Bioceramics, Bioglasses and Gels for Tissue Engineering

Bioceramics, Bioglasses and Gels for Tissue Engineering

Editors

Arish Dasan
Ashokraja Chandrasekar

MDPI

Basel • Beijing • Wuhan • Barcelona • Belgrade • Novi Sad • Cluj • Manchester

Editors

Arish Dasan	Ashokraja Chandrasekar
Department of Glass Processing	Department of Coating Processes
FunGlass, Alexander Dubček	Alexander Dubček University
University of Trenčín	of Trenčín
Trenčín	Trenčín
Slovakia	Slovakia

Editorial Office
MDPI
St. Alban-Anlage 66
4052 Basel, Switzerland

This is a reprint of articles from the Special Issue published online in the open access journal *Gels* (ISSN 2310-2861) (available at: www.mdpi.com/journal/gels/special_issues/gels_tissue_engineering).

For citation purposes, cite each article independently as indicated on the article page online and as indicated below:

Lastname, A.A.; Lastname, B.B. Article Title. *Journal Name* **Year**, *Volume Number*, Page Range.

ISBN 978-3-0365-9495-8 (Hbk)
ISBN 978-3-0365-9494-1 (PDF)
doi.org/10.3390/books978-3-0365-9494-1

Contents

About the Editors

Arish Dasan

Dr. Arish Dasan is currently working as a researcher at FunGlass (European Horizon 2020 project), FunGlass, Alexander Dubček University of Trenčín, Slovakia. Dr. Arish Dasan, after completing his Ph.D in chemistry from Manonmaniam Sundaranar University, Tamilnadu, India, served as a faculty member in the Department of Chemistry at PSN College of Engineering and Technology, Tamilnadu, India, from 2011 to 2012. After that, he was a postdoctoral research associate in the Ceramic Matrix Products Division lab at Vikram Sarabhai Space Center, ISRO, Thiruvananthapuram, India, for three years. He also completed a one-year postdoctoral fellowship at the European Ceramic Center, SPCTS, Limoges, France. Furthermore, he spent more than a year at the University of Padova, Padova, Italy, as a visiting scientist. His research interests include additive manufacturing (3D printing) of glass and ceramic materials, biologically active metal complexes, polymer-derived ceramics, ceramic matrix composites, glass ceramics, ceramic suspensions, the surface modification of ceramic nanoparticles, and ceramic core/shell nanocomposites.

Ashokraja Chandrasekar

Dr. Ashokraja Chandrasekar is a researcher at Alexander Dubcek University in Trencin, Slovakia. His work focuses on the development of bioactive and degradable coatings applied in procedures involving degradable implants. His most recent discoveries involve the development of polyol-bioactive hybrid sol–gel-based coatings on AZ31B magnesium alloys for use in applications involving degradable implants. He has experience in the fields of biotechnology, medical bionanotechnology, biomaterials, and coatings used for biomedical purposes utilizing methods such as sol–gel deposition and plasma electrolytic oxidation. He has previous experience working with bioactive glass nanorods derived from sol–gel, as well as the electrophoretic deposition and anodization of titanium. He has had 12 papers published, and together they have more than 140 citations. He has acted in the reviewer capacity for a variety of notable publications, including IOP, Elsevier, and MDPI, among others.

Preface

Welcome to the Special Issue "Bioceramics, Bioglasses and Gels for Tissue Engineering". In the realm of regenerative medicine, the convergence of materials science and biomedical engineering has paved the way for groundbreaking advancements in tissue engineering. This collection of articles represents a comprehensive exploration of the latest developments, challenges, and innovations in the synthesis and applications of bioceramics, bioglasses, and gels for tissue regeneration. As editors of this Special Issue, we are delighted to present a compilation of cutting-edge research and insights contributed by leading experts and researchers in this field. We extend our sincere appreciation to the authors whose dedicated work has enriched this Special Issue. Their commitment to advancing our understanding of biomaterials in the context of tissue engineering is evident in the quality and depth of the contributions presented here. Furthermore, we would like to express our gratitude to the reviewers whose expertise and thoughtful assessments have played a crucial role in maintaining the scholarly rigor of this compilation. It is our belief that this collection will serve as a valuable resource for researchers, practitioners, and students alike, fostering continued exploration efforts and advancements in this exciting and dynamic field.

Thank you for joining us on this intellectual journey.

Arish Dasan and Ashokraja Chandrasekar
Editors

gels

Editorial

Special Issue: Bioceramics, Bioglasses, and Gels for Tissue Engineering

Arish Dasan * and **Ashokraja Chandrasekar**

FunGlass, Alexander Dubček University of Trenčín, 911 50 Trenčín, Slovakia; ashokraja.chandrasekar@tnuni.sk
* Correspondence: arish.dasan@tnuni.sk or arishd82@gmail.com

Undoubtedly, biomaterials such as bioceramics, bioactive glasses, and gels have attracted a wide range of research interest in the field of tissue engineering (TE), as they facilitate the essential support and environment for cells to grow, differentiate, and, specifically, regenerate new tissues [1–5]. Orthopaedic and dental implants are increasingly being used to treat patients of all ages who are missing bones or teeth. Biocompatibility, mechanical stability, corrosion resistance, and antimicrobial resistance are just some of the criteria that must be met for a material to be used as a biomaterial or an implant [6,7]. Although biomaterials offer many benefits, challenges still exist, for instance, achieving the desirable mechanical properties, controlling their degradation properties, and ensuring their long-term stability. From this perspective, many researchers working on advancements in biomaterials by adopting novel processing and manufacturing techniques such as 3D printing continue to address these challenges and expand the possibilities for their applications in TE [8–10]. Bioceramics, such as hydroxyapatite (HA) and tricalcium phosphate (TCP), are commonly used in bone TE applications [11,12]. High-strength and aesthetic bioceramics such as zirconia find extensive use in dental applications [13,14]. Bioceramics are also used as a coating on commercial metallic implants, thereby enhancing their biocompatibility, bone integration, and improving the lifetime of the implant [15,16]. Apart from hard tissue, recently, bioceramics have been explored in soft tissue engineering applications, for instance, skin regeneration, the regeneration of periodontal tissues, the cure of articular cartilage, myocardial necrosis, and neovascularization growth [1–3,17]. Further investigations, by modifying their composition and surface properties, for example, are likely to lead to the potentiality of regenerating soft tissues.

One of the biomaterials that has transformed contemporary biomaterial-driven regenerative medicine is bioactive glass, which has created novel applications in biomedicine such as soft tissue repair and drug delivery [6]. Bioglass®, pioneered by Hench, is just one example of how much progress has been made in biomaterials in the last few decades [18]. It would be too modest to acknowledge that, among man-made materials, it is a marvel for its ability to both chemically bond to the host bone and promote cell proliferation. Bioactive glasses' main benefit is the controlled release of therapeutic ions from their silica matrix, which stimulates protein and cellular attachment and aids in the repair of damaged bones by the means of cell proliferation [19–21]. Many varieties of bioactive glasses, including the typical 45S5 silicate glass (45S5 or Bioglass®), antibacterial bioactive glasses (S53P4 or BonAlive®), and borate-based glasses (13-93B3 bioactive glass), have been produced and marketed [22]. Recent advances in the development of bioactive glasses for bone regeneration have relied on porous scaffolds that can serve as 3D representations of bone structures. In addition to conventional foaming procedures and template-based methods, a variety of additive manufacturing techniques are currently being employed to construct scaffolds from melt and sol-gel-derived glasses. These techniques permit greater control over the pore structure and formation of the scaffolds [22].

Gels are widely used in TE because of their unique properties, such as high water content, softness, and the provision of a three-dimensional network environment, which

Citation: Dasan, A.; Chandrasekar, A. Special Issue: Bioceramics, Bioglasses, and Gels for Tissue Engineering. *Gels* **2023**, *9*, 586. https://doi.org/10.3390/gels9070586

Received: 4 July 2023
Accepted: 16 July 2023
Published: 21 July 2023

make them suitable for cell growth, nutrient diffusion, and tissue regeneration [23–25]. Hydrogels formed from natural components such as collagen, fibrin, and gelatin are the most used in TE due to their ability to mimic the native extracellular matrix (ECM). Hydrogels can serve as delivery vehicles for potential therapeutic agents, such as drugs and growth factors, enabling their localized and sustained release [25]. In addition, the inclusion of potential therapeutic inorganic biomaterials within a gel network can stimulate by delivering the physical and biochemical cues necessary for tissue development [25]. Considering their limitations, the lack of mechanical strength that restricts their use in load-bearing applications, achieving a detailed microstructure, and controlling the degradation rate are the main challenges. However, it should be noted that the plusses and constraints can vary depending on the gel materials, formulations, and applications. Research teams are continually working to improve reliable gel systems and minimize their limitations.

In recent years, significant research attention has been paid to utilizing biomaterials as feedstock in AM technology [26–28]. The advent of such technology has enabled new possibilities for producing high-performance components directly from customized digital models, which is not feasible using conventional fabrication techniques. It is also possible to manufacture optimal pore sizes with interconnection. Recent emerging approaches, such as omics-based approaches, can allow for a comprehensive study of the regenerative potential of biomaterials [1]. Another hot topic is to develop engineered biomaterials that are (i) bio-instructive (designed to provide chemical and physical cues to guide cellular behavior), (ii) biomimetic (aimed at replicating the structure, including the microstructure and properties, including the chemical and physical ones of natural tissues or organs), and (iii) bioresponsive (designed to exhibit specific responses when exposed to biological stimuli or signals).

Advancements in interdisciplinary fields such as chemistry, physics, material science, nanotechnology, manufacturing technology, and bioengineering have led to the development of a wide range of biomaterial components. These materials are being increasingly explored for a wide range of applications, including regenerative medicine, biosensors, bioelectronics, and personalized implants, with the aim of improving human health and well-being.

This Special Issue collection of articles represents the keen and diverse research ensuing toward innovative functionalities and technologies in the biomedical sector. Atkinson I et al. [29] fabricated composite scaffolds consisting of poly methyl methacrylate (PMMA) Cerium-doped mesoporous bioactive glass (MBG), by means of the phase separation method. In addition, they studied the effect of ceria addition and thereby the property changes, particularly the crystallization behavior, of the SiO_2–CaO–P_2O_5–CeO_2 system [30]. Yergeshov AA et al. [31] investigated the in vivo and in vitro effects of metal ion (Cu, Co, and Zn)-doped biodegradable macroporous cryogels. The research performed by Dascalu LM et al. [32] revealed the adjuvant effect of natural photosensitizers, based on curcuma extract and oregano essential oil, on induced periodontal diseases. Mosas KKA et al. [15] apprized the recent developments in biomaterials and coatings for different biomedical implants. Another review article [33] discussed soft tissue repair using nanoparticle-modified biocompatible polymers including hydrogels. Schrade S et al. [34] found that gelatin nanoparticles can be used for targeted dual drug release out of aleginate-di-aldehyde-gelatin gels. Calcium magnesium silicate (CaO-MgO-SiO_2)-based bioceramics, containing bioactive phases of diopside, akermanite, and merwinite, were prepared by Alecu AE et al. [35] using the sol-gel method followed by thermal treatment. Reyes-Peces MV and their research group [36] fabricated bioactive 3D scaffolds based on silica, gelatin, and β-tricalcium phosphate by the means of robocasting additive manufacturing technology and laser micromachining. Phetcharat P et al. [37] studied the influence of the addition of copper iodide nanoparticles on poly(vinyl alcohol) (PVA) liquid bandages for wound healing applications. Another interesting review article by Kishani M and their team [38] summarized the recent advancements in scaffolds, particularly those fabricated from bio-based natural materials.

Miyamoto Y [39] extensively reviewed cell sheets, vitrified hydrogel membranes, and their cryopreservation applications in regenerative and cellular medicine.

Author Contributions: Conceptualization, A.D.; writing—original draft preparation, A.D. and A.C.; writing—review and editing, A.D. and A.C.; visualization, A.D. and A.C.; supervision, A.D.; project administration, A.D. All authors have read and agreed to the published version of the manuscript.

Conflicts of Interest: The authors declare no conflict of interest.

References

1. Gaharwar, A.K.; Singh, I.; Khademhosseini, A. Engineered biomaterials for in situ tissue regeneration. *Nat. Rev. Mater.* **2020**, *5*, 686–705. [CrossRef]
2. Koons, G.L.; Diba, M.; Mikos, A.G. Materials design for bone-tissue engineering. *Nat. Rev. Mater.* **2020**, *5*, 584–603. [CrossRef]
3. Rodrigo-Navarro, A.; Sankaran, S.; Dalby, M.J.; del Campo, A.; Salmaeron-Sanchez, M. Engineered living biomaterials. *Nat. Rev. Mater.* **2021**, *6*, 1175–1190. [CrossRef]
4. Mochi, F.; Scatena, E.; Rodriguez, D.; Ginebra, M.P.; Gaudio, C.D. Scaffold-based bone tissue engineering in microgravity: Potential, concerns and implications. *NPJ Microgravity* **2022**, *8*, 45. [CrossRef]
5. Morneau, D. Engineered hydrogels for mechanobiology. *Nat. Rev. Methods Primers* **2022**, *2*, 99. [CrossRef]
6. Crush, J.; Hussain, A.; Seah, K.T.M.; Khan, W.S. Bioactive Glass: Methods for Assessing Angiogenesis and Osteogenesis. *Front. Cell. Dev. Biol.* **2021**, *9*, 643781. [CrossRef]
7. Chitra, S.; Mathew, N.K.; Jayalakshmi, S.; Balakumar, S.; Rajeshkumar, S.; Ramya, R. Strategies of Bioceramics, Bioactive Glasses in Endodontics: Future Perspectives of Restorative Dentistry. *Biomed. Res. Int.* **2022**, *2022*, 2530156. [CrossRef]
8. Dasan, A.; Kraxner, J.; Grigolato, L.; Savio, G.; Elsayed, H.; Galusek, D.; Bernardo, E. 3D Printing of Hierarchically Porous Lattice Structures Based on Åkermanite Glass Microspheres and Reactive Silicone Binder. *J. Funct. Biomater.* **2022**, *13*, 8. [CrossRef]
9. Urciuolo, A.; Giobbe, G.G.; Dong, Y.; Michielin, F.; Brandolino, L.; Magnussen, M.; Gagliano, O.; Selmin, G.; Scattolini, V.; Raffa, P.; et al. Hydrogel-in-hydrogel live bioprinting for guidance and control of organoids and organotypic cultures. *Nat. Commun.* **2023**, *14*, 3128. [CrossRef]
10. Kaliaraj, G.S.; Shanmugam, D.K.; Dasan, A.; Mosas, K.K.A. Hydrogels—A Promising Materials for 3D Printing Technology. *Gels* **2023**, *9*, 260. [CrossRef]
11. Re, F.; Borsani, E.; Rezzani, R.; Sartore, L.; Russo, D. Bone Regeneration Using Mesenchymal Stromal Cells and Biocompatible Scaffolds: A Concise Review of the Current Clinical Trials. *Gels* **2023**, *9*, 389. [CrossRef]
12. Dibazar, Z.E.; Nie, L.; Azizi, M.; Nekounam, H.; Hamidi, M.; Shavandi, A.; Izadi, Z.; Delattre, C. Bioceramics/Electrospun Polymeric Nanofibrous and Carbon Nanofibrous Scaffolds for Bone Tissue Engineering Applications. *Materials* **2023**, *16*, 2799. [CrossRef]
13. Abu-Naba'a, L.A. A Narrative Review of Recent Finite Element Studies Reporting References for Elastic Properties of Zirconia Dental Ceramics. *Ceramics* **2023**, *6*, 898–911. [CrossRef]
14. Branco, A.C.; Colaço, R.; Figueiredo-Pina, C.G.; Serro, A.P. Recent Advances on 3D-Printed Zirconia-Based Dental Materials: A Review. *Materials* **2023**, *16*, 1860. [CrossRef]
15. Amirtharaj Mosas, K.K.; Chandrasekar, A.R.; Dasan, A.; Pakseresht, A.; Galusek, D. Recent Advancements in Materials and Coatings for Biomedical Implants. *Gels* **2022**, *8*, 323. [CrossRef]
16. Fatma, K.; Tripathy, J. Bioceramic coatings for tissue engineering. In *Advanced Ceramic Coatings for Emerging Applications*; A volume in Elsevier Series in Advanced Ceramic Materials; Elsevier: Amsterdam, The Netherlands, 2023; pp. 291–309. [CrossRef]
17. Kumar, R.; Pattanayak, I.; Dash, P.A.; Mohanty, S. Bioceramics: A review on design concepts toward tailor-made (multi)-functional materials for tissue engineering applications. *J. Mater. Sci.* **2023**, *58*, 3460–3484. [CrossRef]
18. Ramadoss, R.; Padmanaban, R.; Subramanian, B. Role of Bioglass in Enamel Remineralization: Existing Strategies and Future Prospects-A Narrative Review. *J. Biomed. Mater. Res. Part B Appl. Biomater.* **2022**, *110*, 45–66. [CrossRef]
19. Moeini, A.; Chinijani, H.T.; Khachatourian, M.A.; Fook, V.L.M.; Baino, F.; Montazerian, M. A Critical Review of Bioactive Glasses and Glass–Ceramics in Cancer Therapy. *Int. J. Appl. Glass. Sci.* **2023**, *14*, 69–87. [CrossRef]
20. Obata, A.; Lee, S.; Kasuga, T. Bioactive Glass Materials for Tissue Regeneration. *J. Ceram. Soc. Jpn.* **2022**, *130*, 595–604. [CrossRef]
21. Taye, M.B. Biomedical Applications of Ion-Doped Bioactive Glass: A Review. *Appl. Nanosci.* **2022**, *12*, 3797–3812. [CrossRef]
22. Cheah, C.W.; Al-Namnam, N.M.; Lau, M.N.; Lim, G.S.; Raman, R.; Fairbairn, P.; Ngeow, W.C. Synthetic Material for Bone, Periodontal, and Dental Tissue Regeneration: Where Are We Now, and Where Are We Heading Next? *Materials* **2021**, *14*, 6123. [CrossRef]
23. Zhang, Z.; Bi, F.; Guo, W. Research Advances on Hydrogel-Based Materials for Tissue Regeneration and Remineralization in Tooth. *Gels* **2023**, *9*, 245. [CrossRef]
24. Li, J.; Song, W.; Li, F. Polymeric DNA Hydrogels and Their Applications in Drug Delivery for Cancer Therapy. *Gels* **2023**, *9*, 239. [CrossRef]
25. Zhong, R.; Talebian, S.; Mendes, B.B.; Wallace, G.; Langer, R.; Conde, J.; Shi, J. Hydrogels for RNA delivery. *Nat. Mater.* **2023**, *22*, 818–831. [CrossRef]

26. Naranjo-Alcazar, R.; Bendix, S.; Groth, T.; Gallego Ferrer, G. Research Progress in Enzymatically Cross-Linked Hydrogels as Injectable Systems for Bioprinting and Tissue Engineering. *Gels* **2023**, *9*, 230. [CrossRef]
27. Liu, H.; Gong, Y.; Zhang, K.; Ke, S.; Wang, Y.; Wang, J.; Wang, H. Recent Advances in Decellularized Matrix-Derived Materials for Bioink and 3D Bioprinting. *Gels* **2023**, *9*, 195. [CrossRef]
28. Zhang, B.; Li, S.; Zhang, Z.; Meng, Z.; He, J.; Ramakrishna, S.; Zhang, C. Intelligent biomaterials for micro and nanoscale 3D printing. *Curr. Opin. Biomed.* **2023**, *26*, 100454. [CrossRef]
29. Atkinson, I.; Seciu-Grama, A.; Mocioiu, O.; Mocioiu, A.; Predoana, L.; Voicescu, M.; Cusu, J.; Grigorescu, R.; Ion, R.; Craciunescu, O. Preparation and Biocompatibility of Poly Methyl Methacrylate (PMMA)-Mesoporous Bioactive Glass (MBG) Composite Scaffolds. *Gels* **2021**, *7*, 180. [CrossRef]
30. Anghel, E.; Petrescu, S.; Mocioiu, O.; Cusu, J.; Atkinson, I. Influence of Ceria Addition on Crystallization Behavior and Properties of Mesoporous Bioactive Glasses in the SiO_2–CaO–P_2O_5–CeO_2 System. *Gels* **2022**, *8*, 344. [CrossRef]
31. Yergeshov, A.; Zoughaib, M.; Ishkaeva, R.; Savina, I.; Abdullin, T. Regenerative Activities of ROS-Modulating Trace Metals in Subcutaneously Implanted Biodegradable Cryogel. *Gels* **2022**, *8*, 118. [CrossRef]
32. Dascalu (Rusu), L.; Moldovan, M.; Sarosi, C.; Sava, S.; Dreanca, A.; Repciuc, C.; Purdoiu, R.; Nagy, A.; Badea, M.; Paun, A.; et al. Photodynamic Therapy with Natural Photosensitizers in the Management of Periodontal Disease Induced in Rats. *Gels* **2022**, *8*, 134. [CrossRef] [PubMed]
33. Yudaev, P.; Mezhuev, Y.; Chistyakov, E. Nanoparticle-Containing Wound Dressing: Antimicrobial and Healing Effects. *Gels* **2022**, *8*, 329. [CrossRef]
34. Schrade, S.; Ritschl, L.; Suss, R.; Schilling, P.; Seidenstuecker, M. Gelatin Nanoparticles for Targeted Dual Drug Release out of Alginate-di-Aldehyde-Gelatin Gels. *Gels* **2022**, *8*, 365. [CrossRef]
35. Alecu, A.; Costea, C.; Surdu, V.; Voicu, G.; Jinga, S.; Busuioc, C. Processing of Calcium Magnesium Silicates by the Sol—Gel Route. *Gels* **2022**, *8*, 574. [CrossRef] [PubMed]
36. Reyes-Peces, M.; Felix, E.; Martinez-Vazquez, F.; Fernandez-Montesinos, R.; Bomati-Miguel, O.; Mesa-Diaz, M.; Alcantara, R.; Vilches-Perez, J.; Salido, M.; De la Rosa-Fox, N.; et al. Robocasting and Laser Micromachining of Sol-Gel Derived 3D Silica/Gelatin/β-TCP Scaffolds for Bone Tissue Regeneration. *Gels* **2022**, *8*, 634. [CrossRef]
37. Phetcharat, P.; Sangsanoh, P.; Choipang, C.; Chaiarwut, S.; Suwantong, O.; Chuysinuan, P.; Supaphol, P. Curative Effects of Copper Iodide Embedded on Gallic Acid Incorporated in a Poly(vinyl alcohol) (PVA) Liquid Bandage. *Gels* **2023**, *9*, 53. [CrossRef]
38. Krishani, M.; Shin, W.; Suhaimi, H.; Sambudi, N. Development of Scaffolds from Bio-Based Natural Materials for Tissue Regeneration Applications: A Review. *Gels* **2023**, *9*, 100. [CrossRef] [PubMed]
39. Miyamoto, Y. Cryopreservation of Cell Sheets for Regenerative Therapy: Application of Vitrified Hydrogel Membranes. *Gels* **2023**, *9*, 321. [CrossRef] [PubMed]

gels

Review

Recent Advancements in Materials and Coatings for Biomedical Implants

Kamalan Kirubaharan Amirtharaj Mosas [1,*], Ashok Raja Chandrasekar [1,*], Arish Dasan [1], Amirhossein Pakseresht [1] and Dušan Galusek [1,2,*]

[1] Centre for Functional and Surface-Functionalized Glass, Alexander Dubcek University of Trencín, 911 50 Trencín, Slovakia; arish.dasan@tnuni.sk (A.D.); amir.pakseresht@tnuni.sk (A.P.)
[2] Joint Glass Centre of the IIC SAS, TnUAD, and FChFT STU, FunGlass, Alexander Dubcek University of Trencín, 911 50 Trencín, Slovakia
* Correspondence: kamalan.mosas@tnuni.sk (K.K.A.M.); ashokraja.chandrasekar@tnuni.sk (A.R.C.); dusan.galusek@tnuni.sk (D.G.)

Abstract: Metallic materials such as stainless steel (SS), titanium (Ti), magnesium (Mg) alloys, and cobalt-chromium (Co-Cr) alloys are widely used as biomaterials for implant applications. Metallic implants sometimes fail in surgeries due to inadequate biocompatibility, faster degradation rate (Mg-based alloys), inflammatory response, infections, inertness (SS, Ti, and Co-Cr alloys), lower corrosion resistance, elastic modulus mismatch, excessive wear, and shielding stress. Therefore, to address this problem, it is necessary to develop a method to improve the biofunctionalization of metallic implant surfaces by changing the materials' surface and morphology without altering the mechanical properties of metallic implants. Among various methods, surface modification on metallic surfaces by applying coatings is an effective way to improve implant material performance. In this review, we discuss the recent developments in ceramics, polymers, and metallic materials used for implant applications. Their biocompatibility is also discussed. The recent trends in coatings for biomedical implants, applications, and their future directions were also discussed in detail.

Keywords: bioceramics; metallic implants; biomedical applications; coatings; surface modification; biocompatibility; Mg-alloys; Ti-alloys

Citation: Amirtharaj Mosas, K.K.; Chandrasekar, A.R.; Dasan, A.; Pakseresht, A.; Galusek, D. Recent Advancements in Materials and Coatings for Biomedical Implants. *Gels* **2022**, *8*, 323. https://doi.org/10.3390/gels8050323

Received: 2 May 2022
Accepted: 18 May 2022
Published: 21 May 2022

1. Introduction

Bioimplants are defined as engineered medical devices that are developed to replace the non-functional or broken biological structural parts of the human body, providing support to the given host. Biomaterial surface modification plays a key role in determining the outcome of the interaction between human biology and materials. Substantial development in research in the field of biomaterials has increased the scope of use for a wide range of orthopedic and dental implants that include total bone replacement, fracture fixation, dental screws, joint arthrodesis, and so on [1]. Essentially, the success of bioimplants depends not only on their bulk properties but also on the properties of their surfaces, which interact with human body tissues. As a result, the evolution of bioimplants has reached a level of choice of materials based on specific properties on the basis of selected specific materials [2]. Though alloys and metallic substances meet many of the biomedical requirements, their interfacial bonding between the surrounding tissue or bone and the metallic surface ranges from poor to virtually absent. The failure of the metallic implant originates at the implant-tissue interface due to poor bonding at the interface, which leads to the formation of a nonadherent layer and movement at the tissue-implant interface [3].

Corrosion in biometallic implants can affect the surface and biocompatible behavior that induce tissue reactions, which lead to the release of corrosion byproducts from the implant surface and result in premature failure. A minimum durability of 15 to 20 years for older patients and more than 20 years for younger patients is expected from a bioimplant [4].

However, there are problems associated with the use of metallic implants due to the lack of poor implant fixation, lack of osteoconductivity, corrosion, and wear resistance leading to the formation of wear debris and release of corrosive ions [5–7]. These problems are mostly associated with the surface of the metallic implants. In view of this, the surface of the bioimplant plays a major role in the biological environment because the reactions occur directly on the surface of the implant after implant fixation. Hence, it is necessary to modify the surface of the metallic substrate with specific properties that are different from those in bulk [8,9]. This modification is required to accomplish good bone formability and desired biological interactions. In some applications, biocompatibility, wear, and corrosion resistance are also required.

Surface modifications of bioimplants are explored intensively with many bioactive materials to avoid adverse effects such as lack of biocompatibility, post-surgery infections, long-term survivability, and risks related to implant surface corrosion [10,11]. At first, the research in this field was focused on the improvement in biomechanical properties of metallic implants, but in recent days, it has turned towards improvement in the biological properties of these biomedical devices [12,13]. By applying the appropriate modification on the surface of the material, one can tailor and improve the biocompatibility, cell interactions, and adhesion [14]. Thus, the development and design of biomaterials rely on surface modification. For that, it is necessary to develop techniques for functionalization of the surface of metallic implants through changing the materials' surface composition, morphology, and structure without losing their mechanical properties. By adopting this, the service life and performance of orthopedic and dental implants can be significantly increased. This can be achieved by applying suitable biocompatible coatings with a unique combination of properties.

In view of reliability and performance, the best way to functionalize the implants in direct contact with bones and tissues is ceramic coatings owing to their excellent osteoconductive properties and high stability [15,16]. Surface modification by coating can enhance the antibacterial activity of a bioimplant. The coated surfaces facilitate grafting of cell-binding peptides, directed mutations of the cellular host, protein of extracellular matrix (ECM), and growth of tissues to improve the acceptance of a bioimplant further. Ceramic coatings on bioimplants show promising results in orthopedics with improved bone regeneration and repair [17]. The overview of applications of ceramic coatings used for metallic implants is listed in Table 1.

Table 1. Ceramic coatings used for biomedical applications [18].

Coatings	Applications	Advantages
Oxides (TiO_2, ZrO_2)	Oral implant application Maxillofacial reconstruction Ophthalmic implants	Good regenerative capability Corrosion resistance Antibacterial activities
Nitrides (TiN, ZrN, TiCN, ZrCN, TiAlN) and Oxynitrides (TiON, ZrON)	Dental implants Fracture fixation devices Components of joint endoprostheses	Resistance to corrosion Low frictional coefficient Better adhesion to the substrates
Carbon Based Coatings (a-C, DLC, NCD, carbides, and carbontirides)	Artificial heart valves Orthopedic fixation devices Sensors Artificial ligaments	Low frictional coefficient Excellent biocompatibility High blood compatibility Hydrophobicity
Calcium phosphates (CaP, HAp) and bioactive glass	Spinal implants Orthopedic implants Maxillofacial reconstruction Skull plates	High osteointegration capability Excellent biocompatibility Bioactivity

The major requirements for the selection of coating materials are (a) biocompatibility and nondetrimental effects such as allergy, inflammation, and toxicity, (b) adequate fracture toughness, fatigue, and mechanical strength to withstand the forces, and (c) resistance to corrosion in the human body fluid atmosphere, which contains many constituents such as amino acids, chlorine, water, proteins, sodium, and plasma acids. The choice of coatings, by considering their degeneration and surface properties, plays a major role in terms of reliability and performance of bioimplants. The coatings for biomedical applications can be subdivided into three groups: (a) bioinert, (b) bioactive, and (c) bioresorbable coatings [19]. The coatings having a minimum interaction with the surrounding tissues after implantation in the human body are considered as bioinert coatings. The typical examples of bioinert coatings are metal oxides, nitrides, carbides, carbonitrides, and oxynitrides. Transition metal nitrides (TiN, ZrN, TiAlN, NbN), carbides (TiC), oxides (ZrO_2, Al_2O_3, TiO_2), or oxynitride (TiON) coatings find a wide range of applications in bioceramic coatings due to their remarkable properties such as wear, tear, hardness, biocompatibility, and corrosion resistance [20,21].

The current review incorporates a description of the biomaterials and coatings that are commonly used in the manufacturing of different orthopedic and dental implants.

2. Biomaterials for Biomedical Applications

Biomaterials are used to make devices that interact with the biological systems in the human body and coexist for a long time with minimal failure. The type of material used in implant applications shows specific properties that make them primary candidates for specific applications. The key requirements for the selection of biometallic materials consist of (a) cost effectiveness, (b) mechanical behavior equal to that of the human skull and bones, and (c) their biocompatibility [22,23]. In addition, the major requirement for the bioimplant materials is that it should be compatible with the human body, i.e., it should integrate with the human body without negative impacts. Moreover, it must possess corrosion and wear resistance in the human body environment. These properties will determine the effectiveness of the implant materials.

If a metallic material experiences wear and corrosion, the surrounding tissues present at the implant area can become inflamed, causing unfavorable biological reactions within the human body [24]. The ions and toxins released from the metallic substrates as a byproduct may be potentially harmful and can cause life threatening diseases and increase the risk of using metallic implants. Therefore, it is important to choose correct material for correct applications while performing bioimplants. In addition to that, the mechanical performance of the biomaterial should be close to that of the replacing material where it must sustain complicated and varying mechanical loading cycles [25]. Typical examples for implanting areas are teeth, knee joints, and hips. The selection of biomaterial based on mechanical properties is important to ensure no implant failures within the body when subjected to numerous loading cycles during service life. Moreover, the material should be biocompatible with the surrounding tissues and economically viable. Finally, it is essential that the choice of material should be cost effective, efficient, and able to integrate with the human body. Based on the requirements defined above, several materials were developed in recent years to be used as biomaterials for implant applications. Still, it is hard for a single metallic material to fulfill the desired properties. Biomaterials used for biomedical applications are broadly classified into ceramics, polymers, and metallic systems.

2.1. Ceramics

Ceramics are inorganic compounds formed at high temperatures. Typical examples are bioactive glass (BG), zirconium oxide (ZrO_2), aluminum oxide (Al_2O_3), hydroxyapatite (HAp), and other calcium and silica-based ceramics. These ceramics are noted for their great biocompatibility, which makes them an excellent candidate for biomedical implant applications. Depending on the reactivity with the human body, ceramic implants are classified into three categories: (a) bioactive, (b) bioinert, and (c) bioresorbable ceramics [26]. Bioactive ceramics are used to interact with the surrounding cells and exhibit a higher level of reactivity within the implant sites. Typical examples for bioactive ceramics are HA and fluorapatites [27]. In an opposite trend, bioinert ceramics do not show any reactivity with the host tissues at the implant sites but form a physical bonding when implanted [28].

Bioresorbable ceramics exhibit a low level of reactivity with the host body tissues [29]. After implantation, these ceramics are gradually resorbed and finally replaced with the bone tissue. These bioresorbable ceramics are widely used in orthopedics and dentistry due to their better biocompatibility and chemical interactions [30].

No risk of transmitting disease plus immunogenicity after implantation are the major advantages of the ceramics [31]. Other remarkable advantages are higher resistance to compressive force, low toxicity, good corrosion resistance, and promotion of the formation of new hard tissues. For example, hydroxyapatite-based ceramics exhibit higher Ca/P ratios, which are desirable due to similar chemical properties of bone and teeth hard tissues [32–35]. Due to these attractive properties, ceramics are increasingly utilized for bioimplant applications.

Ceramics are known for their high hardness and stress-shielding effects due to their high elastic moduli, and slow initiation of crack growth over time, which significantly decrease the reliability of the implants [36]. In addition, brittleness, fracture toughness, and fabrication issues limit their use as bioimplants. The ceramics share the brittleness factor, which limits the performance in terms of load-bearing applications (hip implants). If the difference in mechanical properties of ceramic and bone is large, the load will not be transmitted through the bone, thus leading to failure of the bone [37].

Ceramic composite materials provide superior properties compared to single materials. The inferior mechanical properties of monolithic ceramics can be overcome by composite ceramics while diminishing the limitations of each component. The remarkable properties of composites such as the weight to strength ratio enable them to be used extensively for the restoration of bones, ligaments, and dental fillings [38]. Moreover, the composites prepared through the combination of bioactive and bioinert ceramics show better bioactivity and mechanical strength [39]. Typical examples are HA and Al_2O_3 composites which show better osteointegration with bone, good bioactivity, and high yield strength [40,41].

2.2. Polymers

The most widely used materials in biomedical applications are polymers. Polymers are the building blocks of small repeating units' monomers and are classified into two categories called biodegradable and non-biodegradable. Typical examples for biodegradable polymers are polyacetal, chitosan (CS), alginate, polylactide, and polycaprolactone, whereas non-biodegradable polymers include polypropylene, polytetrafluoroethylene, polyethylene terephthalate, polymethylmethacrylate, etc. Polymer implants are mostly used in replacing heart valves, kidneys, bone, skin, contact lens, and artificial blood vessels, in addition as pacemakers [42]. Among biodegradable polymers, CS shows remarkable properties such as biocompatibility, biodegradability, wound healing, and antibacterial activity [43]. It is also environmentally friendly and hence acts as a capping agent [44,45]. Polymers show lower strength and elastic moduli as compared to metals and ceramics. Therefore, they are not generally used for load-bearing applications such as joint and knee prostheses. The polymers are also degraded in the body environment due to biochemical factors.

Polymer implants are quite interesting as bioimplants due to their low cost while offering sufficient mechanical properties. For example, Polyether ether ketone (PEEK), composed of 20% TiO_2 particles and an additional ketone group results in 80% higher compressive strength and better fatigue properties than pure PEEK [46]. Depending upon the replacement anatomy to which the polymer is being applied, a wide variety of polymers can be applied. Polymers have the advantage of complete degradation over time, leaving no signs of their presence at the implant locations in a body. This was possible with the subsequent research and development in biodegradable polymer materials, where the proteins and extracellular matrix mimic the cell signaling functions of the surrounding tissue, permitting better bio-integration [47].

Though polymers show exceptional properties and are cost-efficient and easy to manufacture, they show different forms of cytotoxicity: depending on the host body conditions, inflammatory reactions can occur within the implant region. This will induce bone degeneration, abnormalities, rapid rate of corrosion, and decreases in mechanical properties over time. Moreover, the elastic modulus of polymers is extremely low compared to human bone (between 10 and 30 GPa) [48]. This will create an impact while applying load. Another major issue that is being faced is that the polymer implant degrades as the bone heals. If the process is too fast, the neighboring tissues feel more stress, which causes potential discomfort. These limitations prevent them from being widely used as bioimplants.

Polymeric Gels

Natural polymers such as collagen are the main components of natural bone due to their hydrophilic nature, enabling the formation of hydrogels with aqueous solutions that exhibit several desirable characteristics for bone-tissue engineering [49]. Polymeric gels are often referred to as hydrogels owing to their ability to hold water inside their networks [50]. These hydrogels swell upon water intake and shrink upon drying [51]. Taking advantage of this property, water soluble drugs, growth factors, and other biological entities such as proteins and even live cells can be incorporated into these hydrogels [52]. These gels can be designed for delivery systems based on certain external stimuli such as pH [53,54], temperature, or the presence of specific chemicals or target molecules [55]. Many researchers choose collagen because it is the most important organic component of human bone [56–58].

Hydrogels are attractive soft biomaterials because of their soft consistency (stiffness and viscoelasticity are essential in directing the immune response), high water content, porosity, and biocompatibility [59]. They are widely used in 3D cell cultures for modeling the biological extracellular matrix or as coatings for promoting cell attachment. Other natural polymer-based hydrogels used as bone tissue engineering (BTE) materials include polysaccharides (e.g., cellulose) and polypeptides (e.g., alginate). Compared with natural polymeric gels, synthetic polymeric gels offer more possibilities for molecular alterations that facilitate tailoring the candidate properties to specific requirements, i.e., tuning mechanical properties and biophysical and biochemical cues. For instance, Poly(ethylene glycol) (PEG) hydrogels, modified with adhesion ligand arginine–glycine–aspartic acid (RGD), offer tunable mechanical properties as well as improved cell attachment and cell differentiation [60]. However, generally, the poor mechanical strength of hydrogels limits their usage and needs further improvement for bone regeneration. Recent emerging technologies such as 3D printing in the manufacturing of hydrogel-based components may offer entirely new possibilities for addressing the challenges [61].

2.3. Metals and Alloys

Even though ceramics show excellent biocompatible performance, they have poor fracture toughness and exhibit brittle behavior, and their use in load-bearing applications is limited. Thus, metals and alloys are generally used for implants where high strength and load-bearing capacity are required. Most medical industrial segments rely on metallic implants. They are generally used to replace some load-bearing applications such as the hip, plates, knee prostheses, pins, dental materials, screws, and cardiovascular applications [62]. Though metals show high strength and durability, they can lose their properties under physiological conditions with a potential release of various ions and debris which may trigger a biological response. Most of the alloys release metal ions to the plasma in the blood [63]. The excessive release of ions in the blood has a high risk of accumulation in organs such as the spleen and liver that later form particulates, affecting the normal functioning of these organs. This phenomenon leads to cytotoxicity followed by organ failure upon prolonged accumulation.

Metallic materials are not fully accepted by the human body, and the tissue growth is impaired because of inadequate attachment of the implant, leading to discomfort or pain in the implant region [64]. As compared to ceramic materials, the risk of infection is higher, and the healing time is slower in the case of metallic implants. Although metallic implants have some limitations, preference should be given based on their corrosion resistance, cost effectiveness, and mechanical strength. The chemically inert platinum and gold do not show any corrosion in situ, and these materials can be used as bioimplants, but they are expensive. Hence, recent biomedical industries use Ti-based alloys and Mg-based alloys due to their better biocompatibility and good mechanical strength under human body conditions [65]. The widely used metallic materials used as biomedical devices are stainless steel and Ti- and Co-based alloys [66,67].

2.3.1. Stainless Steels (SS)

In India, SS 304 and 316L are the most used implant materials for biomedical applications due to their cost effectiveness, wide resource availability, reliability, and ease of fabrication as compared to Ti- and Co-based alloys. Among various grades of SS, the primary recommended grade for implant applications is AISI type 316L SS. The presence of chromium (minimum content of 10.5 wt. %) yields a thin and passive oxide layer and protects the implant surface against corrosion [68]. The presence of carbon (min. 0.03 wt. %) in SS increases its mechanical properties, especially fracture toughness, corrosion resistance, and tribological performance of the implants. Their load-bearing capability makes them a suitable orthopedic implant material [69]. However, almost 90% of 316L grade SS implants lose their properties due to a pitting corrosion attack and the release of nickel and chromium ions, which cause allergic reactions in the implant region. Hence, a small addition of molybdenum (2 to 4 wt. %) improves the corrosion resistance and strengthens the 316L SS grade.

The 316L SS used in biomedical devices is classified into two categories: (a) conventional SS and (b) Ni-free stainless steels [70]. The primary use of conventional stainless steels is to provide a load-bearing property to the implanted surfaces: they are often used as fracture plates, nails, screws, and stents in the implant process. In addition, the Ni-free SS provides higher corrosion resistance and biocompatibility [71]. When compared to other bioimplants, the chemical composition of SS alloys offers an advantage when good mechanical properties are desired. Moreover, they have a high cost-to-benefit ratio and exhibit a linear relationship with the manufacturing processes and final structure/properties.

Its elastic modulus (200 GPa), which is higher than that of the human bone (10–30 GPa), results in high stress-shielding effect at the tissue/implant interface leading to the failure of the implanted SS [72–74]. In recent days, SS was modified with hydroxyapatite (HAp) which improves its bio-integration and osteointegration properties. Typical implanted materials are screws, pins, sutures, bone plates, steel threads, and medullary nails, which are used in fracture fixation. However, the corrosion resistance, biocompatibility, and osseointegration of SS are lower compared to Ti-based alloys, where implant success rates are much higher [75].

2.3.2. Co-Cr Alloys

Co-based alloys are considered as one of the most successful materials used for implant applications. This alloy was first used in the early 1900s, where it was used as an implant material for hip replacement. Co-based alloys show better corrosion, wear, and mechanical properties and are used in bioimplant applications. The in vivo and in vitro studies confirmed that Co-based alloys show better biocompatibility and can be used for the manufacturing of surgical implants such as in the hip, knee, shoulder, and fractured bone surfaces [76,77]. The most widely used combination of Co alloys are Co-Cr-Mo owing to their unique combination of strength and ductility. By comparing with other metallic implants, this alloy shows a better elastic modulus, density as well as stiffness, becoming an ideal material for the implant process [78]. This alloy is primarily focused on permanent implant fixation procedures because these alloys maintain their initial properties for a long time after implantation. The cumulative likelihood of endurance reached 96% at 12 years for patients aged above 60 years [79]. A Co-Cr-Mo alloy combined with ultra-high molecular weight polyethylene (UHMWPE) is used in artificial ankles and knees [80,81].

Other major alloying elements of Co-based alloys include Ni, Mo, and Cr. These elements were proven to be toxic to the human body when leached out from the metal surface to the body fluid during corrosion of Co alloys and can lead to skin-related diseases. An excessive leaching of these trace elements leads to damage to organs such as the liver, kidney, blood cells, and lungs [82,83]. The addition of nickel into Co-Cr-Mo improves corrosion resistance and mechanical properties, but due to the cytotoxicity of Ni, the use of this alloy in bioimplants is limited [84]. The elastic modulus (200–250 GPa) and ultimate tensile strength (400–1000 GPa) of Co-based alloys are 10 times higher than those of the human bone. The use of these implants manufactured from Co-based alloys thus results in a stress-shielding effect at the tissue/implant interface. The surface modification of Co-Cr-Mo alloys under plasma treatment improves hardness, wear, and corrosion resistance [85–87]. However, they are still not recommended for joint fixtures due to their inferior frictional and tensile properties. Apart from their biocompatibility and corrosion behavior, Co-based alloys are not ideal materials for bearing and joint surfaces due to their sub-par frictional properties [88].

2.3.3. Ti Alloys

Commercially pure titanium (Ti) and its alloys (Ti-6Al-4V, Ti-6Al-7Nb, Ti-5Al-6Nb, and Ti-13Nb-13Zr) have become major assets in the biomedical field owing to their superior biocompatibility, low density, and suitable mechanical properties. At first, it was intended to be used for aerospace applications, but later in the 1970s, the discovery of its biocompatibility led to a demand for Ti and Ti alloys in biomedical applications. If commercial pure titanium (Cp Ti) is used to replace its alloys, the mechanical properties lost due to alloying elements must be compensated for [89,90]. The alloys of Ti show enhanced mechanical and biocompatibility properties in comparison to pure titanium. Depending on the presence of the iron and oxygen content in the Ti alloy, four different grades of alloys are used. The most widely used Ti alloy is Ti-6Al-4V, comprising an estimated 50% of total titanium alloys' usage for bioimplants of this grade [91,92]. By comparing with other grades of Ti alloys, it offers excellent corrosion resistance, biocompatibility, formability, structural stability, and a better weight to strength ratio. The applications of Ti alloys as bioimplants include heart valves, dental prostheses, osteosynthesis, artificial joints, and bone replacements [93].

Biomedical grade titanium alloys are generally categorized as alpha (α, Ti-6Al-4V), near-α, α-β, and metastable β (Ti-6Al-7Nb) [94,95]. These alloys are widely used as biometallic implants, but they cause stress shielding issues at the implant-tissue interface due to their high elastic modulus values. The elastic modulus of Ti and α-β Ti-alloys (100–110 GPa) is higher than that of human bone which limits its usage in joints. The presence of vanadium and aluminum compounds results in the release of toxic ions of vanadium (oxidovanadium (IV) and vanadate (V)) and aluminum (Al^{3+}) under the physiological environment, leading to adverse health issues [96–98]. Therefore, much interest has been paid to β alloys in combination with Zr, Nb, Ta, or Mo to replace V and Al in the alloy. Such alloys possess better mechanical properties, ductility, good structural stability, higher wear resistance, a lower elastic modulus, and improved corrosion resistance [99–101].

One of the disadvantages of using Ti alloys is their below par tribological properties, due to their high friction and abrasive wear nature [102,103]. Moreover, the formation of TiO_2 during exposure protects the surface of the Ti alloy, which hinders the bioimplant-tissue relationship. The formation of titanium compounds around the surrounding tissues of the implant causes failure of the implant [104].

2.3.4. Mg Alloys

Metal-based biodegradable orthopedic implants nullify the complications associated with the long-term existence of implants inside the human body. In recent days, biodegradable metallic implants were investigated as biomedical implants [105]. Magnesium (Mg) is present in the human body as the fourth most abundant cation and is essential to the human metabolism. Mg corrodes faster in the chloride containing physiological environment; thus, it has emerged as biocompatible and biodegradable material for use as implants [106]. Moreover, Mg and its alloys have received much attention in the category of biodegradable alloys due to their leading properties such as low density, an elastic modulus close to that of bones, light weight, biocompatibility, and excellent mechanical properties [107,108]. The revision surgeries performed to remove hardware components in implants such as screws and plates from the implanted site after healing are often discomforting and expensive for the patients. The revision surgery can also lead to complications such as nosocomial infection and delay the patient's recovery to a normal lifestyle. Mg-based biodegradable metallic implant components can overcome the revision surgery by degrading in situ, thus also eliminating the need for the procedure to remove the implant components after healing [109].

The high mechanical strength of metallic materials limits the use as bioimplants, whereas the Mg implant shows a reduced elastic modulus and prevents the mismatch between a bone and the Mg-based implant. This leads to the reduction in stress shielding at the bone/implant interface. Their mechanical and corrosion properties can be enhanced by alloying with Al, Zn, and other elements [110,111]. Current research is focused on the development of Mg-based alloys with zero or low cytotoxicity. Alloying Mg with other metals must be selected carefully to avoid metal-related toxic issues and corrosion. Different type grades of Mg alloys such as Mg-Ca and Mg-Y-Nd were studied as biodegradable bioimplants for orthopedic applications [112].

The major limitation associated with Mg and Mg-based alloys is their rapid corrosion in physiological conditions. Rapid corrosion results in quick release of byproducts such as hydrogen gases due to fast in vivo degradation. This indicates the necessity for surface modification. To overcome the rapid corrosion, alloying with various elements has been explored. For example, elements such as calcium (Ca), zinc (Zn), silver (Ag), aluminum (Al), zirconium (Zr), yttrium (Y), and Neodymium (Nd) were added to Mg to enhance the corrosion and mechanical properties [113–117]. Typical examples are Mg-Ca, Mg-Zn, and Mg-Zn-Ca. By carefully selecting a suitable element and its composition, the microstructure can be tailored to meet mechanical properties such as bone. This makes them ideal for bone replacement. Table 2 shows the overall comparison of materials used for biomedical applications and their applications.

Table 2. The pros and cons of various biomaterials used in the biomedical industry [118].

Materials	Advantages	Disadvantages	Applications
Polymers	Good performance in cyclic load applications, degrade completely over time.	Different cytotoxicity mechanism, inflammatory reactions, bone degradation, show higher corrosion rate.	Bearing surfaces [119]
Ceramics	Zero risk of transmitting diseases/immunogenicity, compression force resistance, corrosion resistance.	Low mechanical properties, high stress-shielding effects, lower rate of biodegradation, fracture toughness is poor.	Bearing surfaces
Stainless Steels	Better mechanical strength, high ductility, flexibility in bending, low manufacturing cost.	High stress-shielding effects, low resistance to corrosion, less osseointegration, biocompatibility issue.	Bone plates, pins, nails, screws, threads, steel threads, and sutures
Co-Cr based alloys	High strength, ductility, elastic modulus, stiffness, and density.	Higher modulus than bones, stress-shielding effects, not ideal for bearing surfaces in a joint, low frictional properties.	Orthopedic implants for knee, ankle, hip, shoulder, and fracture fixation devices
Titanium and its alloys	Good corrosion resistance, light weight, low density, good mechanical strength.	Poor tribological performance, high frictional coefficient, adhesive wear, and low abrasion resistance.	Total knee, hip replacement, bone plates, and screws for fixation and maxillofacial applications
Mg and its alloys	Low Young's modulus, no stress shielding, biodegradable.	Biocompatibility issue, corrosion resistance, low mechanical integrity.	Mesh cage for segmental defects in bone, 3D scaffold design for better bone regeneration

3. Need for Surface Modification of Bioimplants

In an implant operation, any material inserted into the human body is treated as a foreign substance. If the foreign substances are not biocompatible, layers of fibrous tissues, also known as scar tissues, begin to develop between the tissue and implant. Eventually, due to scar tissue development, the implant fails to osteointegrate with the host bone, leading to implant failure. Therefore, the primary requirement for the successful implant process is to have a complete integration between bioimplants and human body tissues [120]. The biological responses of biomedical devices to the lifespan and performance are better controlled by their surface morphology and chemistry. To achieve better biocompatibility and osteoconductivity, surface modification on biometallic materials has been recommended to achieve the desired properties (Figure 1) to increase the success rate of implants. When the surface is effectively modified, the bulk functionality and properties of the biomedical implant device will remain unaffected for a long time [121,122]. With the advantage of bio-integration and the load-bearing capability of biomaterials, the success rate for bioimplants can be greatly increased.

In recent years, researchers tried to enhance the bio-integration of implants by modifying the implant surface that is in contact with the body environment. Two approaches are considered for modifying the surface of the implants. The first approach is to deposit organic/inorganic-based coatings on the metallic surface without modifying the implant substrate [123]. The second approach is to use conversion coatings or surface modified layers, where the chemical surface modification of a substrate results in a slight increase in thickness [124]. In this case, the substrate elements are involved in developing conversion coatings. For conversion coating, surface preparation by grinding and polishing is required to improve the surface roughness for better mechanical interlocking of coatings. This process is critical, and surface modification by depositing an overlay coating is recommended [125]. Recently, a combination of both surface modification and deposition of thin films was performed to achieve the synergy of both properties.

In a modern biomedical implant industry, surface modification of metallic implants with an appropriate coating material is used to enhance biocompatibility, corrosion resistance, antimicrobial behavior, and mechanical properties. Although there are many methods for the deposition of bioactive surface coatings, an optimal coating technique for biomedical applications has not been developed yet. Currently, the coatings on implant materials are deposited by one of the deposition techniques such as physical vapor deposition (PVD), chemical vapor deposition (CVD), electrophoretic deposition (EPD), electrodeposition (ED), or sol-gel methods [2]. Among these, PVD is recommended to deposit metal/ceramic materials over the implant surface and provide exact stoichiometry, excellent adhesion, high density, and good uniformity. Another method for surface modification other than coating methods is chemical etching to prevent bacterial adhesion and improve osseointegration [126].

The success of an implant is dependent on the stability of the coating, which provides better biocompatibility. This section is focused on the recent advancements in various types of ceramic and polymer coatings to improve bioimplant performance and reliability.

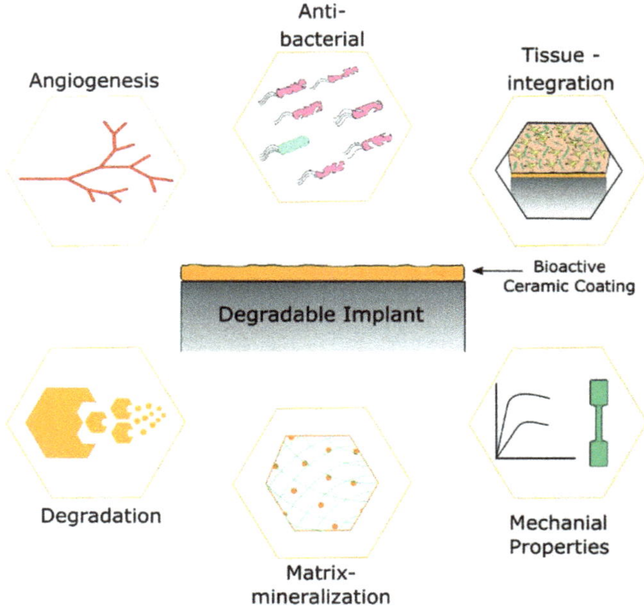

Figure 1. The role of bioactive coated metallic implants as a potential implant material [127]. The qualities of coated implants are superior to those of uncoated metallic implants.

3.1. Polyether Ether Ketone (PEEK)

PEEK is a thermoplastic material that shows a combination of excellent stiffness, chemical and physical properties, and toughness and offers a wide range of applications [128]. Therefore, it is widely used as a bone substitute in orthopedic and dental implants, and in clamps for removable dental prostheses [129]. The PEEK coated substrates show better tribological properties, which are useful for the development of coatings on light weight alloys which lack tribological performance. Most of the sliding and bearing implant materials are coated with PEEK due to its better wear resistance and thermal stability [130,131]. Generally, PEEK coating and its composites are prepared using thermal spraying or electrophoretic processes [132–135]. PEEK coating (70–90 μm thick) deposited through electrophoretic deposition on the Ti-13Nb-13Zr titanium alloy showed excellent wear resistance, 200 times higher than the uncoated alloy [136].

PEEK in combination with other bioactive materials shows better antibacterial activity than PEEK alone [137]. Many authors reported on PEEK-based composite coatings on metallic substrates. These coatings enhance bioactivity and electrochemical corrosion resistance, especially for implant structural components. Typical examples for the composite coatings are TiO_2/PEEK [138], sol-gel glass/PEEK [139], bioactive glass/PEEK [140], h-BN/PEEK [141], Ag/bioactive glass/PEEK [142], and h-BN/bioactive glass/PEEK coatings [137]. A combination of bioactive glass embedded in a polymeric matrix of PEEK makes it an interesting material for orthopedic applications as it meets biological and biomechanical requirements for the application. A cold sprayed Bioglass/PEEK composite prepared by Garrido et al. [143] showed an increase in wear resistance by more than 70%, higher hardness, and a lower coefficient of friction compared to pure PEEK. Coatings based on Bioglass/PEEK on porous Ti substrates resulted in higher adhesion between Bioglass/PEEK coating and Ti substrates [144].

Flame sprayed hexagonal boron nitride (h-BN) incorporated PEEK coating on low-carbon steel substrate increased the hardness and decreased wear and frictional coefficient values for the composite coating containing 8 wt. % h-BN due to its self-lubrication properties [145]. The coefficient of the friction value can also be reduced by the addition of alumina. The Al_2O_3/PEEK composite coating deposited on a Ti alloy using electrophoretic deposition showed increased corrosion resistance and significantly improved wear resistance under dry sliding conditions. The viability test revealed that the Al_2O_3/PEEK coating was found to be cytocompatible with MG-63 osteoblast cells [146]. The scratch resistance of PEEK coatings can be increased with the addition of amorphous Si_3N_4 nanoparticles. Tomasz et al. [147] performed the electrophoretic deposition of the PEEK/Si_3N_4 nanocomposite using a chitosan stabilizer: the coating showed higher scratch resistance than PEEK coating alone. This suggests that PEEK-based nanocomposite coatings potentially improve the bioactive as well as bio-tribological performance of Ti-based alloys used in biomedical applications. The use of PEEK with HAp as a coating can reduce the stress shielding effect. The combination of PEEK/HAp offers similar stiffness to that of the bone tissue. Recent studies suggest that the incorporation of HAp into PEEK coating improves bioactivity and mechanical properties [148]. PEEK coating prepared by different methods and their properties are summarized in Table 3.

Table 3. Methods and properties of PEEK-based composite coatings

S. No.	Coatings	Deposition Method	Significance	Ref.
1	PEEK coating on Ti alloy (Ti-13Nb-13Zr)	Electrophoretic deposition (EPD)	Excellent wear resistance Very good adhesion Low frictional coefficients	[136]
2	HAp/PEEK composite coating on PEEK substrate	Cold Spray coating	Better biocompatibility and osseointgration for clinical applications	[149]
3	SiC/PEEK composite coating on SS	electrostatic spray coating method	Scratch resistance Hardness increases	[150]
4	h-BN/bioactive glass/PEEK coating on SS 316L	Electrophoretic deposition (EPD)	Good adhesion strength Wetting behavior	[137]
5	PEEK/HAp on 316L SS	Electrophoretic deposition (EPD)	Good antibacterial activity	[151]
6	PEEK coating on Ti implant	Thermal spraying	Improved stability and fracture resistance Abrasion resistance	[152]
7	PEEK/ Bioglass composite coating on PEEK substrates	Cold gas spray	Better wear resistance Biomechanical performance	[143]
8	ZrO_2/PEEK coating on Ti6Al4V substrates	Thermal spraying	Improved wettability Blood compatibility Great potential for medical applications	[153]
9	Al_2O_3/PEEK, SiO_2/PEEK coatings on Ti6Al4V substrates	Thermal spraying	High hardness Optimum tribological properties Potential candidate for bearing material	[154]

3.2. *Titanium Dioxide (TiO_2)*

TiO_2 coatings are the most important materials in biomedical applications that are known for their antibacterial properties along with good mechanical properties. The applications of TiO_2 coating include drug delivery systems [155], orthopedic [156], and dental applications [157]. It also shows high catalytic activity, antibacterial activity, and long-term stability under photo and chemical corrosion [158]. TiO_2 promotes the formation of bone-like apatite or calcium phosphate on its surface. This property makes it a suitable candidate for reconstruction and bone replacement [159].

TiO_2 coated metallic substrates show better antibacterial properties. Gartner et al. [160] observed the same biocidal effect by applying TiO_2 coating on glass substrates by a sol-gel method. Photocatalytic activity of TiO_2 coating received much attention as a potential material for anti-bacterial coatings. The antibacterial effects of TiO_2 coating involve both a reduction in bacteria's viability and their destruction [161]. Park et al. [162] showed that the antibacterial effect against *S. aureus* could be improved by adjusting the nucleation time of TiO_2 film during the deposition process. The antibacterial effect of TiO_2 was explained by the formation of reactive oxygen species. Apart from antibacterial properties, the antiviral properties of the TiO_2 coating are also studied [163]. Table 4 summarizes the use of TiO_2 and its composite coatings for bioimplant applications.

Yetim [164] prepared TiO_2 coating with different concentrations of Ag using the sol-gel process on the commercially pure titanium substrate. Electrochemical corrosion properties obtained from electrochemical impedance spectroscopy measurements and potentiodynamic polarization tests in simulated body fluid (SBF) suggest that Ag doped TiO_2 enhances corrosion resistance over that of the bare Ti substrates as well as undoped TiO_2 coated samples [165]. The silver doped TiO_2 (Ag/TiO_2) nanocomposite coated glass substrate with varying Ag content synthesized by the sol-gel route showed antiviral properties against *E. coli*, enterovirus, and influenza A virus (H1N1) [166]. The highest level of photocatalytic degradation under irradiation with either visible or ultraviolet light was observed at an optimum Ag:TiO_2 weight ratio of 1:100. The antibacterial effectiveness was greater than 99.99% against *E. coli* and other infectious diseases after visible light illumination.

Sol-gel derived TiO_2-PTFE nanocomposite coating on stainless steel substrates was prepared by Zhang et al. [164] and their bacterial adherence were tested against two pathogens, namely *S. aureus* and *E. coli*. The bacterial adhesion and bacterial growth studies were evaluated by fluorescence microscopy after 2 h, 6 h, 12 h, and 24 h of incubation (Figure 2a,b). The TiO_2-PTFE coated substrate shows the lowest bacterial adhesion when compared with the uncoated substrate. The bacterial inhibition increases with the increasing TiO_2 concentration (Figure 2c,d). It is also observed that Gram-positive bacteria are less sensitive due to their cell wall thickness.

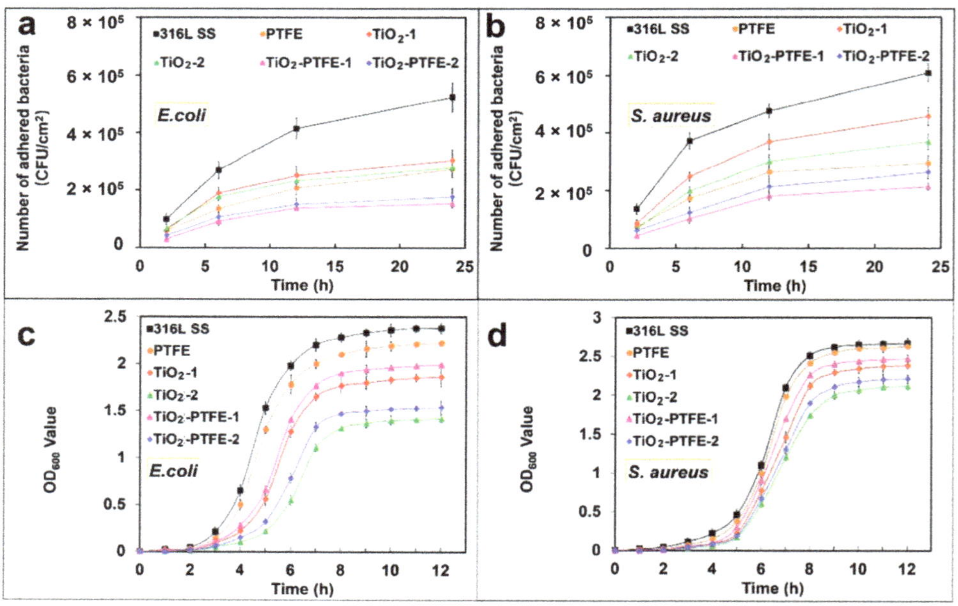

Figure 2. Effect of bacterial adhesion (**a**,**b**) and bacterial growth of *E. coli* and *S. aureus* pathogens on TiO_2-PTFE coated and uncoated substrates [164]. TiO_2-PTFE coated substrates exhibit lower bacterial adherence and a significant reduction in bacterial growth (**c**,**d**) as compared to uncoated substrates.

Table 4. Uses of TiO_2 and its composite coatings in bioimplant applications.

S. No.	Coatings	Deposition Method	Significance	Ref.
1	TiO_2 coating on Ti substrates	Anodic oxidation	Potential rehabilitation to internal bone fracture	[167]
2	TiO_2 coating on PEEK substrate	Dip coating	Recommended for maxillofacial and oral implants applications	[168]
3	TiO_2/$MoSe_2$/chitosan coating on Ti implants	Micro-arc oxidation process	Excellent in vivo and in vitro antibacterial property against *S. mutans* Better biocompatibility and hydrophilicity Better antibacterial properties	[169]
4	Poly(epsilon-caprolactone)/titania (PCL/TiO_2) coating on Ti implants	Electrospinning technique	Good bioactivity against osteoblast cell Superior antibacterial against *S. aureus* Promoting cell attachment	[170]
5	TiO_2 coating on Ti substrates	Direct lithographic anodic oxidation	Corrosion resistant	[171]
6	TiO_2 nano coating	Anodizing oxidation technique	Better cell proliferation and adhesion Better osseointegration	[172]
7	Graphene/TiO_2 coating on Ti substrate	Drop casting method	Better cell adhesion and proliferation behavior	[173]
8	TiO_2/HAp bilayer coating on Ti substrate	MOCVD/Plasma spraying	Better hardness In vitro bioactivity	[174]
9	Y-doped TiO_2 coating on Ti alloy	Plasma electrolytic oxidation method	Better antibacterial activity against *E. coli* and *S. aureus*	[175]
10	Fe_3O_4/TiO_2 composite coating on Ti implants	Micro-arc oxidation process	Prevent inflammatory Better fibroblast response	[176]

3.3. *Transition Metal Nitrides*

Earlier, transition metal nitrides and carbides were widely used to protect the metallic components against wear, tear, and corrosion, potentially offering high-temperature stability. Titanium nitride (TiN) coatings were used as decorative coatings in earlier days. In the last decade, nitride coatings for orthopedic implants were also proposed to protect the implants against wear and tear and to act as a diffusion barrier layer preventing the toxic ion release from the implant metal surfaces to the human body fluids [177–180]. The physical properties of TiN coated substrates show high scratch resistance, hardness, and low frictional coefficients. These properties make them a potential candidate for use as coatings on different metals used in arthroplasty. TiN-based coatings used for orthopedic applications show better biological properties as compared to other nitrides [181]. TiN coatings show better blood tolerability properties with a hemolysis percentage near zero [182]. TiAlN is another biocompatible nitride that has proven to be a promising alternative to TiN in biomedical applications despite its aluminum (Al) content [183].

Transition metal carbonitrides (TiCN, ZrCN) were found to increase the service life of orthopedic implants in terms of wear resistance in biological media [184–186]. Recently, quaternary carbonitrides-based coatings (TiAlCN, TiCrCN, TiNbCN, etc.) were found to show increased anticorrosive, mechanical, and tribological properties compared to ternary carbonitride-based coatings [187–189]. The tribological properties of these carbonitride coatings are very complex. However, the carbon-based carbonitride coatings show good biocompatibility, better wear resistance, and low friction [190]. Much attention has been paid to developing MeSiC-, MeSiCN-, and MeSiN- (where Me is a transition metal, and Si is an alloying element) based hard coatings [191–194]. These types of coatings show high thermal stability, a low frictional coefficient, excellent wear resistance, and good mechanical properties (hardness, Young's modulus). Moreover, in many investigations, TiSi-based carbide and carbonitride coatings proved to be a potential candidate for a metallic implant which combines the mechanical, tribological, and anticorrosive properties of TiN and TiC with the biocompatibility behavior of SiC and SiCN [192,195–197].

TiN coating shows plastic deformation at the coating/surface interfaces due to dissimilarities in the hardness of the substrate and coating [198]. Thus, TiN coating cannot accommodate the fracture and deformation that creates flakes, and defects in the coatings cause deterioration of the coatings from the substrate. Therefore, chromium nitride (CrN) and chromium carbonitride (CrCN) coatings are recommended, which act as a better diffusion barrier for ion release from the alloys. These coatings also exhibit higher toughness, higher cohesive strength, and lower wear debris than TiN coatings [199].

TiN and TiCuN coatings were prepared by the axial magnetic field enhanced arc ion plating (AMFE-AIP) technique, and the in vitro angiogenic response of human umbilical vein endothelial cells was studied by Liu et al. [200]. The TiCuN coating showed better antibacterial activity, and both coatings showed no cytotoxicity to human umbilical vein endothelial cells (HUVECs). TiCuN coatings promote early cell apoptosis, which is important for vascular tissue modeling (Figure 3).

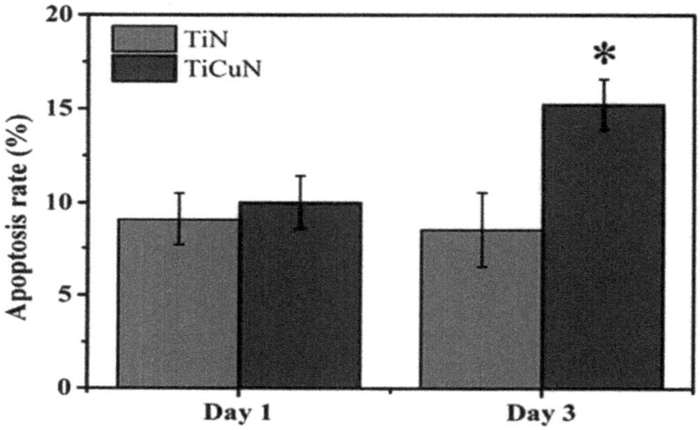

Figure 3. Apoptosis rate of TiN and TiCuN coatings tested for Day 1 and Day 3. Annexin V-FITC/PI double staining kit was used to evaluate the apoptosis rate of these coatings [200]. TiCuN coating promoted the early cell apoptosis rate more than TiN coating. *: Denotes TiCuN coating superior performance.

Transition metal oxynitrides have been considered as interesting materials due to their known mechanical properties, chemical stability, and corrosion resistance in simulated body fluid. Zirconium oxynitride (ZrON) and titanium oxynitride (TiON) based coatings were recently used in biomedical applications for their better corrosion resistance than TiN coating and their anti-fouling ability [201,202]. The magnetron sputtered ZrON and TiON coated 316L SS specimen show better hardness and wear resistance behavior than the uncoated substrate [203]. In addition, both coatings show better anti-fouling performance against *Pseudomonas aeruginosa* bacterial adhesion than uncoated substrates. The coated substrates also show better corrosion protection with or without the addition of hydrogen peroxide (H_2O_2) in artificial blood plasma (ABP) solution [203].

Surface modified coatings prepared from ternary nitrides such as TiZrN, TiCrN, and TiAlN gained considerable attention because they retain their physiochemical properties, such as oxidation resistance, hardness, corrosion resistance, biocompatibility, and structural stability after implantation [204,205]. Magnetron sputtered TiZrN coated 316L SS substrates showed less bacterial adhesion, increased corrosion protection, and negligible human blood platelets activity than uncoated substrates [206]. Recent developments in binary, ternary, and quaternary systems of transition metal nitrides and carbide coatings are tabulated in Table 5.

Table 5. Recent work on binary, ternary, and quaternary systems of transition metal nitride and carbide coatings for implant applications.

S. No.	Coatings	Deposition Method	Significance	Refs.
1	Nano-TiN coating on Ti-6A1-4V	Magnetron sputtering	Enhanced hardness and anti-wear resistance, good hemocompatibility, and biocompatibility	[207]
2	TiN coating on Ti alloy	Cathodic arc deposition	Better corrosion protection Low wear rate Reduced coefficient of friction	[208,209]
3	TiON coating on 316L SS	Magnetron sputtering	Better adhesion Good resistance to corrosion	[210]
4	TiON coating on Ti substrates	Magnetron sputtering	Better biological activity Highly biocompatible	[211]
5	TiCN coating on Ni-Cr alloy	Magnetron sputtering	Good adhesion of fibroblasts Less cytotoxic	[212]
6	TiZr/a-C coatings on Ti substrate	Cathodic arc deposition	Good compatibility with human skin fibroblast cells Good human skin fibroblast cell viability	[213]
7	TiZrCN, TiNbCN, and TiSiCN coatings on steel substrates	Cathodic arc deposition	Better adhesion to the substrate Corrosion resistance	[214]
8	TiAlN coating	Multi arc ion plating technique	Better tribological performance	[215]
9	Nanolayer CrAlN/TiSiN coating on steel substrates	Magnetron sputtering	Excellent tribological performance	[216]
10	TiCN/TiAlN and TiAlN/TiCN bilayer nitride coatings on cemented carbide substrates	Cathodic arc deposition	Higher hardness High scratch resistance	[217]
11	CoCrMoC/CrN and CrN/CoCrMoC coatings on medical grade SS substrates	Magnetron sputtering	Better tribo-corrosion behavior	[218]

3.4. Carbon Based Coatings

Carbon based materials are categorized under bioinert coatings. These coatings are used in load-bearing applications and wear components to improve elevated corrosion resistance, wear, and frictional effects [219]. Besides, carbon-based coatings show minimum protein adhesion and very good biocompatibility due to the hydrophobic nature of carbon-coated surfaces. Three different types of carbon-based coatings are used for biomedical applications. They are (a) nanocrystalline diamond (NCD), (b) pyrolytic carbon (PyC), and (c) diamond-like carbon (DLC) [220]. Some of the coatings are commercially available, while others are under development.

Most of the PyC coatings in biomedical applications are found in the heart valves due to their thromboresistant qualities and biocompatibility [221]. Most of the artificial heart valves are lined with a thick PyC coating. PyC biocompatibility in heart valves is well established. PyC coatings have also been used in orthopedic applications [222]. By varying the process parameters of the PyC (such as temperature, surface area, gas flow rate, precursor) in the CVD process, a variety of the structures can be produced. The most interesting structure for biomedical applications is lamellar, isotropic, granular, and columnar [223–225]. PyC coated orthopedic implants are used to replace small joints such as wrist joints, knuckles, and arthroplasty of proximal interphalangeal joints [226].

Carbon coatings, including nanocrystalline diamond and DLC coating, show many remarkable biological properties and are considered as coatings for medical implants. NCD coatings deposited by the CVD process consist of sp^3-hybridized carbon bonds and show grain sizes in the range of a few nanometers. NCD coatings generally show very low surface roughness and possess the properties of a diamond, such as hydrophobicity and excellent biocompatibility with blood [227,228]. This makes them an ideal coating choice for wear-resistant implant applications and cardiovascular devices. NCD coating can also be used as hard antibacterial coatings that reduce the risk of infections. The electrically active NCD coating surfaces can establish a chemical bond with the biomolecules in the surrounding environment. Medina et al. [229] observed that the NCD coating surfaces react with the cell wall or membrane of Gram-negative *P. aeruginosa* bacteria and establish a chemical bond that alters the bacteria morphology, hindering bacterial adhesion and colonization on the surface of the coating. The properties of NCD films are utilized in biosensing and neurochemical sensing applications [230].

More experimental studies have been reported on DLC based coatings, which are considered as the most promising materials for bioimplant applications [231–234]. Medical grade PEEK samples were coated with DLC using plasma immersion ion implantation and deposition (PIII & D) technique, and their in vitro cytocompatibility and osteogenesis studies were carried out by Mo et al. using human bone marrow mesenchymal stem cells (hBMSCs) [235]. DLC coated substrates show better surface coverage of cells and show high cell viability on the seventh day, which indicates better biocompatibility of DLC-PEEK coatings than PEEK coating (Figure 4). However, DLC suffers from residual stress arising from the substrate/coating thermal expansion mismatch and lattice misfit, which cause poor substrate adhesion and delamination of the coating from the substrate. Another major concern about DLC coatings is their instability in the aqueous environment, which promotes delamination of the coating [236]. To avoid this issue, it is recommended to use interlayers (called buffer layer) such as CrC, Ti, and Si_3N_4 at the interface of the substrate and DLC coating [237]. Another approach is to dope DLC coating with N, F, Ag, Zr, or Ti to avoid a thermal expansion mismatch and residual stress [238].

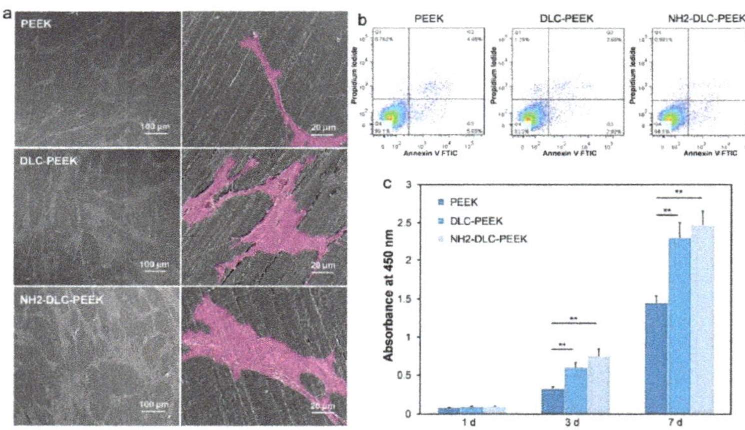

Figure 4. Represents in vitro cytocompatibility of DLC coated PEEK substrates, (**a**) surface morphology of hBMSC cultured on PEEK, DLC-PEEK, and NH2-DLC-PEEK substrates for 1 day, and the enlarged cells are shown in pseudo-color, (**b**) cell viability for 1 day, and (**c**) proliferation of hBMSCs after culturing samples for 1 d, 3 d, and 7 days [235]. ** denotes $p < 0.01$.

The properties of DLC such as chemical inertness, surface smoothness, and hydrophobicity are important for providing better compatibility with blood, reducing platelet activation in contact with the blood, which could trigger thrombosis. DLC can act as a protective coating under the conditions of the human blood environment, which limits the release of nickel ions from metallic implants such as SS 316L. Several studies suggest that DLC coating prepared by various routes is biocompatible and does not induce any inflammation reaction both under in vivo and in vitro conditions [235,239]. Because of these remarkable features, DLC coatings found various applications as coatings in many implant devices such as cardiovascular stents, heart valves, surgery needles, medical wires, contact lenses, etc. DLC coatings can also be used as protective coating in knee replacement because of their high corrosion resistance, hardness, and low wear rate. Generally, DLC films are used to reduce the frictional coefficient and offer better wear resistance [238]. Carbon based coatings and their significance in biomedical field are summarized in Table 6.

Table 6. Different carbon coatings and their properties.

S. No.	Coatings	Deposition Method	Significance	Ref.
1	DLC on Ti alloy	Plasma immersion ion deposition (PIID)	Improvement in tribo-corrosion behavior	[240]
2	Si-DLC on Polyethylene (PE) substrates	Plasma and laser-based processing methods	Improvement in hydrophobicity, lubricity, and electrical conductivity	[241]
3	Carbon coatings on X39CR13 and 316LVM steels	Magnetron sputtering	Improved adhesion and wettability properties	[242]
4	Amorphous carbon/diamond-like carbon (a-C:H) coatings on PEEK substrate	Plasma enhanced chemical vapor deposition	No toxicity issues and better biological performance	[243]
5	DLC with Zr interlayers on Ti alloy	Magnetron sputtering	Reduced coefficient of friction	[244]
6	Si-DLC Coatings on Ti alloy	Magnetron sputtering	High level of biocompatibility due to the presence of Si	[245]
7	a-C:H coating on Co-Cr alloy	PVD/PE-CVD	Excellent mechanical properties, high hardness, and elastic modulus	[246]
8	Si doped DLC on Ti alloy	Magnetron sputtering	Reduced microbial colonization of *E. coli*	[247]
9	DLC on stainless steel	Pulsed DC PE-CVD	Improved biocompatibility and corrosion resistance	[248]
10	DLC with TiO_2 on stainless steel	PE-CVD	Better biocompatibility and antimicrobial activity	[249]

3.5. Calcium Phosphates

Calcium phosphate (CaP) ceramics are widely used as implants since they have a chemical composition similar to the inorganic composition of the bone. By controlling the surface properties such as roughness and porosity of CaP, one can regulate the biomineral formation and cell/protein adhesion. Bioactivity properties are varied depending on the type of calcium phosphates (HAp, tricalcium phosphate (TCP)) because of the differences in crystallinity, solubility, stability, ion release, and mechanical properties. At first, CaP coatings were deposited through the vapor phase process, but in recent years, biomimetic and solution-based methods were developed. Each synthesis approach has its own intrinsic properties, but in general, CaP based coatings are promising to improve implant longevity and biocompatibility. Many studies have been focused on the development of CaP ceramic coatings on metallic substrates to achieve the biological properties identical to a bulk and to enhance the implant durability and fixation [250–253].

Presently, atmospheric plasma spraying (APS) is currently employed to develop CaP coating on implant surfaces [2]. The CaP phases in the coatings exhibit higher solubility in an aqueous medium than HAp which is desirable for activating bone formation. However, faster dissolution reduces the stability and can cause loosening of the implant. A highly crystalline HAp phase dissolves in human physiological conditions at a lower rate which provides long-term stability of the implants. Thus, for the development of implants with required properties, one must control the purity and crystallinity of the coatings. CaP coatings with a denser microstructure lower the risk of delamination of the coating during in vivo tests with human body fluids. Coating surface roughness affects its dissolution and bone apposition and growth. Porous surfaces may enhance cell attachment or formation of the extra-cellular matrix, but the accumulation of macropores at the coating/substrate interface weakens the coating adhesion [254].

CaP in the form of HAp is widely used in implant applications due to its superior biological response. The HAp composition is $Ca_{10}(PO_4)_6(OH)_2$ (Ca/P = 1.67), which resembles the chemical composition of hard tissues such as bone and teeth [255]. Hence, HAp is considered as a primary candidate material due to its exceptional biological properties such as excellent biocompatibility, osteoconductivity, osteoinductivity, and bioactivity [256]. HAp coatings release calcium and phosphate ions and regulate the activation of osteoclasts and osteoblasts, facilitating bone regeneration [257]. The use of HAp ceramics enhances the regeneration of bones, improves osteoconductivity for bone growth, and promotes mineralization through ion release control and encapsulating growth factors. HAp ceramic coating enhances bone apposition in orthopedic implants through the formation of an extremely thin bonding layer with the existing bone. Due to such tissue bonding characteristics, Hap-based ceramics are considered as bioactive-based coatings. The continuous effort to improve the durability of the HAp ceramic coatings has led to development of high-quality HAp coatings and the development of Hap-based composite coatings.

Highly porous or highly crystalline HAp coating shows poor adhesion to the substrate. Sankar et al. [258] studied the corrosion behavior of HAp coatings prepared by electrophoretic deposition (EPD) and the pulsed laser deposition (PLD) method. The corrosion results suggest that the HAp coatings show lower corrosion protection than the coatings prepared by the PLD method due to the formation of denser and pore-free coating [258]. Corrosion protection can also be enhanced by the addition of antimicrobial dopants. For example, Yugeswaran et al. [259] prepared HAp-TiO_2 nanocomposite coatings by APS. The coating shows better corrosion performance in SBF medium than HAp coating without dopants due to its high compactness and the presence of TiO_2 [259]. Silver (Ag) containing HAp coatings prepared by Trujillo et al. [260] show better antibacterial activity than HAp coating alone against *P. aeruginosa* and *S. epidermidis* pathogens due to the antibacterial activity of Ag. The antimicrobial activity of the Ag-doped HAp composite against *E. coli* and *S. aureus* was tested by Lett et al. [261]. The results indicated that the Ag-doped HAp composite has better inhibition of bacterial growth and shows a stronger ability against *S. aureus* bacteria to fight against toxic responses (Figure 5). The absence of Ag in the composite results in lower antibacterial activity of HAp composites. The variation in antibacterial activity was attributed to a thinner cell wall response of *S. aureus* (Figure 5b) to Ag ions than *E. coli* (Figure 5a) [261].

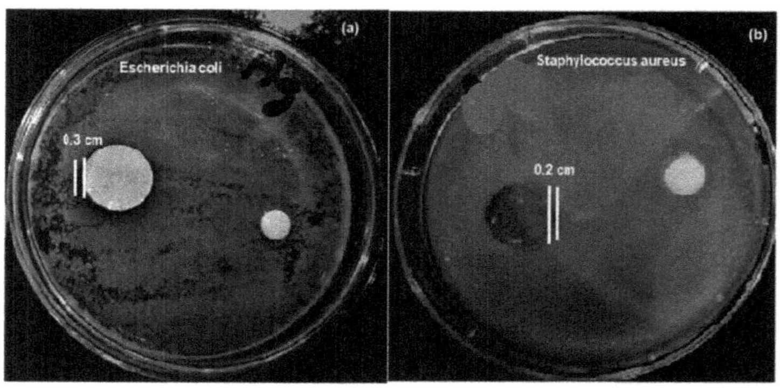

Figure 5. Demonstration of antimicrobial activity of HAp and Ag doped HAp composites against *E. coli* (Gram-negative) (**a**) and *S. aureus* (Gram-positive) (**b**) bacteria [261]. The photograph shows that Ag-doped HAp inhibits *S. aureus* bacteria more effectively than *E. coli*.

In biomedical implants, the major challenge for the performance of implants is bacterial invasion. During surgical operation, the bacteria may enter the surface of the implants through surgical equipment or cross contamination which form a biofilm. Once the surrounding implant is infected, the infection causes implant loosening. To overcome this issue, antimicrobial agents are used as dopants in ceramics, protecting implant material from bacterial invasion and improving their durability. Zinc doped HAp composites prepared by the sol-gel route and annealed at different temperatures (500 °C and 700 °C) show higher antimicrobial activity against *C. albicans* fungal cells and *S. aureus* bacteria [262].

Multiple doping of ions into HAp coatings was also attempted to improve their structural stability, partial dissolution, and biocompatibility. Wang et al. [263] prepared Sr and F^- doped hydroxyapatite and studied the properties of the coating. The addition of the dopant improves the structural stability of the HAp lattice and promotes osteogenic cell differentiation. Moreover, the addition of F^- ions potentially arrests the formation of *S. aureus*. Dopants such as Cu, Zn, Mg, Ag added to HAp enhance antibacterial activity and decrease the toxic effects towards the human body cells [264–267]. For example, Mg-doped HAp shows better osteoblast cell adhesion than pure HAp [268].

The differences in the thermal expansion coefficient of HAp and metallic alloys result in residual thermal stress. The stress accumulation increases with the increase in the coating thickness, which promotes cracking or delamination of the coating. For a thicker coating, the outer layer may detach from the implant, whereas a thin HAp coating can prematurely resorb during bone regeneration. Various HAp composites and their biological properties are summarized in Table 7.

Table 7. Hydroxyapatite and its composites' coatings for implant applications.

S. No.	Coatings	Deposition Method	Significance	Ref.
	HAp nanowire coating on glass substrate	Solvothermal method	Excellent apatite-forming ability	[269]
	Fe doped HAp on Si substrate	Co-precipitation method	Promote better proliferation and adhesion of the osteoblast cells	[270]
	Ce doped HAp/collagen coating on Ti surface	Biomimetic method	Better antibacterial efficacy against *Escherichia coli* and *Staphylococcus aureus* bacteria than HAp coating	[271]
	Si substituted HAp coating on Ti substrate	Precipitation method	Favorable regeneration of crystalline Si-HA layer	[272]
	HAp/$CaSiO_3$/Chitosan Porous coating on Ti substrate	EPD	Improved bioactivity and biocompatibility	[273]
	Bioactive glass/HAp coatings on Ti substrate	Pulsed laser deposition	Significant bioactivity, cytocompatibility, and hemocompatibility	[274]
	PyC/SiC/HAp coating on carbon fibers	Chemical vapor deposition/pulsed electrochemical deposition	Excellent corrosion resistance, induces the nucleation process and growth of bone-like apatite	[275]
	PEEK/HAp composite coating on 316L SS substrate	Electrophoretic deposition	Enhanced in vitro bioactivity	[148]
	Ag/HAp coating on Ti substrate	Sol-gel route	Enhanced antibacterial activity and better corrosion protection	[276]
	TiO_2/HAp coating on Ti substrate	High velocity oxy fuel (HVOF) method	Improved corrosion resistance	[277]
	B_2O_3/Al_2O_3/HAp coating on Ti substrate	High velocity oxy fuel (HVOF) method	Improved adhesion strength	[278]
	TiO_2/HAp nanocomposite coating on 316L SS substrate	Electrophoretic deposition	Excellent corrosion protection under SBF medium	[279]

3.6. Zirconia

Zirconia (ZrO_2) is a ceramic material that can withstand high temperatures as well as higher stresses. It has widespread applications in dental implants and in the coatings on metallic implants to increase their corrosion resistance [280]. ZrO_2 ceramics offer many advantages, including mechanical strength, chemical stability, biocompatibility, good aesthetics, and better wear resistance. Zirconia stabilized with yttria (YSZ) has been used as a dental implant due to its excellent mechanical strength and fracture toughness [281]. YSZ coatings show better hardness and scratch resistance than HAp coating [282]. Gobi Saravanan et al. [283] observed that the YSZ coated Ti substrates show improved hemocompatibility, activating blood platelets with pseudopods. In addition to that, superior in vitro biomineralization behavior was observed and documented through the weight gain on YSZ coating.

Zirconia stabilized with different weight fractions (0, 4, 10 wt. %) of yttria yields different phases (monoclinic, tetragonal, and cubic): zirconia ceramics with tailored mechanical properties and biocompatibility can be thus prepared. Attempts were made to deposit different phases of zirconia (m-ZrO_2, t-ZrO_2, and c-ZrO_2) with the use of electron beam physical vapor deposition (EBPVD) [284]. All the coatings show lower surface roughness than coating prepared through the APS method and reduce pathogen bacterial invasion. Particularly, t-ZrO_2 shows superior hardness over the other two zirconia phases. All the allotropes show better blood plasma protein adhesion and enhanced resistance to corrosion in comparison to uncoated medical grade stainless steel substrates in ABP solution.

Antibacterial activity of ZrO_2 coating can be enhanced by the addition of Ag. Ag-ZrO_2 composite coatings were prepared by Pradhaban et al. [285]. The results suggest that the coating shows antimicrobial activity against *E. coli*. Santos et al. [286] prepared glass ceramic composites with different concentrations of ZrO_2 particles (0–50 vol. %) and carried out a ball-on-plate tribology test. ZrO_2 glass ceramic composite (30 vol. % of ZrO_2) shows optimal wear properties (coefficient of friction is 0.3) and is recommended for load-bearing applications. Bermi et al. [287] deposited YSZ coating through pulsed plasma deposition, and the tribological behavior of the coating in both dry and wet conditions was tested. YSZ coating deposited on a Ti6Al4V alloy ball sliding against the UHMWPE disk shows a reduction in wear rate (17% and 4% in dry and lubricated conditions) than uncoated alloy substrate.

Kaliaraj et al. [288] prepared zirconia coatings on a 316L SS substrate by electron beam physical vapor deposition (EBPVD), and a bacterial adhesion study with *P. aeruginosa* was carried out. Epifluorescence microscopy analysis of live/dead cells after incubation of 1, 2, 3, and 4 days showed a drastic reduction in bacterial adhesion on ZrO_2 coatings, along with retardation in biofilm formation (Figure 6). This observation was attributed to the decrease in surface roughness obtained through coating deposition and the surface chemistry of ZrO_2 that inhibits bacterial adhesion. Electrochemical impedance corrosion results show that ZrO_2 exhibited superior corrosion resistance in the presence of H_2O_2 in an artificial blood plasma electrolyte solution. [288].

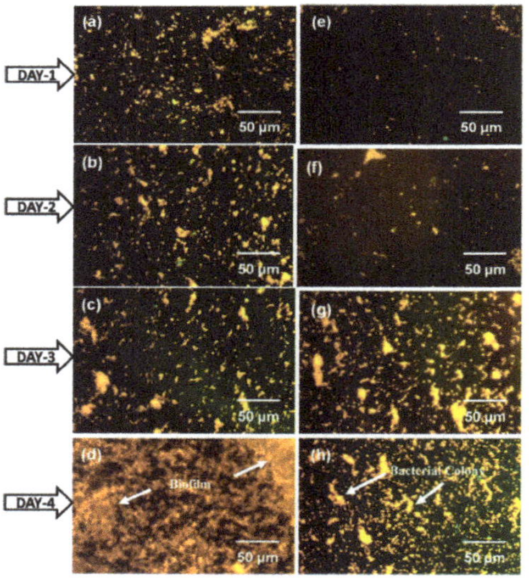

Figure 6. Epifluorescence microscopy analysis of P. aeruginosa bacterial invasion on 316L SS (**a–d**) and ZrO_2 film (**e–h**) after 1, 2, 3, and 4 days incubation [288]. The used acridine orange staining shows orange color for live cells and green color for dead cells. The reduction in bacterial adhesion was seen on ZrO_2 coated substrate compared to uncoated 316L SS.

3.7. Bioactive Glass Coatings

Hench pioneered bioactive materials research and revolutionized the fields of bioactive materials and ceramics with his discovery of bioactive glass (45S5 composition), commercially known as Bioglass [289]. In the wake of Bioglass, various compositions and composites of bioactive glasses or silicates prepared both by melt quench and sol-gel techniques were investigated. Although bioactive glasses exhibit excellent bioactivity, because of their amorphous or semi-crystalline nature, they often fail as an implant material due to their poor mechanical strength. To overcome the shortage in mechanical properties, bioactive glasses are often composited with various metal oxides such as TiO_2, Al_2O_3, ZrO_2, and 2-D materials such as graphene and its derivatives (graphene oxide and reduced graphene oxide) [290]. These composites were reported to improve the corrosion resistance, antibacterial activity, and angiogenic properties of bioactive glass coatings without losing the bioactivity [291]. Similar to many ceramic materials, bioactive glasses can also be prepared in the form of particles of nano and micron size, as mesoporous particles, fibers, 3D scaffolds or monoliths, and thin films or coatings [292].

In this section, various types of coating technologies that can be used for the coating of bioactive glasses and their composites on different types of metals, alloys, and certain specific surfaces are discussed. One of the most simple and economical coating processes is the sol-gel dip-coating process. However, the coatings are often porous because of the solvent evaporation leading to poor corrosion resistance and mechanical properties. Nevertheless, this problem can be solved by incorporating metal oxides such as B_2O_3 as reported by Pinki Dey et al. [293]. According to their report, by replacing the silica weight percentage in the 45S5 system by 1% to 5 wt. %, they were able to decrease the porosity in the particles. Thermal spray coating, an industrial coating process, can also be employed for bioactive glass coating preparation. This process involves the coating of bioactive glasses as fine droplets or as plasma and sprayed over metal surfaces. Porous and non-porous layers with varying coating thicknesses can be achieved by the thermal spray process by tuning the deposition parameters such as velocity, size of the droplets, and temperature of the substrates [294].

Bioactive glasses can also be coated by physical deposition techniques such as radio-frequency magnetron sputtering (RF-MS) and pulsed laser deposition. In a recent study conducted by Qaisar Nawaz et al. [295], silver nanoclusters embedded in a silica matrix were deposited over the PEEK/BG layer using RF co-sputtering. They report a uniform 100 nm of the Ag-SiO_2 layer that showed slower and sustained release of silver ions compared to the electrophoretically deposited coating. Although the physical deposition techniques are very robust and highly reproducible, their shortcoming is often the expensive experimental setup and precursors when compared to wet chemical sol-gel coating techniques. On the other hand, electrophoretic deposition (EPD) combines both the advantages and disadvantages of sol-gel coating and physical deposition methods. EPD is both a versatile and cost-effective method for coating ceramic materials on conducting surfaces.

Ashokraja et al. [296] reported bioactivity in simulated body fluid (SBF) and reactive oxygen production using the XTT assay for reduced graphene oxide (rGO), sol-gel derived bioactive glass rods (BGNR) followed by different methods for developing composites of rGO and BGNR such as under constant stirring (COL), under constant sonication (SOL), and with a simultaneous reduction in graphene oxide-BGNR composites (RED). In their study, they report the role of pH changes in the sol-gel process facilitating one-dimensional rod-shaped bioactive glass formation, and their immersion studies exhibited a 50-micron thick HAp layer on the seventh day for rGO/BG composites [297]. Their work also reports that the different methods employed to prepare the composites influence the HCA formation, antibacterial efficacy, hemocompatibility, and cell proliferation as shown in Figure 7.

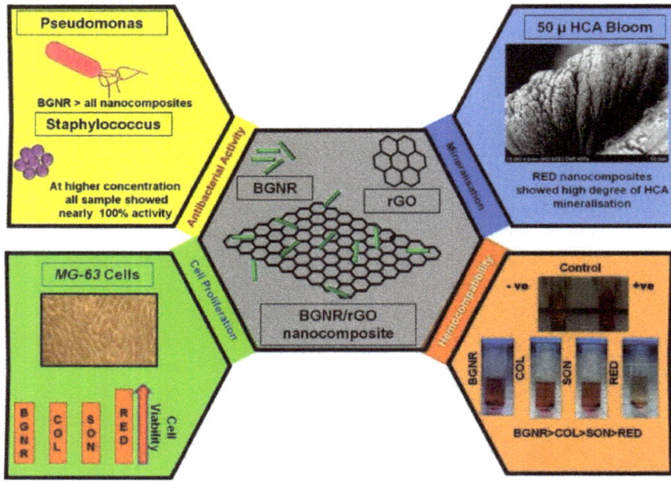

Figure 7. Schematics for HCA formation, antibacterial activity, hemocompatibility, and cell proliferation of bioactive glass rods (BGNR) and their composites with rGO (COL, SON, and RED) [297]. Figure also shows the bioactive behavior of the BGNR-rGO composites. It is noticed that the RED composites showed better HCA layer formation, cell proliferation, and hemocompatibility.

A recent comparative study reported results between pure BG and rGO/BG thin films deposited over the anodized surface of titanium by EPD. The deposited bioactive coatings (both pure and composites) were 2 μm thick and exhibited very good HAp formation in simulated body fluids along with super hydrophilicity in pure bioactive glass coatings [298]. Table 8 summarizes a brief list of bioactive glass coatings, their compositions, coating processes: important features are elucidated.

Table 8. Composition, the substrate used, coating process, and their salient features of bioactive glasses.

S. No.	Coatings	Substrate	Deposition Method	Significance	Ref.
1	Titanium, HAp, Bioactive glass wt.% (57–60 SiO_2, 21–24 CaO, 9–11 Na2O, 2–3P_2O_5, 0.5–1.5 TiO_2, and 2–3B_2O_3)	Ti-alloy—Ti6Al4 V	Laser engineered net shaping	Improved hardness and wear resistance	[299]
2	58S Bioactive glass (molar composition of 35% CaO, 60% SiO_2, and 5% P_2O_5) seeded in HAp	Commercial AISI 316L SS	Cold uniaxial pressing	Seeding of HAp increased the hardness as well as apatite layer formation	[300]
3	Bioglass with silver nanoparticles and Chitosan	Ti-alloy—Ti6Al4 V	Electrophoretic deposition	Increased coating uniformity and nanoscale roughness for bioactivity	[301]
4	(1) 65% SiO_2, 5% P_2O_5, and 30% CaO, (2) 45% SiO_2, 5% P_2O_5, and 50% CaO	Carbon foam	Dip Coating	Compact and dense coating is reported in 65% rather than 45% SiO_2	[302]
5	Manganese modified Bioglass/alginate	316L SS	Electrophoretic deposition	Increase in manganese improves the corrosion resistance in SBF	[303]
6	Bioglass composite with chitosan and iron oxide nanoparticles	Ti-alloy—Ti-13Nb-13Zr	Electrophoretic deposition	Better corrosion resistance, coating adhesion, and hydrophilicity	[304]

Table 8. *Cont.*

S. No.	Coatings	Substrate	Deposition Method	Significance	Ref.
7	Silver incorporated HAp and Bioglass	Nickel titanium alloy	Dip coating	Increased corrosion resistance and coating adhesion	[305]
8	Bioglass	AISI 304L SS	APS	Improved mechanical strength and corrosion resistance	[306]
9	Bioglass, TiO_2, Al_2O_3, and Hap composite with PMMA	Stainless steel 304	Dip coating	PMMA-TiO_2 coating exhibited higher corrosion resistance than other composites coatings	[307]
10	58S bioactive glass-gelatin-polycaprolactone composite	316L SS	Electrospinning	Increase in bioactive glass weight % improved surface Roughness and adhesion strength, exhibited good corrosion resistance, apatite formation and cell viability	[308]
11	58S Bioactive glass	Vitallium alloy	Dip coating	Decreased porosity and increased bioactivity	[309]
12	Bioglass	Ti6Al4V alloy	Electrophoretic deposition	Scratch resistance, hardness, and coating bonding strength	[310]
13	HAp-Bioglass-Iron oxide composite	Ti-alloy—Ti-13Nb-13Zr	Electrophoretic deposition	Corrosion resistance and non-toxic effects	[311]
14	Reduced graphene oxide—Bioglass sol-gel composite	Grade 2 titanium	Electrophoretic deposition	rGO facilitated low hemolysis and improved cell proliferation	[298]

4. Summary and Future Directions

This paper reviews different biomaterials and explains their significant characteristics that influence their bioactivity. Bioimplant manufacturing involves an integrated process of selection of materials, design, fabrication, and surface modification through micro/nano texturing or coating application. Engineering of native metals by converting them into alloys yields desired properties and provides flexibility in designing the needs as per implant requirements. For a long-term application of bioimplants, surface characteristics and their biological functions are considered as key factors. Engineering the surface of the biomaterials by applying suitable coatings provides flexibility in tailoring the properties as per the requirements.

Bioceramic coatings hold great potential by tailoring the biological properties that suit our needs: the choice of the coating depends on the interaction between the cells with the coatings and substrates that are being used. Coatings on metallic implants are invaluable due to their functionality, biocompatibility, durability, and stability. Bioactive coatings are used to enhance the biological fixation between the bone and metallic implant despite their poor tribological and mechanical properties. Hence, they are often improved by developing composites with materials that possesses good mechanical strength. These improved coatings can be also used for durable load-bearing implants. All these properties lead to a better clinical success rate in long-term use in comparison to uncoated metallic implants. The bioactive ceramic coated biodegradable implants provide synergistic properties of both the implants and coating. Thus, these coatings find applications in cardiovascular stents, heart valves, orthopedic applications, tissue engineering, drug delivery, and biosensors. The current trends of ceramic coatings coated metallic implants are more focused on orthopedic applications.

Feasibility studies on complex structures, designing, fabrication of metallic alloys to form complex shapes without losing mechanical properties and surface integrity are a challenging task and should be attempted. The degradation mechanism of coatings on metallic implants changes in the human body environment. Moreover, lattice mismatch and the accumulation of residual stress cause degradation of the implant after implantation. Thus, there is a need to develop mathematical models for the prediction of degradation mechanisms. Another approach to reducing the residual stress is to deposit a functionally graded multi-layered or nanocomposite coating with multifunctional properties.

Author Contributions: Conceptualization, K.K.A.M., A.R.C. and A.D.; validation, A.P. and D.G.; writing—original draft preparation, K.K.A.M., A.R.C. and A.D.; writing—review and editing, A.P. and D.G.; supervision, A.P. and D.G.; project administration, D.G. and A.P.; funding acquisition, D.G. and A.P. All authors have read and agreed to the published version of the manuscript.

Funding: This project received funding from the European Union's Horizon 2020 research and innovation program under grant agreement No 739566. This work was also created in the frame of the project Centre for Functional and Surface Functionalized Glass (CEGLASS); ITMS code is 313011R453, operational program research and innovation, co-funded from the European Regional Development Fund.

Institutional Review Board Statement: Not Applicable.

Informed Consent Statement: Not Applicable.

Data Availability Statement: Not Applicable.

Conflicts of Interest: The authors declare no conflict of interest.

References

1. Tang, G.; Liu, Z.; Liu, Y.; Yu, J.; Wang, X.; Tan, Z.; Ye, X. Recent Trends in the Development of Bone Regenerative Biomaterials. *Front. Cell Dev. Biol.* **2021**, *9*, 1001. [CrossRef] [PubMed]
2. Campbell, A.A. Bioceramics for Implant Coatings. *Mater. Today* **2003**, *6*, 26–30. [CrossRef]
3. Goharian, A. Fundamentals in Loosening and Osseointegration of Orthopedic Implants. *Osseointegration Orthop. Implant.* **2019**, *1*, 1–26. [CrossRef]
4. Rodriguez-Gabella, T.; Voisine, P.; Puri, R.; Pibarot, P.; Rodés-Cabau, J. Aortic Bioprosthetic Valve Durability: Incidence, Mechanisms, Predictors, and Management of Surgical and Transcatheter Valve Degeneration. *J. Am. Coll. Cardiol.* **2017**, *70*, 1013–1028. [CrossRef] [PubMed]
5. Prasad, K.; Bazaka, O.; Chua, M.; Rochford, M.; Fedrick, L.; Spoor, J.; Symes, R.; Tieppo, M.; Collins, C.; Cao, A.; et al. Metallic Biomaterials: Current Challenges and Opportunities. *Materials* **2017**, *10*, 884. [CrossRef]
6. Shayesteh Moghaddam, N.; Taheri Andani, M.; Amerinatanzi, A.; Haberland, C.; Huff, S.; Miller, M.; Elahinia, M.; Dean, D. Metals for Bone Implants: Safety, Design, and Efficacy. *Biomanufacturing Rev.* **2016**, *1*, 1. [CrossRef]
7. Bazaka, O.; Bazaka, K.; Kingshott, P.; Crawford, R.J.; Ivanova, E.P. Chapter 1 Metallic Implants for Biomedical Applications. In *Chemistry of Inorganic Biomaterials*; Royal Society of Chemistry: London, UK, 2021; Volume 8, pp. 1–98. [CrossRef]
8. Liu, W.; Liu, S.; Wang, L. Surface Modification of Biomedical Titanium Alloy: Micromorphology, Microstructure Evolution and Biomedical Applications. *Coatings* **2019**, *9*, 249. [CrossRef]
9. Nedela, O.; Slepicka, P.; Švorcík, V. Surface Modification of Polymer Substrates for Biomedical Applications. *Materials* **2017**, *10*, 1115. [CrossRef]
10. Aherwar, A.; Singh, A.K.; Patnaik, A.; Aherwar, A.; Singh, A.K.; Patnaik, A. Current and Future Biocompatibility Aspects of Biomaterials for Hip Prosthesis. *AIMS Bioeng.* **2015**, *3*, 23–43. [CrossRef]
11. Wang, G.; Zreiqat, H. Functional Coatings or Films for Hard-Tissue Applications. *Materials* **2010**, *3*, 3994. [CrossRef]
12. Bhat, S.; Kumar, A. Biomaterials and Bioengineering Tomorrow's Healthcare. *Biomatter* **2013**, *3*, e24717. [CrossRef] [PubMed]
13. Joyce, K.; Fabra, G.T.; Bozkurt, Y.; Pandit, A. Bioactive Potential of Natural Biomaterials: Identification, Retention and Assessment of Biological Properties. *Signal Transduct. Target. Ther.* **2021**, *6*, 122. [CrossRef]
14. Chandel, A.K.S.; Shimizu, A.; Hasegawa, K.; Ito, T. Advancement of Biomaterial-Based Postoperative Adhesion Barriers. *Macromol. Biosci.* **2021**, *21*, e2000395. [CrossRef] [PubMed]
15. Kazimierczak, P.; Przekora, A. Osteoconductive and Osteoinductive Surface Modifications of Biomaterials for Bone Regeneration: A Concise Review. *Coatings* **2020**, *10*, 971. [CrossRef]
16. Akhtar, M.; Uzair, S.A.; Rizwan, M.; Ur Rehman, M.A. The Improvement in Surface Properties of Metallic Implant via Magnetron Sputtering: Recent Progress and Remaining Challenges. *Front. Mater.* **2022**, *8*, 602. [CrossRef]
17. Ahirwar, H.; Zhou, Y.; Mahapatra, C.; Ramakrishna, S.; Kumar, P.; Nanda, H.S. Materials for Orthopedic Bioimplants: Modulating Degradation and Surface Modification Using Integrated Nanomaterials. *Coatings* **2020**, *10*, 264. [CrossRef]
18. Vladescu, A.; Surmeneva, M.A.; Cotrut, C.M.; Surmenev, R.A.; Antoniac, I.V. Bioceramic Coatings for Metallic Implants. *Handb. Bioceram. Biocomposites* **2016**, 703–733. [CrossRef]
19. Sarian, M.N.; Iqbal, N.; Sotoudehbagha, P.; Razavi, M.; Ahmed, Q.U.; Sukotjo, C.; Hermawan, H. Potential Bioactive Coating System for High-Performance Absorbable Magnesium Bone Implants. *Bioact. Mater.* **2022**, *12*, 42–63. [CrossRef]

20. Pana, I.; Vladescu, A.; Constantin, L.R.; Sandu, I.G.; Dinu, M.; Cotrut, C.M. In Vitro Corrosion and Tribocorrosion Performance of Biocompatible Carbide Coatings. *Coatings* **2020**, *10*, 654. [CrossRef]
21. Probst, J.; Gbureck, U.; Thull, R. Binary Nitride and Oxynitride PVD Coatings on Titanium for Biomedical Applications. *Surf. Coat. Technol.* **2001**, *148*, 226–233. [CrossRef]
22. Dos Santos, G.A. The Importance of Metallic Materials as Biomaterials. *Adv. Tissue Eng. Regen. Med. Open Access* **2017**, 3, 300–302. [CrossRef]
23. Raghavendra, G.M.; Varaprasad, K.; Jayaramudu, T. Biomaterials: Design, Development and Biomedical Applications. *Nanotechnol. Appl. Tissue Eng.* **2015**, 21–44. [CrossRef]
24. Eliaz, N. Corrosion of Metallic Biomaterials: A Review. *Materials* **2019**, *12*, 407. [CrossRef] [PubMed]
25. Prasadh, S.; Wong, R.C.W. Unraveling the Mechanical Strength of Biomaterials Used as a Bone Scaffold in Oral and Maxillofacial Defects. *Oral Sci. Int.* **2018**, *15*, 48–55. [CrossRef]
26. Yamamuro, T. Bioceramics. *Biomech. Biomater. Orthop.* **2004**, 22–33. [CrossRef]
27. Shen, J.Z.; Fäldt, J. Requirements of Bioactive Ceramics for Dental Implants and Scaffolds. *Adv. Ceram. Dent.* **2014**, 279–300. [CrossRef]
28. Piconi, C.; Porporati, A.A. Bioinert Ceramics: Zirconia and Alumina. *Handb. Bioceram. Biocomposites* **2016**, 59–89. [CrossRef]
29. Barnes, D.H.; Moavenian, A.; Sharma, A.; Best, S.M. Biocompatibility of Ceramics. *Mater. Med. Devices* **2012**, 128–134. [CrossRef]
30. Elbadawi, M.; Meredith, J.; Hopkins, L. Ian Reaney Progress in Bioactive Metal and, Ceramic Implants for Load-Bearing Application. In *Advanced Techniques in Bone Regeneration*; InTech: Rijeka, Croatia, 2016. [CrossRef]
31. Whitehouse, M.R.; Blom, A.W. The Use of Ceramics as Bone Substitutes in Revision Hip Arthroplasty. *Materials* **2009**, *2*, 1895. [CrossRef]
32. Huang, C.H.; Yoshimura, M. Direct Ceramic Coating of Calcium Phosphate Doped with Strontium via Reactive Growing Integration Layer Method on α-Ti Alloy. *Sci. Rep.* **2020**, *10*, 10602. [CrossRef]
33. Lu, J.; Yu, H.; Chen, C. Biological Properties of Calcium Phosphate Biomaterials for Bone Repair: A Review. *RSC Adv.* **2018**, *8*, 2015–2033. [CrossRef] [PubMed]
34. Fiume, E.; Magnaterra, G.; Rahdar, A.; Verné, E.; Baino, F. Hydroxyapatite for Biomedical Applications: A Short Overview. *Ceramics* **2021**, *4*, 542–563. [CrossRef]
35. Díaz-Cuenca, A.; Rabadjieva, D.; Sezanova, K.; Gergulova, R.; Ilieva, R.; Tepavitcharova, S. Biocompatible Calcium Phosphate-Based Ceramics and Composites. *Mater. Today Proc.* **2022**. [CrossRef]
36. Kim, M.; An, S.; Huh, C.; Kim, C. Development of Zirconium-Based Alloys with Low Elastic Modulus for Dental Implant Materials. *Appl. Sci.* **2019**, *9*, 5281. [CrossRef]
37. Ridzwan, M.I.Z.; Shuib, S.; Hassan, A.Y.; Shokri, A.A.; Mohammad Ibrahim, M.N. Problem of Stress Shielding and Improvement to the Hip Implant Designs: A Review. *J. Med. Sci.* **2007**, *7*, 460–467. [CrossRef]
38. Fiorillo, L.; Cicciù, M.; Tozum, T.F.; Saccucci, M.; Orlando, C.; Romano, G.L.; D'amico, C.; Cervino, G. Endosseous Dental Implant Materials and Clinical Outcomes of Different Alloys: A Systematic Review. *Materials* **2022**, *15*, 1979. [CrossRef]
39. Abbas, Z.; Dapporto, M.; Tampieri, A.; Sprio, S. Toughening of Bioceramic Composites for Bone Regeneration. *J. Compos. Sci.* **2021**, *5*, 259. [CrossRef]
40. Díez-Pascual, M.; Li, M.; Komasa, S.; Hontsu, S.; Hashimoto, Y.; Okazaki, J. Structural Characterization and Osseointegrative Properties of Pulsed Laser-Deposited Fluorinated Hydroxyapatite Films on Nano-Zirconia for Implant Applications. *Int. J. Mol. Sci.* **2022**, *23*, 2416. [CrossRef]
41. Smargiassi, A.; Bertacchini, J.; Checchi, M.; Cavani, F.; Ferretti, M.; Palumbo, C. Biocompatibility Analyses of Al_2O_3-Treated Titanium Plates Tested with Osteocyte and Fibroblast Cell Lines. *Biomedicines* **2017**, *5*, 32. [CrossRef]
42. Lam, M.T.; Wu, J.C. Biomaterial Applications in Cardiovascular Tissue Repair and Regeneration. *Expert Rev. Cardiovasc. Ther.* **2012**, *10*, 1039. [CrossRef]
43. Ohta, S.; Mitsuhashi, K.; Chandel, A.K.S.; Qi, P.; Nakamura, N.; Nakamichi, A.; Yoshida, H.; Yamaguchi, G.; Hara, Y.; Sasaki, R.; et al. Silver-Loaded Carboxymethyl Cellulose Nonwoven Sheet with Controlled Counterions for Infected Wound Healing. *Carbohydr. Polym.* **2022**, *286*, 119289. [CrossRef] [PubMed]
44. Jiménez-Gómez, C.P.; Cecilia, J.A. Chitosan: A Natural Biopolymer with a Wide and Varied Range of Applications. *Molecules* **2020**, *25*, 3981. [CrossRef] [PubMed]
45. Kołodziejska, M.; Jankowska, K.; Klak, M.; Wszoła, M. Chitosan as an Underrated Polymer in Modern Tissue Engineering. *Nanomaterials* **2021**, *11*, 19. [CrossRef]
46. Cevik, P.; Schimmel, M.; Yilmaz, B. New Generation CAD-CAM Materials for Implant-Supported Definitive Frameworks Fabricated by Using Subtractive Technologies. *BioMed Res. Int.* **2022**, *2022*. [CrossRef] [PubMed]
47. Vigier, S.; Fülöp, T. Exploring the Extracellular Matrix to Create Biomaterials. In *Composition and Function of the Extracellular Matrix in the Human Body*; InTech: Rijeka, Croatia, 2016. [CrossRef]
48. Liao, C.; Li, Y.; Tjong, S.C. Polyetheretherketone and Its Composites for Bone Replacement and Regeneration. *Polymers* **2020**, *12*, 2858. [CrossRef] [PubMed]
49. Koons, G.L.; Diba, M.; Mikos, A.G. Materials Design for Bone-Tissue Engineering. *Nat. Rev. Mater.* **2020**, *5*, 584–603. [CrossRef]

50. Chandel, A.K.S.; Nutan, B.; Raval, I.H.; Jewrajka, S.K. Self-Assembly of Partially Alkylated Dextran-Graft-Poly[(2-Dimethylamino)Ethyl Methacrylate] Copolymer Facilitating Hydrophobic/Hydrophilic Drug Delivery and Improving Conetwork Hydrogel Properties. *Biomacromolecules* **2018**, *19*, 1142–1153. [CrossRef]
51. Nutan, B.; Chandel, A.K.S.; Jewrajka, S.K. Liquid Prepolymer-Based in Situ Formation of Degradable Poly(Ethylene Glycol)-Linked-Poly(Caprolactone)-Linked-Poly(2-Dimethylaminoethyl)Methacrylate Amphiphilic Conetwork Gels Showing Polarity Driven Gelation and Bioadhesion. *ACS Appl. Bio Mater.* **2018**, *1*, 1606–1619. [CrossRef]
52. Tallet, L.; Gribova, V.; Ploux, L.; Vrana, N.E.; Lavalle, P. New Smart Antimicrobial Hydrogels, Nanomaterials, and Coatings: Earlier Action, More Specific, Better Dosing? *Adv. Healthc. Mater.* **2021**, *10*, 2001199. [CrossRef]
53. Paroha, S.; Chandel, A.K.S.; Dubey, R.D. Nanosystems for Drug Delivery of Coenzyme Q10. *Environ. Chem. Lett.* **2018**, *16*, 71–77. [CrossRef]
54. Paroha, S.; Chandel, A.K.S.; Dubey, R.D. *Nanotechnology Delivery Systems of Coenzyme Q10: Pharmacokinetic and Clinical Implications*; Springer: Cham, Switzerland, 2017; pp. 213–228.
55. Chandel, A.K.S.; Kumar, C.U.; Jewrajka, S.K. Effect of Polyethylene Glycol on Properties and Drug Encapsulation-Release Performance of Biodegradable/Cytocompatible Agarose-Polyethylene Glycol-Polycaprolactone Amphiphilic Co-Network Gels. *ACS Appl. Mater. Interfaces* **2016**, *8*, 3182–3192. [CrossRef] [PubMed]
56. Dupin, D.; Deng, Z.; Guo, Y.; Zhao, X.; Du, T.; Zhu, J.; Xie, Y.; Wu, F.; Wang, Y.; Guan, M. Poly(N-Isopropylacrylamide) Based Electrically Conductive Hydrogels and Their Applications. *Gels* **2022**, *8*, 280. [CrossRef]
57. Romischke, J.; Scherkus, A.; Saemann, M.; Krueger, S.; Bader, R.; Kragl, U.; Meyer, J. Swelling and Mechanical Characterization of Polyelectrolyte Hydrogels as Potential Synthetic Cartilage Substitute Materials. *Gels* **2022**, *8*, 296. [CrossRef]
58. Liu, S.; Khan, R.; Zaman, M.; Salawi, A.; Khan, M.A.; Iqbal, M.O.; Riaz, R.; Ahmed, M.M.; Butt, M.H.; Alvi, M.N.; et al. Synthesis of Chemically Cross-Linked PH-Sensitive Hydrogels for the Sustained Delivery of Ezetimibe. *Gels* **2022**, *8*, 281. [CrossRef]
59. Zhao, Y.; Song, S.; Ren, X.; Zhang, J.; Lin, Q.; Zhao, Y. Supramolecular Adhesive Hydrogels for Tissue Engineering Applications. *Chem. Rev.* **2022**, *122*, 5604–5640. [CrossRef]
60. Yue, S.; He, H.; Li, B.; Hou, T. Hydrogel as a Biomaterial for Bone Tissue Engineering: A Review. *Nanomaterials* **2020**, *10*, 1511. [CrossRef] [PubMed]
61. Chang, H.K.; Yang, D.H.; Ha, M.Y.; Kim, H.J.; Kim, C.H.; Kim, S.H.; Choi, J.W.; Chun, H.J. 3D Printing of Cell-Laden Visible Light Curable Glycol Chitosan Bioink for Bone Tissue Engineering. *Carbohydr. Polym.* **2022**, *287*, 119328. [CrossRef] [PubMed]
62. Yang, H.; Jia, B.; Zhang, Z.; Qu, X.; Li, G.; Lin, W.; Zhu, D.; Dai, K.; Zheng, Y. Alloying Design of Biodegradable Zinc as Promising Bone Implants for Load-Bearing Applications. *Nat. Commun.* **2020**, *11*, 1–16. [CrossRef]
63. Cadosch, D.; Chan, E.; Gautschi, O.P.; Filgueira, L. Metal Is Not Inert: Role of Metal Ions Released by Biocorrosion in Aseptic Loosening—Current Concepts. *J. Biomed. Mater. Res. A* **2009**, *91*, 1252–1262. [CrossRef]
64. Gupta, S.; Gupta, H.; Tandan, A. Technical Complications of Implant-Causes and Management: A Comprehensive Review. *Natl. J. Maxillofac. Surg.* **2015**, *6*, 3. [CrossRef]
65. Siti Nur Hazwani, M.R.; LIM, L.X.; Lockman, Z.; Zuhailawati, H. Fabrication of Titanium-Based Alloys with Bioactive Surface Oxide Layer as Biomedical Implants: Opportunity and Challenges. *Trans. Nonferrous Met. Soc. China* **2022**, *32*, 1–44. [CrossRef]
66. Hermawan, H.; Ramdan, D.; Djuansjah, J.R.P. Metals for Biomedical Applications. In *Biomedical Engineering—From Theory to Applications*; InTech: Rijeka, Croatia, 2011. [CrossRef]
67. Hanawa, T. Research and Development of Metals for Medical Devices Based on Clinical Needs. *Sci. Technol. Adv. Mater.* **2012**, *13*. [CrossRef] [PubMed]
68. Ma, L.; Wiame, F.; Maurice, V.; Marcus, P. Origin of Nanoscale Heterogeneity in the Surface Oxide Film Protecting Stainless Steel against Corrosion. *npj Mater. Degrad.* **2019**, *3*, 1–9. [CrossRef]
69. Marcolongo, M.; Sarkar, S.; Ganesh, N. 7.11 Trends in Materials for Spine Surgery. *Compr. Biomater. II* **2017**, 175–198. [CrossRef]
70. Yan, X.; Cao, W.; Li, H. Biomedical Alloys and Physical Surface Modifications: A Mini-Review. *Materials* **2021**, *15*, 66. [CrossRef]
71. Li, M.; Yin, T.; Wang, Y.; Du, F.; Zou, X.; Gregersen, H.; Wang, G. Study of Biocompatibility of Medical Grade High Nitrogen Nickel-Free Austenitic Stainless Steel in Vitro. *Mater. Sci. Eng. C* **2014**, *43*, 641–648. [CrossRef]
72. Niinomi, M.; Liu, Y.; Nakai, M.; Liu, H.; Li, H. Biomedical Titanium Alloys with Young's Moduli Close to That of Cortical Bone. *Regen. Biomater.* **2016**, *3*, 173. [CrossRef]
73. Li, Y.; Yang, C.; Zhao, H.; Qu, S.; Li, X.; Li, Y. New Developments of Ti-Based Alloys for Biomedical Applications. *Materials* **2014**, *7*, 1709. [CrossRef]
74. Noyama, Y.; Miura, T.; Ishimoto, T.; Itaya, T.; Niinomi, M.; Nakano, T. Bone Loss and Reduced Bone Quality of the Human Femur after Total Hip Arthroplasty under Stress-Shielding Effects by Titanium-Based Implant. *Mater. Trans.* **2012**, *53*, 565–570. [CrossRef]
75. Plecko, M.; Sievert, C.; Andermatt, D.; Frigg, R.; Kronen, P.; Klein, K.; Stübinger, S.; Nuss, K.; Bürki, A.; Ferguson, S.; et al. Osseointegration and Biocompatibility of Different Metal Implants—A Comparative Experimental Investigation in Sheep. *BMC Musculoskelet. Disord.* **2012**, *13*, 32. [CrossRef]
76. Garcia-Mendez, M.C.; Urrutia-Baca, V.H.; Cuao-Moreu, C.A.; Lorenzo-Bonet, E.; Alvarez-Vera, M.; Ortiz-Martinez, D.M.; de la Garza-Ramos, M.A. In Vitro Biocompatibility Evaluation of a New Co-Cr-B Alloy with Potential Biomedical Application. *Metals* **2021**, *11*, 1267. [CrossRef]
77. Marti, A. Cobalt-Base Alloys Used in Surgery. *Injury* **2000**, *31*. [CrossRef]

78. Yu, W.; Li, X.; Ma, X.; Xu, X. Biomechanical Analysis of Inclined and Cantilever Design with Different Implant Framework Materials in Mandibular Complete-Arch Implant Restorations. *J. Prosthet. Dent.* **2022**. [CrossRef] [PubMed]
79. Kaivosoja, E.; Tiainen, V.-M.; Takakubo, Y.; Rajchel, B.; Sobiecki, J.; Konttinen, Y.T.; Takagi, M. Materials Used for Hip and Knee Implants. *Wear Orthop. Implant. Artif. Jt.* **2013**, 178–218. [CrossRef]
80. Nine, M.J.; Choudhury, D.; Hee, A.C.; Mootanah, R.; Osman, N.A.A. Wear Debris Characterization and Corresponding Biological Response: Artificial Hip and Knee Joints. *Materials* **2014**, *7*, 980. [CrossRef] [PubMed]
81. González-Mora, V.A.; Hoffmann, M.; Stroosnijder, R.; Espinar, E.; Llamas, J.M.; Fernández-Fairén, M.; Gil, F.J. Influence of Different CoCrMo Counterfaces on Wear in UHMWPE for Artificial Joints. *J. Biomed. Sci. Eng.* **2011**, *4*, 375–382. [CrossRef]
82. Briffa, J.; Sinagra, E.; Blundell, R. Heavy Metal Pollution in the Environment and Their Toxicological Effects on Humans. *Heliyon* **2020**, *6*, e04691. [CrossRef]
83. Jaishankar, M.; Tseten, T.; Anbalagan, N.; Mathew, B.B.; Beeregowda, K.N. Toxicity, Mechanism and Health Effects of Some Heavy Metals. *Interdiscip. Toxicol.* **2014**, *7*, 60. [CrossRef]
84. Genchi, G.; Carocci, A.; Lauria, G.; Sinicropi, M.S.; Catalano, A. Nickel: Human Health and Environmental Toxicology. *Int. J. Environ. Res. Public Health* **2020**, *17*, 679. [CrossRef]
85. Guo, Z.; Pang, X.; Yan, Y.; Gao, K.; Volinsky, A.A.; Zhang, T.Y. CoCrMo Alloy for Orthopedic Implant Application Enhanced Corrosion and Tribocorrosion Properties by Nitrogen Ion Implantation. *Appl. Surf. Sci.* **2015**, *347*, 23–34. [CrossRef]
86. Liu, R.; Li, X.; Hu, X.; Dong, H. Surface Modification of a Medical Grade Co-Cr-Mo Alloy by Low-Temperature Plasma Surface Alloying with Nitrogen and Carbon. *Surf. Coat. Technol.* **2013**, *232*, 906–911. [CrossRef]
87. Lourenço, M.L.; Cardoso, G.C.; Sousa, K.d.S.J.; Donato, T.A.G.; Pontes, F.M.L.; Grandini, C.R. Development of Novel Ti-Mo-Mn Alloys for Biomedical Applications. *Sci. Rep.* **2020**, *10*, 1–8. [CrossRef] [PubMed]
88. Wilson, J. Metallic Biomaterials: State of the Art and New Challenges. *Fundam. Biomater. Met.* **2018**, 1–33. [CrossRef]
89. Sarraf, M.; Rezvani Ghomi, E.; Alipour, S.; Ramakrishna, S.; Liana Sukiman, N. A State-of-the-Art Review of the Fabrication and Characteristics of Titanium and Its Alloys for Biomedical Applications. *Bio-Des. Manuf.* **2021**, *1*, 1–25. [CrossRef] [PubMed]
90. Estrin, Y.; Kim, H.E.; Lapovok, R.; Ng, H.P.; Jo, J.H. Mechanical Strength and Biocompatibility of Ultrafine-Grained Commercial Purity Titanium. *BioMed Res. Int.* **2013**, *2013*. [CrossRef] [PubMed]
91. Farooq, M.U.; Mughal, M.P.; Ahmed, N.; Mufti, N.A.; Al-Ahmari, A.M.; He, Y. On the Investigation of Surface Integrity of Ti6Al4V ELI Using Si-Mixed Electric Discharge Machining. *Materials* **2020**, *13*, 1549. [CrossRef]
92. Singh, N.; Hameed, P.; Ummethala, R.; Manivasagam, G.; Prashanth, K.G.; Eckert, J. Selective Laser Manufacturing of Ti-Based Alloys and Composites: Impact of Process Parameters, Application Trends, and Future Prospects. *Mater. Today Adv.* **2020**, *8*, 100097. [CrossRef]
93. Elias, C.N.; Lima, J.H.C.; Valiev, R.; Meyers, M.A. Biomedical Applications of Titanium and Its Alloys. *Jom* **2008**, *60*, 46–49. [CrossRef]
94. Lampman, S. Titanium and Its Alloys for Biomedical Implants. *Mater. Med. Devices* **2012**, 223–236. [CrossRef]
95. Martin, J.W. Metals and Alloys. *Mater. Eng.* **2006**, 71–132. [CrossRef]
96. Khadija, G.; Saleem, A.; Akhtar, Z.; Naqvi, Z.; Gull, M.; Masood, M.; Mukhtar, S.; Batool, M.; Saleem, N.; Rasheed, T.; et al. Short Term Exposure to Titanium, Aluminum and Vanadium (Ti 6Al 4V) Alloy Powder Drastically Affects Behavior and Antioxidant Metabolites in Vital Organs of Male Albino Mice. *Toxicol. Rep.* **2018**, *5*, 765–770. [CrossRef] [PubMed]
97. Costa, B.C.; Tokuhara, C.K.; Rocha, L.A.; Oliveira, R.C.; Lisboa-Filho, P.N.; Costa Pessoa, J. Vanadium Ionic Species from Degradation of Ti-6Al-4V Metallic Implants: In Vitro Cytotoxicity and Speciation Evaluation. *Mater. Sci. Eng. C* **2019**, *96*, 730–739. [CrossRef] [PubMed]
98. Han, X.; Ji, X.; Zhao, M.; Li, D. Mg/Ag Ratios Induced in Vitro Cell Adhesion and Preliminary Antibacterial Properties of TiN on Medical Ti-6Al-4V Alloy by Mg and Ag Implantation. *Surf. Coat. Technol.* **2020**, *397*, 126020. [CrossRef]
99. Chlebus, E.; Kuźnicka, B.; Kurzynowski, T.; Dybała, B. Microstructure and Mechanical Behaviour of Ti-6Al-7Nb Alloy Produced by Selective Laser Melting. *Mater. Charact.* **2011**, *62*, 488–495. [CrossRef]
100. Challa, V.S.A.; Mali, S.; Misra, R.D.K. Reduced Toxicity and Superior Cellular Response of Preosteoblasts to Ti-6Al-7Nb Alloy and Comparison with Ti-6Al-4V. *J. Biomed. Mater. Res. Part A* **2013**, *101A*, 2083–2089. [CrossRef] [PubMed]
101. Lyczkowska, E.; Szymczyk, P.; Dybała, B.; Chlebus, E. Chemical Polishing of Scaffolds Made of Ti-6Al-7Nb Alloy by Additive Manufacturing. *Arch. Civ. Mech. Eng.* **2014**, *14*, 586–594. [CrossRef]
102. Niu, Q.L.; Zheng, X.H.; Ming, W.W.; Chen, M. Friction and Wear Performance of Titanium Alloys against Tungsten Carbide under Dry Sliding and Water Lubrication. *Tribol. Trans.* **2013**, *56*, 101–108. [CrossRef]
103. Fellah, M.; Labaïz, M.; Assala, O.; Dekhil, L.; Taleb, A.; Rezag, H.; Iost, A. Tribological Behavior of Ti-6Al-4V and Ti-6Al-7Nb Alloys for Total Hip Prosthesis. *Adv. Tribol.* **2014**, *2014*. [CrossRef]
104. Delgado-Ruiz, R.; Romanos, G. Potential Causes of Titanium Particle and Ion Release in Implant Dentistry: A Systematic Review. *Int. J. Mol. Sci.* **2018**, *19*, 3585. [CrossRef]
105. Shuai, C.; Li, S.; Peng, S.; Feng, P.; Lai, Y.; Gao, C. Biodegradable Metallic Bone Implants. *Mater. Chem. Front.* **2019**, *3*, 544–562. [CrossRef]
106. Harrison, R.; Maradze, D.; Lyons, S.; Zheng, Y.; Liu, Y. Corrosion of Magnesium and Magnesium–Calcium Alloy in Biologically-Simulated Environment. *Prog. Nat. Sci. Mater. Int.* **2014**, *24*, 539–546. [CrossRef]

107. Amukarimi, S.; Mozafari, M. Biodegradable Magnesium-based Biomaterials: An Overview of Challenges and Opportunities. *MedComm* **2021**, *2*, 123. [CrossRef]
108. Tsakiris, V.; Tardei, C.; Clicinschi, F.M. Biodegradable Mg Alloys for Orthopedic Implants—A Review. *J. Magnes. Alloy.* **2021**, *9*, 1884–1905. [CrossRef]
109. Alam, M.E.; Pal, S.; Decker, R.; Ferreri, N.C.; Knezevic, M.; Beyerlein, I.J. Rare-Earth- and Aluminum-Free, High Strength Dilute Magnesium Alloy for Biomedical Applications. *Sci. Rep.* **2020**, *10*, 15839. [CrossRef] [PubMed]
110. Prasad, S.V.S.; Prasad, S.B.; Verma, K.; Mishra, R.K.; Kumar, V.; Singh, S. The Role and Significance of Magnesium in Modern Day Research—A Review. *J. Magnes. Alloy.* **2022**, *10*, 1–61. [CrossRef]
111. Loukil, N. Alloying Elements of Magnesium Alloys: A Literature Review. In *Magnesium Alloys [Working Title]*; InTech: Rijeka, Croatia, 2021. [CrossRef]
112. Sezer, N.; Evis, Z.; Kayhan, S.M.; Tahmasebifar, A.; Koç, M. Review of Magnesium-Based Biomaterials and Their Applications. *J. Magnes. Alloy.* **2018**, *6*, 23–43. [CrossRef]
113. Li, Q.; Wang, Q.; Wang, Y.; Zeng, X.; Ding, W. Effect of Nd and Y Addition on Microstructure and Mechanical Properties of As-Cast Mg-Zn-Zr Alloy. *J. Alloys Compd.* **2007**, *427*, 115–123. [CrossRef]
114. Liu, C.; Ren, Z.; Xu, Y.; Pang, S.; Zhao, X.; Zhao, Y. Biodegradable Magnesium Alloys Developed as Bone Repair Materials: A Review. *Scanning* **2018**, *2018*. [CrossRef]
115. Xu, Y.; Li, J.; Qi, M.; Liao, L.; Gao, Z. Enhanced Mechanical Properties of Mg-Zn-Y-Zr Alloy by Low-Speed Indirect Extrusion. *J. Mater. Res. Technol.* **2020**, *9*, 9856–9867. [CrossRef]
116. Wang, S.; Han, Z.; Nie, Y.; Dai, J.; Li, X.; Cheng, J.; Zhang, X. Modified Mechanical Properties of Mg-Nd-Zn-Ag-Zr Alloy by Solution Treatment for Cardiovascular Stent Application. *Mater. Res. Express* **2019**, *6*, 085416. [CrossRef]
117. Jamel, M.M.; Jamel, M.M.; Lopez, H.F. Designing Advanced Biomedical Biodegradable Mg Alloys: A Review. *Metals* **2022**, *12*, 85. [CrossRef]
118. Ralls, A.; Kumar, P.; Misra, M.; Menezes, P.L. Material Design and Surface Engineering for Bio-Implants. *JOM* **2020**, *72*, 684–696. [CrossRef]
119. Hussain, M.; Askari Rizvi, S.H.; Abbas, N.; Sajjad, U.; Shad, M.R.; Badshah, M.A.; Malik, A.I. Recent Developments in Coatings for Orthopedic Metallic Implants. *Coatings* **2021**, *11*, 791. [CrossRef]
120. Prakasam, M.; Locs, J.; Salma-Ancane, K.; Loca, D.; Largeteau, A.; Berzina-Cimdina, L. Biodegradable Materials and Metallic Implants—A Review. *J. Funct. Biomater.* **2017**, *8*, 44. [CrossRef] [PubMed]
121. Lam, M.; Migonney, V.; Falentin-Daudre, C. Review of Silicone Surface Modification Techniques and Coatings for Antibacterial/Antimicrobial Applications to Improve Breast Implant Surfaces. *Acta Biomater.* **2021**, *121*, 68–88. [CrossRef] [PubMed]
122. Accioni, F.; Vázquez, J.; Merinero, M.; Begines, B.; Alcudia, A. Latest Trends in Surface Modification for Dental Implantology: Innovative Developments and Analytical Applications. *Pharmaceutics* **2022**, *14*, 455. [CrossRef] [PubMed]
123. Sikder, P.; Ren, Y.; Bhaduri, S.B. Synthesis and Evaluation of Protective Poly(Lactic Acid) and Fluorine-Doped Hydroxyapatite–Based Composite Coatings on AZ31 Magnesium Alloy. *J. Mater. Res.* **2019**, *34*, 3766–3776. [CrossRef]
124. Metroke, T.L.; Parkhill, R.L.; Knobbe, E.T. Passivation of Metal Alloys Using Sol–Gel-Derived Materials—A Review. *Prog. Org. Coat.* **2001**, *41*, 233–238. [CrossRef]
125. Bekmurzayeva, A.; Duncanson, W.J.; Azevedo, H.S.; Kanayeva, D. Surface Modification of Stainless Steel for Biomedical Applications: Revisiting a Century-Old Material. *Mater. Sci. Eng. C* **2018**, *93*, 1073–1089. [CrossRef]
126. Quinn, J.; McFadden, R.; Chan, C.W.; Carson, L. Titanium for Orthopedic Applications: An Overview of Surface Modification to Improve Biocompatibility and Prevent Bacterial Biofilm Formation. *iScience* **2020**, *23*, 101745. [CrossRef]
127. Nilawar, S.; Uddin, M.; Chatterjee, K. Surface Engineering of Biodegradable Implants: Emerging Trends in Bioactive Ceramic Coatings and Mechanical Treatments. *Mater. Adv.* **2021**, *2*, 7820–7841. [CrossRef]
128. Friedrich, K.; Zhang, Z.; Schlarb, A.K. Effects of Various Fillers on the Sliding Wear of Polymer Composites. *Compos. Sci. Technol.* **2005**, *65*, 2329–2343. [CrossRef]
129. Almasi, D.; Iqbal, N.; Sadeghi, M.; Sudin, I.; Abdul Kadir, M.R.; Kamarul, T. Preparation Methods for Improving PEEK's Bioactivity for Orthopedic and Dental Application: A Review. *Int. J. Biomater.* **2016**, *2016*. [CrossRef] [PubMed]
130. Lu, Z.P.; Friedrich, K. On Sliding Friction and Wear of PEEK and Its Composites. *Wear* **1995**, *181–183*, 624–631. [CrossRef]
131. Vande Voort, J.; Bahadur, S. The Growth and Bonding of Transfer Film and the Role of CuS and PTFE in the Tribological Behavior of PEEK. *Wear* **1995**, *181–183*, 212–221. [CrossRef]
132. Zhang, G.; Leparoux, S.; Liao, H.; Coddet, C. Microwave Sintering of Poly-Ether-Ether-Ketone (PEEK) Based Coatings Deposited on Metallic Substrate. *Scr. Mater.* **2006**, *55*, 621–624. [CrossRef]
133. Normand, B.; Takenouti, H.; Keddam, M.; Liao, H.; Monteil, G.; Coddet, C. Electrochemical Impedance Spectroscopy and Dielectric Properties of Polymer: Application to PEEK Thermally Sprayed Coating. *Electrochim. Acta* **2004**, *49*, 2981–2986. [CrossRef]
134. Zhang, G.; Liao, H.; Yu, H.; Ji, V.; Huang, W.; Mhaisalkar, S.G.; Coddet, C. Correlation of Crystallization Behavior and Mechanical Properties of Thermal Sprayed PEEK Coating. *Surf. Coat. Technol.* **2006**, *200*, 6690–6695. [CrossRef]
135. Fiołek, A.; Zimowski, S.; Kopia, A.; Łukaszczyk, A.; Moskalewicz, T. Electrophoretic Co-Deposition of Polyetheretherketone and Graphite Particles: Microstructure, Electrochemical Corrosion Resistance, and Coating Adhesion to a Titanium Alloy. *Materials* **2020**, *13*, 3251. [CrossRef]

136. Sak, A.; Moskalewicz, T.; Zimowski, S.; Cieniek, Ł.; Dubiel, B.; Radziszewska, A.; Kot, M.; Łukaszczyk, A. Influence of Polyetheretherketone Coatings on the Ti-13Nb-13Zr Titanium Alloy's Bio-Tribological Properties and Corrosion Resistance. *Mater. Sci. Eng. C* **2016**, *63*, 52–61. [CrossRef]
137. Virk, R.S.; Ur Rehman, M.A.; Boccaccini, A.R. PEEK Based Biocompatible Coatings Incorporating H-BN and Bioactive Glass by Electrophoretic Deposition. *ECS Trans.* **2018**, *82*, 89–95. [CrossRef]
138. Wu, X.; Liu, X.; Wei, J.; Ma, J.; Deng, F.; Wei, S. Nano-TiO_2/PEEK Bioactive Composite as a Bone Substitute Material: In Vitro and in Vivo Studies. *Int. J. Nanomed.* **2012**, *7*, 1215–1225. [CrossRef]
139. Moskalewicz, T.; Zych, A.; Łukaszczyk, A.; Cholewa-Kowalska, K.; Kruk, A.; Dubiel, B.; Radziszewska, A.; Berent, K.; Gajewska, M. Electrophoretic Deposition, Microstructure, and Corrosion Resistance of Porous Sol–Gel Glass/Polyetheretherketone Coatings on the Ti-13Nb-13Zr Alloy. *Metall. Mater. Trans. A Phys. Metall. Mater. Sci.* **2017**, *48*, 2660–2673. [CrossRef]
140. Atiq Ur Rehman, M.; Bastan, F.E.; Haider, B.; Boccaccini, A.R. Electrophoretic Deposition of PEEK/Bioactive Glass Composite Coatings for Orthopedic Implants: A Design of Experiments (DoE) Study. *Mater. Des.* **2017**, *130*, 223–230. [CrossRef]
141. Lebga-Nebane, J.L.; Sankarasubramanian, M.; Chojecki, G.; Ning, B.; Yuya, P.A.; Moosbrugger, J.C.; Rasmussen, D.H.; Krishnan, S. Polyetheretherketone, Hexagonal Boron Nitride, and Tungsten Carbide Cobalt Chromium Composite Coatings: Mechanical and Tribological Properties. *J. Appl. Polym. Sci.* **2021**, *138*, 50504. [CrossRef]
142. Seuss, S.; Heinloth, M.; Boccaccini, A.R. Development of Bioactive Composite Coatings Based on Combination of PEEK, Bioactive Glass and Ag Nanoparticles with Antibacterial Properties. *Surf. Coat. Technol.* **2016**, *301*, 100–105. [CrossRef]
143. Garrido, B.; Albaladejo-Fuentes, V.; Cano, I.G.; Dosta, S. Development of Bioglass/PEEK Composite Coating by Cold Gas Spray for Orthopedic Implants. *J. Therm. Spray Technol.* **2022**, 1–11. [CrossRef]
144. Torres, Y.; Romero, C.; Chen, Q.; Pérez, G.; Rodríguez-Ortiz, J.A.; Pavón, J.J.; Álvarez, L.; Arévalo, C.; Boccaccini, A.R. Electrophoretic Deposition of PEEK/45S5 Bioactive Glass Coating on Porous Titanium Substrate: Influence of Processing Conditions and Porosity Parameters. *Key Eng. Mater.* **2016**, *704*, 343–350. [CrossRef]
145. Tharajak, J.; Palathai, T.; Sombatsompop, N. Morphological and Physical Properties and Friction/Wear Behavior of h-BN Filled PEEK Composite Coatings. *Surf. Coat. Technol.* **2015**, *273*, 20–29. [CrossRef]
146. Moskalewicz, T.; Zimowski, S.; Zych, A.; Łukaszczyk, A.; Reczyńska, K.; Pamuła, E. Electrophoretic Deposition, Microstructure and Selected Properties of Composite Alumina/Polyetheretherketone Coatings on the Ti-13Nb-13Zr Alloy. *J. Electrochem. Soc.* **2018**, *165*, D116–D128. [CrossRef]
147. Moskalewicz, T.; Zych, A.; Kruk, A.; Kopia, A.; Zimowski, S.; Sitarz, M.; Cieniek, Ł. Electrophoretic Deposition and Microstructure Development of Si3N4/Polyetheretherketone Coatings on Titanium Alloy. *Surf. Coat. Technol.* **2018**, *350*, 633–647. [CrossRef]
148. Baştan, F.E.; Atiq Ur Rehman, M.; Avcu, Y.Y.; Avcu, E.; Üstel, F.; Boccaccini, A.R. Electrophoretic Co-Deposition of PEEK-Hydroxyapatite Composite Coatings for Biomedical Applications. *Colloids Surf. B Biointerfaces* **2018**, *169*, 176–182. [CrossRef] [PubMed]
149. Lee, J.H.; Jang, H.L.; Lee, K.M.; Baek, H.R.; Jin, K.; Noh, J.H. Cold-Spray Coating of Hydroxyapatite on a Three-Dimensional Polyetheretherketone Implant and Its Biocompatibility Evaluated by in Vitro and in Vivo Minipig Model. *J. Biomed. Mater. Res.-Part B Appl. Biomater.* **2017**, *105*, 647–657. [CrossRef] [PubMed]
150. Kadiyala, A.K.; Bijwe, J.; Kalappa, P. Investigations on Influence of Nano and Micron Sized Particles of SiC on Performance Properties of PEEK Coatings. *Surf. Coat. Technol.* **2018**, *334*, 124–133. [CrossRef]
151. Abdulkareem, M.H.; Abdalsalam, A.H.; Bohan, A.J. Influence of Chitosan on the Antibacterial Activity of Composite Coating (PEEK /HAp) Fabricated by Electrophoretic Deposition. *Prog. Org. Coat.* **2019**, *130*, 251–259. [CrossRef]
152. Chen, X.; Ma, R.; Min, J.; Li, Z.; Yu, P.; Yu, H. Effect of PEEK and PTFE Coatings in Fatigue Performance of Dental Implant Retaining Screw Joint: An in Vitro Study. *J. Mech. Behav. Biomed. Mater.* **2020**, *103*. [CrossRef]
153. Song, J.; Liao, Z.; Shi, H.; Xiang, D.; Xu, L.; Liu, Y.; Mu, X.; Liu, W. Blood Compatibility of ZrO_2 Particle Reinforced PEEK Coatings on Ti6Al4V Substrates. *Polymers* **2017**, *9*, 589. [CrossRef]
154. Song, J.; Liao, Z.; Wang, S.; Liu, Y.; Liu, W.; Tyagi, R. Study on the Tribological Behaviors of Different PEEK Composite Coatings for Use as Artificial Cervical Disk Materials. *J. Mater. Eng. Perform.* **2016**, *25*, 116–129. [CrossRef]
155. Xie, C.; Li, P.; Liu, Y.; Luo, F.; Xiao, X. Preparation of TiO_2 Nanotubes/Mesoporous Calcium Silicate Composites with Controllable Drug Release. *Mater. Sci. Eng. C Mater. Biol. Appl.* **2016**, *67*, 433–439. [CrossRef]
156. Pradhan, D.; Wren, A.W.; Misture, S.T.; Mellott, N.P. Investigating the Structure and Biocompatibility of Niobium and Titanium Oxides as Coatings for Orthopedic Metallic Implants. *Mater. Sci. Eng. C* **2016**, *58*, 918–926. [CrossRef]
157. Mirak, M.; Alizadeh, M.; Ashtiani, M.N. Characterization, Mechanical Properties and Corrosion Resistance of Biocompatible Zn-HA/TiO_2 Nanocomposite Coatings. *J. Mech. Behav. Biomed. Mater.* **2016**, *62*, 282–290. [CrossRef] [PubMed]
158. Cao, Y.Q.; Zi, T.Q.; Zhao, X.R.; Liu, C.; Ren, Q.; Fang, J.B.; Li, W.M.; Li, A.D. Enhanced Visible Light Photocatalytic Activity of Fe2O3 Modified TiO_2 Prepared by Atomic Layer Deposition. *Sci. Rep.* **2020**, *10*, 1–10. [CrossRef] [PubMed]
159. He, X.; Zhang, G.; Zhang, H.; Hang, R.; Huang, X.; Yao, X.; Zhang, X. Cu and Si Co-Doped Microporous TiO_2 Coating for Osseointegration by the Coordinated Stimulus Action. *Appl. Surf. Sci.* **2020**, *503*, 144072. [CrossRef]
160. Gartner, M.; Trapalis, C.; Todorova, N.; Giannakopoulou, T.; Dobrescu, G.; Anastasescu, M.; Osiceanu, P.; Ghita, A.; Enache, M.; Dumitru, L.; et al. Doped Sol-Gel TiO_2 Films for Biological Applications. *Bull. Korean Chem. Soc.* **2008**, *29*, 1038–1042. [CrossRef]
161. Radeka, M.; Markov, S.; Lončar, E.; Rudić, O.; Vučetić, S.; Ranogajec, J. Photocatalytic Effects of TiO_2 Mesoporous Coating Immobilized on Clay Roofing Tiles. *J. Eur. Ceram. Soc.* **2014**, *34*, 127–136. [CrossRef]

162. Park, S.; Park, J.; Heo, J.; Hong, B.Y.; Hong, J. Growth Behaviors and Biocidal Properties of Titanium Dioxide Films Depending on Nucleation Duration in Liquid Phase Deposition. *Appl. Surf. Sci.* **2017**, *425*, 547–552. [CrossRef]
163. Hamza, R.Z.; Gobouri, A.A.; Al-Yasi, H.M.; Al-Talhi, T.A.; El-Megharbel, S.M. A New Sterilization Strategy Using TiO_2 Nanotubes for Production of Free Radicals That Eliminate Viruses and Application of a Treatment Strategy to Combat Infections Caused by Emerging SARS-CoV-2 during the COVID-19 Pandemic. *Coatings* **2021**, *11*, 680. [CrossRef]
164. Zhang, S.; Liang, X.; Gadd, G.M.; Zhao, Q. Advanced Titanium Dioxide-Polytetrafluorethylene (TiO_2-PTFE) Nanocomposite Coatings on Stainless Steel Surfaces with Antibacterial and Anti-Corrosion Properties. *Appl. Surf. Sci.* **2019**, *490*, 231–241. [CrossRef]
165. Yetim, T. Corrosion Behavior of Ag-Doped TiO_2 Coatings on Commercially Pure Titanium in Simulated Body Fluid Solution. *J. Bionic Eng.* **2016**, *13*, 397–405. [CrossRef]
166. Moongraksathum, B.; Chien, M.-Y.; Chen, Y.-W. Antiviral and Antibacterial Effects of Silver-Doped TiO 2 Prepared by the Peroxo Sol-Gel Method. *J. Nanosci. Nanotechnol.* **2019**, *19*, 7356–7362. [CrossRef]
167. Hua, Z.; Zhang, C.; Xu, Y.; Jia, H.; Chen, X.; Bai, Y.; Wei, B. Efficiently Reduced Heat Rise in TiO 2 Coating Ti-Based Metallic Implants Using Anodic Oxidation Method. *Surf. Coat. Technol.* **2019**, *363*, 75–79. [CrossRef]
168. Xian, P.; Chen, Y.; Gao, S.; Qian, J.; Zhang, W.; Udduttula, A.; Huang, N.; Wan, G. Polydopamine (PDA) Mediated Nanogranular-Structured Titanium Dioxide (TiO_2) Coating on Polyetheretherketone (PEEK) for Oral and Maxillofacial Implants Application. *Surf. Coat. Technol.* **2020**, *401*. [CrossRef]
169. Chai, M.; An, M.; Zhang, X. Construction of a TiO_2/$MoSe_2$/CHI Coating on Dental Implants for Combating Streptococcus Mutans Infection. *Mater. Sci. Eng. C* **2021**, *129*. [CrossRef] [PubMed]
170. Kiran, A.S.K.; Kumar, T.S.S.; Sanghavi, R.; Doble, M.; Ramakrishna, S. Antibacterial and Bioactive Surface Modifications of Titanium Implants by PCL/TiO_2 Nanocomposite Coatings. *Nanomaterials* **2018**, *8*, 860. [CrossRef] [PubMed]
171. Doll, P.W.; Ahrens, R.; Guber, A.E. Etch-Less Microfabrication of Structured TiO_2 Implant Coatings on Bulk Titanium Grade 23 by Direct Lithographic Anodic Oxidation. *J. Micromech. Microeng.* **2021**, *31*. [CrossRef]
172. Zhu, Y.; Zheng, T.; Wen, L.M.; Li, R.; Zhang, Y.B.; Bi, W.J.; Feng, X.J.; Qi, M.C. Osteogenic Capability of Strontium and Icariin-Loaded TiO_2 Nanotube Coatings in Vitro and in Osteoporotic Rats. *J. Biomater. Appl.* **2021**, *35*, 1119–1131. [CrossRef] [PubMed]
173. Long, X.; Duan, L.; Weng, W.; Cheng, K.; Wang, D.; Ouyang, H. Light-Induced Osteogenic Differentiation of BMSCs with Graphene/TiO_2 Composite Coating on Ti Implant. *Colloids Surf. B Biointerfaces* **2021**, *207*. [CrossRef]
174. Visentin, F.; el Habra, N.; Fabrizio, M.; Brianese, N.; Gerbasi, R.; Nodari, L.; Zin, V.; Galenda, A. TiO_2-HA Bi-Layer Coatings for Improving the Bioactivity and Service-Life of Ti Dental Implants. *Surf. Coat. Technol.* **2019**, *378*. [CrossRef]
175. Zhang, B.; Li, B.; Gao, S.; Li, Y.; Cao, R.; Cheng, J.; Li, R.; Wang, E.; Guo, Y.; Zhang, K.; et al. Y-Doped TiO_2 Coating with Superior Bioactivity and Antibacterial Property Prepared via Plasma Electrolytic Oxidation. *Mater. Des.* **2020**, *192*. [CrossRef]
176. Li, K.; Liu, S.; Xue, Y.; Zhang, L.; Han, Y. A Superparamagnetic Fe_3O_4-TiO_2 Composite Coating on Titanium by Micro-Arc Oxidation for Percutaneous Implants. *J. Mater. Chem. B* **2019**, *7*, 5265–5276. [CrossRef]
177. Gobbi, S.J. Orthopedic Implants: Coating with TiN. *Biomed. J. Sci. Tech. Res.* **2019**, *16*. [CrossRef]
178. Pogrebnjak, A.D.; Kong, C.H.; Webster, R.F.; Tilley, R.D.; Takeda, Y.; Oyoshi, K.; Bondar, O.V.; Buranich, V.V.; Konstantinov, S.V.; Baimoldanova, L.S.; et al. Antibacterial Effect of Au Implantation in Ductile Nanocomposite Multilayer (TiAlSiY)N/CrN Coatings. *ACS Appl. Mater. Interfaces* **2019**, *11*, 48540–48550. [CrossRef] [PubMed]
179. Kim, H.; Kim, C.Y.; Kim, D.W.; Lee, I.S.; Lee, G.H.; Park, J.C.; Lee, S.J.; Lee, K.Y. Wear Performance of Self-Mating Contact Pairs of TiN and TiAlN Coatings on Orthopedic Grade Ti-6Al-4V. *Biomed. Mater.* **2010**, *5*, 044108. [CrossRef] [PubMed]
180. Subramanian, B.; Muraleedharan, C.V.; Ananthakumar, R.; Jayachandran, M. A Comparative Study of Titanium Nitride (TiN), Titanium Oxy Nitride (TiON) and Titanium Aluminum Nitride (TiAlN), as Surface Coatings for Bio Implants. *Surf. Coat. Technol.* **2011**, *205*, 5014–5020. [CrossRef]
181. Van Hove, R.P.; Sierevelt, I.N.; van Royen, B.J.; Nolte, P.A. Titanium-Nitride Coating of Orthopaedic Implants: A Review of the Literature. *BioMed Res. Int.* **2015**, *2015*. [CrossRef] [PubMed]
182. Dion, I.; Baquey, C.; Candelon, B.; Monties, J.R. Hemocompatibility of Titanium Nitride. *Int. J. Artif. Organs* **1992**, *15*, 617–621. [CrossRef]
183. Çomaklı, O. Improved Structural, Mechanical, Corrosion and Tribocorrosion Properties of Ti45Nb Alloys by TiN, TiAlN Monolayers, and TiAlN/TiN Multilayer Ceramic Films. *Ceram. Int.* **2021**, *47*, 4149–4156. [CrossRef]
184. Dinu, M.; Pana, I.; Scripca, P.; Sandu, I.G.; Vitelaru, C.; Vladescu, A. Improvement of CoCr Alloy Characteristics by Ti-Based Carbonitride Coatings Used in Orthopedic Applications. *Coatings* **2020**, *10*, 495. [CrossRef]
185. Ul-Hamid, A. Microstructure, Properties and Applications of Zr-Carbide, Zr-Nitride and Zr-Carbonitride Coatings: A Review. *Mater. Adv.* **2020**, *1*, 1012–1037. [CrossRef]
186. Ghufran, M.; Uddin, G.M.; Arafat, S.M.; Jawad, M.; Rehman, A. Development and Tribo-Mechanical Properties of Functional Ternary Nitride Coatings: Applications-Based Comprehensive Review. *Proc. Inst. Mech. Eng. Part J J. Eng. Tribol.* **2020**, *235*, 196–232. [CrossRef]
187. Chen, S.N.; Zhao, Y.M.; Zhang, Y.F.; Chen, L.; Liao, B.; Zhang, X.; Ouyang, X.P. Influence of Carbon Content on the Structure and Tribocorrosion Properties of TiAlCN/TiAlN/TiAl Multilayer Composite Coatings. *Surf. Coat. Technol.* **2021**, *411*, 126886. [CrossRef]

188. Hovsepian, P.E.; Ehiasarian, A.P.; Petrov, I. TiAlCN/VCN Nanolayer Coatings Suitable for Machining of Al and Ti Alloys Deposited by Combined High Power Impulse Magnetron Sputtering/Unbalanced Magnetron Sputtering. *Surf. Eng.* **2013**, *26*, 610–614. [CrossRef]
189. Caicedo, J.C.; Amaya, C.; Yate, L.; Gómez, M.E.; Zambrano, G.; Alvarado-Rivera, J.; Muñoz-Saldaña, J.; Prieto, P. TiCN/TiNbCN Multilayer Coatings with Enhanced Mechanical Properties. *Appl. Surf. Sci.* **2010**, *256*, 5898–5904 [CrossRef]
190. Rajak, D.K.; Kumar, A.; Behera, A.; Menezes, P.L. Diamond-Like Carbon (DLC) Coatings: Classification, Properties, and Applications. *Appl. Sci.* **2021**, *11*, 4445. [CrossRef]
191. Moritz, Y.; Saringer, C.; Tkadletz, M.; Stark, A.; Schell, N.; Letofsky-Papst, I.; Czettl, C.; Pohler, M.; Schalk, N. Oxidation Behavior of Arc Evaporated TiSiN Coatings Investigated by In-Situ Synchrotron X-Ray Diffraction and HR-STEM. *Surf. Coat. Technol.* **2020**, *404*, 126632. [CrossRef]
192. Parau, A.C.; Vitelaru, C.; Balaceanu, M.; Braic, V.; Constantin, L.R.; Braic, M.; Vladescu, A. TiSiC, TiSiC-Zr, and TiSiC-Cr Coatings—Corrosion Resistance and Tribological Performance in Saline Solution. *Tribol. Trans.* **2015**, *59*, 72–79. [CrossRef]
193. Xie, X.; Li, J.; Dong, M.; Zhang, H.; Wang, L. Structure and Properties of TiSiCN Coatings with Different Bias Voltages by Arc Ion Plating. *Surf. Topogr. Metrol. Prop.* **2018**, *6*, 014003. [CrossRef]
194. Lin, J.; Wei, R.; Ge, F.; Li, Y.; Zhang, X.; Huang, F.; Lei, M. TiSiCN and TiAlVSiCN Nanocomposite Coatings Deposited from Ti and Ti-6Al-4V Targets. *Surf. Coat. Technol.* **2018**, *336*, 106–116. [CrossRef]
195. Varghese, V.; Ramesh, M.R.; Chakradhar, D.; Shaik, H. Characterisation and Performance Evaluation of TiSiN &TiAlSiN Coatings by RF Magnetron Sputtering Deposition during End Milling of Maraging Steel. *Mater. Res. Express* **2020**, *6*, 126440. [CrossRef]
196. Vitu, T.; Polcar, T.; Cvrcek, L.; Novak, R.; Macak, J.; Vyskocil, J.; Cavaleiro, A. Structure and Tribology of Biocompatible Ti-C:H Coatings. *Surf. Coat. Technol.* **2008**, *202*, 5790–5793. [CrossRef]
197. Wan, Q.; Liu, N.; Yang, B.; Liu, H.; Chen, Y. Influence of Si Content on Properties of Ti(1-x)SixN Coatings. *J. Wuhan Univ. Technol.-Mater. Sci. Ed.* **2019**, *34*, 774–780. [CrossRef]
198. Yuan, Z.; Han, Y.; Zang, S.; Chen, J.; He, G.; Chai, Y.; Yang, Z.; Fu, Q. Damage Evolution Behavior of TiN/Ti Multilayer Coatings under High-Speed Impact Conditions. *Surf. Coat. Technol.* **2021**, *426*, 127807. [CrossRef]
199. Zhu, L.; Feng, C.; Zhu, S.; Wang, F.; Yuan, J.; Wang, P. Comparison of CrN, AlN and TiN Diffusion Barriers on the Interdiffusion and Oxidation Behaviors of Ni+CrAlYSiN Nanocomposite Coatings. *Crystals* **2021**, *11*, 1333. [CrossRef]
200. Liu, H.; Zhang, X.; Jin, S.; Zhao, Y.; Ren, L.; Yang, K. Effect of Copper-Doped Titanium Nitride Coating on Angiogenesis. *Mater. Lett.* **2020**, *269*, 127634. [CrossRef]
201. Castro, J.D.; Lima, M.J.; Carvalho, I.; Henriques, M.; Carvalho, S. Cu Oxidation Mechanism on Cu-Zr(O)N Coatings: Role on Functional Properties. *Appl. Surf. Sci.* **2021**, *555*. [CrossRef]
202. Pana, I.; Braic, V.; Dinu, M.; Massima Mouele, E.S.; Parau, A.C.; Petrik, L.F.; Braic, M. In Vitro Corrosion of Titanium Nitride and Oxynitride-Based Biocompatible Coatings Deposited on Stainless Steel. *Coatings* **2020**, *10*, 710. [CrossRef]
203. Kaliaraj, G.S.; Kumar, N. Oxynitrides Decorated 316L SS for Potential Bioimplant Application. *Mater. Res. Express* **2018**, *5*, 036403. [CrossRef]
204. Lee, D.B.; Lee, Y.C.; Kwon, S.C. High Temperature Oxidation of TiCrN Coatings Deposited on a Steel Substrate by Ion Plating. *Surf. Coat. Technol.* **2001**, *141*, 232–239. [CrossRef]
205. Chiu, K.A.; Fu, C.W.; Fang, Y.S.; Do, T.H.; Shih, F.H.; Chang, L. Heteroepitaxial Growth and Microwave Plasma Annealing of DC Reactive Sputtering Deposited TiZrN Film on Si (100). *Surf. Coat. Technol.* **2020**, *394*, 125873. [CrossRef]
206. Kaliaraj, G.S.; Vishwakarma, V.; Ramadoss, A.; Ramachandran, D.; Rabel, A.M. Corrosion, Haemocompatibility and Bacterial Adhesion Behavior of TiZrN-Coated 316L SS for Bioimplants. *Bull. Mater. Sci.* **2015**, *38*, 951–955. [CrossRef]
207. Cui, W.; Qin, G.; Duan, J.; Wang, H. A Graded Nano-TiN Coating on Biomedical Ti Alloy: Low Friction Coefficient, Good Bonding and Biocompatibility. *Mater. Sci. Eng. C* **2017**, *71*, 520–528. [CrossRef] [PubMed]
208. Hussein, M.A.; Adesina, A.Y.; Kumar, A.M.; Sorour, A.A.; Ankah, N.; Al-Aqeeli, N. Mechanical, in-Vitro Corrosion, and Tribological Characteristics of TiN Coating Produced by Cathodic Arc Physical Vapor Deposition on Ti20Nb13Zr Alloy for Biomedical Applications. *Thin Solid Films* **2020**, *709*. [CrossRef]
209. Uddin, G.M.; Jawad, M.; Ghufran, M.; Saleem, M.W.; Raza, M.A.; Rehman, Z.U.; Arafat, S.M.; Irfan, M.; Waseem, B. Experimental Investigation of Tribo-Mechanical and Chemical Properties of TiN PVD Coating on Titanium Substrate for Biomedical Implants Manufacturing. *Int. J. Adv. Manuf. Technol.* **2019**, *102*, 1391–1404. [CrossRef]
210. Beshchasna, N.; Ho, A.Y.K.; Saqib, M.; Kraśkiewicz, H.; Wasyluk, Ł.; Kuzmin, O.; Duta, O.C.; Ficai, D.; Trusca, R.D.; Ficai, A.; et al. Surface Evaluation of Titanium Oxynitride Coatings Used for Developing Layered Cardiovascular Stents. *Mater. Sci. Eng. C* **2019**, *99*, 405–416. [CrossRef]
211. Banakh, O.; Moussa, M.; Matthey, J.; Pontearso, A.; Cattani-Lorente, M.; Sanjines, R.; Fontana, P.; Wiskott, A.; Durual, S. Sputtered Titanium Oxynitride Coatings for Endosseous Applications: Physical and Chemical Evaluation and First Bioactivity Assays. *Appl. Surf. Sci.* **2014**, *317*, 986–993. [CrossRef]
212. Banaszek, K.; Klimek, L.; Zgorzynska, E.; Swarzynska, A.; Walczewska, A. Cytotoxicity of Titanium Carbonitride Coatings for Prostodontic Alloys with Different Amounts of Carbon and Nitrogen. *Biomed. Mater.* **2018**, *13*. [CrossRef]
213. Huang, H.L.; Chang, Y.Y.; Liu, J.X.; Tsai, M.T.; Lai, C.H. Antibacterial Activity and Cell Compatibility of TiZrN, TiZrCN, and TiZr-Amorphous Carbon Coatings. *Thin Solid Films* **2015**, *596*, 111–117. [CrossRef]

214. Constantin, L.; Braic, M.; Dinu, M.; Balaceanu, M.; Braic, V.; Farcau, C.; Vladescu, A. Effects of Zr, Nb, or Si Addition on the Microstructural, Mechanical, and Corrosion Resistance of TiCN Hard Coatings. *Mater. Corros.* **2016**, *67*, 929–938. [CrossRef]
215. Zheng, L.; Zhao, L.; Xiong, W. Tribological Properties of TiAlN-Coated Cermets. *Rare Met.* **2009**, *28*, 57–62. [CrossRef]
216. Miletić, A.; Panjan, P.; Čekada, M.; Kovačević, L.; Terek, P.; Kovač, J.; Dražič, G.; Škorić, B. Nanolayer CrAlN/TiSiN Coating Designed for Tribological Applications. *Ceram. Int.* **2021**, *47*, 2022–2033. [CrossRef]
217. Sampath Kumar, T.; Vinoth Jebaraj, A.; Shankar, E.; Tamiloli, N.; Sivakumar, K. Metallurgical and Mechanical Characterization of TiCN/TiAlN and TiAlN/TiCN Bilayer Nitride Coatings. *Surf. Interfaces* **2019**, *15*, 256–264. [CrossRef]
218. Chetcuti, R.; Dearnley, P.A.; Mazzonello, A.; Buhagiar, J.; Mallia, B. Tribocorrosion Response of Duplex Layered CoCrMoC/CrN and CrN/CoCrMoC Coatings on Implant Grade 316LVM Stainless Steel. *Surf. Coat. Technol.* **2020**, *384*. [CrossRef]
219. Rathmann, L.; Rusche, T.; Hasselbruch, H.; Mehner, A.; Radel, T. Friction and Wear Characterization of LIPSS and TiN / DLC Variants. *Appl. Surf. Sci.* **2022**, *584*, 152654. [CrossRef]
220. Islam, M.; Díaz Lantada, A.; Mager, D.; Korvink, J.G.; Islam, M.; Mager, D.; Korvink, J.G.; Lantada, A.D. Carbon-Based Materials for Articular Tissue Engineering: From Innovative Scaffolding Materials toward Engineered Living Carbon. *Adv. Healthc. Mater.* **2022**, *11*, 2101834. [CrossRef] [PubMed]
221. More, R.B.; Haubold, A.D.; Bokros, J.C. Pyrolytic Carbon for Long-Term Medical Implants. In *Biomaterials Science: An Introduction to Materials*, 3rd ed.; Academic Press: Cambridge, MA, USA, 2013; pp. 209–222. [CrossRef]
222. Ross, M.; Williams, D.; Couzens, G.; Klawitter, J. Pyrocarbon for Joint Replacement. *Jt. Replace. Technol.* **2021**, 145–163. [CrossRef]
223. Victoria Cabañas, M. Bioceramic Coatings for Medical Implants. *Bio-Ceram. Clin. Appl.* **2014**, 249–289. [CrossRef]
224. Ren, J.; Lv, C.; Duan, Y.; Zhang, Y.; Zhang, J. Microstructure and Ablation Performance of HfC/PyC Core-Shell Structure Nanowire-Reinforced Hf1-XZrxC Coating. *J. Eur. Ceram. Soc.* **2021**, *41*, 7450–7463. [CrossRef]
225. Feng, S.L.; Yang, Y.G.; Bai, S.; Xu, L.; Yang, X.M.; Xia, H.H.; Zhou, X.T. Microstructure of a Pyrolytic Carbon Coating on a Nuclear Graphite Substrate IG-110. *Xinxing Tan Cailiao/New Carbon Mater.* **2015**, *30*, 275–281. [CrossRef]
226. Adkinson, J.M.; Chung, K.C. Advances in Small Joint Arthroplasty of the Hand. *Plast. Reconstr. Surg.* **2014**, *134*, 1260. [CrossRef]
227. Nistor, P.A.; May, P.W. Diamond Thin Films: Giving Biomedical Applications a New Shine. *J. R. Soc. Interface* **2017**, 14. [CrossRef]
228. Catledge, S.A.; Thomas, V.; Vohra, Y.K. Nanostructured Diamond Coatings for Orthopaedic Applications. *Woodhead Publ. Ser. Biomater.* **2013**, *2013*, 105. [CrossRef] [PubMed]
229. Medina, O.; Nocua, J.; Mendoza, F.; Gómez-Moreno, R.; Ávalos, J.; Rodríguez, C.; Morell, G. Bactericide and Bacterial Anti-Adhesive Properties of the Nanocrystalline Diamond Surface. *Diam. Relat. Mater.* **2012**, *22*, 77–81. [CrossRef]
230. Siddiqui, S.; Dutta, G.; Tan, C.; Arumugam, P.U. Nanocrystalline Diamond Electrodes: Enabling Electrochemical Microsensing Applications with High Reliability and Stability. *IEEE Nanotechnol. Mag.* **2016**, *10*, 12. [CrossRef] [PubMed]
231. Roy, R.K.; Lee, K.R. Biomedical Applications of Diamond-like Carbon Coatings: A Review. *J. Biomed. Mater. Res. B Appl. Biomater.* **2007**, *83*, 72–84. [CrossRef]
232. Peng, F.; Lin, Y.; Zhang, D.; Ruan, Q.; Tang, K.; Li, M.; Liu, X.; Chu, P.K.; Zhang, Y. Corrosion Behavior and Biocompatibility of Diamond-like Carbon-Coated Zinc: An in Vitro Study. *ACS Omega* **2021**, *6*, 9843–9851. [CrossRef]
233. Bean, P.A.; Evans, M.D.; Bendavid, A. Biomineralization of Osteoblasts on DLC Coated Surfaces for Bone Implants. *Biointerphases* **2018**, *13*, 041002. [CrossRef]
234. Liu, J.; Wang, X.; Wu, B.J.; Zhang, T.F.; Leng, Y.X.; Huang, N. Tribocorrosion Behavior of DLC-Coated CoCrMo Alloy in Simulated Biological Environment. *Vacuum* **2013**, *92*, 39–43. [CrossRef]
235. Mo, S.; Zhao, F.; Gao, A.; Wu, Y.; Liao, Q.; Xie, L.; Pan, H.; Tong, L.; Chu, P.K.; Wang, H. Simultaneous Application of Diamond-like Carbon Coating and Surface Amination on Polyether Ether Ketone: Towards Superior Mechanical Performance and Osseointegration. *Smart Mater. Med.* **2021**, *2*, 219–228. [CrossRef]
236. Park, S.J.; Lee, K.R.; Ahn, S.H.; Kim, J.G. Instability of Diamond-like Carbon (DLC) Films during Sliding in Aqueous Environment. *Diam. Relat. Mater.* **2008**, *17*, 247–251. [CrossRef]
237. Lung, B.H.; Chiang, M.J.; Hon, M.H. Effect of Gradient A-SiCx Interlayer on Adhesion of DLC Films. *Mater. Chem. Phys.* **2001**, *72*, 163–166. [CrossRef]
238. Kumar, P.; Babu, P.D.; Mohan, L.; Anandan, C.; Grips, V.K.W. Wear and Corrosion Behavior of Zr-Doped DLC on Ti-13Zr-13Nb Biomedical Alloy. *J. Mater. Eng. Perform.* **2013**, *22*, 283–293. [CrossRef]
239. Hauert, R. DLC Films in Biomedical Applications. *Tribol. Diam.-Like Carbon Film. Fundam. Appl.* **2008**, 494–509. [CrossRef]
240. Hatem, A.; Lin, J.; Wei, R.; Torres, R.D.; Laurindo, C.; Soares, P. Tribocorrosion Behavior of DLC-Coated Ti-6Al-4V Alloy Deposited by PIID and PEMS + PIID Techniques for Biomedical Applications. *Surf. Coat. Technol.* **2017**, *332*, 223–232. [CrossRef]
241. Joshi, P.; Riley, P.R.; Denning, W.; Shukla, S.; Khosla, N.; Narayan, J.; Narayan, R. Laser-Patterned Carbon Coatings on Flexible and Optically Transparent Plastic Substrates for Advanced Biomedical Sensing and Implant Applications. *J. Mater. Chem. C* **2022**. [CrossRef]
242. López, R.; Menéndez, M.; Fernández, C.; Chmiela, A.; Bernardo-Sánchez, A. The Influence of Carbon Coatings on the Functional Properties of X39CR13 and 316LVM Steels Intended for Biomedical Applications. *Metals* **2019**, *9*, 815. [CrossRef]
243. Awaja, F.; Cools, P.; Lohberger, B.; Nikiforov, A.Y.; Speranza, G.; Morent, R. Functionalized, Biocompatible, and Impermeable Nanoscale Coatings for PEEK. *Mater. Sci. Eng. C* **2017**, *76*, 865–870. [CrossRef] [PubMed]
244. Choudhury, D.; Lackner, J.; Fleming, R.A.; Goss, J.; Chen, J.; Zou, M. Diamond-like Carbon Coatings with Zirconium-Containing Interlayers for Orthopedic Implants. *J. Mech. Behav. Biomed. Mater.* **2017**, *68*, 51–61. [CrossRef]

245. Bociaga, D.; Sobczyk-Guzenda, A.; Komorowski, P.; Balcerzak, J.; Jastrzebski, K.; Przybyszewska, K.; Kaczmarek, A. Surface Characteristics and Biological Evaluation of Si-DLC Coatings Fabricated Using Magnetron Sputtering Method on Ti6Al7Nb Substrate. *Nanomaterials* **2019**, *9*, 812. [CrossRef]
246. Rothammer, B.; Neusser, K.; Marian, M.; Bartz, M.; Krauß, S.; Böhm, T.; Thiele, S.; Merle, B.; Detsch, R.; Wartzack, S. Amorphous Carbon Coatings for Total Knee Replacements—Part I: Deposition, Cytocompatibility, Chemical and Mechanical Properties. *Polymers* **2021**, *13*, 1952. [CrossRef]
247. Bociaga, D.; Sobczyk-Guzenda, A.; Szymanski, W.; Jedrzejczak, A.; Jastrzebska, A.; Olejnik, A.; Swiatek, L.; Jastrzebski, K. Diamond like Carbon Coatings Doped by Si Fabricated by a Multi-Target DC-RF Magnetron Sputtering Method—Mechanical Properties, Chemical Analysis and Biological Evaluation. *Vacuum* **2017**, *143*, 395–406. [CrossRef]
248. Derakhshandeh, M.R.; Eshraghi, M.J.; Hadavi, M.M.; Javaheri, M.; Khamseh, S.; Sari, M.G.; Zarrintaj, P.; Saeb, M.R.; Mozafari, M. Diamond-like Carbon Thin Films Prepared by Pulsed-DC PE-CVD for Biomedical Applications. *Surf. Innov.* **2018**, *6*, 167–175. [CrossRef]
249. Wachesk, C.C.; Seabra, S.H.; dos Santos, T.A.T.; Trava-Airoldi, V.J.; Lobo, A.O.; Marciano, F.R. In Vivo Biocompatibility of Diamond-like Carbon Films Containing TiO2 Nanoparticles for Biomedical Applications. *J. Mater. Sci. Mater. Med.* **2021**, *32*. [CrossRef] [PubMed]
250. Su, S.; Chen, W.; Zheng, M.; Lu, G.; Tang, W.; Huang, H.; Qu, D. Facile Fabrication of 3D-Printed Porous Ti6Al4V Scaffolds with a Sr-CaP Coating for Bone Regeneration. *ACS Omega* **2022**, *7*, 8391–8402. [CrossRef] [PubMed]
251. Taranu, B.-O.; Ianasi, P.; Rus, S.F.; Bucur, A.I. Simultaneous Precipitation and Electrodeposition of Hydroxyapatite Coatings at Different Temperatures on Various Metal Substrates. *Coatings* **2022**, *12*, 288. [CrossRef]
252. Feddes, B.; Vredenberg, A.M.; Wehner, M.; Wolke, J.C.G.; Jansen, J.A. Laser-Induced Crystallization of Calcium Phosphate Coatings on Polyethylene (PE). *Biomaterials* **2005**, *26*, 1645–1651. [CrossRef] [PubMed]
253. Azem, F.A.; Delice, T.K.; Ungan, G.; Cakir, A. Investigation of Duty Cycle Effect on Corrosion Properties of Electrodeposited Calcium Phosphate Coatings. *Mater. Sci. Eng. C* **2016**, *68*, 681–686. [CrossRef] [PubMed]
254. Heimann, R.B.; Lehmann, H.D. Deposition, Structure, Properties and Biological Function of Plasma-Sprayed Bioceramic Coatings. *Bioceram. Coat. Med. Implant.* **2015**, 253–308. [CrossRef]
255. Kaliaraj, G.S.; Siva, T.; Ramadoss, A. Surface Functionalized Bioceramics Coated on Metallic Implants for Biomedical and Anticorrosion Performance—A Review. *J. Mater. Chem. B* **2021**, *9*, 9433–9460. [CrossRef]
256. Al-Amin, M.; Abdul Rani, A.M.; Abdu Aliyu, A.A.; Bryant, M.G.; Danish, M.; Ahmad, A. Bio-Ceramic Coatings Adhesion and Roughness of Biomaterials through PM-EDM: A Comprehensive Review. *Mater. Manuf. Processes* **2020**, *35*, 1157–1180. [CrossRef]
257. Jeong, J.; Kim, J.H.; Shim, J.H.; Hwang, N.S.; Heo, C.Y. Bioactive Calcium Phosphate Materials and Applications in Bone Regeneration. *Biomater. Res.* **2019**, *23*, 1–11. [CrossRef]
258. Sankar, M.; Suwas, S.; Balasubramanian, S.; Manivasagam, G. Comparison of Electrochemical Behavior of Hydroxyapatite Coated onto WE43 Mg Alloy by Electrophoretic and Pulsed Laser Deposition. *Surf. Coat. Technol.* **2017**, *309*, 840–848. [CrossRef]
259. Yugeswaran, S.; Kobayashi, A.; Ucisik, A.H.; Subramanian, B. Characterization of Gas Tunnel Type Plasma Sprayed Hydroxyapatite–Nanostructure Titania Composite Coatings. *Appl. Surf. Sci.* **2015**, *347*, 48–56. [CrossRef]
260. Trujillo, N.A.; Oldinski, R.A.; Ma, H.; Bryers, J.D.; Williams, J.D.; Popat, K.C. Antibacterial Effects of Silver-Doped Hydroxyapatite Thin Films Sputter Deposited on Titanium. *Mater. Sci. Eng. C* **2012**, *32*, 2135–2144. [CrossRef]
261. Lett, J.A.; Sagadevan, S.; Paiman, S.; Mohammad, F.; Schirhagl, R.; Léonard, E.; Alshahateet, S.F.; Oh, W.C. Exploring the Thumbprints of Ag-Hydroxyapatite Composite as a Surface Coating Bone Material for the Implants. *J. Mater. Res. Technol.* **2020**, *9*, 12824–12833. [CrossRef]
262. Predoi, D.; Iconaru, S.; Predoi, M.; Buton, N.; Motelica-Heino, M. Zinc Doped Hydroxyapatite Thin Films Prepared by Sol–Gel Spin Coating Procedure. *Coatings* **2019**, *9*, 156. [CrossRef]
263. Wang, Q.; Li, P.; Tang, P.; Ge, X.; Ren, F.; Zhao, C.; Fang, J.; Wang, K.; Fang, L.; Li, Y.; et al. Experimental and Simulation Studies of Strontium/Fluoride-Codoped Hydroxyapatite Nanoparticles with Osteogenic and Antibacterial Activities. *Colloids Surf. B Biointerfaces* **2019**, *182*, 110359. [CrossRef]
264. Sutha, S.; Dhineshbabu, N.R.; Prabhu, M.; Rajendran, V. Mg-Doped Hydroxyapatite/Chitosan Composite Coated 316L Stainless Steel Implants for Biomedical Applications. *J. Nanosci. Nanotechnol.* **2015**, *15*, 4178–4187. [CrossRef]
265. Ofudje, E.A.; Adeogun, A.I.; Idowu, M.A.; Kareem, S.O. Synthesis and Characterization of Zn-Doped Hydroxyapatite: Scaffold Application, Antibacterial and Bioactivity Studies. *Heliyon* **2019**, *5*, e01716. [CrossRef]
266. Chi, W.; Zou, J.; Ai, F.; Lin, Y.; Li, W.; Cao, C.; Yang, K.; Zhou, K. Research of Cu-Doped Hydroxyapatite Microbeads Fabricated by Pneumatic Extrusion Printing. *Materials* **2019**, *12*, 1769. [CrossRef]
267. Prodan, A.M.; Iconaru, S.L.; Predoi, M.V.; Predoi, D.; Motelica-Heino, M.; Turculet, C.S.; Beuran, M. Silver-Doped Hydroxyapatite Thin Layers Obtained by Sol-Gel Spin Coating Procedure. *Coatings* **2019**, *10*, 14. [CrossRef]
268. Mishra, V.K.; Bhattacharjee, B.N.; Parkash, O.; Kumar, D.; Rai, S.B. Mg-Doped Hydroxyapatite Nanoplates for Biomedical Applications: A Surfactant Assisted Microwave Synthesis and Spectroscopic Investigations. *J. Alloys Compd.* **2014**, *614*, 283–288. [CrossRef]
269. Sun, T.W.; Zhu, Y.J. Solvothermal Growth of Ultralong Hydroxyapatite Nanowire Coating on Glass Substrate. *Chem. Lett.* **2019**, *48*, 1462–1464. [CrossRef]

270. Predoi, D.; Iconaru, S.L.; Ciobanu, S.C.; Predoi, S.A.; Buton, N.; Megier, C.; Beuran, M. Development of Iron-Doped Hydroxyapatite Coatings. *Coatings* **2021**, *11*, 186. [CrossRef]
271. Ciobanu, G.; Harja, M. Cerium-Doped Hydroxyapatite/Collagen Coatings on Titanium for Bone Implants. *Ceram. Int.* **2019**, *45*, 2852–2857. [CrossRef]
272. Golovanova, O.A.; Zaits, A.V. Biomimetic Coating of a Titanium Substrate with Silicon-Substituted Hydroxyapatite. *Inorg. Mater.* **2018**, *54*, 1124–1130. [CrossRef]
273. Zhang, L. Surface Modification of Titanium by Hydroxyapatite/CaSiO3/Chitosan Porous Bioceramic Coating. *Int. J. Electrochem. Sci.* **2020**, *15*. [CrossRef]
274. Dhinasekaran, D.; Kaliaraj, G.S.; Jagannathan, M.; Rajendran, A.R.; Prakasarao, A.; Ganesan, S.; Subramanian, B. Pulsed Laser Deposition of Nanostructured Bioactive Glass and Hydroxyapatite Coatings: Microstructural and Electrochemical Characterization. *Mater. Sci. Eng. C* **2021**, *130*. [CrossRef] [PubMed]
275. Zhang, L.; Pei, L.; Li, H.; Zhu, F. Design and Fabrication of Pyrolytic Carbon-SiC-Fluoridated Hydroxyapatite-Hydroxyapatite Multilayered Coating on Carbon Fibers. *Appl. Surf. Sci.* **2019**, *473*, 571–577. [CrossRef]
276. Batebi, K.; Abbasi Khazaei, B.; Afshar, A. Characterization of Sol-Gel Derived Silver/Fluor-Hydroxyapatite Composite Coatings on Titanium Substrate. *Surf. Coat. Technol.* **2018**, *352*, 522–528. [CrossRef]
277. Melero, H.C.; Sakai, R.T.; Vignatti, C.A.; Benedetti, A.V.; Fernández, J.; Guilemany, J.M.; Suegama, P.H. Corrosion Resistance Evaluation of HVOF Produced Hydroxyapatite and TiO_2-Hydroxyapatite Coatings in Hanks' Solution. *Mater. Res.* **2018**, *21*. [CrossRef]
278. Evcin, A.; Buyukleblebici, B. Ti6A14V Coating with B_2O_3 and Al_2O_3 Containing Hydroxyapatite by HVOF Technique. *Sci. Iran.* **2019**, *26*, 1980–1989. [CrossRef]
279. Farrokhi-Rad, M. Electrophoretic Deposition of Fiber Hydroxyapatite/Titania Nanocomposite Coatings. *Ceram. Int.* **2018**, *44*, 622–630. [CrossRef]
280. Ionescu, R.N.; Totan, A.R.; Imre, M.M.; Țâncu, A.M.C.; Pantea, M.; Butucescu, M.; Farcașiu, A.T. Prosthetic Materials Used for Implant-Supported Restorations and Their Biochemical Oral Interactions: A Narrative Review. *Materials* **2022**, *15*, 1016. [CrossRef] [PubMed]
281. Treccani, L.; Yvonne Klein, T.; Meder, F.; Pardun, K.; Rezwan, K. Functionalized Ceramics for Biomedical, Biotechnological and Environmental Applications. *Acta Biomater.* **2013**, *9*, 7115–7150. [CrossRef] [PubMed]
282. Fu, L.; Khor, K.A.; Lim, J.P. Yttria Stabilized Zirconia Reinforced Hydroxyapatite Coatings. *Surf. Coat. Technol.* **2000**, *127*, 66–75. [CrossRef]
283. Kaliaraj, G.S.; Muthaiah, B.; Alagarsamy, K.; Vishwakarma, V.; Kirubaharan, A.M.K. Role of Bovine Serum Albumin in the Degradation of Zirconia and Its Allotropes Coated 316L SS for Potential Bioimplants. *Mater. Chem. Phys.* **2021**, *258*, 123859. [CrossRef]
284. Kaliaraj, G.S.; Vishwakarma, V.; Alagarsamy, K.; Kamalan Kirubaharan, A.M. Biological and Corrosion Behavior of M-ZrO2 and t-ZrO2 Coated 316L SS for Potential Biomedical Applications. *Ceram. Int.* **2018**, *44*, 14940–14946. [CrossRef]
285. Pradhaban, G.; Kaliaraj, G.S.; Vishwakarma, V. Antibacterial Effects of Silver–Zirconia Composite Coatings Using Pulsed Laser Deposition onto 316L SS for Bio Implants. *Prog. Biomater.* **2014**, *3*, 123–130. [CrossRef]
286. Santos, R.L.P.; Buciumeanu, M.; Silva, F.S.; Souza, J.C.M.; Nascimento, R.M.; Motta, F.V.; Henriques, B. Tribological Behavior of Zirconia-Reinforced Glass–Ceramic Composites in Artificial Saliva. *Tribol. Int.* **2016**, *103*, 379–387. [CrossRef]
287. Berni, M.; Lopomo, N.; Marchiori, G.; Gambardella, A.; Boi, M.; Bianchi, M.; Visani, A.; Pavan, P.; Russo, A.; Marcacci, M. Tribological Characterization of Zirconia Coatings Deposited on Ti_6Al_4V Components for Orthopedic Applications. *Mater. Sci. Eng. C* **2016**, *62*, 643–655. [CrossRef]
288. Kaliaraj, G.S.; Vishwakarma, V.; Kirubaharan, A.M.K. Biocompatible Zirconia-Coated 316 Stainless Steel with Anticorrosive Behavior for Biomedical Application. *Ceram. Int.* **2018**, *44*, 9780–9786. [CrossRef]
289. Sergi, R.; Bellucci, D.; Cannillo, V. A Review of Bioactive Glass/Natural Polymer Composites: State of the Art. *Materials* **2020**, *13*, 5560. [CrossRef] [PubMed]
290. Pachaiappan, R.; Rajendran, S.; Show, P.L.; Manavalan, K.; Naushad, M. Metal/Metal Oxide Nanocomposites for Bactericidal Effect: A Review. *Chemosphere* **2021**, *272*. [CrossRef] [PubMed]
291. Crush, J.; Hussain, A.; Seah, K.T.M.; Khan, W.S. Bioactive Glass: Methods for Assessing Angiogenesis and Osteogenesis. *Front. Cell Dev. Biol.* **2021**, *9*. [CrossRef]
292. Cannio, M.; Bellucci, D.; Roether, J.A.; Boccaccini, D.N.; Cannillo, V. Bioactive Glass Applications: A Literature Review of Human Clinical Trials. *Materials* **2021**, *14*, 5440. [CrossRef]
293. Dey, P.; Pal, S.K.; Banerjee, I.; Sarkar, R. Effect of Addition of B_2O_3 to the Sol-Gel Synthesized 45S5 Bioglass. *J. Aust. Ceram. Soc.* **2020**, *56*, 1309–1322. [CrossRef]
294. Henao, J.; Poblano-Salas, C.; Monsalve, M.; Corona-Castuera, J.; Barceinas-Sanchez, O. Bio-Active Glass Coatings Manufactured by Thermal Spray: A Status Report. *J. Mater. Res. Technol.* **2019**, *8*, 4965–4984. [CrossRef]
295. Nawaz, Q.; Fastner, S.; Rehman, M.A.U.; Ferraris, S.; Perero, S.; di Confiengo, G.G.; Yavuz, E.; Ferraris, M.; Boccaccini, A.R. Multifunctional Stratified Composite Coatings by Electrophoretic Deposition and RF Co-Sputtering for Orthopaedic Implants. *J. Mater. Sci.* **2021**, *56*, 7920–7935. [CrossRef]

296. Ashok raja, C.; Balakumar, S.; Durgalakshmi, D.; George, R.P.; Anandkumar, B.; Kamachi Mudali, U. Reduced Graphene Oxide/Nano-Bioglass Composites: Processing and Super-Anion Oxide Evaluation. *RSC Adv.* **2016**, *6*, 19657–19661. [CrossRef]
297. Balakumar, S.; Bargavi, P.; Rajashree, P.; Anandkumar, B.; George, R.P. Decoration of 1-D Nano Bioactive Glass on Reduced Graphene Oxide Sheets: Strategies and in Vitro Bioactivity Studies. *Mater. Sci. Eng. C* **2018**, *90*, 85–94. [CrossRef]
298. Ashok raja, C.; Balakumar, S.; Anandkumar, B.; George, R.P.; Kamachi Mudali, U. Formation of Bioactive Nano Hybrid Thin Films on Anodized Titanium via Electrophoretic Deposition Intended for Biomedical Applications. *Mater. Today Commun.* **2020**, *25*, 101666. [CrossRef]
299. Chalisgaonkar, V.; Das, M.; Balla, V.K. Laser Processing of Ti Composite Coatings Reinforced with Hydroxyapatite and Bioglass. *Addit. Manuf.* **2018**, *20*, 134–143. [CrossRef]
300. López-Cuevas, J.; Rendón-Angeles, J.C.; Méndez-Nonell, J.; Barrientos-Rodríguez, H. In Vitro Bioactivity of AISI 316L Stainless Steel Coated with Hydroxyapatite-Seeded 58S Bioglass. *MRS Adv.* **2019**, *4*, 3133–3142. [CrossRef]
301. Estrada-Cabrera, E.; Torres-Ferrer, L.R.; Aztatzi-Aguilar, O.G.; de Vizcaya-Ruiz, A.; Meraz-Rios, M.A.; Zarate-Triviño, D.G.; Arizmendi-Morquecho, A.; de Luna Bugallo, A.; Prokhorov, E.; Luna-Barcenas, G. Chitosan-Bioglass Coatings on Partially Nanostructured Anodized Ti-6Al-4V Alloy for Biomedical Applications. *Surf. Coat. Technol.* **2019**, *375*, 468–476. [CrossRef]
302. Nino, D.; Bayona, M.; Güiza, V.; Córdoba, E. Approach for Fabricating Bioglass Coatings on Reticulated Vitreous Carbon Foams for Tissue Engineering Applications. *J. Phys. Conf. Ser.* **2019**, *1159*, 012007. [CrossRef]
303. Hosseini, S.; Farnoush, H. Characterization and in Vitro Bioactivity of Electrophoretically Deposited Mn-Modified Bioglass-Alginate Nanostructured Composite Coatings. *Mater. Res. Express* **2019**, *6*. [CrossRef]
304. Singh, S.; Singh, G.; Bala, N. Analysis of in Vitro Corrosion Behavior and Hemocompatibility of Electrophoretically Deposited Bioglass-Chitosan-Iron Oxide Coating for Biomedical Applications. *J. Mater. Res.* **2020**, *35*, 1749–1761. [CrossRef]
305. Say, Y.; Aksakal, B. Enhanced Corrosion Properties of Biological NiTi Alloy by Hydroxyapatite and Bioglass Based Biocomposite Coatings. *J. Mater. Res. Technol.* **2020**, *9*, 1712–1749. [CrossRef]
306. Rojas, O.; Prudent, M.; López, M.E.; Vargas, F.; Ageorges, H. Influence of Atmospheric Plasma Spraying Parameters on Porosity Formation in Coatings Manufactured from 45S5 Bioglass® Powder. *J. Therm. Spray Technol.* **2020**, *29*, 185–198. [CrossRef]
307. Li, X.; Zhitomirsky, I. Deposition of Poly(Methyl Methacrylate) and Composites Containing Bioceramics and Bioglass by Dip Coating Using Isopropanol-Water Co-Solvent. *Prog. Org. Coat.* **2020**, *148*. [CrossRef]
308. Shafiee, B.M.; Torkaman, R.; Mahmoudi, M.; Emadi, R.; Derakhshan, M.; Karamian, E.; Tavangarian, F. Surface Modification of 316l Ss Implants by Applying Bioglass/Gelatin/Polycaprolactone Composite Coatings for Biomedical Applications. *Coatings* **2020**, *10*, 1220. [CrossRef]
309. Haftbaradaran-Esfahani, M.; Ahmadian, M.; Nassajpour-Esfahani, A. Fabrication and Characterization of Porous Biomedical Vitallium Alloy with 58S Bioglass Coating Prepared by Sol-Gel Method. *Appl. Surf. Sci.* **2020**, *505*. [CrossRef]
310. Azzouz, I.; Faure, J.; Khlifi, K.; Larbi, A.C.; Benhayoune, H. Electrophoretic Deposition of 45s5 Bioglass® Coatings on the Ti6al4v Prosthetic Alloy with Improved Mechanical Properties. *Coatings* **2020**, *10*, 1192. [CrossRef]
311. Singh, S.; Singh, G.; Bala, N. Characterization, Electrochemical Behavior and in Vitro Hemocompatibility of Hydroxyapatite-Bioglass-Iron Oxide-Chitosan Composite Coating by Electrophoretic Deposition. *Surf. Coat. Technol.* **2021**, *405*. [CrossRef]

gels

Review

Nanoparticle-Containing Wound Dressing: Antimicrobial and Healing Effects

Pavel Yudaev, Yaroslav Mezhuev and Evgeniy Chistyakov *

Mendeleev University of Chemical Technology of Russia, 125047 Moscow, Russia;
yudaevpavel5@gmail.com (P.Y.); valsorja@mail.ru (Y.M.)
* Correspondence: ewgenijj@rambler.ru

Abstract: The dressings containing nanoparticles of metals and metal oxides are promising types of materials for wound repair. In such dressings, biocompatible and nontoxic hydrophilic polymers are used as a matrix. In the present review, we take a look at the anti-microbial effect of the nanoparticle-modified wound dressings against various microorganisms and evaluate their healing action. A detailed analysis of 31 sources published in 2021 and 2022 was performed. Furthermore, a trend for development of modern antibacterial wound-healing nanomaterials was shown as exemplified in publications starting from 2018. The review may be helpful for researchers working in the areas of biotechnology, medicine, epidemiology, material science and other fields aimed at the improvement of the quality of life.

Keywords: wound; wound dressing; nanoparticle; hydrogel; nanofiber

Citation: Yudaev, P.; Mezhuev, Y.; Chistyakov, E. Nanoparticle-Containing Wound Dressing: Antimicrobial and Healing Effects. *Gels* **2022**, *8*, 329. https://doi.org/10.3390/gels8060329

Academic Editors: Arish Dasan, Ashokraja Chandrasekar and Nupur Kohli

Received: 22 April 2022
Accepted: 22 May 2022
Published: 24 May 2022

1. Introduction

The large surface area and the ultra-fine size, low toxicity and biocompatibility provide the high potential for the use of nanoparticles in the biomedical field [1,2], in dental restorative materials [3,4], in pharmacology as carriers for targeted drug delivery [5,6] including cancer drugs [7]. At the same time, many nanoparticles also show antibacterial action. This has to do with the fact that nanoparticles are much smaller compared with the size of a bacterial cell and can diffuse into the bacterial cell walls with further destruction [8–10].

Nanoparticles have a high potential as components of wound dressings being an alternative to antibiotics thanks to the fact that unlike the former, the latter cause fewer side effects and are not prone to causing microbial resistance. This makes it possible to use them for inhibiting the growth of drug-resistant bacteria [11].

It is known that in order to be used in clinical practice the skin regenerating and healing wound dressings must have a range of properties such as biocompatibility, nontoxicity, porosity, as well as air and vapor permeability. They must also sustain a moist environment on the wound surface and absorb exudates, be capable of imitating the extracellular matrix structure and have an antibacterial effect [12]. Such properties can be found in nanofibers and hydrogels based on natural and synthetic polymers modified with nanoparticles. Unlike roller bandage and gauze, hydrogels create a moist wound environment and provide a cooling effect relieving painful sensations in patients [13].

A number of review articles are known dealing with various wound healing materials, i.e., nanofibers modified with silver nanoparticles [14], electroconductive films, membranes, hydrogels [15], nanocomposites based on polylactic acid and zinc oxide nanoparticles [16], polysaccharide antibacterial hydrogels [17,18], hydrogels with copper nanoparticles [19], membranes with gold or silver nanoparticles [20] and others. However, the aforementioned papers do not contain a detailed analysis of antibacterial and healing properties of wound dressings.

The aim of the present paper was to provide an outlook on the use of nanoparticles of metals and metal oxides for production of polymer composite wound dressings through studying their antibacterial and healing action.

2. Materials and Methods

The Scopus database was used to conduct the literature survey. The "wound dressing nanoparticles" search returned 1747 results, of which 146 were dated 2022 and 324 were dated 2021, as well as 565 articles from 2018 to 2020. For the purposes of the present analysis 31 most relevant articles published in 2021–2022 were selected, of which 18 articles dedicated to the silver nanoparticles (AgNPs), gold (AuNPs) and copper (CuNPs) nanoparticles. 13 articles dedicated to metal oxides nanoparticles as follows: 7 articles dealing with zinc oxide nanoparticles (ZnO NPs), 2 articles related to iron oxide nanoparticles (FeO NPs or Fe_3O_4 NPs), 2 articles on cerium dioxide nanoparticles (CeO_2 NPs), article dedicated to titanium dioxide nanoparticles with multi-walled carbon nanotubes (MWCNT_TiO_2), and 1 article on copper oxide nanoparticles (CuO NPs).

3. Results

3.1. Nanoparticles of Metals

3.1.1. Silver Nanoparticles

The silver nanoparticles obtained by green methods are efficient against pathogens in case of acute, chronic wounds and burns. It is known from the literature [21] that the formation of free radical forms of oxygen is the main mechanism of AgNPs interaction with bacteria leading to damage of the bacteria cell walls.

It is noteworthy that the introduction of AgNPs into the hydrogel matrices is often complicated by low stability of the colloids in aqueous media, which may lead to loss of bioactivity, formation of toxic structure and side effects in patients. Therefore, the biopolymers are normally used for stabilization. For instance, the authors of [22] synthesized AgNPs in the presence of stabilizers in the form of polysaccharides, i.e., ulvan and cellulose. The AgNPs retained its stability even upon loading of a hyaluronic acid-based hydrogel. It was determined that polysaccharides stabilized the AgNPs through the formation of a polyanionic outer layer around the inorganic core attributable to the presence of the hydroxyl, sulphate and carboxyl groups leading to electrostatic repulsion of AgNPs and a high negative zeta-potential value. The addition of polysaccharides did not affect the antibacterial action of AgNPs against *E. coli*, *P. aeruginosa* and *S. aureus*.

Antezana et al. [23] studied the antibacterial activity and biocompatibility of hydrogels based on collagen and spherical AgNPs ranging from 10 to 15 nm. The growth inhibiting effect of the hydrogels on the *S. aureus* and *P. aeruginosa* was established, as evidenced by the presence of the zones of inhibition vs. the nanoparticle-free hydrogel (Figure 1). The CFU mL^{-1} values for the *S. aureus* and *P. aeruginosa* bacteria significantly decreased within 7 days for the hydrogels containing 6.7 mg g^{-1} and 67 mg g^{-1} of AgNPs. However, for the hydrogels with the lower AgNPs content (0.67 mg g^{-1}) the decrease in CFU mL^{-1} was observed only for 3 days followed by a later rise. Unfortunately, the authors did not provide an explanation for this fact.

The viability of the MDCK dog kidney epithelial cells decreased by 80–90% after 48 h of observation for all silver-containing hydrogels. According to the authors this has to do with the release of the toxic silver ions. In order to create biocompatible hydrogels, the *Cannabis sativa* plant extract was used, which provided an antioxidant effect. The MDCK cell proliferation increase by 40–60% was discovered as compared to the extract-free hydrogel. This fact was explained by the authors through the reduction in oxidative stress caused by O_2, the generation of which is stimulated by AgNPs.

Nešović et al. [24] studied an AgNPs-containing hydrogel with 40–60 nm particles based on polyvinyl alcohol. The germ-kill effect against *S. aureus* (the TL strain) and *E. coli* (the ATCC 25922 strain) was observed within 24 h. In that publication the role of the stabilizing agent was played by polyvinyl alcohol, which prevented the agglomeration and the bioactivity loss of AgNPs. It was determined that the release rate of silver (in the phosphate buffer medium with pH = 7.4) from the hydrogel can be accurately controlled, which is important for clinical use of such wound repair materials and stable protection against infections. According to the authors, the silver-containing hydrogel can be applied for wound healing for a prolonged period of time since it releases the silver slowly and maintains the required concentration thereof for 28 days. However, the authors of paper [24] did not study the mechanical properties of the hydrogel.

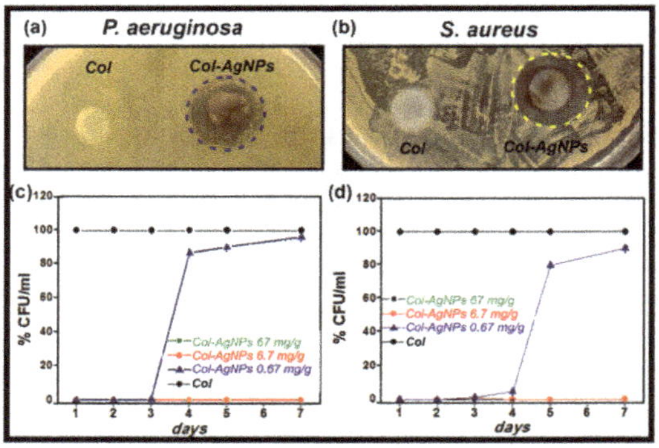

Figure 1. The antimicrobial activity of hydrogels against the *P. aeruginosa* (**a**,**c**) and *S. aureus* (**b**,**d**) bacteria (green and red lines match).

Santiago-Castillo et al. [25] the antibacterial properties of nanofibers based on high molecular weight polyvinyl alcohol (146–186 kDa), chitosan (190–310 kDa) and cube-shaped AgNPs were investigated. A combination of chitosan with polyvinyl alcohol was used since the solutions of chitosan are highly viscous at room temperature due to the polycationic nature of chitosan in solutions and due to the rigid structure, which complicated the production of fiber by means of electrospinning. The nanofibers demonstrated antibacterial activity against the *E. coli* (the ATCC 25922 strain) and the *S. aureus* bacteria (the ATCC 25923 strain). The diameters of the zones of inhibition were equal to 22 mm and 20 mm for *E. coli* and *S. aureus*, respectively. The antibacterial activity of the obtained nanofibers was better compared to the nanofibers containing ZnO NPs and copper nanoparticles. The hardness increase in the silver-containing fiber from 32 hPa to 152 hPa compared with the polyvinyl alcohol-chitosan fiber was established.

In contrast to Santiago-Castillo et al. [25], the authors of [26] used castor oil for loading of the composite instead of chitosan in order to increase the antibacterial properties of the polyvinyl alcohol and AgNPs hydrogels. The silver nanoparticles at that were synthesized in the presence of the *Mentha piperita* leaf extract and then dispersed in the polymer matrix. The nanoparticles were equidistributed in the polymer matrix without aggregation. The films were obtained by green method using environmentally friendly raw materials, water and ethanol as solvents. The formulated films displayed the growth inhibition zone diameters of 8.33 mm and 9 mm for the *S. aureus* and *P. aeruginosa* bacteria, respectively. At the same time, the obtained values were lower than in [25] for *S. aureus* and lower than that of commercially available antibiotics amoxicillin (13.7 for *S. aureus*) and amikacin (19.7 for *P. aeruginosa*).

A burn-repair hydrogel was developed, in which oxidized dextran, adipic dihydrazide grafted hyaluronic acid and quaternized chitosan were used as the polymer matrix and the spherical 50–100 nm AgNPs was used as the filler [27]. The reduction in the silver ions Ag^+ to AgNPs was performed by placing the gel into 0.1 M aqueous solution of silver nitrate. The hydrogel showed antibacterial effect against *E. coli*, *S. aureus* and *P. aeruginosa*. The diameters of the zones of inhibition growth for *E. coli*, *S. aureus* and *P. aeruginosa* were 16, 20 and 17 mm, respectively. The in vivo study of the healing effect showed a decrease in the burn area in the SD male rats after 7 days of observation. The wound was completely healed after 14 days of observation, whereas in the control group (AgNPs-free gel) the healing occurred only on the 21st day. Furthermore, the histological examination showed full re-epithelization and a pronounced deposition of collagen after 21 days of observation and the immunohistochemistry assay demonstrated the expression reduction in the inflammatory cytokines IL-6, IL-1β and TNF-α on the 14th day.

Since the obtained hydrogel had a pronounced antibacterial, healing and anti-inflammatory effect it can be used for fast healing of burns. However, the study of cytotoxicity on the L929 mice fibroblast cells showed less than 70% cell viability, which is indicative of the toxicity of the silver-containing hydrogel.

The authors of paper [28] produced the antibacterial nontoxic hybrid nanofibers based on quaternized chitin, tannic acid, polylactic acid, and polyurethane with the addition of AgNPs using the electrospinning method. The obtained nanofibers inhibited the growth of the Gram-positive *S. aureus* bacteria and the Gram-negative *E. coli*. Noteworthy, the authors chose to express the diameter of the inhibition zone in nm and not in mm, which does not match the indications on the Petri dishes. This does not allow for a high-quality data interpretation.

Bozkaya et al. [29] studied the antibacterial and wound-healing effects of the AgNPs-containing (particle size 14.8 nm) polycaprolactone and polyethylene oxide fibers. The water-methanol extract of the *Centella asiatica* plant was used in the role of the stabilizer and the reductive agent. The fibers had the sufficient mechanical properties (tensile strength min. 2.5 MPa) for wound-dressing applications, porosity (pore size min 1000 mm), water absorption, water vapor transfer rate (in the range of 2000–2500 g m^{-2} day^{-1}), air permeability, antibacterial effect (against *S. aureus, E. coli, C. Albicans*) and biocompatibility. The diameters of the zones of inhibition for the resulting fibers grew along with the rise of the AgNPs concentration. The inhibition zone diameters after 24 h of incubation were 24, 21 and 21 mm for *S. aureus, E. coli*, and *C. albicans*, respectively. The L929 cell viability was above 85%, which indicates that the fibers are nontoxic. The authors plan to use the obtained nanomaterial for healing of burns. That being said, the in vivo studies of the fibers' healing and anti-inflammatory effects are needed for their introduction into clinical practice.

The authors of [30] manufactured a high-swelling (4000%) cryogel based on gelatin and AgNPs (particle size 10–20 nm). The cryogel was active against the methicillin-resistant *S. aureus* (MRSA) and *P. aeruginosa* and promoted regeneration of the burn wound tissue in the Kunming female mice weighing 30–35 g. The wound area decreased after 2 weeks. The wound contraction was 96%, whereas that of the commercially available Tegaderm wound dressing was only 56%. The expression of the TNF-α cytokines was decreased compared to Tegaderm within 1 week of treatment, which indicates better anti-inflammatory properties of the cryogel. The cryogel also possessed the ability to absorb blood and showed haemostatic action since it was able to decrease the blood loss from 350 mg down to 81 mg. The authors proposed the use of the cryogel for the production of burn wound dressings.

Since gelatin yields high-swelling gels it can increase the same parameter for other polymers as well. Thus, Sethi et al. [31] synthesized a hybrid hydrogel containing spherical (particle size 4–19 nm) and quasispherical (particle size 4–58 nm) AgNPs based on starch and gelatin. N,N'–methylenebisacrylamide was used for chemical crosslinking of the polymers, whereas for physical crosslinking polyacrylic acid was grafted to starch and gelatin. The maximum swelling degree in blood, water, 0.9% aqueous solutions of $MgCl_2$ and NaCl for the hybrid gel were by 12–27% higher compared to that of the starch-based gel and the compressive modulus grew by 201%. The hybrid hydrogel possessed sufficient porosity, antibacterial activity against *E. coli* and *S. aureus* and was non-cytotoxic towards the human skin fibroblasts (cell viability 89%). The release of silver into the phosphate-buffered solution was fast within 2 days, whereas later it became nearly stable (Figure 2). The amount of the released silver varied in the range of 21–51 mcg per 0.5 g of gel, which is acceptable for use in the wound-repair application.

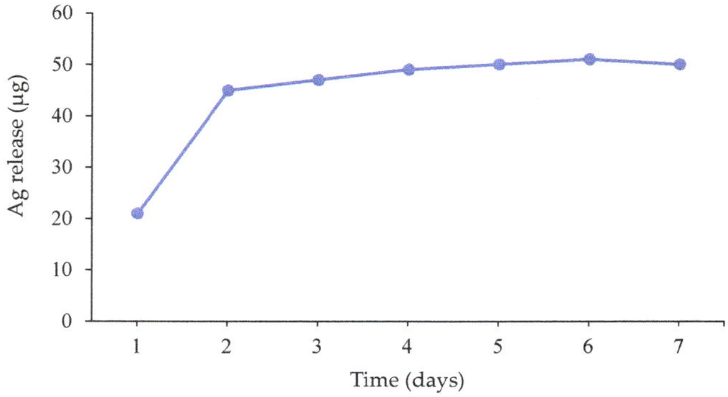

Figure 2. The release of silver from the hybrid hydrogel in vitro.

It is often the case that starch requires other types of modification. Due to low solubility in organic solvents, it is difficult to obtain starch-based nanofibers by the electrospinning method. For that reason, the starch is chemically transformed into hydroxypropyl starch and for mechanical strengthening of the nanofibers it is blended with synthetic polymers. For instance, El-Hefnawy et al. [32] obtained nanofibers for wound dressings on the basis of hydroxypropyl starch, polyurethane, and AgNPs with particle size below 5 nm. The AgNPs was synthesized by green method in the presence of the *Nerium oleander* leaf extract as the stabilizer and the reductive agent. Four polymer compositions were prepared for electrospinning containing 0 (AgNPs-0@NFs), 1 (AgNPs-1@NFs), 2 (AgNPs-2@NFs) and 3 (AgNPs-3@NFs) ml of AgNPs water dispersion. The full composition of the formulations is shown in Table 1.

Table 1. The lists of the utilized volumes of polymers and AgNPs used in the preparation of electrospinning solutions (PU—polyurethane (12 wt.% solution in DMF), HPS—hydroxypropyl starch (15 wt.% solution in DMSO)).

Composition Code	PU Volume (mL)	HPS Volume (mL)	Water Dispersion AgNPs (mL)	Total Volume (mL)
AgNPs-0@NFs	10	5	0	15
AgNPs-1@NFs	9	5	1	15
AgNPs-2@NFs	8	5	2	15
AgNPs-3@NFs	7	5	3	15

The obtained nanofibers showed an antibacterial effect against wound-present pathogen including the drug-resistant Gram-negative *P. aeruginosa* bacteria. The zones of inhibition diameters values (ZOI) are presented in Table 2. It was determined that the highest ZOI values can be observed for the AgNPs-3@NFs nanofibers, whereas the lowest were seen in the case of AgNPs-1@NFs. Furthermore, 2 the ZOI values for AgNPs-3@NFs were much higher than that of the commercially available ciprofloxacin antibiotic.

A composite based on the cellulose nanofibers and AgNPs was developed and studied in terms of its antimicrobial properties [33]. The zone of inhibition diameters for *E. coli*, *P. aeruginosa*, *S. aureus*, *B. subtilis*, *P. mirabilis* and *C. albicans* was 8.7, 8.0, 10.7, 11.0, 10.7 and 10.3 mm, respectively. The cellulose nanofibers prevented the aggregation of AgNPs and the loss of their bioactivity. The silver ions Ag^+ were adsorbed on the negatively charged surface of the nanofibers, and thanks to the carbonyl groups on the surface thereof, were reduced to silver nanoparticles. The obtained composite was more active against the abovementioned microorganisms compared with the commercially available gentamycin antibiotic. It is planned to use the composite as a dressing for repairing wounds of various origin.

Table 2. The antimicrobial activity and ZOI diameters for AgNPs-0@NFs, AgNPs-1@NFs, AgNps-2@NFs, and AgNPs-3@NFs against human-associated pathogens.

Composition Code	ZOI Diameters (mm)			
	P. aeruginosa	*E. faecalis*	*C. albicans*	*A. niger*
AgNPs-0@NFs	0 ± 0	0 ± 0	0 ± 0	0 ± 0
AgNPs-1@NFs	15 ± 0.20	13 ± 0.18	10 ± 0.16	11 ± 0.23
AgNPs-2@NFs	21 ± 0.17	19 ± 0.12	17 ± 0.20	15 ± 0.15
AgNPs-3@NFs	26 ± 0.23	24 ± 0.25	23 ± 0.23	21 ± 0.17
Ciprofloxacin	12 ± 0.19	11 ± 0.25	8 ± 0.21	7 ± 0.14

Rao et al. [34] obtained a gel based on carboxymethyl chitosan and AgNPs stabilized by tannic acid. The silver nanoparticles were of spherical shape 5 nm in size. The hydrogel displayed antibacterial action against *E. coli* and *S. aureus*, good adhesion and cell proliferation; it was nontoxic to the CCDK skin fibroblast cells. The diameters of the zones of growth inhibition for *E. coli* and *S. aureus* were 19.2 mm and 17.5 mm, respectively. Unfortunately, the publication [34] does not contain a study of the gel's mechanical properties, whereas the gel must be sufficiently strong to withstand the external forces during motion of the human body.

Yan et al. [35] studied the mechanical properties of the hydrogel membrane on the basis of AgNPs-containing calcium alginate-polydopamine-carboxymethyl chitosan. The membrane itself was prepared by immersion of the calcium alginate-polydopamine-carboxymethyl chitosan film into a silver nitrate water solution. It was determined that the tensile strength and the elongation at break were by 342.99% and 13.84% higher vs. the calcium alginate membrane and were equal to 91.92 MPa and 2.55%, respectively. It was also reported that the hydrogel membrane showed antibacterial activity against *E. coli* and *S. aureus*. The diameters of the growth inhibition zones for the *E. coli* and *S. aureus* bacteria were 14 mm and 13 mm at the silver nitrate concentration 2 mM. The membrane was nontoxic at the silver nitrate concentration of 2 mM since the viability of the HSF human skin fibroblast cells was over 70%. However, as the silver nitrate concentration increased from 2 mM to 4 mM the cell viability dropped down to 10%, which is indicative of the membrane's cytotoxicity. This means that in order for the membrane to be applied as a wound dressing the $AgNO_3$ concentrations must be low.

The polyacrylonitrile nanofibers containing curcumin, tannic acid and AgNPs (average particle size 19.1 nm) are active against *E. coli* and *S. aureus* and are biocompatible with the MSCs mesenchymatous stem cells [36]. The tannic acid played the role of the reducing agent for the silver ions in the production of AgNPs.

The maximum diameters of the growth inhibition zones for the *E. coli* and *S. aureus* bacteria were 1.1 and 1.2 cm. It was also established that the silver is released from the fibers for a prolonged period of time (more than 6 days).

In view of this, the AgNPs are a promising component for the wound dressings. The materials based on polymers and AgNPs have an antibacterial, antifungal, healing and anti-inflammatory effect and demonstrate a lasting prolonged release of silver from the polymer matrix [29,31]. The high-water retention capacity of the materials containing AgNPs ensures the absorption of the wound exudates. It is attributable to the hydrogen bonds and the dipole–dipole interactions between the polar (-OH, -COOH) groups within the AgNPs structure stabilized by the plant extracts and water [29].

Noteworthy, the silver nanoparticles applied to the skin in the dispersion form did not cause aggressive skin rashes in rabbits and swine [37]. However, the AgNPs may come into contact with keratinocytes upon the skin barrier damage, e.g., in case of burns and chronic wounds. In publication [38] the toxic effect of AgNPs towards the human skin keratinocytes (the HaCaT cell line) was reported, i.e., the main epidermal cells. Additional in vitro and in vivo studies of AgNPs are needed in order to avoid side effects for application in wound dressings.

3.1.2. Nanoparticles of Gold

Zhang et al. [39] developed a nanoporous hydrogel dressing based on heparin and polyvinyl alcohol containing spherical AuNPs with the particle size below 100 nm. The dressing inhibited the growth of the *S. aureus* and *E. coli* bacteria, improved healing of wounds in the Kunming mice; it was nontoxic towards the HFFF2 human dermal fibroblasts (cell viability over 90%). It was suggested to use the dressing for burn repair. The spherical gold nanoparticles of various sizes are nontoxic to human skin [40]. However, the AuNPs-containing polycaprolactone nanofibers showed a minor antimicrobial effect against the *S. aureus, E. coli, P. aeruginosa, C. albicans* (the zone of inhibition values below 3 mm) [41].

3.1.3. Nanoparticles of Copper

In paper [42], a double-layer nanofiber was synthesized. The first layer was made of polyvinyl alcohol and chitosan containing CuNPs, whereas the second layer consisted of poly-N-vinyl pyrrolidone. The high antibacterial activity against the Gram-positive *S. aureus* (15.6 $\pm$ 1.1 mm) and *B. cereus* (29.6 $\pm$ 0.42 mm) and the Gram-negative *E. coli* (13.3 $\pm$ 0.8 mm) and *P. aeruginosa* (10 $\pm$ 1 mm) bacteria was established. The authors also showed the faster healing in the albino Wistar rats on the 3rd day of observation vs. the controls (nanoparticles-free gel). On the 16th day, the wounds healed completely. The nontoxicity of the CuNPs as compared to the AgNPs towards the human skin keratinocytes is advantageous [43]. However, the activity of the copper-containing nanofiber against the *S. aureus, E. coli* and *P. aeruginosa* microorganisms was lower even than that of the tetracycline antibiotic [42].

3.2. Nanoparticles of Metal Oxides

3.2.1. Wound Dressings Containing Zinc Oxide Nanoparticles

ZnO NPs inhibit the growth of wound microbial population, improve healing, promote tissue regeneration and wound contraction. Bandeira et al. [44] synthesized nanofibers based on polyacrylic acid and polyallylamine hydrochloride modified with spherical ZnO NPs 18 $\pm$ 5 nm in diameter by means of electrospinning. The nanoparticles were synthesized by green method using the *Ilex paraguariensis* leaf extract. The ZnO NPs content in the fiber was 11.2 wt.%. The nanofibers showed the morphology similar to that of the skin extracellular matrix and inhibited the growth of both Gram-negative *E. coli* (the ATCC 35218 strain) and the Gram-positive *S. aureus* (the ATCC 25923 strain). The nanofibers decreased the viability of the *S. aureus* and *E. coli* bacterial cells by 65% and 10%, respectively. Thus, the Gram-negative *E. coli* are more resistant to the zinc oxide nanoparticles as compared to the *S. aureus* bacteria.

The paper [45] demonstrated that the hydrocolloid patch with ZnO NPs from CGBio (Seoul, South Korea) improves wound healing in the Sprague Dawley rats weighing 200–300 g. A decrease in the wound area was observed in comparison to the control group after 10 days. The wound healed by 98% on the 10th day vs. day 0. The microscopic study showed a thicker outer skin and granulation tissue layers as well as higher collagen on the 10th day compared to the control. The IF staining showed a decrease in the CD68 anti-inflammatory cytokines concentration by ca. 30%, the IL-8 by 50%, the TNF-α by 50%, the MCP-1 by 50%, the IL-6 by 90% and the IL-1β by 20% in the 10th day. Moreover, higher levels of the α-CMA, TGF-β3 fibroblasts biomarkers, as well as those of vimentin and M2 were determined vs. controls, which is critical during the wound healing process. Therefore, the obtained patch had the necessary healing and anti-inflammatory effects. Furthermore, the authors plan to investigate the cytotoxicity of the produced material. However, the authors of [45] did not name the polymer from which the hydrocolloid patch was made and what materials and solvents were used for its preparation, which is not entirely meaningful from a scientific standpoint.

In publications [46,47] biocomposite chitosan-based films modified with ZnO NPs were prepared that were active against wound-present pathogens, i.e., *E. coli*, *S. aureus*, *K. pneumoniae*, *B. subtilis*. The work [46] dealt with the analysis of a film based on chitosan (mol. weight 800 kDa), glycyrrhizinic acid and ZnO NPs/palygorskite nanorods. The film had a high degree of swelling of 472.49%, which indicates the fact that the film can easily absorb the wound exudates. The agar diffusion method was used for qualitative evaluation of the antibacterial effect of the films against the pathogens such as *E. coli* and *S. aureus* as well as the drug-resistant β-lactamase-producing *E. coli* (ESBL-*E. coli*) and MRSA; for the quantitative evaluation the colony count method was used. At 5 wt.% content of nanorods the film inhibited 99.5% *E. coli*, 99.8% *S. aureus*, 99.6% ESBL-*E. coli* and 99.8% MRSA. The diameters of the growth inhibition zones for *E. coli*, *S. aureus*, ESBL-*E. coli* and MRSA were 15.04 $\pm$ 1.07 mm, 14.50 $\pm$ 0.42 mm, 13.28 $\pm$ 0.31 mm and 13.64 $\pm$ 0.29 mm, respectively. The ZnO NPs release the zinc ions, which bind to the negatively charged of the bacterial cell membrane causing lysis. It is necessary to conduct an in vivo study of the wound surface evolution over time.

In manuscript [48], the authors developed a wound dressing on the basis of cotton pads and a biocomposite consisting of chitosan, glycogen and 30–80 nm ZnO NPs. Contrary to publications [46,47], in [48] the healing effect of the material was studied as compared to sterile gauze. A fast and almost complete healing of back wounds in the Wistar rats was demonstrated (males weighing 180–200 g), as well as excellent epithelization, granulation, tissue generation and collagen deposition. In 3 days, the wound diameter decreased by 11.7%, whereas the wound area decreased by 22.5%, which is more than double of that in the control group (sterile gauze). After 17 days of observation, the size of the wounds decreased by 99.7% vs. 89% in the control group. Additionally, the dressing had a significant antibacterial effect against the wound bacteria *P. aeruginosa*, *S. aureus*, *S. epidermidis* and a fungicidal effect against the causal fungus *C. albicans*. The decrease percentage of the CFU for the Gram-negative *P. aeruginosa* bacteria and the *C. albicans* fungus was lower than that for the Gram-positive bacteria *S. epidermidis* and *S. aureus* and was equal to 35%, 63%, 97% and 85%, respectively. The antibacterial action was explained by the authors through the electrostatic attraction between the positively charged dressing and the negatively charged bacterial cell walls. According to the authors, the prepared wound dressing can be used in clinical practice for treatment of chronic wounds and diabetic foot sores. However, the paper did not study the material's cytotoxicity and other important parameters applicable to wound dressings.

The antibacterial and healing effects of the hydrogel of another mixed polymer composition, i.e., polyvinyl alcohol, chitosan and starch with addition of ZnO NPs (particle size below 30 nm) was studied [49]. *E. coli* (the ATCC 25922 strain) and *S. aureus* (the ATCC 25923 strain) were selected as the test microorganisms. The healing effect was studied in vivo on the male Wistar rats (200–250 g). The minimum inhibiting concentration values against *E. coli* and *S. aureus* were 200 μg mL^{-1} and 50 μg mL^{-1}, respectively. On the 7th day of the in vivo study the wound contraction percentage was lower than in the control group, whereas 100% wound contraction was observed on the 14th day. The microscopic wound repair investigation showed 60% collagen fiber formation after 7 and 14 days, which is necessary for the regeneration of blood vessels, whereas in the control group it was under 50%. It was established that the viability of the L-929 and HDF cells was over 75%, which suggests the nontoxicity of the dressing. Additionally, the hydrogel had a porous structure (pore size 20 $\pm$ 8 μm), the optimal degree of swelling (swelling ratio 5.5–6.7) and the water vapor transmission rate (in the range of 2000–3000 g m^{-2} day^{-1}). Furthermore, the addition of ZnO NPs improved the hydrogel's tensile strength. The formulated material had great prospects in terms of the wound dressing production.

With that said, the wound dressings containing the ZnO NPs are a promising type of materials for chronic wound repair and diabetic foot sores treatment thanks to their improved healing effect and antimicrobial action compared to cotton gauze. However, high concentrations of ZnO NPs (50 μg mL^{-1} and above) did inhibit the proliferation of the HGF-1 human gingival fibroblast cells [50].

It should be noted that along with all of its positive properties, the ZnO NPs are potentially genotoxic towards the human epidermal cells even at low concentrations (below 1 μg mL^{-1}) [51]. In the review paper [52] an analysis of the liver toxicity, lung toxicity, neurotoxicity and immunotoxicity studies of the ZnO NPs was performed. It was established that the ZnO NPs toxicity depends on the concentration/dose, administration route, exposure time and the particle size. This means that further toxicity study of the ZnO NPs is required, and they should be applied with caution.

3.2.2. Wound Dressings Containing Nanoparticles of Iron Oxides

In publication [53], a porous nanocomposites based on chitosan, polyvinyl alcohol and the FeO iron oxide nanoparticles (particle size under 50 nm) was obtained. The nanoparticles were prepared using the *Pinus densiflora* leaf extract. The nanocomposite containing 0.01 wt.% of nanoparticles had a large zone of bacterial growth inhibition with respect to *B. cereus* (22 mm), *S. aureus* (21 mm), *E. coli* (20 mm), *S. enterica* (22 mm). The inclusion of the nanoparticles into the polymer matrix increased proliferation of the HEK923 cells (the cell line derived from human embryonic cells) in comparison with chitosan, which improved the healing of wounds. It was suggested by the authors to use the obtained composite for diabetic sores repair. However, a further in vivo study is needed.

Paydayesh et al. [54] synthesized a polyhydroxyethyl methacrylate hydrogel with addition of the 20–40 nm Fe_3O_4 iron oxide nanoparticles. The swelling index of the gel decreased along with the nanoparticles content growth, which has to do with the nanoparticles acting as crosslinking centers. The gel containing 15 wt.% iron oxide nanoparticles inhibited the growth of the *E. coli* and *S. aureus* bacteria. Additionally, the obtained nanocomposite hydrogel was nontoxic towards the HFFF2 fibroblast cells, which makes its application in the wounds dressing area possible. However, the Fe_3O_4 iron oxide nanoparticles decreased the viability of the human keratinocytes at concentrations exceeding 50 μg mL^{-1} [55].

3.2.3. Wound Dressings Containing Cerium Dioxide Nanoparticles

Zamani et al. [56] established that the gelatin-polycaprolactone nanofibers containing spherical cerium dioxide nanoparticles (max. 20 m) show an antibacterial effect against *P. aeruginosa*. The minimum bactericide concentration was equal to 50 μg mL^{-1}. It was also reported that the expression of the *shv, kpc, imp* genes found in the resistant *P. aeruginosa* strains was decreased in the presence of the nanofibers. The nanofiber containing 200 μg mL^{-1} of CeO_2 NPs was nontoxic since 97% of the human fibroblast cells (cell line HU 2) survived. The developed fiber was proposed for use in production of dressings for skin infection treatment.

The CeO_2 cerium dioxide nanoparticles were used for preparation of the nanofiber on the basis of poly-L-lactic acid and gelatin [57]. The fibers demonstrated no cytotoxicity towards the NIH 3T3 mice fibroblast cells. The healing effect of the membrane was evaluated via a model back skin burn on the SD rats weighing 200 g. On the 10th day of observation, the open area decreased considerably. On the 21st day the wounds healed contrary to the control group. The scars were minimal, and the healing effect was better compared to the control (medical patch). It was suggested to use the developed fibers as a cheap material for wound dressings. This being said, the antimicrobial effect of the nanofiber was not studied in the present paper. Respective additional research is required for application of the material in the medical field.

The cerium oxide nanoparticles showed a low toxicity towards the HaCaT human skin keratinocytes [58]. However, the provided toxicological information is insufficient and the genotoxicity and the apoptosis studies for the CeO_2 NPs are required.

3.2.4. Wound Dressings Containing Titanium Dioxide Nanoparticles

David et al. [59] manufactured a nanocomposite on the basis of multi-walled carbon nanotubes with spherical titanium dioxide nanoparticles (ca. 15 nm in diameter) on their surface (MWCNT_TiO_2). The nanocomposite was loaded into a cellulose acetate-collagen porous film. The antimicrobial action, biocompatibility and cytotoxicity of the obtained films were investigated. By means of a diffusion test it was determined that the films inhibit the growth of the *S. aureus* and *E. coli* bacteria, as well as the *C. Albicans* fungi. The largest inhibition zone was observed for the Gram-negative *E. coli* bacteria (Figure 3).

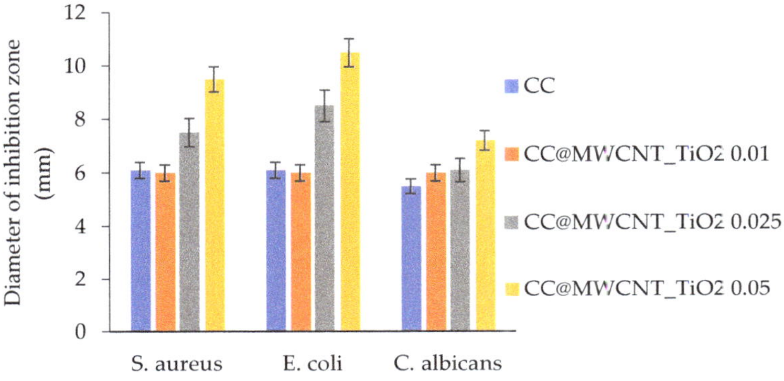

Figure 3. The antimicrobial activity of the films against the *S. aureus*, *E. coli* bacteria and the *C. Albicans* fungi. CC—cellulose acetate-collagen, MWCNT—multi-walled carbon nanotubes, 0.01, 0.025, 0.05—the weight (g) of MWCNT_TiO_2 added to the solution of cellulose acetate and collagen in acetic acid and water.

The diameters of the inhibition zones grew along with the growth of the MWCNT_TiO_2 content in the film. Fluorescent microscopy showed that the obtained films were nontoxic towards the HDFn human dermal fibroblast cells, which participate in the wound healing process. However, the paper [59] does not contain any in vivo studies of the obtained films.

It should be noted that the titanium dioxide nanoparticles of various sizes (10 nm, 21 nm, 32 nm) did not show any statistically significant effect on the viability of the HaCaT human keratinocyte cell line [60,61].

3.2.5. Wound Dressings Containing Copper Oxide Nanoparticles

In publication [62], the antibacterial properties of the nanofibers based on polycaprolactone and gelatin containing 1 wt.% CuO NPs were studied. It was determined that the nanofibers have a strong antibacterial activity against the Gram-positive *S. aureus* (51 $\pm$ 1.2 mm), the multidrug resistant *S. aureus* (40 $\pm$ 1.7 mm) and the Gram-negative *P. aeruginosa* (31 $\pm$ 0.5 mm) and the *E. coli* (30.5 $\pm$ 0.3 mm) bacteria, which was explained by the authors through generation of the reactive oxygen intermediates that destroy bacterial cells by cell membrane oxidation. At that, the CuO NPs may be toxic to NIH3T3 [63].

Moreover, it was established that the CuO NPs were toxic towards the HaCaT cells at concentrations exceeding 5 μg mL^{-1} after 24 h of exposure as they damage the cell membranes and are genotoxic [64], which limits their use as part of wound dressings.

4. Discussion

From the reviewed publications, it may be concluded that both natural (hyaluronic acid, chitosan, carboxymethyl chitosan, cellulose, cellulose acetate, collagen, gelatin, starch, hydroxypropyl starch, sodium alginate) and synthetic polymers (polyvinyl alcohol, polycaprolactone, polyethylene glycol, polyacrylonitrile, polyurethane, polylactic acid, polyhydroxy ethyl methacrylate, and polyacrylic acid) or blends thereof are used for the production of modern wound-repair materials. The polymers provide biocompatibility, swelling, porosity, as well as favorable conditions for tissue regeneration. However, the antimicrobial action of these polymers is oftentimes insufficient. Therefore, nanoscale fillers are additionally introduced into the polymer matrix that provide the antibacterial effect, such as silver, gold, copper, zinc oxide, cerium dioxide, iron oxides (FeO, Fe_3O_4) and titanium dioxide nanoparticles with multi-walled carbon nanotubes. Table 3 shows the characteristics studied for these composites.

Table 3. The characteristics of the wound dressings studied in the presented publications.

Nanofiller	The Studied Characteristics of the Wound Dressings	Ref.
AgNPs	Antimicrobial activity	[22–36]
	Cytotoxicity analysis	[22,27,29,31,34,35]
	Biocompatibility	[22,28,30]
	Silver release measurements	[24,29,31,36]
	In vivo wound healing activity	[27,30]
	Swelling degree	[27,30,31]
	Porosity	[29,31]
	Mechanical properties	[25,28,29,31,35,36]
	Immunohistochemical analysis	[27,30]
	Water absorption	
	Water vapor transmission rate	[29]
	Air permeability	
	Hemocompatibility	[30,31]
	In vitro cell compatibility	[36]
AuNPs	Antimicrobial activity	[39,41]
	Cytotoxicity analysis	[41]
	Biocompatibility	[39]
	In vivo wound healing activity	[39]
	Swelling degree	[39]
	Mechanical properties	[39,41]
CuNPs	Antimicrobial activity	
	In vivo wound healing activity	[42]
	Copper ion release tesr	
ZnO NPs	Antimicrobial activity	[44,47–49]
	Cytotoxicity analysis	[49,50]
	In vivo wound healing activity	[45,48,49]
	Swelling degree	[46,49]
	Porosity	[49]
	Mechanical properties	[46,47]
	Water vapor transmission rate	[49]
	Hemocompatibility	[46]
FeO NPs	Antimicrobial activity	
	Porosity	
	Water absorption	[53]
	Iron release	
	Antidiabetic activity	
Fe_3O_4 NPs	Antimicrobial activity	
	Cytotoxicity analysis	
	Biocompatibility	
	Swelling degree	[54]
	Porosity	
	Mechanical properties	
	Water vapor transmission rate	

Table 3. *Cont.*

Nanofiller	The Studied Characteristics of the Wound Dressings	Ref.
CeO_2 NPs	Antimicrobial activity Cytotoxicity analysis Evaluation of resistance genes expression in *P. aeruginosa*	[56]
	In vivo wound healing activity Mechanical properties Water absorption In vitro cell proliferation test	[57]
MWCNT_TiO_2	Antimicrobial activity Biocompatibility Mechanical properties	[59]
CuO NPs	Mechanical properties Antimicrobial activity Biocompatibility	[62]

The antimicrobial activity was studied in most of the reviewed publications (25 paper) as well as cytotoxicity and biocompatibility (16 papers), and mechanical properties (13 papers) of the materials. However, the in vivo wound healing activity studies of the materials can be found only in 7 articles. Additionally, only 5 publications were dedicated to the water vapor transmission rate, the water absorption and air permeability studies that are necessary for the optimal balance of moisture and oxygen in the wound. More systematic studies of the materials in terms of their healing effect need to be carried out on large animal models (e.g., horses), as well as air and vapor permeability and water absorption must be investigated in order to evaluate the medical application potential thereof

When applying nanoparticles in the wound-repair products, their activity and the antimicrobial action should be considered since the particles may also affect various tissue cells in patients leading to all kinds of complications (Table 4).

Table 4. The action, the advantages and disadvantages of the nanoparticles used for wound dressing production.

Nanofiller	Effect on Cells	Advantage (+) Disadvantage (−)
AgNPs	Oxidative stress. Superoxide and hydroxyl radical generation	Strong antibacterial action (+) May cause allergies (−) Toxic to human skin keratinocyte cells (−)
AuNPs	Presumably cause oxidative damage to bacteria	Nontoxic to human skin keratinocyte cells (+) Weak antimicrobial effect (−)
CuNPs	Presumably copper ions bind the DNA molecules of a bacterial cell	Nontoxic to human skin keratinocyte cells (+) Lower antimicrobial effect vs. antibiotics (−)
ZnO NPs	Electrostatic attraction of zinc ions to the bacterial cell membrane followed by release of the cell contents	Strong antibacterial action (+) Genotoxic to human epidermal cells (−)
FeO NPs, Fe_3O_4 NPs	Penetrate through cell membrane and prevent transmembrane electron trnsfer	Strong antibacterial action (+) Toxic to human skin keratinocyte cells (−)
CeO_2 NPs	Oxidative stress on lipids and/or proteins in the plasma membrane through reduction in Ce^{4+} to Ce^{3+}.	Strong antibacterial action (+) Low toxicity towards human skin keratinocyte cells (+)
TiO_2 NPs	Oxidative stress. Generation of two reactive oxygen intermediates—OH and H_2O_2.	Nontoxic to human skin keratinocyte cells (+) Weak antimicrobial effect (−)
CuO NPs	Oxidative stress. Generation of four reactive oxygen intermediates—the superoxide oxygen radical, ·OH, H_2O_2, the singlet oxide.	Strong antibacterial action (+) Toxic to human skin keratinocyte cells (−)

An additional survey of the literature published on the subject starting from 2018 was conducted in order to evaluate the application trends for the nanoparticles as antimicrobial preparations in wound dressings. The results are presented in Table 5.

Table 5. The application trends for nanoparticles in the preparation of antimicrobial wound dressings.

Nanoparticle	Year	No. of Publications	References
AgNPs	2018	6	[65–70]
	2019	4	[71–74]
	2020	12	[75–86]
	2021	2	[23,27]
	2022	13	[22,24–26,28–36]
AuNPs	2020	2	[87,88]
	2021	2	[39,41]
CuNPs	2020	1	[89]
	2021	1	[42]
ZnO NPs	2018	5	[70,90–93]
	2019	4	[94–97]
	2020	2	[85,98]
	2021	2	[44,46]
	2022	5	[45,47–50]
FeO NPs	2021	1	[53]
Fe_3O_4 NPs	2022	1	[54]
CeO_2 NPs	2019	1	[99]
	2021	1	[56]
	2022	1	[57]
TiO_2 NPs	2020	1	[100]
	2022	1	[59]
CuO NPs	2021	1	[62]
Cu_2O NPs	2018	1	[101]
Lignin NPs	2018	1	[102]
Silver zeolite NPs	2018	1	[103]
ZrO_2 NPs	2020	1	[104]

From the results presented in Table 4, it can be concluded that the largest share of the research is dedicated to the preparation of the dressings containing silver nanoparticles, i.e., 37 papers. At the same time, the distribution of the research over a five-year period is not always uniform. Nevertheless, in recent years the number of publications on wound-repair AgNPs-based nanomaterials had increased. The materials based on zinc oxide nanoparticles are relatively actively studied (18 articles). Other nanoparticles are much less widely used, which, based on the results, has to do with the weaker antimicrobial action vs. AgNPs and ZnO NPs. Furthermore, the AgNPs and ZnO NPs are less toxic at low concentrations towards the live mammalian cells. At that the silver nanoparticles are more promising in terms of the practical application than ZnO NPs since the latter are potentially genotoxic at elevated concentrations. It must be taken into account that in order to achieve the antimicrobial effect while not harming the patient at the same time is hardly possible by only varying merely the concentration of the nanoparticles. Therefore, the evaluation of the trends in the field of application of polymers for antibacterial wound dressing matrices is equally important because the polymer defines many characteristics of the obtained nanomaterial. The literature survey results are presented in Table 6.

Table 6. The application trends for polymers in the preparation of antimicrobial wound dressings.

Polymer	Year	No. of Publications	References
Chitosan and derivatives thereof	2018	3	[70,93,102]
	2019	4	[71,72,74,94]
	2020	8	[75,76,81,82,86,88,89,100]
	2021	4	[27,42,46,53]
	2022	6	[25,34,35,47–49]
Polyvinyl alcohol	2018	4	[65,91,92,102]
	2019	3	[73,94,96]
	2020	5	[81,83,84,87,88]
	2021	3	[39,42,53]
	2022	4	[24–26,49]
Cellulose and derivatives thereof	2018	3	[67,69,101]
	2020	2	[85,104]
	2022	2	[33,59]
Polycaprolactone	2020	1	[104]
	2021	3	[41,56,62]
	2022	1	[29]
Gelatin	2020	1	[80]
	2021	2	[56,62]
	2022	3	[30,31,57]
Starch and derivatives thereof	2020	1	[83]
	2022	3	[31,32,49]
Konjac glucomannan	2018	1	[68]
	2020	3	[75,82,85]
Collagen	2020	2	[76,85]
	2022	2	[23,59]
Silk fibroin	2018	2	[91,93]
	2019	1	[71]
Sodium alginate and calcium alginate	2019	1	[97]
	2020	1	[80]
	2022	1	[35]
Hyaluronic acid and derivatives thereof	2020	1	[89]
	2021	1	[27]
	2022	1	[22]
Polyalkylene glycols	2018	1	[69]
	2019	1	[72]
	2020	2	[98,100]
	2022	1	[29]

Table 6. *Cont.*

Polymer	Year	No. of Publications	References
Keratin	2019	1	[95]
	2020	1	[79]
Polylactic acid	2022	2	[28,57]
κ-carrageenan	2020	2	[85,87]
Polyurethane	2022	2	[28,32]
Oxidized dextran	2021	1	[27]
Polyvinylpyrrolidone	2018	1	[69]
Agar	2018	1	[69]
Poly(acrylic acid-co-itaconic acid)	2018	1	[90]
Nylon 66	2018	1	[66]
Nylon 4/6 copolymer	2018	1	[103]
Galacto-xyloglucan	2020	1	[77]
Gum acacia and carbopol	2020	1	[78]
HBV	2019	1	[99]
Polyacrylonitrile	2022	1	[36]
Heparin	2021	1	[39]
Polyacrylic acid and polyallylamine hydrochloride	2021		[44]
Glycogen	2022	1	[48]
Polyhydroxyethyl methacrylate	2022	1	[54]
Quaternized chitin	2022	1	[28]

As it can be seen from the table, among the polymer matrices for wound-repair nanomaterials over the past 5 years the most publications were dedicated to biopolymers (66 papers), whereas 44 papers were dedicated to synthetic polymers. The polysaccharides were leading among the biopolymers with the number of publications reaching 50. Chitosan also had a stable leading position. The number of studies dedicated to the use of cellulose had decreased in recent years, whereas that of starch had on the contrary increased. In 2021–2022, some publications emerged dealing with wound dressings based on heparin, glycogen and quaternized chitin. Gelatin is another example of a biopolymer with growing interest.

Polyvinyl alcohol is the synthetic polymer most frequently used in wound-repair nanomaterials. It was mentioned in 19 articles over the abovementioned time period, whereas the distribution from one year to another varied insignificantly. Polycaprolactone was used in the studies to a lesser extent but with similar consistency. The newest articles published in 2022 are presenting the research dedicated to polymers such as polyacrylonitrile, polyhydroxyethyl methacrylate and polylactic acid, among which the latter being the most promising.

5. Conclusions

Of note, the hydrogels, fibers and membranes containing silver nanoparticles have been studied the most in terms of their antibacterial, fungicidal and cytotoxic properties. The AgNPs have a pronounced antibacterial action and often surpass certain antibiotics with respect to their efficiency against the gram-positive *S. aureus* and the gram-negative *P. aeruginosa* bacteria being the main pathogens present in wounds. This makes it possible to use AgNPs in the composition of wound dressings, mainly for burn repair. However, the silver nanoparticles are prone to aggregation, which may lead the change of scale and the loss of antimicrobial action. For this reason, the development of materials in which the silver is uniformly distributed in the polymer matrix is a hot topic for further research [105].

As far as the wound dressing matrices are concerned, despite a wide range of polymers analyzed in many studies, only two have the potential for practical application, i.e., chitosan, the biopolymer, and the synthetic polyvinyl alcohol.

The low molecular weight chitosan is nontoxic; it has healing, haemostatic, pain-relieving and antimicrobial effects [106]. The polymer is rather easy to obtain while the raw material for its production is abundant in nature including food production by-products. Additionally, no skin reactions to chitosan dressings in patients allergic to mollusks were reported [107]. However, there were no studies of potential side effects and allergic reactions to the polymer materials in the reviewed publications, such as allergic contact dermatitis or toxicodermatosis.

The polyvinyl alcohol is also relatively cheap, easy to produce, nontoxic and allows for obtaining a material with a preset molecular weight; it does not contain any toxic monomer admixtures and solvents, which may be found in other polymers. What is especially important, polyvinyl alcohol is easy to modify, which allows regulating the parameters of a PVA-based gel and grafting various functional groups, substances and preparations to its polymer chain. Thanks to modification, this polymer can provide a wide range of properties and it can be used under various conditions.

Thus, the antimicrobial wound-repair materials based on polyvinyl alcohol or chitosan modified with silver nanoparticles have the broadest practical application potential.

Author Contributions: Conceptualization, E.C. and P.Y.; methodology, E.C.; validation, P.Y., Y.M. and E.C.; writing—original draft preparation, P.Y.; writing—review and editing, E.C.; visualization, P.Y.; supervision, E.C.; project administration, E.C.; funding acquisition, E.C. All authors have read and agreed to the published version of the manuscript.

Funding: The survey was funded by the Russian Science Foundation grant No. 22-25-20078, https://rscf.ru/project/22-25-20078/ (accessed on 20 April 2022).

Institutional Review Board Statement: Not applicable.

Informed Consent Statement: Not applicable.

Conflicts of Interest: The authors declare no conflict of interest.

References

1. Varaprasad, K. Co-assembled ZnO (shell)—CuO (core) nano-oxide materials for microbial protection. *Phosphorus Sulfur Silicon Relat. Elem.* **2018**, *193*, 74–80. [CrossRef]
2. Varshney, S.; Nigam, A.; Pawar, S.J.; Mishra, N. An overview on biomedical applications of versatile silica nanoparticles, synthesized via several chemical and biological routes: A review. *Phosphorus Sulfur Silicon Relat. Elem.* **2022**, *197*, 72–88. [CrossRef]
3. Yudaev, P.; Chuev, V.; Klyukin, B.; Kuskov, A.; Mezhuev, Y.; Chistyakov, E. Polymeric dental nanomaterials: Antimicrobial action. *Polymers* **2022**, *14*, 864. [CrossRef] [PubMed]
4. Nikolaidis, A.K.; Koulaouzidou, E.A.; Gogos, C.; Achilias, D.S. Synthesis of novel dental nanocomposite resins by incorporating polymerizable, quaternary ammonium silane-modified silica nanoparticles. *Polymers* **2021**, *13*, 1682. [CrossRef]
5. Yilmaz, B.; Ozay, O. Synthesis of antibiotic-modified silica nanoparticles and their use as a controlled drug release system with antibacterial properties. *Phosphorus Sulfur Silicon Relat. Elem.* **2022**, in press. [CrossRef]
6. Razuvaeva, E.V.; Kalinin, K.T.; Sedush, N.G.; Nazarov, A.A.; Volkov, D.S.; Chvalun, S.N. Structure and cytotoxicity of biodegradable poly(d,l-lactide-co-glycolide) nanoparticles loaded with oxaliplatin. *Mendeleev Commun.* **2021**, *31*, 512–514. [CrossRef]
7. Al Sawaftah, N.M.; Awad, N.S.; Pitt, W.G.; Husseini, G.A. pH-responsive nanocarriers in cancer therapy. *Polymers* **2022**, *14*, 936. [CrossRef]
8. Wang, L.; Hu, C.; Shao, L. The antimicrobial activity of nanoparticles: Present situation and prospects for the future. *Int. J. Nanomed.* **2017**, *12*, 1227–1249. [CrossRef]
9. Gudkov, S.V.; Burmistrov, D.E.; Serov, D.A.; Rebezov, M.B.; Semenova, A.A.; Lisitsyn, A.B. A mini review of antibacterial properties of ZnO nanoparticles. *Front. Phys.* **2021**, *9*, 641481. [CrossRef]
10. Thao, N.T.; Wijerathna, H.M.S.M.; Kumar, R.S.; Choi, D.; Dananjaya, S.H.S.; Attanayake, A.P. Preparation and characterization of succinyl chitosan and succinyl chitosan nanoparticle film: In vitro and in vivo evaluation of wound healing activity. *Int. J. Biol. Macromol.* **2021**, *193*, 1823–1834. [CrossRef]

11. Sharifiaghdam, M.; Shaabani, E.; Asghari, F.; Faridi-Majidi, R. Chitosan coated metallic nanoparticles with stability, antioxidant, and antibacterial properties: Potential for wound healing application. *J. Appl. Polym. Sci.* **2022**, *139*, 51766. [CrossRef]
12. Deng, P.; Liang, X.; Chen, F.; Chen, Y.; Zhou, J. Novel multifunctional dual-dynamic-bonds crosslinked hydrogels for multi-strategy therapy of MRSA-infected wounds. *Appl. Mater. Today* **2022**, *26*, 101362. [CrossRef]
13. Xie, T.; Ding, J.; Han, X.; Jia, H.; Yang, Y.; Liang, S.; Wang, W.; Liu, W.; Wang, W. Wound dressing change facilitated by spraying zinc ions. *Mater. Horiz.* **2020**, *7*, 605–614. [CrossRef]
14. Alven, S.; Buyana, B.; Feketshane, Z.; Aderibigbe, B.A. Electrospun nanofibers/nanofibrous scaffolds loaded with silver nanoparticles as effective antibacterial wound dressing materials. *Pharmaceutics* **2021**, *13*, 964. [CrossRef] [PubMed]
15. Yu, R.; Zhang, H.; Guo, B. Conductive biomaterials as bioactive wound dressing for wound healing and skin tissue engineering. *Nanomicro Lett.* **2022**, *14*, 1–46. [CrossRef]
16. Chong, W.J.; Shen, S.; Li, Y.; Trinchi, A.; Pejak, D.; Kyratzis, I.L.; Sola, A.; Wen, C. Additive manufacturing of antibacterial PLA-ZnO nanocomposites: Benefits, limitations and open challenges. *J. Mater. Sci. Technol.* **2022**, *111*, 120–151. [CrossRef]
17. Zhang, R.; Yu, B.; Tian, Y.; Pang, L.; Xu, T.; Cong, H.; Shen, Y. Diversified antibacterial modification and latest applications of polysaccharide-based hydrogels for wound healthcare. *Appl. Mater. Today* **2022**, *26*, 101396. [CrossRef]
18. Yuan, Y.; Ding, L.; Chen, Y.; Chen, G.; Zhao, T.; Yu, Y. Nano-silver functionalized polysaccharides as a platform for wound dressings: A review. *Int. J. Biol. Macromol.* **2022**, *194*, 644–653. [CrossRef]
19. Salvo, J.; Sandoval, C. Role of copper nanoparticles in wound healing for chronic wounds: Literature review. *Burn. Trauma* **2022**, *10*, tkab047. [CrossRef]
20. Kwiatkowska, A.; Drabik, M.; Lipko, A.; Grzeczkowicz, A.; Stachowiak, R.; Marszalik, A.; Granicka, L.H. Composite Membrane Dressings System with Metallic Nanoparticles as an Antibacterial Factor in Wound Healing. *Membranes* **2022**, *12*, 215. [CrossRef]
21. Vijayakumar, V.; Samal, S.K.; Mohanty, S.; Nayak, S.K. Recent advancements in biopolymer and metal nanoparticle-based materials in diabetic wound healing management. *Int. J. Biol. Macromol.* **2019**, *122*, 137–148. [CrossRef]
22. Massironi, A.; Franco, A.R.; Babo, P.S.; Puppi, D.; Chiellini, F.; Reis, R.L.; Gomes, M.E. Development and Characterization of Highly Stable Silver NanoParticles as Novel Potential Antimicrobial Agents for Wound Healing Hydrogels. *Int. J. Mol. Sci.* **2022**, *23*, 2161. [CrossRef]
23. Antezana, P.E.; Municoy, S.; Pérez, C.J.; Desimone, M.F. Collagen Hydrogels Loaded with Silver Nanoparticles and Cannabis Sativa Oil. *Antibiotics* **2021**, *10*, 1420. [CrossRef] [PubMed]
24. Nešović, K.; Mišković-Stanković, V. Silver/poly (vinyl alcohol)/graphene hydrogels for wound dressing applications: Understanding the mechanism of silver, antibacterial agent release. *J. Vinyl Addit. Technol.* **2022**, *28*, 196–210. [CrossRef]
25. Santiago-Castillo, K.; Torres-Huerta, A.M.; del Ángel-López, D.; Domínguez-Crespo, M.A.; Dorantes-Rosales, H.; Palma-Ramírez, D.; Willcock, H. In Situ Growth of Silver Nanoparticles on Chitosan Matrix for the Synthesis of Hybrid Electrospun Fibers: Analysis of Microstructural and Mechanical Properties. *Polymers* **2022**, *14*, 674. [CrossRef] [PubMed]
26. Mojally, M.; Sharmin, E.; Obaid, N.A.; Alhindi, Y.; Abdalla, A.N. Polyvinyl alcohol/corn starch/castor oil hydrogel films, loaded with silver nanoparticles biosynthesized in Mentha piperita leaves' extract. *J. King Saud Univ. Sci.* **2022**, *34*, 101879. [CrossRef]
27. Chen, X.; Zhang, H.; Yang, X.; Zhang, W.; Jiang, M.; Wen, T.; Wang, J.; Guo, R.; Liu, H. Preparation and Application of Quaternized Chitosan- and AgNPs-Base Synergistic Antibacterial Hydrogel for Burn Wound Healing. *Molecules* **2021**, *26*, 4037. [CrossRef]
28. Zhou, A.; Zhang, Y.; Zhang, X.; Deng, Y.; Huang, D.; Huang, C.; Qu, Q. Quaternized chitin/tannic acid bilayers layer-by-layer deposited poly (lactic acid)/polyurethane nanofibrous mats decorated with photoresponsive complex and silver nanoparticles for antibacterial activity. *Int. J. Biol. Macromol.* **2022**, *201*, 448–457. [CrossRef]
29. Bozkaya, O.; Arat, E.; Gök, Z.G.; Yiğitoğlu, M.; Vargel, İ. Production and characterization of hybrid nanofiber wound dressing containing Centella asiatica coated silver nanoparticles by mutual electrospinning method. *Eur. Polym. J.* **2022**, *166*, 111023. [CrossRef]
30. Huang, Y.; Bai, L.; Yang, Y.; Yin, Z.; Guo, B. Biodegradable gelatin/silver nanoparticle composite cryogel with excellent antibacterial and antibiofilm activity and hemostasis for Pseudomonas aeruginosa-infected burn wound healing. *J. Colloid Interface Sci.* **2022**, *608*, 2278–2289. [CrossRef]
31. Sethi, S.; Thakur, S.; Kaith, B.S.; Sharma, N.; Ansar, S.; Pandey, S.; Kuma, V. Biopolymer starch-gelatin embedded with silver nanoparticle–based hydrogel composites for antibacterial application. *Biomass Convers. Biorefin.* **2022**, 1–22. [CrossRef]
32. El-Hefnawy, M.E.; Alhayyani, S.; El-Sherbiny, M.M.; Sakran, M.I.; El-Newehy, M.H. Fabrication of Nanofibers Based on Hydroxypropyl Starch/Polyurethane Loaded with the Biosynthesized Silver Nanoparticles for the Treatment of Pathogenic Microbes in Wounds. *Polymers* **2022**, *14*, 318. [CrossRef] [PubMed]
33. Madivoli, E.S.; Kareru, P.G.; Gachanja, A.N.; Makhanu, D.S.; Mugo, S.M. Cellulose-Based Hybrid Nanoarchitectonics with Silver Nanoparticles: Characterization and Antimicrobial Potency. *J. Inorg. Organomet. Polym. Mater.* **2022**, *32*, 854–863. [CrossRef]
34. Rao, K.M.; Suneetha, M.; Zo, S.; Won, S.Y.; Kim, H.J.; Han, S.S. Injectable nanocomposite hydrogel as wound dressing agent with tunable multifunctional property. *Mater. Lett.* **2022**, *307*, 131062. [CrossRef]
35. Yan, M.; Shi, J.; Tang, S.; Zhou, G.; Zeng, J.; Zhang, Y.; Zhang, H.; Yu, Y.; Guo, J. The construction of a seaweed-based antibacterial membrane loaded with nano-silver based on the formation of a dynamic united dual network structure. *New J. Chem.* **2022**, *46*, 511–520. [CrossRef]

36. Chen, P.; Chai, M.; Mai, Z.; Liao, M.; Xie, X.; Lu, Z.; Zhang, W.; Zhao, H.; Dong, X.; Fu, X.; et al. Electrospinning polyacrylonitrile (PAN) based nanofibrous membranes synergic with plant antibacterial agent and silver nanoparticles (AgNPs) for potential wound dressing. *Mater. Today Commun.* **2022**, *31*, 103336. [CrossRef]
37. Hadrup, N.; Sharma, A.K.; Loeschner, K. Toxicity of silver ions, metallic silver, and silver nanoparticle materials after in vivo dermal and mucosal surface exposure: A review. *Regul. Toxicol. Pharmacol.* **2018**, *98*, 257–267. [CrossRef]
38. Ahlberg, S.; Meinke, M.C.; Werner, L.; Epple, M.; Diendorf, J.; Blume-Peytavi, U.; Lademann, J.; Vogt, A.; Rancan, F. Comparison of silver nanoparticles stored under air or argon with respect to the induction of intracellular free radicals and toxic effects toward keratinocytes. *Eur. J. Pharm. Biopharm.* **2014**, *88*, 651–657. [CrossRef]
39. Zhang, K.; Zhao, G. An Effective Wound Healing Material Based on Gold Incorporation into a Heparin-Polyvinyl Alcohol Nanocomposite: Enhanced In Vitro and In Vivo Care of Perioperative Period. *J. Clust. Sci.* **2021**, 1–11. [CrossRef]
40. Wang, S.; Lu, W.; Tovmachenko, O.; Rai, U.S.; Yu, H.; Ray, P.C. Challenge in understanding size and shape dependent toxicity of gold nanomaterials in human skin keratinocytes. *Chem. Phys. Lett.* **2008**, *463*, 145–149. [CrossRef]
41. El-Naggar, M.E.; Abd-Al-Aleem, A.H.; Abu-Saied, M.A.; Youssef, A.M. Synthesis of environmentally benign antimicrobial dressing nanofibers based on polycaprolactone blended with gold nanoparticles and spearmint oil nanoemulsion. *J. Mater. Res.* **2021**, *15*, 3447–3460. [CrossRef]
42. Lemraski, E.G.; Jahangirian, H.; Dashti, M.; Khajehali, E.; Sharafinia, S.; Rafiee-Moghaddam, R.; Webster, T.J. Antimicrobial double-layer wound dressing based on chitosan/polyvinyl alcohol/copper: In vitro and in vivo assessment. *Int. J. Nanomed.* **2021**, *16*, 223. [CrossRef] [PubMed]
43. Alizadeh, S.; Seyedalipour, B.; Shafieyan, S.; Kheime, A.; Mohammadi, P.; Aghdami, N. Copper nanoparticles promote rapid wound healing in acute full thickness defect via acceleration of skin cell migration, proliferation, and neovascularization. *Biochem. Byophys. Res. Commun.* **2019**, *517*, 684–690. [CrossRef] [PubMed]
44. Bandeira, M.; Chee, B.S.; Frassini, R.; Nugent, M.; Giovanela, M.; Roesch-Ely, M.; Crespo, J.d.S.; Devine, D.M. Antimicrobial PAA/PAH Electrospun Fiber Containing Green Synthesized Zinc Oxide Nanoparticles for Wound Healing. *Materials* **2021**, *14*, 2889. [CrossRef] [PubMed]
45. Le, V.A.T.; Trinh, T.X.; Chien, P.N.; Giang, N.N.; Zhang, X.-R.; Nam, S.-Y.; Heo, C.-Y. Evaluation of the Performance of a ZnO-Nanoparticle-Coated Hydrocolloid Patch in Wound Healing. *Polymers* **2022**, *14*, 919. [CrossRef] [PubMed]
46. Zhang, Q.; Zhang, H.; Hui, A.; Ding, J.; Liu, X.; Wang, A. Synergistic Effect of Glycyrrhizic Acid and ZnO/Palygorskite on Improving Chitosan-Based Films and Their Potential Application in Wound Healing. *Polymers* **2021**, *13*, 3878. [CrossRef]
47. Kalemtas, A.; Kocer, H.B.; Aydin, A.; Terzioglu, P.; Aydin, G. Mechanical and antibacterial properties of ZnO/chitosan bio-composite films. *J. Polym. Eng.* **2022**, *42*, 35–47. [CrossRef]
48. Hasanin, M.; Swielam, E.M.; Atwa, N.A.; Agwa, M.M. Novel design of bandages using cotton pads, doped with chitosan, glycogen and ZnO nanoparticles, having enhanced antimicrobial and wounds healing effects. *Int. J. Biol. Macromol.* **2022**, *197*, 121–130. [CrossRef]
49. Joorabloo, A.; Khorasani, M.T.; Adeli, H.; Milan, P.B.; Amoupour, M. Using artificial neural network for design and development of PVA/chitosan/starch/heparinized nZnO hydrogels for enhanced wound healing. *J. Ind. Eng. Chem.* **2022**, *108*, 88–100. [CrossRef]
50. Chen, F.C.; Huang, C.M.; Yu, X.W.; Chen, Y.Y. Effect of nano zinc oxide on proliferation and toxicity of human gingival cells. *Hum. Exp. Toxicol.* **2022**, *41*, 1–10. [CrossRef]
51. Sharma, V.; Shukla, R.K.; Saxena, N.; Parmar, D.; Das, M.; Dhawan, A. DNA damaging potential of zinc oxide nanoparticles in human epidermal cells. *Toxicol. Lett.* **2009**, *185*, 211–218. [CrossRef]
52. Keerthana, S.; Kumar, A. Potential risks and benefits of zinc oxide nanoparticles: A systematic review. *Crit. Rev. Toxicol.* **2020**, *50*, 47–71. [CrossRef] [PubMed]
53. Sathiyaseelan, A.; Saravanakumar, K.; Mariadoss, A.V.A.; Wang, M.-H. Antimicrobial and Wound Healing Properties of FeO Fabricated Chitosan/PVA Nanocomposite Sponge. *Antibiotics* **2021**, *10*, 524. [CrossRef] [PubMed]
54. Paydayesh, A.; Heleil, L.; Sh Dadkhah, A. Preparation and application of poly (hydroxyl ethyl methacrylate) nanocomposite hydrogels containing iron oxide nanoparticles as wound dressing. *Polym. Compos.* **2022**, *30*, 09673911211063106. [CrossRef]
55. Moac, E.A.; Farcaş, C.; Coricovac, D.; Avram, S.; Mihali, C.V.; Drâghici, G.A.; Felicia, L.; Cornelia, P.; Dehelean, C. Oleic Acid Double Coated Fe_3O_4 Nanoparticles as Anti-Melanoma Compounds with a Complex Mechanism of Activity-In Vitro and In Ovo Assessment. *J. Biomed. Nanotechnol.* **2019**, *15*, 893–909. [CrossRef] [PubMed]
56. Zamani, K.; Allah-Bakhshi, N.; Akhavan, F.; Yousefi, M.; Golmoradi, R.; Ramezani, M.; Ramezani, F. Antibacterial effect of cerium oxide nanoparticle against Pseudomonas aeruginosa. *BMC Biotechnol.* **2021**, *21*, 1–11. [CrossRef] [PubMed]
57. Lv, Y.; Xu, Y.; Sang, X.; Li, C.; Liu, Y.; Guo, Q.; Ramakrishna, S.; Wang, C.; Hu, P.; Nanda, H.S. PLLA-gelatin composite fiber membranes incorporated with functionalized CeNPs as a sustainable wound dressing substitute promoting skin regeneration and scar remodelling. *J. Mater. Chem. B.* **2022**, *10*, 1116–1127. [CrossRef]
58. Ngoc, L.T.N.; Bui, V.K.H.; Moon, J.Y.; Lee, Y.C. In-vitro cytotoxicity and oxidative stress induced by cerium aminoclay and cerium oxide nanoparticles in human skin keratinocyte cells. *J. Nanosci. Nanotechnol.* **2019**, *19*, 6369–6375. [CrossRef]
59. David, M.E.; Ion, R.M.; Grigorescu, R.M.; Iancu, L.; Holban, A.M.; Iordache, F.; Nicoara, A.I.; Alexandrescu, E.; Somoghi, R.; Teodorescu, S.; et al. Biocompatible and Antimicrobial Cellulose Acetate-Collagen Films Containing MWCNTs Decorated with TiO_2 Nanoparticles for Potential Biomedical Applications. *Nanomaterials* **2022**, *12*, 239. [CrossRef]

60. Zhang, L.W.; Monteiro-Riviere, N.A. Toxicity assessment of six titanium dioxide nanoparticles in human epidermal keratinocytes. *Cutan. Ocul. Toxicol.* **2019**, *38*, 66–80. [CrossRef]
61. Wright, C.; Iyer, A.K.V.; Wang, L.; Wu, N.; Yakisich, J.S.; Rojanasakul, Y.; Azad, N. Effects of titanium dioxide nanoparticles on human keratinocytes. *Drug Chem. Toxicol.* **2017**, *40*, 90–100. [CrossRef]
62. Karuppannan, S.K.; Ramalingam, R.; Khalith, S.M.; Musthafa, S.A.; Dowlath, M.J.H.; Munuswamy-Ramanujam, G.; Arunachalam, K.D. Copper oxide nanoparticles infused electrospun polycaprolactone/gelatin scaffold as an antibacterial wound dressing. *Mater. Lett.* **2021**, *294*, 129787. [CrossRef]
63. Hashmi, M.; Ullah, S.; Kim, I.S. Copper oxide (CuO) loaded polyacrylonitrile (PAN) nanofiber membranes for antimicrobial breath mask applications. *Curr. Biotechnol.* **2019**, *1*, 1–10. [CrossRef]
64. Alarifi, S.; Ali, D.; Verma, A.; Alakhtani, S.; Ali, B.A. Cytotoxicity and genotoxicity of copper oxide nanoparticles in human skin keratinocytes cells. *Int. J. Toxicol.* **2013**, *32*, 296–307. [CrossRef] [PubMed]
65. Yahyaei, B.; Manafi, S.; Fahimi, B.; Arabzadeh, S.; Pourali, P. Production of electrospun polyvinyl alcohol/microbial synthesized silver nanoparticles scaffold for the treatment of fungating wounds. *Appl. Nanosci.* **2018**, *8*, 417–426. [CrossRef]
66. Allafchian, A.; Jalali, S.A.H.; Kabirzadeh, N. Characterisation and investigation of antibacterial properties of nylon 66/TPS/Ag NPs nanofibre membranes. *Micro Nano Lett.* **2018**, *13*, 1747–1751. [CrossRef]
67. Capanema, N.S.; Mansur, A.A.; Carvalho, S.M.; Mansur, L.L.; Ramos, C.P.; Lage, A.P.; Mansur, H.S. Physicochemical properties and antimicrobial activity of biocompatible carboxymethylcellulose-silver nanoparticle hybrids for wound dressing and epidermal repair. *J. Appl. Polym. Sci.* **2018**, *135*, 45812. [CrossRef]
68. Chen, H.; Lan, G.; Ran, L.; Xiao, Y.; Yu, K.; Lu, B.; Dai, F.; Wu, D.; Lu, F. A novel wound dressing based on a Konjac glucomannan/silver nanoparticle composite sponge effectively kills bacteria and accelerates wound healing. *Carbohydr. Polym.* **2018**, *183*, 70–80. [CrossRef]
69. de Lima, G.G.; de Lima, D.W.; de Oliveira, M.J.; Lugão, A.B.; Alcântara, M.T.; Devine, D.M.; de Sá, M.J. Synthesis and in vivo behavior of PVP/CMC/agar hydrogel membranes impregnated with silver nanoparticles for wound healing applications. *ACS Appl. Bio Mater.* **2018**, *1*, 1842–1852. [CrossRef]
70. Ficai, D.; Ardelean, I.L.; Holban, A.M.; Ditu, L.M.; Gudovan, D.; Sönmez, M.A.R.I.A.; Trusca, R.; Kaya, A.; Ficai, A.; Andronescu, E. Manufacturing nanostructured chitosan-based 2D sheets with prolonged antimicrobial activity. *Rom. J. Morphol. Embryol.* **2018**, *59*, 517–525.
71. Wang, Q.; Qian, Z.; Liu, B.; Liu, J.; Zhang, L.; Xu, J. In vitro and in vivo evaluation of new PRP antibacterial moisturizing dressings for infectious wound repair. *J. Biomater. Sci. Polym. Ed.* **2019**, *30*, 462–485. [CrossRef]
72. Masood, N.; Ahmed, R.; Tariq, M.; Ahmed, Z.; Masoud, M.S.; Ali, I.; Asghar, R.; Andleeb, A.; Hasan, A. Silver nanoparticle impregnated chitosan-PEG hydrogel enhances wound healing in diabetes induced rabbits. *Int. J. Pharm.* **2019**, *559*, 23–36. [CrossRef] [PubMed]
73. Aktürk, A.; Taygun, M.E.; Güler, F.K.; Goller, G.; Küçükbayrak, S. Fabrication of antibacterial polyvinyl alcohol nanocomposite mats with soluble starch coated silver nanoparticles. *Colloids Surf. A Physicochem. Eng. Asp.* **2019**, *562*, 255–262. [CrossRef]
74. Hernández-Rangel, A.; Silva-Bermudez, P.; Espana-Sanchez, B.L.; Luna-Hernández, E.; Almaguer-Flores, A.; Ibarra, C.; Garcia-Perez, V.I.; Velasquillo, C.; Luna-Barcenas, G. Fabrication and in vitro behavior of dual-function chitosan/silver nanocomposites for potential wound dressing applications. *Mater. Sci. Eng. C* **2019**, *94*, 750–765. [CrossRef] [PubMed]
75. Wang, Y.; Xie, R.; Li, Q.; Dai, F.; Lan, G.; Shang, S.; Lu, F. A self-adapting hydrogel based on chitosan/oxidized konjac glucomannan/AgNPs for repairing irregular wounds. *Biomater. Sci.* **2020**, *8*, 1910–1922. [CrossRef]
76. Sionkowska, A.; Walczak, M.; Michalska-Sionkowska, M. Preparation and characterization of collagen/chitosan composites with silver nanoparticles. *Polym. Compos.* **2020**, *41*, 951–957. [CrossRef]
77. Preethi, G.U.; Unnikrishnan, B.S.; Sreekutty, J.; Archana, M.G.; Anupama, M.S.; Shiji, R.; Pillai, R.; Joseph, M.M.; Syama, H.P.; Sreelekha, T.T. Semi-interpenetrating nanosilver doped polysaccharide hydrogel scaffolds for cutaneous wound healing. *Int. J. Biol. Macromol.* **2020**, *142*, 712–723. [CrossRef]
78. Lekalakala, R.; Aderibigbe, B.A.; Owonubi, S.J.; Sadiku, E.R.; Fonkui, Y.T.; Ndinteh, D.T.; Ray, S.S. Gum Acacia/Carbopol-Based Biocomposites Loaded with Silver Nanoparticles as Potential Wound Dressings. *Int. J. Nanosci. Nanotechnol.* **2020**, *16*, 219–231.
79. Konop, M.; Czuwara, J.; Kłodzińska, E.; Laskowska, A.K.; Sulejczak, D.; Damps, T.; Zielenkiewicz, U.; Brzozowska, I.; Sureda, A.; Kowalkowski, T.; et al. Evaluation of keratin biomaterial containing silver nanoparticles as a potential wound dressing in full-thickness skin wound model in diabetic mice. *J. Tissue Eng. Regen. Med.* **2020**, *14*, 334–346. [CrossRef]
80. Diniz, F.R.; Maia, R.C.A.; Rannier Andrade, L.; Andrade, L.N.; Vinicius Chaud, M.; da Silva, C.F.; Correa, C.B.; de Albuquerque Junior, R.L.; da Costa, L.P.; Shin, S.R.; et al. Silver nanoparticles-composing alginate/gelatine hydrogel improves wound healing in vivo. *Nanomaterials* **2020**, *10*, 390. [CrossRef]
81. Kumar, A.; Kaur, H. Sprayed in-situ synthesis of polyvinyl alcohol/chitosan loaded silver nanocomposite hydrogel for improved antibacterial effects. *Int. J. Biol. Macromol.* **2020**, *145*, 950–964. [CrossRef]
82. Jiang, Y.; Huang, J.; Wu, X.; Ren, Y.; Li, Z.; Ren, J. Controlled release of silver ions from AgNPs using a hydrogel based on konjac glucomannan and chitosan for infected wounds. *Int. J. Biol. Macromol.* **2020**, *149*, 148–157. [CrossRef] [PubMed]
83. Ounkaew, A.; Kasemsiri, P.; Jetsrisuparb, K.; Uyama, H.; Hsu, Y.I.; Boonmars, T.; Artchayasawat, A.; Knijnenburg, J.T.; Chindaprasirt, P. Synthesis of nanocomposite hydrogel based carboxymethyl starch/polyvinyl alcohol/nanosilver for biomedical materials. *Carbohydr. Polym.* **2020**, *248*, 116767. [CrossRef] [PubMed]

84. Cobos, M.; De-La-Pinta, I.; Quindós, G.; Fernández, M.J.; Fernández, M.D. Synthesis, physical, mechanical and antibacterial properties of nanocomposites based on poly (vinyl alcohol)/graphene oxide–silver nanoparticles. *Polymers* **2020**, *12*, 723. [CrossRef] [PubMed]
85. Li, X.X.; Dong, J.Y.; Li, Y.H.; Zhong, J.; Yu, H.; Yu, Q.Q.; Lei, M. Fabrication of Ag–ZnO@ carboxymethyl cellulose/K-carrageenan/graphene oxide/konjac glucomannan hydrogel for effective wound dressing in nursing care for diabetic foot ulcers. *Appl. Nanosci.* **2020**, *10*, 729–738. [CrossRef]
86. Li, R.; Xu, Z.; Jiang, Q.; Zheng, Y.; Chen, Z.; Chen, X. Characterization and biological evaluation of a novel silver nanoparticle-loaded collagen-chitosan dressing. *Regen. Biomater.* **2020**, *7*, 371–380. [CrossRef]
87. Zepon, K.M.; Marques, M.S.; Hansen, A.W.; Pucci, C.D.; Morisso, F.D.; Ziulkoski, A.L.; do Nascimento, J.H.; Magnago, R.F.; Kanis, L.A. Polymer-based wafers containing in situ synthesized gold nanoparticles as a potential wound-dressing material. *Mater. Sci. Eng. C* **2020**, *109*, 110630. [CrossRef]
88. Menazea, A.A.; Ahmed, M.K. Wound healing activity of Chitosan/Polyvinyl Alcohol embedded by gold nanoparticles prepared by nanosecond laser ablation. *J. Mol. Struct.* **2020**, *1217*, 128401. [CrossRef]
89. Yang, Y.; Dong, Z.; Li, M.; Liu, L.; Luo, H.; Wang, P.; Zhang, D.; Yang, X.; Zhou, K.; Lei, S. Graphene oxide/copper nanoderivatives-modified chitosan/hyaluronic acid dressings for facilitating wound healing in infected full-thickness skin defects. *Int. J. Nanomed.* **2020**, *15*, 8231. [CrossRef]
90. Carrillo-Rodríguez, J.C.; Meléndez-Ortiz, H.I.; Puente-Urbina, B.; Padron, G.; Ledezma, A.; Betancourt-Galindo, R. Composite based on poly (acrylic acid-co-itaconic acid) hydrogel with antibacterial performance. *Polym. Compos.* **2018**, *39*, 171–180. [CrossRef]
91. Patil, P.P.; Meshram, J.V.; Bohara, R.A.; Nanaware, S.G.; Pawar, S.H. ZnO nanoparticle-embedded silk fibroin–polyvinyl alcohol composite film: A potential dressing material for infected wounds. *New J. Chem.* **2018**, *42*, 14620–14629. [CrossRef]
92. Swaroop, K.; Somashekarappa, H.M. In vitro biocompatibility and antibacterial activity of gamma ray crosslinked ZnO/PVA hydrogel nanocomposites. *Mater. Today Proc.* **2018**, *5*, 21314–21321. [CrossRef]
93. Salama, A.H.M.E.D. Chitosan/Silk Fibroin/Zinc Oxide Nanocomposite as a Sustainable and Antimicrobial Biomaterial. *Cell. Chem. Technol.* **2018**, *52*, 903–907.
94. Khorasani, M.T.; Joorabloo, A.; Adeli, H.; Mansoori-Moghadam, Z.; Moghaddam, A. Design and optimization of process parameters of polyvinyl (alcohol)/chitosan/nano zinc oxide hydrogels as wound healing materials. *Carbohydr. Polym.* **2019**, *207*, 542–554. [CrossRef] [PubMed]
95. Villanueva, M.E.; Cuestas, M.L.; Pérez, C.J.; Dall, V.C.; Copello, G.J. Smart release of antimicrobial ZnO nanoplates from a pH-responsive keratin hydrogel. *J. Colloid Interface Sci.* **2019**, *536*, 372–380. [CrossRef] [PubMed]
96. Khalilipour, A.; Paydayesh, A. Characterization of polyvinyl alcohol/ZnO nanocomposite hydrogels for wound dressings. *J. Macromol. Sci. Part B* **2019**, *58*, 371–384. [CrossRef]
97. Dodero, A.; Scarfi, S.; Pozzolini, M.; Vicini, S.; Alloisio, M.; Castellano, M. Alginate-based electrospun membranes containing ZnO nanoparticles as potential wound healing patches: Biological, mechanical, and physicochemical characterization. *ACS Appl. Mater. Interfaces* **2019**, *12*, 3371–3381. [CrossRef]
98. Nosrati, H.; Khodaei, M.; Banitalebi-Dehkordi, M.; Alizadeh, M.; Asadpour, S.; Sharifi, E.; Ai, J.; Soleimannejad, M. Preparation and characterization of poly (ethylene oxide)/zinc oxide nanofibrous scaffold for chronic wound healing applications. *Polym. Med.* **2020**, *50*, 41–51. [CrossRef]
99. Augustine, R.; Hasan, A.; Patan, N.K.; Dalvi, Y.B.; Varghese, R.; Antony, A.; Unni, R.N.; Sandhyarani, N.; Moustafa, A.E.A. Cerium oxide nanoparticle incorporated electrospun poly (3-hydroxybutyrate-co-3-hydroxyvalerate) membranes for diabetic wound healing applications. *ACS Biomater. Sci. Eng.* **2019**, *6*, 58–70. [CrossRef]
100. Ulu, A.; Birhanlı, E.; Köytepe, S.; Ateş, B. Chitosan/polypropylene glycol hydrogel composite film designed with TiO2 nanoparticles: A promising scaffold of biomedical applications. *Int. J. Biol. Macromol.* **2020**, *163*, 529–540. [CrossRef]
101. Hu, Y.; Ke, Q.; Li, Z.; Han, W.; Yan, Z. In situ synthesis of cuprous oxide/cellulose nanofibers gel and antibacterial properties. *Comput. Mater. Contin.* **2018**, *56*, 517–527.
102. Yang, W.; Fortunati, E.; Bertoglio, F.; Owczarek, J.S.; Bruni, G.; Kozanecki, M.; Kenny, J.M.; Torre, L.; Visai, L.; Puglia, D. Polyvinyl alcohol/chitosan hydrogels with enhanced antioxidant and antibacterial properties induced by lignin nanoparticles. *Carbohydr. Polym.* **2018**, *181*, 275–284. [CrossRef] [PubMed]
103. Han, H.; Kim, J.H. Preparation and properties of nylon 4/6 copolymer nanofibers containing silver-zeolite nanoparticles. *Fiber. Polym.* **2018**, *19*, 350–356. [CrossRef]
104. Karimian, R.; Mehrabani, M.G.; Mehramuz, B.; Ganbarov, K.; Ejlali, L.; Tanomand, A.; Kamounah, F.S.; Rezaee, M.A.; Yoursefi, M.; Sheykhsaran, E.; et al. Poly (ε-Caprolactone)/cellulose nanofiber blend nanocomposites containing ZrO_2 nanoparticles: A new biocompatible wound dressing bandage with antimicrobial activity. *Adv. Pharm. Bull.* **2020**, *10*, 577.
105. Yudaev, P.A.; Maslennikova, V.V.; Konkova, A.A.; Butorova, I.A.; Chistyakov, E.M. Silver-containing hydrogel based on polyvinyl alcohol modified with nanoscale cyclotriphosphazene. *Pub. Health Tox.* **2021**, *1*, A23. [CrossRef]
106. Bano, I.; Arshad, M.; Yasin, T.; Ghauri, M.A.; Younus, M. Chitosan: A potential biopolymer for wound management. *Int. J. Biol. Macromol.* **2017**, *102*, 380–383. [CrossRef]
107. Waibel, K.H.; Haney, B.; Moore, M.; Whisman, B.; Gomez, R. Safety of chitosan bandages in shellfish allergic patients. *Mil. Med.* **2011**, *176*, 1153–1156. [CrossRef]

gels

MDPI

Review

Development of Scaffolds from Bio-Based Natural Materials for Tissue Regeneration Applications: A Review

Murugiah Krishani [1,*], Wong Yen Shin [1], Hazwani Suhaimi [1,*] and Nonni Soraya Sambudi [2]

1 Faculty of Integrated Technologies, Universiti Brunei Darussalam, Jalan Tungku Link, Gadong BE1410, Brunei
2 Department of Chemical Engineering, Universitas Pertamina, Simprug, Jakarta 12220, Indonesia
* Correspondence: 20h8400@ubd.edu.bn (M.K.); hazwani.suhaimi@ubd.edu.bn (H.S.)

Abstract: Tissue damage and organ failure are major problems that many people face worldwide. Most of them benefit from treatment related to modern technology's tissue regeneration process. Tissue engineering is one of the booming fields widely used to replace damaged tissue. Scaffold is a base material in which cells and growth factors are embedded to construct a substitute tissue. Various materials have been used to develop scaffolds. Bio-based natural materials are biocompatible, safe, and do not release toxic compounds during biodegradation. Therefore, it is highly recommendable to fabricate scaffolds using such materials. To date, there have been no singular materials that fulfill all the features of the scaffold. Hence, combining two or more materials is encouraged to obtain the desired characteristics. To design a reliable scaffold by combining different materials, there is a need to choose a good fabrication technique. In this review article, the bio-based natural materials and fine fabrication techniques that are currently used in developing scaffolds for tissue regeneration applications, along with the number of articles published on each material, are briefly discussed. It is envisaged to gain explicit knowledge of developing scaffolds from bio-based natural materials for tissue regeneration applications.

Keywords: tissue engineering; scaffold; fabrication techniques; tissue regeneration

Citation: Krishani, M.; Shin, W.Y.; Suhaimi, H.; Sambudi, N.S. Development of Scaffolds from Bio-Based Natural Materials for Tissue Regeneration Applications: A Review. *Gels* **2023**, *9*, 100. https://doi.org/10.3390/gels9020100

Academic Editors: Arish Dasan, Ashokraja Chandrasekar and Nupur Kohli

Received: 30 December 2022
Revised: 19 January 2023
Accepted: 19 January 2023
Published: 23 January 2023

1. Introduction

Tissue regeneration is a dynamic process in which the cells and their surrounding matrix interplay. Further, this process is encouraged by designing biomaterials that adapt to the local cellular signals [1]. Transplantation is the conventional method for tissue regeneration, but donor availability, pain, and risks related to graft rejection and infectious disease are some concerns [2]. Tissue engineering is a modern field that promotes tissue replacement and regeneration substitutes. It is a multidisciplinary field in which a biomaterial such as a scaffold, cells, and growth factors are combined to form a new tissue [3,4]. It also helps to overcome the problems faced during autologous and allogeneic tissue repair, such as inadequacy, donor site dejection, and unbidden immune responses [5]. The scaffold acts as a template in which cells and growth factors are implanted to imitate the extracellular matrix to maintain and restore tissue function. High porosity, pore interconnectivity, biocompatibility, biodegradability, and mechanical properties are indispensable properties that must be considered when designing the scaffold [6]. Besides blood cells, most tissue cells reside in a solid matrix known as the extracellular matrix (ECM). The ECM is an anchor for maintaining a proper structure and providing the tissue with mechanical properties and signaling molecules. Hence, the scaffold selected for engineered tissue should mimic the ECM of that specific tissue [7]. Selecting appropriate cells, isolating and expanding targeted cells, and selecting suitable biomaterial for scaffold designing are factors that thrive in tissue engineering [8]. However, a solitary polymer cannot achieve every single property of a scaffold, so the desired property can be attained by mixing it with a variety of polymers [9]. Along with the selection of material, process technique or fabrication method also provide a more significant impact on the features of the resultant scaffold [10]. This paper provides detailed information on bio-based natural materials and the fabrication techniques currently used to develop scaffolds for tissue regeneration applications.

2. Tissue Engineering

Tissue engineering (TE) is a relatively new, unique, multidisciplinary field. It offers new hope to patients by integrating clinical medicine, materials science, cell biology and genetics, and mechanical engineering to design bio-artificial tissues or biological substitutes that restore or regenerate, preserve, and improve damaged tissue or organs [3]. The three essential parameters in tissue engineering, biomaterial scaffolds, cells, and growth-stimulating signals, are known as the "tissue-engineering triad," as mentioned in Figure 1.

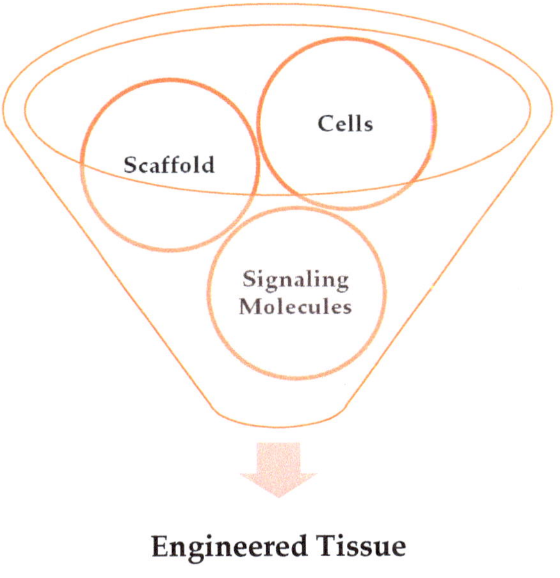

Figure 1. Vital elements of tissue engineering (simplified diagrammatic representation of the basic concept of tissue engineering, i.e., scaffold, cells, and growth-stimulating factors are the three essential parameters responsible in tissue engineering for forming new functional tissue).

The bioreactor uses this triad to imitate a natural environment to reproduce and grow new functional tissues or cellular components. Figure 2 shows an illustration of the basic principle of TE.

Firstly, cells are isolated from a biopsy (allogenic, syngeneic, xenogeneic, or autologous source) and allowed to grow and expand in vitro, in a cell culture system, or in a bioreactor. The expanded cells are then seeded onto a nutrient and growth factors-rich matrix or carrier (scaffold) for structural support. Here, the cells grow, differentiate, and proliferate to form new tissues, then migrate to the carrier to replace the old tissues. Lastly, this TE product will be grafted into the patient to replace the damaged tissues [11].

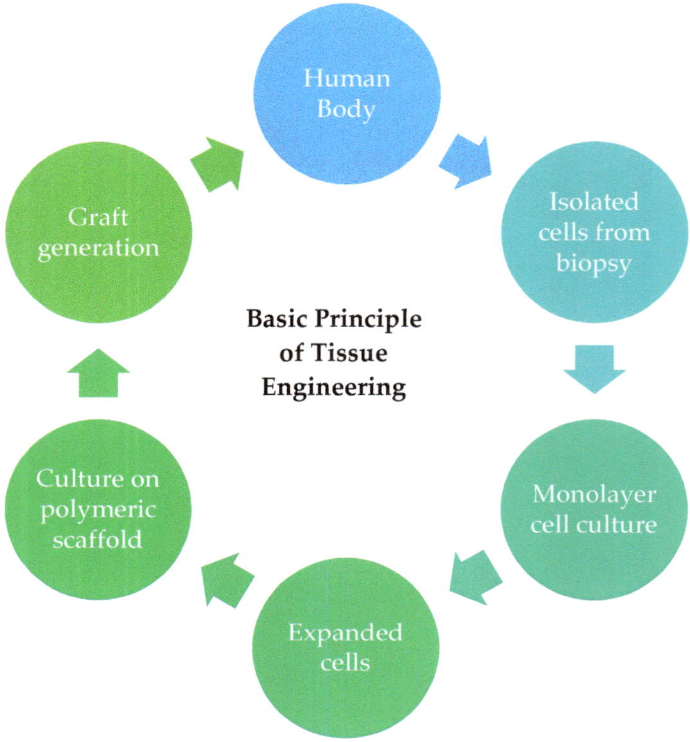

Figure 2. An illustration of the basic principle of TE, which includes cell isolation, cell culture, cell expansion, and tissue grafting into the patient's body.

2.1. Key Elements of Tissue Engineering

2.1.1. Cells

The cell is a structural and functional unit of life in all living organisms. Cells performing the same function are grouped to form tissues and create a body system. While designing a TE product, especially for clinical applications, cell source selection becomes a crucial issue, as it determines the success of the tissue generation step. Essentially, the cells isolated for TE applications should fulfill the essential requirement of combining themselves with the selected tissue with different growth factors and cytokines that activate the endogenous tissue regeneration program. However, natural cells have difficulty reproducing the same particular cell type in large quantities. A promising cell source called stem cells is then developed as an alternative. Stem cells can be categorized into embryonic (ESCs), adult (ASCs), and induced pluripotent stem cells (iPSCs). ESCs are pluripotent cells that can differentiate into any desired lineage, but are ethically controversial and have a shortage in teratoma production [12]. ASCs are multipotent cells and are considered more appropriate for TE applications than ESCs. Though ASCs have more limitations in cell differentiation, they are believed to be less prone to rejections after transplantations. Therefore, ASCs are commonly used to isolate tissues such as bone marrow, muscle, adipose tissue, and umbilical cord [13]. iPSCs, on the other hand, are somatic cells in the pluripotent state that exhibit autologous characteristics and fulfill differentiation capacity [12]. Nevertheless, iPSCs are yet to be extensively used due to the need for precise characterizations of reprogramming the somatic cells before clinical applications [14].

2.1.2. Growth Factors

Growth-stimulating signals include growth factors (GFs), which are a heterogeneous group of polypeptides bonded to specific receptors on the cell surface that regulates a heterogeneous group of polypeptides bonded to specific receptors on the cell surface that regulate cellular responses such as cell proliferation and cell survival, as well as the growth of targeted tissues [15]. Some GFs that have been used in TE applications include bone morphogenetic proteins, vascular epithelial growth factor (VEGF), and transforming growth factor-β (TGF-β) [16].

2.1.3. Scaffolds

Scaffolds play an important role in TE applications, serving as a temporary platform or template for providing guidance and structural support to develop new tissues [17]. Scaffolds refer to a three-dimensional (3D) porous biomaterial that provides a favorable environment for cells to repair and regenerate tissues and organs [3]. It serves as a template for tissue defect reconstruction while promoting cell attachment, proliferation, extracellular matrix regeneration, and restoration of nerves, muscles, and bones. In addition, scaffolds can transport bioactive materials such as drugs, inhibitors, and cytokines as a mechanical barrier against the infiltrating native tissues, which may disturb tissue restoration and regeneration [11].

2.2. Requirements of Scaffold

2.2.1. Microarchitecture

The microarchitecture of the scaffold includes the porosity and pore size and the interconnectivity between the pores. Firstly, the pore size must be adequate for cell migration and attachment onto scaffolds. This also ensures proper mass transfer of nutrients and waste materials into and out of the cells and tissue or vascularization and infiltration. As suggested by Perić Kačarević et al., a smaller pore size is favorable, between 75 and 100 μm in vitro, while the maximum pore size should lie between 200 and 500 μm in vivo to allow optimal tissue penetration and vascularization [18,19]. Moreover, an interconnecting porous system is required to provide a larger scaffold surface area for cell attachment. In addition, having a higher porosity helps to maximize cell-to-cell interactions, thereby promoting the integration of the engineered tissues with the native tissues [20]. Alonzo et al. suggested a pore network comprising more than 60 percent of pores with pore diameters ranging between 150 and 400 μm and at least 20 percent smaller than 20 μm [21].

2.2.2. Biodegradability

As scaffolds only act as a temporary platform for developing cells or tissues, they should be chemically or enzymatically broken down over time when grafted into living organisms. The rate at which the scaffold materials are broken down is known as biodegradability [3]. Ideally, the biodegradation rate of the scaffold should be proportional to the rate of new bone formation or tissue regeneration. When new tissues are successfully engineered and integrated with host bone, they will replace the biomaterial scaffolds via a "creeping substitution" step [22]. The non-toxic products of the scaffold will then be recycled as metabolites in other biochemical reactions or exit the body without interference with other organs and surrounding tissues [18,20].

2.2.3. Biocompatibility

Furthermore, the scaffold should be highly biocompatible for cell adhesion and proliferation. There should be negligible chronic immune responses to prevent severe inflammatory reactions that might affect healing or cause rejection in the body. Even when inflammatory reactions occur, they should be recovered in no more than two weeks [23,24].

2.2.4. Bioactivity

Scaffold bioactivity refers to its ability to interact with the surrounding cellular components of the engineered tissues. Unlike traditional passive biomaterials, which generally pose low or no interactions with the environment, bioactive scaffolds are designed to enhance proper cell migration or differentiation, tissue regeneration or neoformation, and integration in the host, thereby avoiding processes such as scarring [19]. Moreover, the scaffolds may be attached to cell-adhesive ligands to promote cell attachment, or to physical indicators such as topography to enhance cell morphology and alignment. In addition, bioactive scaffolds may serve as a transporter or reservoir for growth-stimulating signals such as GFs to enhance tissue regeneration [7].

2.2.5. Mechanical Properties

Furthermore, the scaffold materials should pose similar intrinsic mechanical properties as native bones or tissues in the anatomical site of implantation. The mechanical properties of tissue vary in nature, as listed in Table 1. It provides structural support and shape stability and, at the same time, helps to minimize the risk of stress shielding, implant-related osteopenia, and subsequent re-fracture. Moreover, the scaffold should also be strong enough to allow surgical handling during transplantations. Some examples of mechanical properties include elastic modulus, tensile strength, fracture toughness, fatigue, and elongation percentage [7,18,19,25].

Table 1. Young's modulus of various tissues.

Tissue	Young's Modulus	Reference
Bone	1–20 GPa	[26,27]
Cardiac	30–400 KPa	[27,28]
Cartilage	10–20 KPa	[27,29]
Endothelium	1–7 KPa	[27,30]
Liver	0.3–0.8 KPa	[27,31]
Lung	1–5 KPa	[27]
Nerve	0.1–2 KPa	[32]
Skin	4.6–20.0 MPa	[33]
Skeletal Muscle	20–100 KPa	[34]

Tensile testing and compressive testing are the conventional methods used to characterize the mechanical properties of a scaffold. Compressive/tensile strength, toughness, and Young's modulus are the important obtained parameters. No limitations for the geometrical structure of the specimen is the biggest advantage of compressive testing over tensile testing. Atomic force microscopy (AFM), dynamic mechanical analysis (DMA), rheometry, and micro indentation are the alternative methods for the characterization of mechanical properties [25]. Elasticity (Young's modulus), shear strength, and viscoelasticity measurement are some significant mechanical properties in cardiac tissue engineering. Due to its thin geometric structure (μm thickness), it is inadequate for DMA. Hence, viscoelasticity measurement for the cardiac scaffold is incorporated only in a few studies [35]. For the healing process, to endure osteogenic loads, adequate compressive strength is needed in bone tissue engineering [36]. Compared to other tissues, neural tissues have low mechanical stiffness with the range of 0.1 KPa for Young's modulus [37]. Mechanical properties play a crucial role in skin tissue engineering to resist physiological forces such as nerve bundles, vascular networks, and collagen deposition during the wound healing process [38]. Figure 3 depicts the requirements to be considered while developing the scaffold.

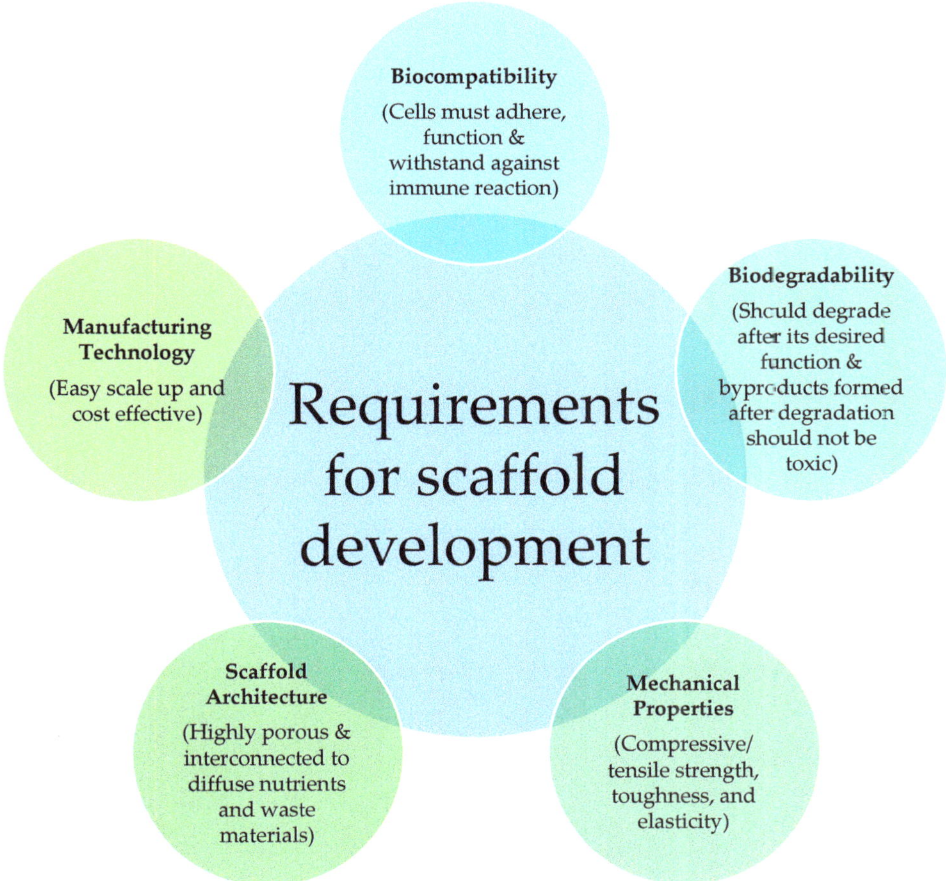

Figure 3. The necessary ideal scaffold requirements include biocompatibility, biodegradability, mechanical properties, scaffold architecture, and manufacturing technology.

2.2.6. Manufacturing Technologies

As stated by Place et al., TE products must be both productive and cost-effective, introducing a potential dichotomy between the need for sophistication and ease of production [39]. While ensuring scaffold efficiency, it is also essential to consider the cost and availability, ensuring scale-up production of the scaffolds is feasible when required. Another key factor to consider is delivering and packaging the scaffolds to the clinicians. Even though clinicians usually prefer off-the-shelf availability to lessen waiting time before implantations, it may not be possible for some tissue types [40]. Therefore, this should be considered while implementing a TE strategy.

2.3. Materials Used for Developing Scaffold

The material source for scaffolds should depend on the patient's status. For instance, patients with cancer or osteoporosis generally experience low bone metabolism; hence, the scaffold material should be non-resorbable. Nevertheless, the material source would come under biomaterials. According to the European Society for Biomaterials (ESB), a biomaterial is a material meant to interface with biological systems to treat, evaluate, augment or replace any tissue, organ or function of the body [18,40]. The four major biomaterials typically used in the fabrication of scaffolds are polymers, bio-ceramics, metals, and carbon-based nanomaterials. As each group has specific advantages and disadvantages, scaffolds may comprise more than one of these biomaterial types [40,41]. Natural polymers, synthetic polymers, bio-ceramics, biodegradable metals, and carbon-based nanomaterials are currently used in scaffold development [1].

2.3.1. Polymer

A polymer is a long-chained macromolecule built up by repeated monomers, and polymer-based biomaterials are considered a good choice for fabricating a scaffold [42]. Polymers are a good candidate in TE applications for their great versatility and flexibility in providing a wide range of mechanical, chemical, and physical properties. They show good biocompatibility, are light in weight, and are resistant to biochemical attack. Moreover, polymers are highly available at a reasonable cost and quickly processed into desired shapes. In addition, the inertness of polymers towards host tissues makes them an eligible candidate for a drug delivery system. Some biomedical applications involving polymers include artificial organs and blood vessels, breast implants, contact lenses, coatings for pharmaceutical tablets and capsules, external and internal ear repairs, cardiac assist devices, and joint replacements [43]. Polymeric biomaterials have been obtained from natural and synthetic polymers, each having pros and cons [44,45]. Carbohydrates such as chitin, cellulose, starch, alginate, and hyaluronic acid and proteins such as collagen, elastin, keratin, gelatin, and fibrin fall under natural polymers, where polyesters such as poly ε-caprolactone (PCL), polylactic acid (PLA), and polyglycolic acid (PGA) and Polyurethanes come under synthetic polymers [44].

1. Natural polymers: Biopolymers are toxic-free, highly biocompatible, easily adhere to cells, and improve proliferation and differentiation. Nevertheless, they have poor mechanical strength and are highly sensitive to elevated temperatures [46]. Biopolymers are also known as natural polymers. Natural polymers are materials that can be obtained from natural sources. They can be categorized into protein-based biomaterials (naturally occurring polymers in the human body such as collagen, fibrin, and elastin) and polysaccharides-based biomaterials (such as silk, chitosan, alginate, and gelatin). They exhibit similar characteristics to soft tissues, showing bioactivity, excellent cell adhesion and growth, and fulfilling biodegradability and biocompatibility. Moreover, they are also known for their wide availability, ecological safety, and modifiability to suit different applications. However, natural sources indicate the requirement of a purification step to avoid foreign immunological responses after implantation. In addition, natural polymers typically show poor physical and mechanical stability, limiting their applications in the load-bearing orthopaedic field [17,43].
2. Synthetic polymers: In contrast to natural polymers, synthetic polymers have good mechanical properties. However, they also have a high risk of immune rejection, and toxic substances such as carbon dioxides are released during degradation, leading to cell damage [40]. Synthetic polymers serve as a more predictable biomaterial providing a wide range of mechanical and physical properties such as degradation rates. If they are synthesized under controlled conditions, they do not pose any immunological risks, and desired characteristics can be brought together. One common synthetic polymer used for BTE applications is aliphatic polyesters, including poly (ε-caprolactone) (PCL) and polylactide (PLA). PCL is a semi-crystalline, biodegradable, and non-toxic polyester that shows hydrophobicity and slow degradation rates of more than 24 months. These problems can be addressed by blending with other polymers or producing composites. In contrast, the porous PLA exhibits high biocompatibility, but shows slow degradation rates of 3–5 years. Thus, PLA is combined with hydroxyapatite (HAp) to improve its mechanical and physical strength [18,43].

2.3.2. Bio-Ceramics

Ceramic materials obtained from natural products, termed bio-ceramics, have been widely used in dental and bone tissue engineering. Bio-ceramics are organic, non-metallic solids with good compatibility, bio-inertness, bioactivity, osteoconductivity, and mechanical strength [18,43]. In addition, bio-ceramics can promote new bone generation and the osteo-potential of scaffolds. However, bio-ceramics are low in elasticity with a brittle surface, limiting their use in implants. Thus, they are usually blended or coated with other materials to improve their elasticity and strength. Among the biomaterials, bio-ceramic scaffolds have been proven to be more successful in treating minor bone defects, such as orthopaedic implants and bone-filling applications. There are three main types of ceramics: bioinert, bioactive, and bioresorbable. Bioinert ceramics include alumina (Al_2O_3), Zirconia (ZrO_2), and pyrolytic carbon; bioactive ceramics include bioglasses (BG) and glass ceramics, while bioresorbable ceramic contains calcium phosphates. Of all three types, the most commonly used ceramics in BTE applications are HAp, tricalcium phosphates (TCP), and their composites [3]. Human bone and teeth are composed of an inorganic compound known as hydroxyapatite, which constitutes calcium, phosphate, and OH radicals with high tensile strength and quickly adheres to host tissues. Many studies revealed that HAp is non-toxic and lacks inflammatory and pyrogenetic response [47].

HAp is a naturally occurring calcium phosphate-based mineral with the chemical formula $Ca_{10}(PO_4)_6(OH)_2$. It is structurally similar to a biological apatite in the human body, known as bone mineral, which makes up approximately 60–70% of human bone tissues on a dry weight basis. It shows similar chemical and physical properties to human bone and dental tissues [48,49]. Hence, some hydroxyapatite-rich natural products such as shells, corals, algae, fish scales, and animal bones are used to develop scaffolds [50,51]. HAp exhibits excellent biocompatibility and bioactivity, high osteoinductivity and osteoconductivity, non-toxicity, and non-inflammatory characteristics. Moreover, it vitalizes growth factors and promotes cell growth and proliferation. HAp is therefore considered a highly potential implant material and bone substitute. Nevertheless, HAp also shows poor mechanical properties, slow resorption and remodeling rates, and slow degradation rates in vivo, making it unsuitable for all BTE applications. Thus, HAp is usually synthesized with other natural or synthetic polymers to create more effective composite scaffolds [19,52].

HAp can be obtained through extractions from natural sources or chemical syntheses, divided into three categories: high-temperature methods, wet methods, and dry methods. Dry methods include mechanochemical methods and solid-state reactions. Here, dry precursors of calcium and phosphate are mixed without any precisely controlled conditions to synthesize HAp. According to Sadat-Shojai et al., this is to ease the mass production of HAp powders. Wet methods include sol-gel, chemical/wet/co-precipitation, hydrothermal, hydrolysis, sonochemical method, and emulsion method. With this method, the morphology and average powder size can be controlled. However, HAp yielded usually exhibits low crystallinity due to low operating temperatures. High-temperature processes include combustion and pyrolysis methods, where samples undergo thermal decompositions. These two methods are rarely used for HAp synthesis due to poor control over the operating conditions [53].

HAp can also be extracted from food wastes or biological sources such as aquatic or marine sources, mammalian bones shell sources, and plant sources [54]. This method is relatively safe, more sustainable, and economical to fabricate HAp, thereby contributing to the economy, environment, and general health. However, it is notable that natural HAp is non-stoichiometric, either calcium or phosphorus-deficient. Generally, calcium positions would be the vacancy, where cations such as Na^+, Mg^{2+}, and Al^{3+} are substituted into the vacant space. Likewise, carbonate ions would replace phosphate or hydroxyl ions, and fluoride ions would substitute in place of hydroxyl ions. These trace elements present in the natural HAp resemble the apatite in human bone, which is crucial in accelerating bone formation and regeneration. For instance, blending 3–5 mol% silicon with synthetic HAp can boost cell growth density, enhancing osteoblast growth. Another example is adding

1–10% of strontium ions in synthetic HAp, which improves osteoblast activity and material differentiation [55]. Calcium carbonate is abundantly found in the exoskeleton of most marine organisms such as corals, sea urchins, and some algae. HAp produced from these exoskeletons are highly porous, have good vascularization and blood supply, and help to form new tissue [56]. Over the past 20 years, extensive research has been done to constantly improve the synthesis methods and introduce new technologies, aiming to develop an ideal HAp composite or scaffold that fulfills all the desired specifications.

2.3.3. Metals

Metals such as stainless steel, cobalt–chromium–molybdenum alloy, aluminium, lead, silver, and titanium alloys have been considered good load-bearing implants because of their excellent quality electrical and thermal conductivity, appropriate mechanical properties, corrosion resistance, biocompatibility, and reasonable cost. However, metals are non-biodegradable. Therefore, researchers introduced the use of biodegradable metals [18]. Biodegradable metals are metals having controlled corrosion properties. They can be grouped into pure biodegradable metals (Mg^{-}- and Fe^{-}-based), biodegradable alloys, and biodegradable metal matrix composites [43]. Pure biodegradable metal implants have similar mechanical properties to stainless steel and bone and are non-toxic. However, they show slow degradation rates and are incompatible with MRI (Magnetic Resonance Imaging). These problems can be addressed through newly-developed fabrication methods such as casting, electroforming, powder metallurgy, and inkjet 3D printing. Moreover, it is essential to note that the patients implanted with biodegradable metals should not have an iron-related disease, and the patient's intestines can absorb only Fe^{2+}. Thus, any Fe^{3+} released should be first reduced to Fe^{2+} before being absorbed [43].

Biodegradable porous metal scaffolds have attracted researchers in scaffold development by their high compressive strength. Biodegradable metals overcome problems such as innate immune rejection and have good load-bearing capacity during bone healing. However, biodegradable metals such as Mg and their alloys have a high corrosion rate. Recently, scientists have concentrated on the Zn-based alloy system to produce biodegradable metal scaffolds [57,58].

2.3.4. Carbon-Based Nanomaterials

Researchers developed carbon-based nanomaterial scaffolds by combining tissue engineering and nanotechnology to enhance the scaffold's features. Carbon nanotubes (CNTs), graphene oxide (GO), carbon dots (CDs), fullerenes, and nanodiamonds are some carbon nanomaterials used as scaffolds in tissue engineering. Biocompatibility, mechanical stability, low cytotoxicity, facilitating cell communication, and nutrition delivery are advantages of carbon-based nanomaterials that pull down to use in scaffold development. However, limited biodegradability and potential cytotoxicity are significant drawbacks [59].

Carbon-based nanomaterials, including graphene oxide (GO), carbon nanotubes (CNTs), fullerenes, carbon dots (CDs), nanodiamonds (NDs), and their derivatives, are highly potential scaffold materials for bone restoration applications. They are biocompatible, mechanically stable, and commercially available. In addition to that, they show essential qualities such as good biodegradability, efficient cell proliferation and osteogenic differentiations, significant cell growth stimulations, proper mass transfer of nutrients in the scaffold microenvironment, improved cell distributions, and appropriate cell bioactivity. Yet, further studies regarding the low cytotoxicity and the adverse environmental effects of carbon-based nanomaterials are to be conducted before they can be clinically tested and brought into application [1,60]. The materials used for developing the scaffold are summarized in Figure 4.

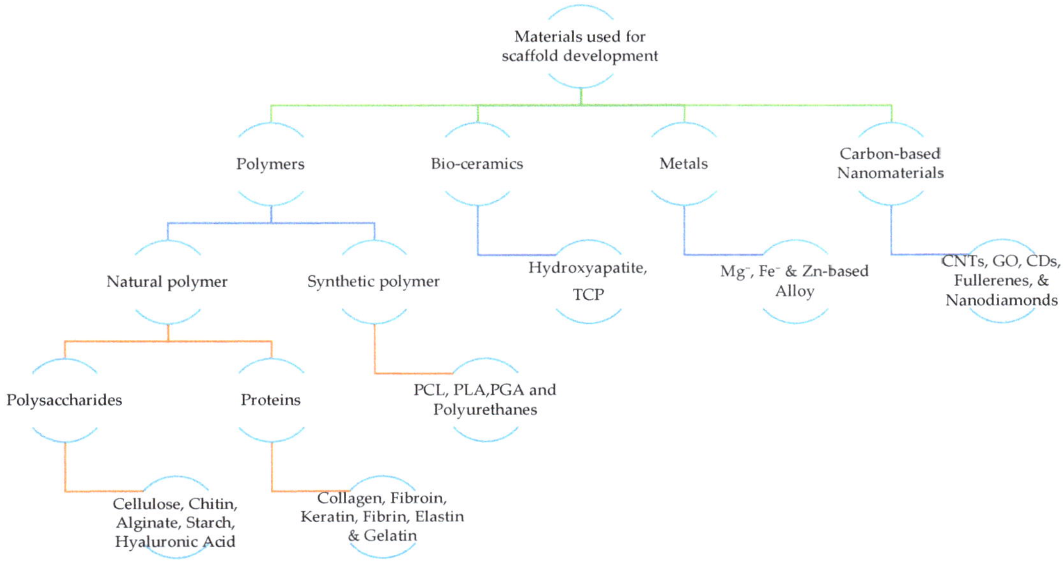

Figure 4. Materials used for scaffold development. (Materials are divided into four broad categories such as polymers, bio-ceramics, metals, and carbon nanomaterials. Classification with a few examples is summarized).

In recent years, some innovative synthetic scaffolds based on natural products have been developed by applying recombinant DNA technology and advanced genetic engineering. Elastin-like recombinant [ELR] and elastin-like peptides [ELP] are a few scaffolds developed by obtaining the principles of advanced genetic engineering techniques. The Arginine, Glycine, and Aspartic acid (RGD) sequence is an integrin-binding sequence in the ELP scaffolds, which helps in cell adhesion and proliferation. ELR scaffolds constitute a fibronectin domain that helps cell adhesion, particularly in vascular regeneration [61,62]. B. Gurumurthy et al. developed a collagen-based scaffold by incorporating it with an elastin-like polypeptide obtained from genetically modified *Escherichia coli* bacteria and bioglass to examine the osteogenic differentiation [63]. Repeated sequences of elastin and silk blocks are recombinantly combined to form silk-elastin-like protein polymers (SELPs). The hydrogels of SELPs play a vital role in wound healing [62].

On the whole, bio-based polymers have good features such as compatibility, versatility, and adaptability, and are also abundant in nature; they can be obtained from various agricultural resources and biodegradable waste materials. Hence, the processing and synthesis cost is low and environmentally friendly [50,64]. Many researchers have tried to produce scaffolds from natural polymers by keeping this in mind by modifying and enhancing their stability using various fabrication methods [65].

2.4. Common Natural Polymers Used in Tissue Regeneration Applications

2.4.1. Cellulose

Cellulose is a fundamental structural unit of the plant cell wall. It is also found in red, green, and brown algae, some fungi, and as an extracellular component in bacteria [66]. Cellulose is a homopolysaccharide composed of D-glucose units connected by β-(1→4) glycosidic bonds [67]. Cellulose is an ideal material for tissue growth. It has several features such as biocompatibility, biodegradability, and cheap cost. It is already used as a scaffolding material in wound repair, cartilage tissue regeneration, differentiating endothelial cells, and bone tissue engineering [60]. Scaffolds developed based on bacterial cellulose are widely used in various biomedical applications [68]. Based on recent research, the performance of the material toward cell growth or biocompatibility is mentioned in Table 2.

Table 2. Recent research on the performance of the materials toward cell growth or biocompatibility.

Natural Material, Application, and References	Cells	Assay	Result
Cellulose in Bone Tissue Engineering [69]	Human osteoblast cells	MTT assay	Significant increase in cell viability of scaffold with 0.5 weight% bacterial cellulose
Cellulose in Cartilage Tissue Engineering [70]	Chondrocytes	Presto Blue™ assay	Chondrocyte viability percentage in scaffold was found to be greater than 70%
Cellulose in Cardiac Tissue Engineering [71]	H9C2 rat cardiac myoblasts	MTT assay	Excellent biocompatibility in which scaffold exhibited cell proliferation and retention over the time frame of the study
Cellulose in Nerve Tissue Engineering [72]	Rat PC12 cells	Dojindo's cell counting kit-8 (CCK-8) assay	An increase in cell viability was observed when the concentration of Poly(3-hexylthiophene)—an organic voltaic material, was up to 0.15M for the respective scaffold
Cellulose in Skin Tissue Engineering [73]	L929 mouse fibroblast	MTT assay	In vitro studies show low cell viability due to MTT assay, which is unreliable in calculating the number of cells settled inside the scaffold, but in vivo studies with Wistar rats revealed it is a promising material for diabetic wound healing
Chitosan in Bone Tissue Engineering [74]	MC3T3-E1 cells (Mouse calvaria pre-osteoblast)	Dojindo's cell counting kit-8 (CCK-8) assay	Cell attachment, viability, and proliferation in regenerated cellulose nanofibers into chitosan hydrogel is more excellent than pure chitosan hydrogel
Chitosan in Skin Tissue Engineering [75]	Human dermal fibroblast	MTS assay	The presence of chitosan along with gelatin helps in the cellular behavior of substrates and enhances the proliferation rate of fibroblasts
Chitosan in Nerve Tissue Engineering [76]	Schwann cells	MTT assay	The rate of Schwann cell proliferation was increased after the introduction of gold nanoparticles
Chitosan in Cartilage Tissue Engineering [77]	ATDC5 (Chondrocytes)	Live/dead kit (Invitrogen)	The cells migrated toward the edges of the scaffold, and the cell population at the edges became higher. From this result, necessary modifications were carried out to develop smooth strands without any slope to encourage the cells to spread on the whole surface of scaffold
Chitosan in Cardiac Tissue Engineering [78]	H9C2 (Rat cardiac myoblast cells) and HUVEC (Human umbilical vein endothelial cells)	Alamar Blue assay	For HUVEC, more cell viability was seen in the scaffold combined with polyurethane, chitosan, and carbon nanotube than in the polyurethane scaffold and control, since polyurethane is hydrophobic and lacks enough surface of cell recognition sites, while chitosan is hydrophilic. H9C2 revealed that the developed scaffold is promising for infarcted myocardium

Table 2. *Cont.*

Natural Material, Application, and References	Cells	Assay	Result
Chitosan in Liver Tissue Engineering [79]	HepG2 (Human hepatic carcinoma cells)	MTT assay	The P-value greater than 0.05 in all cases indicates that the scaffold is suitable for liver tissue engineering and in vivo tests
Alginate, Cellulose, and Gelatin in Bone Tissue Engineering [80]	hBMSC (Human bone marrow stromal cells)	WST-1 assay	No evidence of side effects after the cell seeding in the scaffold revealed its biocompatibility, and rapid bone regeneration was observed in the in vivo model three weeks after transplantation
Alginate in Skin Tissue Engineering [81]	Fibroblast L929 cell line	MTT assay	The resulting scaffold showed good cell adhesion based on cell concentration in the scaffold as well as the assessment of cell growth
Alginate in Cartilage Tissue Engineering [82]	AMSC (Mesenchymal stem cells derived from adipose tissue)	MTT assay	Alginate helps AMSC for chondrogenesis differentiation without any aid of exogenous differential agents
Alginate in Nerve Tissue Engineering [83]	Olfactory ecto-mesenchymal stem cells	Resazurin assay and live/dead viability assay	From the live/dead viability assay, hydrogels with 5 µm and 25 µm magnetic short fibers (MSF) and alginate ease neural-like cell proliferation. From the Resazurin assay, the cell proliferation is higher in MSF-containing hydrogel than in pure alginate
Hyaluronic Acid in Skin Tissue Engineering [84]	HDF (Human dermal fibroblast)	MTT assay	Enhanced cell proliferation was seen on the nanocomposite scaffold along with cell viability. Not only the cell proliferation but also the drug delivery was exhibited by the MTT assay
Hyaluronic Acid in Neural Tissue Engineering [85]	SH–SY5Y (Human neuroblastoma cell line)	MTT assay	The rate of SH–SY5Y proliferation is accelerated by the optimal amount of hyaluronic acid
Starch in Bone Tissue Engineering [86]	MG-63 (Human osteoblast cells)	MTT assay	Cell viability of all samples was found to be greater than 94%, which shows good cytocompatibility
Collagen in Cartilage Tissue Engineering [87]	Articular chondrocytes from new-born Sprague Dawley (SD) rats	Live/dead cell viability assay	The proportion of live cells in the collagen and sodium alginate scaffold was found to be greater than in the sodium alginate and agarose scaffold
Collagen in Corneal Tissue Engineering [88]	hBM-MSCs (Human bone marrow mesenchymal stem cells)	Cell Counting Kit-8 assay	Hydrogel combined with gelatin and collagen showed increased cell viability and proliferation with time than gelatin hydrogel
Collagen in Bone Tissue Engineering [89]	MC3T3-E1 cells from mice	Cell Counting Kit-8 assay	Collagen I proteins are relatively expressed at a higher level, promoting cell differentiation. The constructed porous microsphere had excellent biocompatibility and effectively enhanced cell adhesion, proliferation, and differentiation

Table 2. *Cont.*

Natural Material, Application, and References	Cells	Assay	Result
Collagen in Skin Tissue Engineering [90]	ATCCR PCS-201-012 (Normal adult human dermal fibroblasts) and ATCCR PCS-200-011 (Normal primary human adult epidermal keratinocytes)	MTT assay	The scaffold made up of collagen and elastin promotes cell adhesion and proliferation
Collagen in Oral Mucosa Tissue Engineering [91]	Human primary oral fibroblast and keratinocyte cells	PrestoBlue assay	The comparative study revealed that the biological properties of the collagen-based hydrogel are superior to gelatin methacryloyl in terms of growth of oral fibroblast within the scaffold and epithelial cell differentiation and adhesion on the engineered substrate surface
Fibroin in Bone Tissue Engineering [92]	HADMSC (Human adipose-derived mesenchymal stem cells)	Live/dead assay and MTT assay	The MTT assay revealed that PRP (platelet-rich plasma)-treated composite scaffold showed a greater cell proliferation rate than untreated scaffold. The live/dead assay revealed that cells were active after day 14 on both PRP-treated and untreated scaffolds
Fibroin in Skin Tissue Engineering [93]	L929 cells	Cell Counting Kit-8 assay	In the total of 7 days, the cell proliferation rate was found to be lowest on day 3, and the cell proliferation rate increased significantly on days 5 and 7
Fibroin in Cartilage Tissue Engineering [94]	Human chondrocyte	Cell Counting Kit-8 assay and Live/dead assay	The CCK-8 assay revealed that significant cell growth was noticed from 7–14 days. From the live/dead assay, the cell viability was detected from 5–14 days
Fibroin in Corneal Tissue Engineering [95]	The limbal cells (Isolated from corneal limbus)	MTT assay	Vigorous cell adhesion and proliferation were seen on the surface of the scaffold
Fibroin in Musculoskeletal System Tissue Engineering [96]	Bone marrow-derived mesenchymal stem cells from rabbit	Live/dead assay	Silk fibroin/pullulan hydrogels contain 90% live cells after seven days of culture, which explicitly shows its good cytocompatibility
Fibroin in Neural Tissue Engineering [97]	SH-SY5Y (Human neuroblastoma cell line)	Cell Counting Kit-8 assay	Silk fibroin scaffold showed good cell survival with an increased number over time
Keratin in Bone Tissue Engineering [98]	MG-63 cells	MTT assay	Hydroxyapatite-containing scaffold showed higher cell viability
Keratin in Nerve Tissue Engineering [99]	L929 mouse lung fibroblasts, human skin fibroblasts, human Schwann cells, and human pulmonary microvascular endothelial cells	MTS assay & Alamar Blue assay	Cells seeded on keratin combined chitosan membrane showed more significant cell adhesion and metabolic activity than plain chitosan membrane

Table 2. *Cont.*

Natural Material, Application, and References	Cells	Assay	Result
Keratin in Vascular Tissue Engineering [100]	HUVEC (Human umbilical vein endothelial cells) and HUASMC (Human umbilical arterial smooth muscle cells)	MTT assay	The developed mat had good biocompatibility, including prolonged activated partial thromboplastin time (APTT), cytocompatibility, and lower platelet adhesion. Moreover, these mats could speed up the nitric oxide generation from the donor in the blood, which accelerates endothelial cell growth, reduces smooth muscle cell proliferation, and inhibits platelet adhesion
Keratin in Skin Tissue Engineering [101]	L929 cells from mouse fibroblast	MTT assay	In vitro studies revealed cell adhesion and proliferation, whereas in vivo studies revealed wound healing
Keratin in Urethral Tissue Engineering [102]	Smooth muscle cells from rabbit	Live/dead assay	Scaffold containing calcium peroxide (CPO) displayed greater cell viability (92–94%) than scaffold without CPO (88–93%)
Elastin in Cardiac Tissue Engineering [103]	Cardiac progenitor cells from rats	Dil Cell Labeling	Quantitative evaluation of Dil-labelled cells occupying a fractional area after 72 h of seeding in the scaffold confirms the cell viability by in vitro studies. From the detection of immunofluorescence in the myocardium, after ten days of implant, the cell viability by in vivo study is revealed
Elastin in Vascular Tissue Engineering [104]	hAd-MSCs (Human adipose-derived mesenchymal stem cells)	MTT assay	Cell viability and proliferation were confirmed by the MTT assay, and reverse transcription-polymerase chain reaction ensures cell differentiation
Elastin in Cartilage Tissue Engineering [105]	Chondrocytes (Cartilaginous tissues of the bovine knee from calves)	XTT (2,3-bis(2-methoxy-4-nitro-5-sulfophenyl)-2H-tetrazolium-5-carboxanilide)	During implantation, the surface of the scaffold affected the cell, resulting in decreased cell activity. Then, better cell proliferation was seen after the surface modification with elastin and other materials
Gelatin in Bone Tissue Engineering [106]	MC3T3-E1 osteoblasts	Histology assay	The biocompatibility of the scaffold was determined by comparing it with Gelfoam. The cell number on the gelatin scaffold is significantly higher than on Gelfoam
Gelatin in Skin Tissue Engineering [107]	HSF (Human skin fibroblast)	MTT assay	Cells activity was not affected by the scaffold material, and it was well-suited for cell proliferation and adhesion
Gelatin in Cartilage Tissue Engineering [108]	Articular cartilage progenitor cell line	Resazurin assay	Not only the hydrophilic character of gelatin but also the presence of Arginylglycylaspartic acid (RGD)—a cell recognition domain, in its structure facilitates cell attachment
Gelatin in Nerve Tissue Engineering [109]	L929 cells from mouse fibroblast	MTT assay	Axons and neuronal dendrites formed on day 14 confirm cell differentiation along with cell viability and proliferation

Table 2. *Cont.*

Natural Material, Application, and References	Cells	Assay	Result
Fibrin in Cartilage Tissue Engineering [110]	Human hyaline-derived chondrocytes	WST-1 assay	An increase in cellular metabolic activity with time, along with a decrease in the biomaterial volume
Fibrin in Liver Tissue Engineering [111]	HepG2 cell lines	MTT assay	From the MTT assay, the quantitative assessment of cell viability was found to be 86.75 ± 1.7%
Fibrin in Retinal Tissue Engineering [112]	ahRPE cells (Adult human retinal pigment epithelial cells)	MTT assay	Proper ahRPE cell encapsulation was done by a 84 mg/dL concentration of fibrin glue
Fibrin in Neural Tissue Engineering [113]	hEnSC (Human endometrial stem cells)	MTT assay	Novel hydrogel fabricated with fibrin, polyurethane, and multiwall carbon nanotube showed more significant cell viability and proliferation than fibrin

2.4.2. Chitin and Chitosan

Chitin is the second most common polysaccharide globally, followed by cellulose. It exists in the exoskeleton of arthropods such as crabs, shrimps, lobsters, insects, prawns, and fungal cell walls. Chitin comprises repeated units of 2-(acetylamino)-2-deoxy-D-glucose. Chitin and chitosan are differentiated by a degree of deacetylation. Chitin has various biomedical applications in tissue engineering due to its outstanding properties such as non-toxicity, biocompatibility, biodegradability, and chelating of metal ions. It also supports cell adhesion, differentiation, and migration. Chitin also has structural similarity with N-glycosaminoglycans, essential components of connective tissues; hence, it is a good option for skin tissue regeneration. Further, it is also used in dental, bone, and cartilage implants [3,114,115]. Mokhtari et al. have developed a scaffold hydrogel by combining chitosan with collagen and aldehyde-modified nanocrystalline cellulose loaded with gold nanoparticles, showing a potential application in tissue engineering [116].

2.4.3. Alginate

Alginate is a seaweed-derived polysaccharide extracted from Phaeophyceae-brown algae. Alginate comprises β-(1–4)-d-mannuronic acid and α-(1,4)-l-guluronic acid connected as repeated linear chains [66,117]. Alginate displays biocompatibility, biodegradability, a simple production process, and tunable mechanical properties, leaping to join in developing scaffolds in cartilage tissue engineering [118]. Moreover, alginate is hydrophilic, so it is used in wound dressing to absorb the pus and help it heal. It is also used in cell growth scaffolds, supporting blood vessels' formation, healing bone injuries, cartilage regeneration, and drug delivery systems [119]. Alginate-based scaffolds are widely used in various tissues or organs, including skeletal muscles, pancreas, nerve, liver, and dental tissue engineering [117]. For cardiac repair, Rosellini. E. et al. produced a scaffold using alginate, elastin, and gelatin, which successfully attained the desired cellular response [103]. The molecular structure of some polysaccharides is shown in Figure 5.

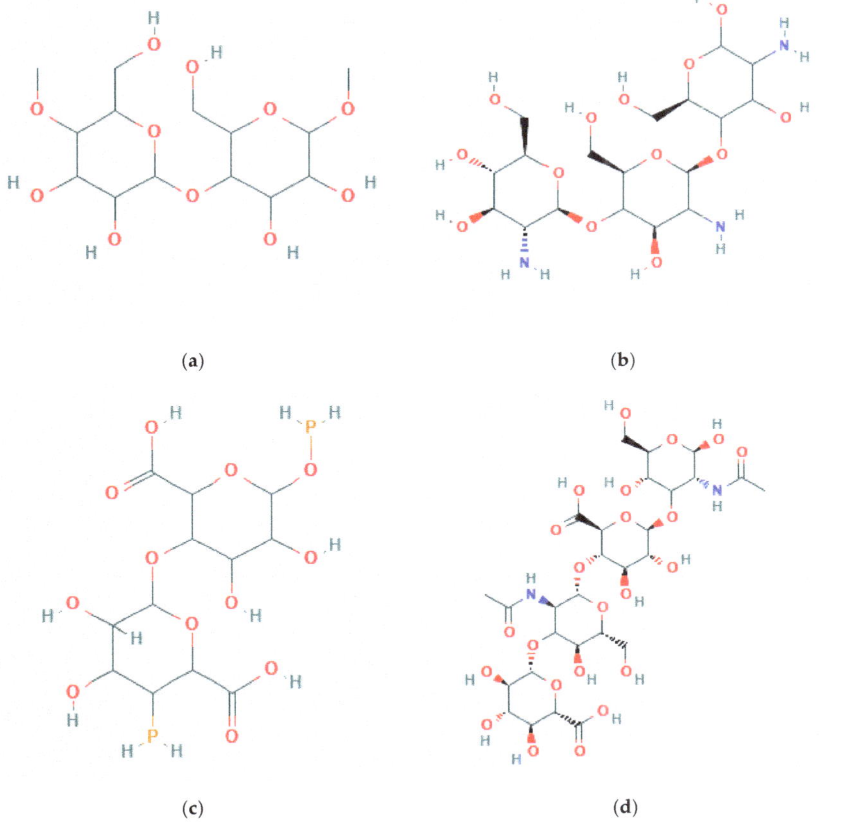

Figure 5. Molecular structure of polysaccharides: (**a**) cellulose-microcrystalline [120]; (**b**) chitosan [121]; (**c**) alginate [122]; (**d**) hyaluronic acid [123].

2.4.4. Starch

Starch is a popular polysaccharide produced by plants for energy storage. It consists of amylose and amylopectin. Amylose (a linear polymer linked by α (1–4) linkages) is connected to amylopectin (highly branched polymer) by α (1–6) linkages. Starch is highly porous and allows cells to penetrate vascularization and tissue growth. Biocompatibility, biodegradation, osteoconduction, and osteo production are some characteristics that display starch to apply in tissue engineering [124].

A study revealed that starch membrane, collagen, and chitosan enhance epithelial tissue regeneration during wound healing, clearly showing that starch-based scaffolds have more significance in wound healing [125]. It also helps in cell adhesion, growth, proliferation, and differentiation. Starch generally shows poor mechanical properties in aqueous media and is easily dissolved. Starch was incorporated with bio-additives to attain good mechanical properties to overcome this problem. For instance, researchers combined starch-based scaffolds with bio-additives such as citric acid, cellulose nanofibers, and hydroxyapatite to obtain the desired result. Many in vivo and in vitro assessments certified that starch-based scaffolds are better for bone regeneration [86]. The molecular structure of starch is shown in Figure 6.

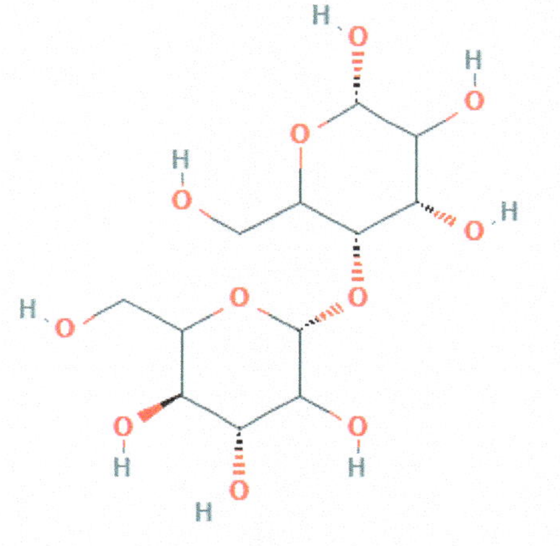

Figure 6. Molecular structure of starch soluble [126]. (Starch is a polysaccharide mainly found in plant cells.)

2.4.5. Hyaluronic Acid

Hyaluronic acid is a glycosaminoglycan that deficits sulfate bonds commonly secreted by chondrocytes and fibroblasts. It comprises repeated β-1,4-D-glucuronic acid and β-1,3-N-acetyl-D-glucosamine disaccharide units. It is mainly present in the synovial fluid, connective tissues of the dermis, the vitreum, and the dental pulp matrix. It maintains the viscoelasticity of ECM by acting as a lubricant [127]. It plays a vital role in the cell's structural maintenance, keeps tissue hydrated, and helps cell signaling and wound repair. It is highly biocompatible, biodegradable, and can be easily modified chemically. Therefore, it is widely used as scaffolds in various forms such as sponges, cryogels, hydrogels, and injectable hydrogels [128–130]. A combination of collagen and hyaluronic acid scaffold material was used in cartilage regeneration, which plays a significant role in tissue repair. Mohammadi et al. prepared a scaffold by combining hyaluronic acid and collagen loaded with prednisolone to make a proper dosage form for cartilage repair [130]. According to Sieni et al., scaffolds based on hyaluronic acid show several more valuable features than collagen scaffolds in breast cancer treatment [131]. The advantages, disadvantages, and applications of each polysaccharide are mentioned in Table 3.

Table 3. Advantages, disadvantages, and applications of polysaccharides in various tissue regeneration applications.

Polysaccharide	Advantages	Disadvantages	Applications
Cellulose [132–136]	Excellent bioactivity and biocompatibility, having high mechanical properties, depends on the chosen source	Non biodegradability	Bone, tendons, cartilage, cardiovascular, muscle, neural, and skin
Chitosan [137–143]	Easy digestion, biocompatibility, biodegradability, antibacterial activity, and hemostatic activity	Low mechanical resistance, stiff and brittle	Bone, cartilage, skin, cardiac, muscle, liver, and nervous tissue engineering

Table 3. *Cont.*

Polysaccharide	Advantages	Disadvantages	Applications
Alginate [144–148]	Bioactivity, biocompatibility, biodegradability, non-immunogenicity, and non-antigenicity	Toughness, limited strength, and difficulty in controlled gelation	Bone, cartilage, Skin, and neural regeneration
Hyaluronic acid [149–156]	Bioactivity, biocompatibility, biodegradability, and easy chemical modification	Rapid degradation and poor mechanical properties	Skin and neural regeneration
Starch [124,157,158]	Biocompatibility, biodegradability, cheap, pertinent porosity	Low mechanical strength, high water uptake, difficult to process, and unstable in long-term application	Bone cement in bone defects and dental cavities

2.4.6. Guar Gum

Guar gum is a galactomannan gum, a polysaccharide obtained from the seed of a leguminous plant, namely, guar beans, commonly known as cluster beans (*Cyamopsis teteragonolobha*). Easy accessibility, biodegradability, biocompatibility, non-toxicity, and non-immunogenicity are attractive features that tempt many researchers to develop scaffolds from guar gum [159].

2.4.7. Pullulan

Pullulan is a polysaccharide made up of repeated maltotriose units connected by alpha (1–6) linkages obtained from fungi known as *Aureobasidium*. Pullulan plays a vital role in tissue engineering due to its adjustable property, biocompatibility, biodegradability, and adhesive nature. Oxidized pullulan was cross-linked with collagen, and scaffolds were produced for various biomedical applications [160–162].

2.4.8. Collagen

Collagen is the critical protein in the connective tissues of animals, mainly in mammals. It is a protein with high biocompatibility and biodegradability. Therefore, it is applied in the medical field in various forms, such as a scaffold, drug carrier, and wound dressing [163]. The latest research shows that collagen obtained from marine organisms is used in multiple biomedical applications [164]. Collagen-based scaffolds are widely used in myocardial tissue engineering [137], cartilage tissue engineering [165], neural tissue engineering [166], musculoskeletal tissue engineering [167], and bone tissue engineering [48]. Massimino et al. developed a collagen-based scaffold obtained from bovine tendon for dermal regeneration applications [49]. Pericardial bovine and porcine tissue underwent TRICOL decellularization (detergent-based treatment), and decellularized pericardial scaffold containing collagen and elastin was considered a potential biomaterial for tissue replacement [52].

2.4.9. Fibroin

Fibroin is protein silk produced by some larvae such as spiders, silkworms, mites, scorpions, and flies. The silk obtained from *Bombyx mori* (silkworm) and spiders such as *Araneus diadematus* and *Nephila clavipes* are widely used commercially [50]. Due to its excellent structural integrity and mechanical properties, silk fibroin-based biomaterial is used in musculoskeletal tissue engineering [168]. Hadisi et al. developed a silk fibroin-based scaffold composed of hardystonite loaded with gentamicin as an antibiotic agent to evaluate the in vitro and in vivo studies on bone tissue engineering applications [169]. According to Zakeri-Siavashani et al., fibroin-based scaffold containing keratin and vanillin particles acts as a potential antibacterial agent in skin tissue engineering [170].

2.4.10. Keratin

Keratin is a fibrous protein rich in cysteine and is widely present in hair, nails, wool, feathers, and horns [171]. The flexible transverse bonds in the keratin molecular chain provide suitable mechanical properties to its fibrous protein structure [172]. Keratin is insoluble, highly durable, chemically unreactive, and has binding factors that help cell adhesion and growth [173]. Keratin-based scaffolds are widely used in skin, bone, and nerve regeneration [100]. Wan et al. developed a biocomposite mat that constitutes poly (ε-caprolactone), keratin, heparin, and vascular endothelial growth factor, which acts as a well-suited scaffold in vascular tissue engineering [174]. The molecular structure of some protein molecules is shown in Figure 7.

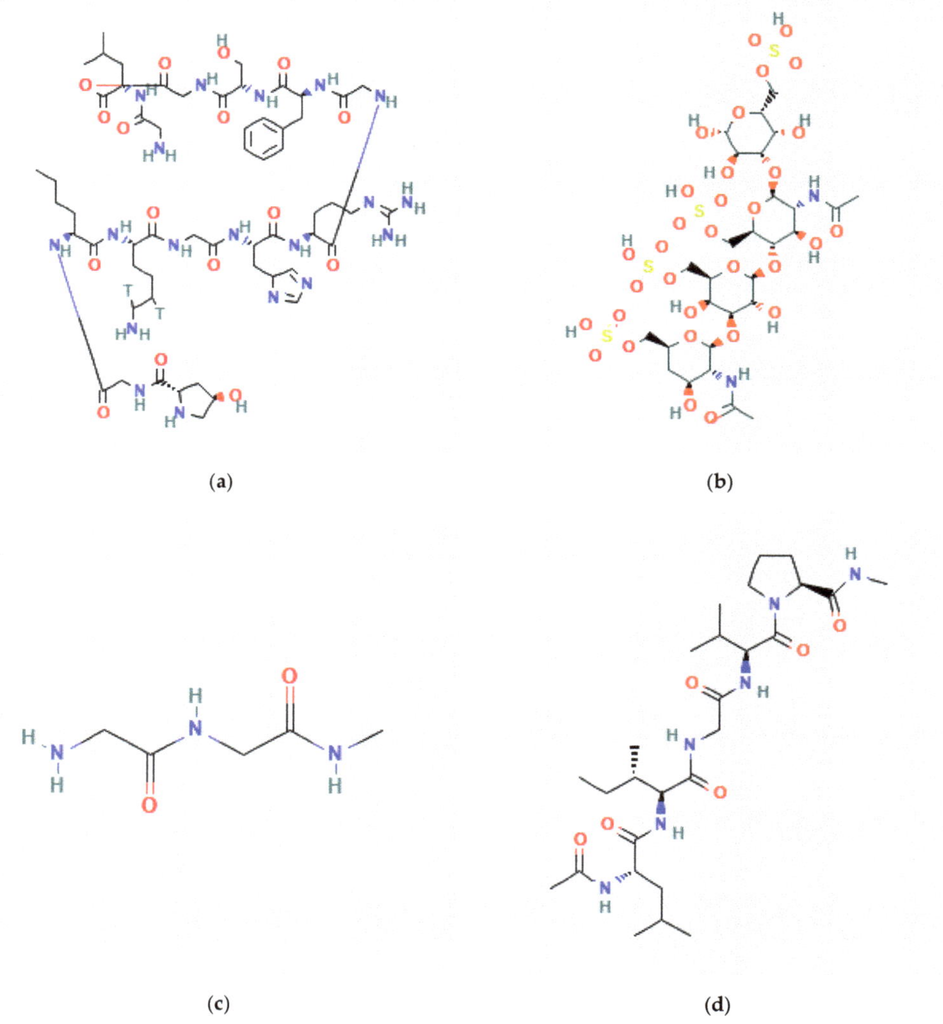

Figure 7. Molecular structure of some protein molecules: (**a**) collagen I [175]; (**b**) keratin [176]; (**c**) fibrin [177]; (**d**) elastin [178].

2.4.11. Elastin

Elastin is a structural protein with elastic properties widely found in connective tissue and other load-bearing tissues. Elastin is in the collagen network in many organs, including the lungs, skin, and blood vessels [179]. In vascular tissue engineering, the successful development of elastin-based vascular graft materials helps to facilitate arterial regeneration and helps to understand the macrophage-mediated immune response created after implantation [180]. Rodrigues I. C. P. et al. stated that adding elastin and collagen to his polyurethane-based scaffold improves cellular response and wettability [181]. Matriderm and glyaderm are some dermal substitutes used in wound healing made up of elastin combined with collagen, whereas matriderm constitutes bovine collagen [60].

2.4.12. Fibrin

Fibrin is a protein molecule formed during blood clotting by polymerizing thrombin and fibrinogen in blood plasma. Easy fabrication, rapid biodegradability, and good biocompatibility are some properties that make fibrin used in tissue regeneration applications. It is mainly used in nerve tissue engineering, skin tissue engineering, musculoskeletal tissue engineering, and cardiac tissue engineering [182,183]. According to Bluteau et al., the low thrombin concentration increased the rate of osteoblastic marker expression. It brought out the increased angiogenic response of osteoblasts by vascular endothelial growth factor (VEGF) expression. Thus, fibrin also helps in bone tissue engineering [184].

2.4.13. Gelatin

Gelatin is a protein molecule obtained by the hydrolysis of collagen, and it constitutes the Arg–Gly–Asp (RGD) peptide sequence, which helps in cell adhesion, proliferation, and differentiation [185]. The primary source of gelatin production is extracted from mammals, especially bovine hides and porcine skin [186]. Scaffold coated with gelatin inhibits complement system and opsonization. Thus, it reduces their immunogenicity [187]. In vitro studies show that scaffolds based on gelatin can control cell differentiation and gene expression [188]. Dehghan M. et al. combined gelatin, polycaprolactone, and polydimethylsiloxane to produce a scaffold, and further investigations on tests regarding biocompatibility, biodegradability, and mechanical properties gave a positive result [9].

Singh S. et al. used gelatin as a fabricating material for a cellulose-based scaffold produced from cotton to improve cell adhesion [189]. Goudarzi Z. M. et al. concluded that a poly (ε-caprolactone) and gelatin composite scaffold incorporated with acetylated cellulose nanofiber is an ideal scaffold for soft tissue engineering [190]. The list of advantages, disadvantages, and applications of each protein is mentioned in Table 4.

Table 4. Advantages, disadvantages, and applications of proteins in tissue regeneration applications.

Protein and References	Advantages	Disadvantages	Applications
Collagen [46,150,191–195]	Bioactive, biocompatible, biodegradable, poorly immunogenic, and mimics ECM	Poor mechanical properties	Bone, skin, dental, cornea, vascular, and cartilage regeneration
Fibroin [196–201]	Bioactivity, biocompatibility, biodegradability, low immunogenic, good mechanical properties, high tensile strength, excellent structural integrity, water-based processing, and cheap	Weak and brittle as a scaffold	Bone, skin, vascular, cartilage, tendon, hepatic, cornea, and neural regeneration, and musculoskeletal tissue engineering

Table 4. *Cont.*

Protein and References	Advantages	Disadvantages	Applications
Keratin [100,171,174]	Biocompatibility, biodegradability, mechanical durability	Poor mechanical properties and brittle	Skin, bone, and nerve regeneration, urinary tract and vascular tissue engineering
Elastin [202–208]	Bioactivity, biocompatibility, good biomechanical and biophysical properties	Difficult in sourcing, water-insoluble, difficult to manipulate in vitro, risk of contamination and inflammation	Cartilage, skin, tendon, and cardiovascular regeneration
Gelatin [209–214]	Bioactive, biocompatible, biodegradable, ECM mimicking, low immunogenic, water-soluble, and cheap	Poor mechanical properties, low solubility in concentrated aqueous media, and speed enzymatic degradation,	Bone, skin, cartilage, adipose, and neural regeneration
Fibrin [215–219]	Biocompatible, biodegradable, low immunogenic, and mimics ECM	Risk of contamination, expensive, poor mechanical properties, and rapid degradation rate	Cartilage, liver, retina, vascular, and neural regeneration

Figures 8 and 9 depict the number of publications on polysaccharides and proteins in tissue engineering applications. It can be seen from both figures that there is an apparent increase in terms of publications in research involving the usage of polysaccharides and proteins as a natural ingredient in developing suitable scaffolds for tissue engineering applications

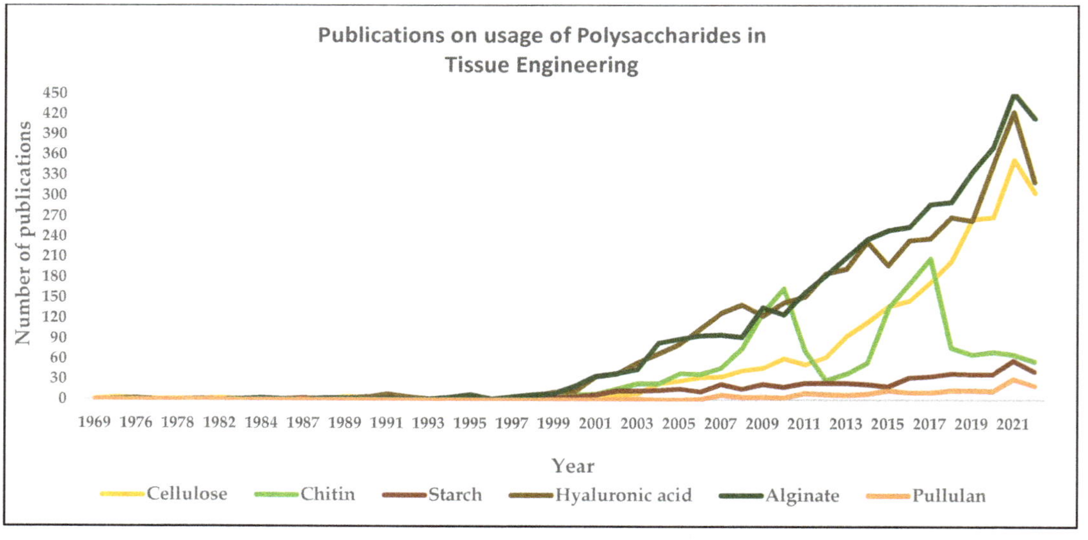

Figure 8. Publications on the usage of polysaccharides in tissue engineering. (The number of papers published on individual polysaccharides such as cellulose, chitin, alginate, starch, hyaluronic acid, and pullulan is drawn based on the year and the respective total number of papers published, using search engine: www.scopus.com, accessed on 15 January 2023.)

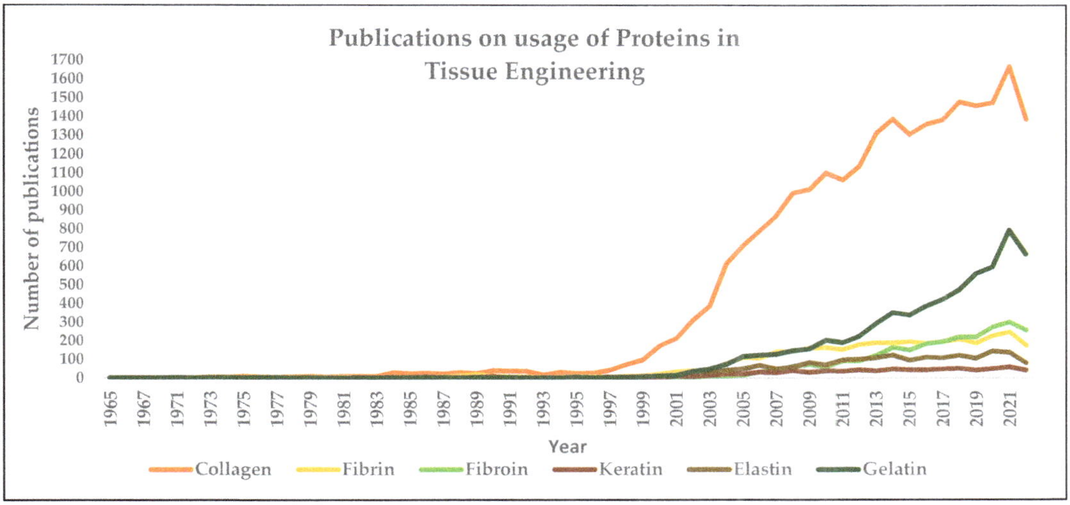

Figure 9. Publications on the usage of proteins in tissue engineering. (The number of papers published on individual proteins such as collagen, fibrin, fibroin, keratin, elastin, and gelatin is drawn based on the year and the respective total number of papers published, using search engine: www.scopus.com, accessed on 15 January 2023).

2.5. Scaffold Fabrication Techniques

Usually, the tissue comprises repeated 3D units such as islets that act as a base for coordinating multicellular processes, maintaining mechanical properties, and integrating various organs through the circulation process. Hence, while designing the scaffold for tissue repair, we must remember that tissue substitutes should have desired mechanical properties and facilities for transporting nutrients and wastes [2]. Fabrication techniques are needed to create a proper scaffold with good mechanical properties, interconnected pores, 3D porous structure, and uniform distribution [220]. The scaffold architectural design is characterized into three levels (nano, micro, and macro) to maintain scaffold parameters such as anatomical features, cell–matrix interactions, and nutritional transportation. The nano-level architecture includes surface modification, including attachment of signaling molecules for cell adhesion, proliferation, and differentiation. Micro-level architecture constitutes pore size, porosity, interconnected pores, and spatial arrangements. The anatomical features and organ and patient specificity include macro-level architecture [221].

Fabrication techniques are classified into two categories: conventional and rapid prototyping. Techniques such as freeze drying, solvent casting, particle leaching, electrospinning, gas foaming, and thermal-induced phase separation come under conventional fabrication techniques. These techniques are suitable for constructing porous scaffolds, but the main limitation is the lack of tunable properties to control shapes and internal architecture. In other words, achieving complex micro- and macro-level architecture is difficult in conventional fabrication techniques. Rapid prototyping is developed to overcome the drawbacks caused by conventional fabrication techniques. Rapid prototyping is known as solid free-form fabrication (SFF) and additive manufacturing (AM). It is the fastest fabrication method for assembling the desired item by using computer generation tools such as computer-aided design (CAD), magnetic resonance imaging (MRI), and computer tomography (CT). Nearly 30 rapid prototyping technologies were applied in various fields, of which 20 were used for biomedical applications [222]. Stereolithography, bioprinting, selective laser sintering, solvent-based extrusion-free forming, and fused deposition modeling are standard rapid prototyping methods used in tissue engineering for scaffold fabrication.

Usually, the primary protocol includes forming and slicing a virtual computer model ensured by layer-by-layer fabrication steps that are similar in all the various rapid prototyping techniques. Initially, a CAD model is captured or formulated from a physical unit by digital method, and then the obtained model is converted into a stereolithography file for virtual slicing. Further, it allows for digital slicing to gain cross-sectional layers. This process is termed pre-processing. Then, rapid prototyping starts to print the layer of the prototype. The post-processing steps, including surface treatment and hardening, are applied. It entirely depends on the purpose and manufacturing techniques. The desired complex micro- and macro-level architecture can also be achieved by using rapid prototyping [223].

2.5.1. Freeze Drying

The freeze-drying technique is otherwise known as lyophilization or ice templating. This technique includes three steps: dissolution, solidification or freezing, and sublimation. At first, the chosen polymer is dissolved in a solvent. Secondly, the solution is loaded into a mold and placed in the freezer for solidification or freezing. It is then allowed to cool down using chemicals such as dry ice in aqueous methanol, liquid nitrogen, or mechanical refrigeration. Care should be taken at this step to maintain temperature, or else it will result in the formation of large crystals, which may affect the properties of the scaffold later. Thirdly, the sublimation process is carried out to remove water and other solvent molecules in the frozen component. This technique is highly suitable for producing scaffolds with high porosity, which provides vascularization and helps in cell proliferation and differentiation. The lyophilization method can be combined with salt leaching, gas foaming, gel casting, and liquid dispensing practices to improve the scaffold's properties. No involvement of heat is the primary advantage of this method, so heat-sensitive molecules such as proteins or growth factors can be incorporated into it without hesitation. However, it consumes a longer time and high energy, and the cost of a freeze dryer is expensive, which are some of the drawbacks [224]. C. M. Brougham et al. developed a heart valve-shaped tissue engineering scaffold using collagen and glycosaminoglycan copolymer and fabricated it using the freeze-drying method [225]. During electro-spinning, toxic substances from organic solvents may involve scaffold preparation. Moreover, it can cause damage to the biological activity of cells. To avoid this situation, A. Izadyari Aghmiuni et al. combined freeze-drying and electro-spinning methods to develop a scaffold for tissue engineering [226].

2.5.2. Solvent Casting and Particle Leaching

3D specimens with thin walls or membranes were produced using solvent casting and particle leaching methods. These thin membranes are prepared by adding salt particles to the solvent polymeric solution. Then, the solvent is allowed to evaporate, and the resulting membrane is washed with distilled water to leach out the salt. The main advantages of solvent casting and particle leaching methods are high porosity, cheapness, and straightforwardness. This technique's usage of toxic solvents, poor interconnectivity, and irregularly shaped pores are limitations [227,228]. N. Thadavirul et al. developed a polycaprolactone porous scaffold using solvent casting and particle leaching techniques for bone tissue engineering [228]. To enhance the mechanical properties, researchers incorporate hydroxyapatite into blends of the biodegradable polymer [229].

2.5.3. Gas Foaming

The gas foaming technique was introduced to avoid using organic cytotoxic solvents and high temperatures. However, the resultant material obtained had closed pores, which limited its usage, especially in cell transplantation. In this method, chosen polymer was mixed with salt particles and molded to form solid disks. Then, disks were exposed to inert gas foaming agents such as nitrogen gas or carbon dioxide with high pressure for saturation. Then, gas was decreased to ambient pressure to create thermodynamic instability, resulting in nucleation and facilitating carbon dioxide pores between polymer matrices. Finally, the salt was removed by leaching the polymer using distilled water [2,230].

2.5.4. Electrospinning

It is a simple technique in which solutions produce fibers by applying high-voltage electricity. The main principle behind this technique is the interaction between electrostatic repulsion and surface tension of charging liquid that receives high voltage droplets. This machine consists of four major parts: a power supply unit, a syringe pump, a metallic needle, and a grounded collector [2,231], as shown in Figure 10. Usually, this technique is widely used for producing nano-fibrous scaffolds. The liquid is injected into the capillary tube of the syringe pump. The muscle power of the electric field from a high-voltage power supply increases the surface tension of liquid extruding from the nozzle of the metallic needle.

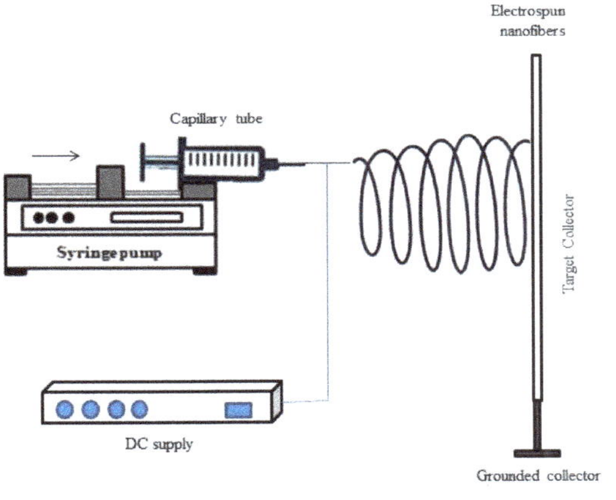

Figure 10. Simplified diagram of an electrospinning device, which consists of four main components such as power supply, syringe pump, metallic needle, and grounded collector. Reprinted with permission from Ref. [2]. Copyright 2019 Abdalla Eltom et al.

Further, the liquid jet is continuously whipped due to electrostatic repulsion and is collected in the form of fibers in the grounded collector. Electrospinning techniques help produce scaffolds with good porosity, patterned architecture, and aligned fibers, which further help cellular response and enhance tissue regeneration. Precise control over fiber formation, homogeneous cell distribution, and lack of cellular infiltration are drawbacks of the electrospinning method [231]. Cellulose nano fiber (CNF) scaffolds developed using potato peel waste promote the adhesion and proliferation of BALB-3T3 fibroblasts cells [232].

2.5.5. Thermal-Induced Phase Separation Method

This method is widely used to fabricate microcellular foams or microporous membranes. This technique de-mixes the homogenous polymer solution into polymer-rich and poor phases by applying variant temperatures. Further, lyophilization of phase-separated polymer solution helps produce microcellular structure [233] Adjustment of pore size can be practically made possible in this method by allowing drugs and fillers. Moreover, these particles are also homogeneously distributed within the pore size. Inadequate resolution and usage of limited materials for fabrications are the main drawbacks of this method. The phase separation technique plays a vital role in fabricating a 3D nanofibrous scaffold, and it can be highly recommended to use along with another fabricating technique such as solid free form [2]. The advantages and disadvantages of various fabrication techniques are mentioned in Table 5.

Table 5. Advantages and disadvantages of various fabrication techniques.

S.No	Techniques	Advantages	Disadvantages
1.	Freeze drying	Capability to do away with high temperatures, applicable in a variety of purposes, and pore size can be manageable to be controlled by changing the freeze-drying method	Long time consumption, high energy consumption, usage of cytotoxic solvents, and irregular pore size
2.	Solvent casting and practical leaching	Cheap and high porosity	Usage of toxic solvents
3.	Gas foaming	Absence of caustic solvents	Poor interconnectivity, low reproducibility, and structural uniformity
4.	Electrospinning	Porosity, control over morphology, and usage of simple equipment	Limited control of pore structure, use of toxic solvents, and many variables involved in the process
5.	Thermal-induced phase separation	Fast, controllable, scalable, and formation of intrinsically interconnected pores	Only used for thermoplastic
6.	Stereolithography	High resolution, fast processing, and smoother surface	Expensive, high temperature, and toxic uncured resin
7.	Selective laser sintering	Fast processing, high resolution, and no support is needed during manufacturing	High temperature, rough surface finish
8.	Fused deposition model	No requirement for solvents and good mechanical properties	Filament requirement and high temperature
9.	Solvent-based extrusion 3D printing method	Applicable to precise control of micron-level scaffold structure, suitable for ceramic and metals too	Temperature extrusion
10.	Bioprinting method	Cheap and higher accuracy	Depends on cell existence

2.5.6. Stereolithography

Stereolithography is considered the first rapid prototype technique commercially available in the fabrication process—an aqueous photo-curable polymer was used as a raw material. An ultraviolet laser beam was used as a light source to irradiate the material surface for solidification where the untreated region remains liquid. Once the solidification of one layer is completed, the lifting table starts to move to the next layer. Subsequently, the solidified layer is recoated with new liquid resin. This photo-polymerization process is repeated until the remaining layer is done. This technique's scaffold material has enhanced cell growth and adhesion. High resolution and uniformity in pore interconnectivity are this method's main advantages [234]. The process involved in stereolithography is shown in Figure 11.

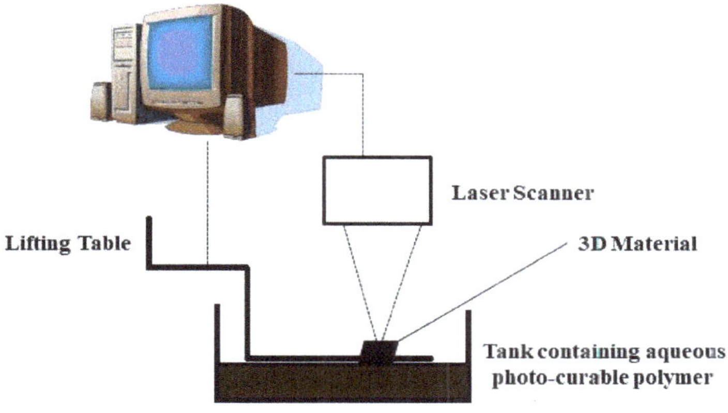

Figure 11. Simplified diagram of stereolithography, which consists of a tank, lifting table, laser scanner, and a computer.

2.5.7. Selective Laser Sintering

It is an additive manufacturing technique in which a high-intensity laser beam fabricates a scaffold layer-wise using computer-aided design models. Usually, materials are used in powder, and this technique can be applied to produce various materials such as ceramic, polymer, and metals. The laser beam is used to heat powder particles to glass transition temperature (near their melting point). The material was sintered to form a solid model directly without permitting the melting phase. Then, the workstation moves down layer by layer. At the same time, fresh powder is spread on the sintered object with the help of a roller, and the process is repeated until the completion of a 3D material. The scaffolds from this method provide excellent compressive strength, fracture toughness, osteoconduction, and osteoinduction. However, the high operating temperature limits the resolution, and additional procedures such as removing injected powder after processing the phase spin are some drawbacks of this method [2,234,235]. The process involved in selective laser sintering is shown in Figure 12.

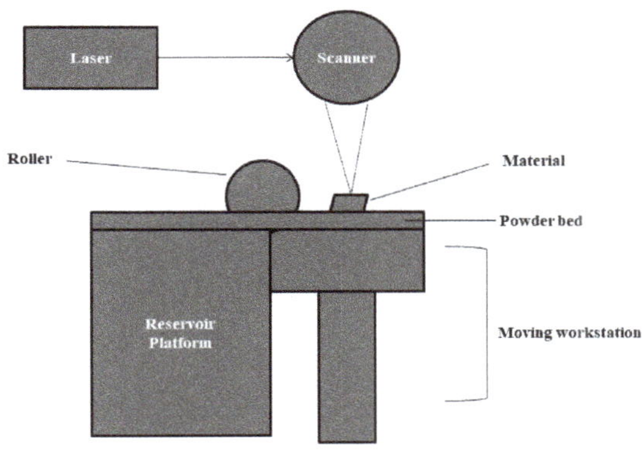

Figure 12. Simplified diagram of selective laser sintering, which consists of a reservoir platform, moving workstation, roller, and a scanner.

2.5.8. Fused Deposition Model

According to Xia et al., the fused deposition model is a filament-based additive manufacturing method. Plastic materials are used in the form of filament. They are inserted into a heating nozzle, where the filament is melted, extruded, and deposited into a plate to produce a 3D structure, layer-by-layer manner, with the help of computer-based devices. This technique is simple, cheap, versatile, and has wide applications. However, some significant deficiencies are there, too, such as difficulty in microporosity establishment, which results in a lack of cell growth and vascularization. The processing time is too long, and the heating process hinders the integration of biomolecules into the scaffold, resulting in a smooth surface unsuitable for cell adhesion, which needs further coating. Many experiments were carried out to overcome these problems, and some series were developed based on the fusion deposition model. Low-temperature deposition modeling is one of the series created, and it also gave positive responses such as better biocompatibility, biodegradability, and all required properties for bone tissue engineering [2,234,236].

2.5.9. Solvent-based Extrusion 3D Printing Method

The solvent-based extrusion 3D printing method keeps biomaterials in solvents to produce inks. Then, obtained inks are extruded from the nozzle in filament to create a scaffold structure in a layer-wise manner. Natural polymers, synthetic polymers, and ceramics are the biomaterials currently being used to produce ink. This technique was widely applied to fabricate scaffolds for cartilage tissue, bone tissue, blood vessel, heart valve tissue, and skin tissue. Difficulty in obtaining appropriate levels of filament uniformity, lack of ink feasibility, and poor fidelity between the structure of computer models and printed scaffold structures are some disadvantages [237].

2.5.10. Bioprinting Method

Bioprinting technology is a promising fabrication technique to develop highly mimicked tissue with digital control. A typical bioprinting method consists of pre-processing, processing, and post-processing phases. At first, in the pre-processing step, the tissue blueprint is created using computer-aided design (CAD). The vital information regarding histological structure and composition, anatomy, and human organ topology for the design can be extracted using imaging approaches. Moreover, parameters for biomaterials are also finalized during this stage. A suitable bioprinter prints the desired structure in the processing step. The bio-ink used for the bioprinter plays a crucial role in delivering the desired scaffold. Finally, post-processing steps are carried out to maturate the obtained scaffold before host implantation. Using an ideal bioreactor for the scale-up process is also under this category. Computer-aided design (CAD) and computer-aided manufacturing (CAM) are used in all three phases and play a crucial role.

Bio-CAD mimics the 3D internal structure, differentiates heterogeneous tissue types, and creates desired models. Bio-CAM is used to predict the feasibility of the fabrication process. The combination of Bio-CAD and Bio-CAM helps accelerate the bioprinting process and enhance the quality of printed tissues. The biomaterials used in this process should be printable, non-toxic, and biodegradable in vivo. Inkjet bioprinting, extrusion bioprinting, laser-assisted bioprinting, and stereolithography are the widely applied bioprinting approaches. Due to their advantages, low cost, accuracy, and high speed, bioprinting technologies have already marked their footprints in cartilage, skin, aortic valve, bone, vascular, and kidney tissues. Dependence on existing cells is the main drawback of this method [238].

2.5.11. Aerosol Jet Printing

The focused airstream is used as ink instead of liquid droplets in aerosol jet printing. Either organic or inorganic materials can be used for this printing technique. A composite suspension is atomized into an aerosol using an ultrasonic or pneumatic atomizer. Then, it is transported to the deposition head by nitrogen gas, which acts as a carrier gas, and jets onto the substrate to form a 3D structure in a layer-wise manner. Polymer, ceramic, and metals can be used for aerosol jet printing. Scaffolds developed from aerosol jet printing show better cytocompatibility in in vitro studies, and it is a low-temperature process, so it is suitable for biomanufacturing too [239]. Some research on natural polymers used to fabricate scaffolds for various tissue regeneration applications is explained in Table 6.

Table 6. Natural polymers used to fabricate scaffolds for various tissue regeneration applications.

Natural Polymer Used and Reference	Fabrication Technique	Cell Type	Applications
Cellulose and starch [240]	Selective laser sintering	-	Drug delivery and tissue engineering
Alginate [241]	Freeze drying	MG-63 human osteosarcoma cell line	Bone tissue engineering
Chitin and chondroitin sulphate [242]	Freeze drying	Human dermal fibroblast	Skin tissue engineering
Hyaluronic acid and collagen [243]	Electrospinning	Schwann cells	Nerve regeneration
Collagen [244]	Electrospinning followed by cross-linking	H9C2 cell line from embryonic rat heart tissue	Cardiac cell therapy
Chitosan [245]	Stereolithography	Human mesenchymal stem cells	Cartilage tissue engineering
Fibrin and chitosan [246]	Electrospinning	MG-63 cell line	Bone tissue engineering
Silk fibroin [247]	3D printing	Human bronchial epithelial cell line (BEAS-2B)	Tracheal epithelial regeneration
Gelatin [248]	3D printing	MG-63 cell line	Hard tissue regeneration
Chitosan [249]	Thermal-induced phase separation	Mouse bone marrow stromal cells	Bone tissue engineering
Corn starch [250]	Solvent casting and particulate leaching	-	Bone tissue engineering
Keratin [251]	Electrospinning	Human umbilical vein endothelial cells	Vascular tissue engineering
Silk fibroin [252]	Freeze drying	Human umbilical vein endothelial cells	Skin tissue engineering
Elastin [253]	Electrospinning	MG-63 osteosarcoma cell line	Bone tissue engineering

3. Conclusions

Scaffolds based on natural products have gained more importance than synthetic products. The research in developing scaffolds from natural-based biomaterials for tissue regeneration applications is rapidly growing due to their outstanding properties such as promoting cell adhesion, proliferation, migration, biocompatibility, biodegradability, porosity, ease of production, inexpensive, and non-toxic. However, natural-based biomaterials have poor mechanical properties. They can be fabricated with suitable materials and used in various biomedical applications, including tissue engineering. The selection of suitable materials is crucial in tissue engineering. In that way, this paper provides a clear idea about the natural-based materials that are currently used in tissue engineering applications. In addition to that, the applications of fabrication techniques in scaffold development have been illustrated. Each technique has its respective benefits and drawbacks, and, as mentioned, appropriate selection to satisfy the need for the tissue to be repaired plays a vital role.

Author Contributions: M.K. and W.Y.S. performed the review, analysis, and editing of the manuscript; H.S. and N.S.S. performed the structure conceptualization, review, and editing of the manuscript. All authors have read and agreed to the published version of the manuscript.

Funding: This research and the APC were funded by Universiti Brunei Darussalam Research Grant No: UBD/RSCH/1.3/FICBF(b)/2020/005.

Institutional Review Board Statement: Not applicable.

Informed Consent Statement: Not applicable.

Data Availability Statement: No new data were created or analyzed in this study. Data sharing is not applicable to this article.

Conflicts of Interest: The authors declare no conflict of interest.

References

1. Koons, G.L.; Diba, M.; Mikos, A.G. Materials design for bone-tissue engineering. *Nat. Rev. Mater.* **2020**, *5*, 584–603. [CrossRef]
2. Eltom, A.; Zhong, G.; Muhammad, A. Scaffold Techniques and Designs in Tissue Engineering Functions and Purposes: A Review. *Adv. Mater. Sci. Eng.* **2019**, *2019*, 3429527. [CrossRef]
3. Chocholata, P.; Kulda, V.; Babuska, V. Fabrication of Scaffolds for Bone-Tissue Regeneration. *Materials* **2019**, *12*, 568. [CrossRef]
4. Lauritano, D.; Limongelli, L.; Moreo, G.; Favia, G.; Carinci, F. Nanomaterials for Periodontal Tissue Engineering: Chitosan-Based Scaffolds. A Systematic Review. *Nanomaterials* **2020**, *10*, 605. [CrossRef] [PubMed]
5. Tan, H.-L.; Kai, D.; Pasbakhsh, P.; Teow, S.-Y.; Lim, Y.-Y.; Pushpamalar, J. Electrospun cellulose acetate butyrate/polyethylene glycol (CAB/PEG) composite nanofibers: A potential scaffold for tissue engineering. *Colloids Surf. B Biointerfaces* **2020**, *188*, 110713. [CrossRef] [PubMed]
6. Biswal, T. Biopolymers for tissue engineering applications: A review. *Mater. Today: Proc.* **2020**, *41*, 397–402. [CrossRef]
7. Chan, B.P.; Leong, K.W. Scaffolding in tissue engineering: General approaches and tissue-specific considerations. *Eur. Spine J.* **2008**, *17*, 467–479. [CrossRef]
8. Junior, A.L.; Pinheiro, C.C.G.; Fernandes, T.L.; Bueno, D.F. The use of human dental pulp stem cells for in vivo bone tissue engineering: A systematic review. *J. Tissue Eng.* **2018**, *9*, 2041731417752766. [CrossRef]
9. Dehghan, M.; Mehrizi, M.K.; Nikukar, H. Modeling and optimizing a polycaprolactone/gelatin/polydimethylsiloxane nanofiber scaffold for tissue engineering: Using response surface methodology. *J. Text. Inst.* **2021**, *112*, 482–493. [CrossRef]
10. Dutta, R.C.; Dey, M.; Dutta, A.K.; Basu, B. Competent processing techniques for scaffolds in tissue engineering. *Biotechnol. Adv.* **2017**, *35*, 240–250. [CrossRef]
11. Mathew, A.; Augustine, R.; Kalarikal, N.; Thomas, S. Tissue Engineering: Principles, Recent Trends and the Future. In *Nanomedicine and Tissue Engineering*; Taylor & Francis: Abingdon, UK, 2016; pp. 31–82. [CrossRef]
12. Castells-Sala, C.; Alemany-Ribes, M.; Fernández-Muiños, T.; Recha-Sancho, L.; López-Chicón, P.; Aloy-Reverté, C.; Caballero-Camino, J.; Márquez-Gil, A.; Semino, C.E. Current applications of tissue engineering in biomedicine. *J. Biochips Tiss. Chips* **2013**, *S2*, 1. [CrossRef]
13. Dufey, V.; Tacheny, A.; Art, M.; Becken, U.; De Longueville, F. Expansion of human bone marrow-derived mesenchymal stem cells in BioBLU 0.3 c single-use bioreactors. *Appl. Note* **2016**, *305*, 1–8. Available online: https://www.eppendorf.com/fileadmin/Community/Cell_Handling/Bioprocess/PDF/Application-Note_305_BioBLU-0.3c_DASbox_Expansion-of-Human-B.pdf (accessed on 29 December 2022).
14. Ye, L.; Swingen, C.; Zhang, J. Induced Pluripotent Stem Cells and Their Potential for Basic and Clinical Sciences. *Curr. Cardiol. Rev.* **2013**, *9*, 63–72. [CrossRef] [PubMed]
15. Damjanov, I. Inflammation and Repair. In *Pathology Secrets*, 3rd ed.; Damjanov, I., Ed.; Mosby: Maryland Heights, MO, USA, 2009; pp. 19–37.
16. Akter, F. Principles of Tissue Engineering. In *Tissue Engineering Made Easy*; Akter, F., Ed.; Academic Press: London, UK, 2016; pp. 3–16.
17. Alaribe, F.N.; Manoto, S.L.; Motaung, S.C.K.M. Scaffolds from biomaterials: Advantages and limitations in bone and tissue engineering. *Biologia* **2016**, *71*, 353–366. [CrossRef]
18. Perić Kačarević, Ž.; Rider, P.; Alkildani, S.; Retnasingh, S.; Pejakić, M.; Schnettler, R.; Gosau, M.; Smeets, R.; Jung, O.; Barbeck, M. An introduction to bone tissue engineering. *Int. J. Artif. Organs* **2019**, *43*, 69–86. [CrossRef] [PubMed]
19. Roseti, L.; Parisi, V.; Petretta, M.; Cavallo, C.; Desando, G.; Bartolotti, I.; Grigolo, B. Scaffolds for Bone Tissue Engineering: State of the art and new perspectives. *Mater. Sci. Eng. C* **2017**, *78*, 1246–1262. [CrossRef] [PubMed]
20. Verma, P.; Verma, V. Concepts of Tissue Engineering. In *Animal Biotechnology*, 2nd ed.; Academic Press: Cambridge, MA, US, 2020; pp. 295–307.
21. Alonzo, M.; Primo, F.A.; Kumar, S.A.; Mudloff, J.A.; Dominguez, E.; Fregoso, G.; Ortiz, N.; Weiss, W.M.; Joddar, B. Bone tissue engineering techniques, advances, and scaffolds for treatment of bone defects. *Curr. Opin. Biomed. Eng.* **2020**, *17*, 100248. [CrossRef]
22. Roberts, T.T.; Rosenbaum, A.J. Bone grafts, bone substitutes and orthobiologics: Bone grafts, bone substitutes and orthobiologics the bridge between basic science and clinical advancements in fracture healing. *Organogenesis* **2012**, *8*, 114–124. [CrossRef]
23. Wong, Y.S.; Tay, C.Y.; Wen, F.; Venkatraman, S.S.; Tan, L.P. Engineered Polymeric Biomaterials for Tissue Engineering. *Curr. Tissue Eng.* **2012**, *1*, 41–53. [CrossRef]

24. Nikolova, M.P.; Chavali, M.S. Recent advances in biomaterials for 3D scaffolds: A review. *Bioact. Mater.* **2019**, *4*, 271–292. [CrossRef]
25. Gurumurthy, B.; Janorkar, A.V. Improvements in mechanical properties of collagen-based scaffolds for tissue engineering. *Curr. Opin. Biomed. Eng.* **2021**, *17*, 100253. [CrossRef]
26. Pal, S. Mechanical Properties of Biological Materials. In *Design of Artificial Human Joints & Organs*; Springer: New York, NY, USA, 2014; pp. 23–40.
27. Salati, M.A.; Khazai, J.; Tahmuri, A.M.; Samadi, A.; Taghizadeh, A.; Taghizadeh, M.; Zarrintaj, P.; Ramsey, J.D.; Habibzadeh, S.; Seidi, F.; et al. Agarose-Based Biomaterials: Opportunities and Challenges in Cartilage Tissue Engineering. *Polymers* **2020**, *12*, 1150. [CrossRef] [PubMed]
28. Ogneva, I.V.; Lebedev, D.; Shenkman, B.S. Transversal Stiffness and Young's Modulus of Single Fibers from Rat Soleus Muscle Probed by Atomic Force Microscopy. *Biophys. J.* **2010**, *98*, 418–424. [CrossRef] [PubMed]
29. Nguyen, B.V.; Wang, Q.G.; Kuiper, N.J.; El Haj, A.J.; Thomas, C.R.; Zhang, Z. Biomechanical properties of single chondrocytes and chondrons determined by micromanipulation and finite-element modelling. *J. R. Soc. Interface* **2010**, *7*, 1723–1733. [CrossRef]
30. Saha, K.; Keung, A.J.; Irwin, E.F.; Li, Y.; Little, L.; Schaffer, D.V.; Healy, K.E. Substrate Modulus Directs Neural Stem Cell Behavior. *Biophys. J.* **2008**, *95*, 4426–4438. [CrossRef] [PubMed]
31. Chen, E.; Novakofski, J.; Jenkins, W.; O'Brien, W. Young's modulus measurements of soft tissues with application to elasticity imaging. *IEEE Trans. Ultrason. Ferroelectr. Freq. Control.* **1996**, *43*, 191–194. [CrossRef]
32. Spedden, E.; White, J.D.; Naumova, E.N.; Kaplan, D.L.; Staii, C. Elasticity Maps of Living Neurons Measured by Combined Fluorescence and Atomic Force Microscopy. *Biophys. J.* **2012**, *103*, 868–877. [CrossRef]
33. Tran, T.; Hamid, Z.; Cheong, K. A Review of Mechanical Properties of Scaffold in Tissue Engineering: Aloe Vera Composites. *J. Phys. Conf. Ser.* **2018**, *1082*, 012080. [CrossRef]
34. Collinsworth, A.M.; Zhang, S.; Kraus, W.E.; Truskey, G.A. Apparent elastic modulus and hysteresis of skeletal muscle cells throughout differentiation. *Am. J. Physiol. Physiol.* **2002**, *283*, C1219–C1227. [CrossRef]
35. Nguyen-Truong, M.; Li, Y.; Wang, Z. Mechanical Considerations of Electrospun Scaffolds for Myocardial Tissue and Regenerative Engineering. *Bioengineering* **2020**, *7*, 122. [CrossRef]
36. Zhang, X.-Y.; Fang, G.; Zhou, J. Additively Manufactured Scaffolds for Bone Tissue Engineering and the Prediction of their Mechanical Behavior: A Review. *Materials* **2017**, *10*, 50. [CrossRef] [PubMed]
37. Sharma, R.; Kirsch, R.; Valente, K.; Perez, M.; Willerth, S. Physical and Mechanical Characterization of Fibrin-Based Bioprinted Constructs Containing Drug-Releasing Microspheres for Neural Tissue Engineering Applications. *Processes* **2021**, *9*, 1205. [CrossRef]
38. Ezhilarasu, H.; Ramalingam, R.; Dhand, C.; Lakshminarayanan, R.; Sadiq, A.; Gandhimathi, C.; Ramakrishna, S.; Bay, B.H.; Venugopal, J.R.; Srinivasan, D.K. Biocompatible Aloe vera and Tetracycline Hydrochloride Loaded Hybrid Nanofibrous Scaffolds for Skin Tissue Engineering. *Int. J. Mol. Sci.* **2019**, *20*, 5174. [CrossRef]
39. Place, E.S.; Evans, N.D.; Stevens, M.M. Complexity in biomaterials for tissue engineering. *Nat. Mater.* **2009**, *8*, 457–470. [CrossRef]
40. O'Brien, F.J. Biomaterials & scaffolds for tissue engineering. *Mater. Today* **2011**, *14*, 88–95. [CrossRef]
41. Francois, E.L.; Yaszemski, M.J. *Preclinical Bone Repair Models in Regenerative Medicine*; Elsevier: Amsterdam, The Netherlands, 2019.
42. Khan, F.; Tanaka, M. Designing Smart Biomaterials for Tissue Engineering. *Int. J. Mol. Sci.* **2018**, *19*, 17. [CrossRef]
43. Sultana, N.; Hassan, M.I.; Lim, M.M. *Scaffolding Biomaterials*; Springer: Cham, Switzerland, 2015.
44. Asadi, N.; Del Bakhshayesh, A.R.; Davaran, S.; Akbarzadeh, A. Common biocompatible polymeric materials for tissue engineering and regenerative medicine. *Mater. Chem. Phys.* **2020**, *242*, 122528. [CrossRef]
45. Gunathilake, T.M.S.U.; Ching, Y.C.; Ching, K.Y.; Chuah, C.H.; Abdullah, L.C. Biomedical and Microbiological Applications of Bio-Based Porous Materials: A Review. *Polymers* **2017**, *9*, 160. [CrossRef]
46. Del Bakhshayesh, A.R.; Mostafavi, E.; Alizadeh, E.; Asadi, N.; Akbarzadeh, A.; Davaran, S. Fabrication of Three-Dimensional Scaffolds Based on Nano-biomimetic Collagen Hybrid Constructs for Skin Tissue Engineering. *ACS Omega* **2018**, *3*, 8605–8611. [CrossRef]
47. Sabir, A.; Abbas, H.; Amini, A.Y.; Asmal, S. Characterization of duck egg shells and bioceramic materials in making denture applications. *IOP Conf. Ser. Mater. Sci. Eng.* **2021**, *1088*, 012116. [CrossRef]
48. Umeyama, R.; Yamawaki, T.; Liu, D.; Kanazawa, S.; Takato, T.; Hoshi, K.; Hikita, A. Optimization of culture duration of bone marrow cells before transplantation with a β-tricalcium phosphate/recombinant collagen peptide hybrid scaffold. *Regen. Ther.* **2020**, *14*, 284–295. [CrossRef] [PubMed]
49. Massimino, L.C.; Martins, V.D.C.A.; Vulcani, V.A.S.; De Oliveira, É.L.; Andreeta, M.B.; Bonagamba, T.J.; Klingbeil, M.F.G.; Mathor, M.B.; de Guzzi Plepis, A.M. Use of collagen and auricular cartilage in bioengineering: Scaffolds for tissue regeneration. *Cell Tissue Bank.* **2020**, 1–12. [CrossRef] [PubMed]
50. Barua, E.; Deoghare, A.B.; Deb, P.; Das Lala, S. Naturally derived biomaterials for development of composite bone scaffold: A review. *IOP Conf. Ser. Mater. Sci. Eng.* **2018**, *377*, 012013. [CrossRef]
51. Awasthi, S.; Pandey, S.K.; Arunan, E.; Srivastava, C. A review on hydroxyapatite coatings for the biomedical applications: Experimental and theoretical perspectives. *J. Mater. Chem. B* **2021**, *9*, 228–249. [CrossRef]

52. Zouhair, S.; Sasso, E.D.; Tuladhar, S.R.; Fidalgo, C.; Vedovelli, L.; Filippi, A.; Borile, G.; Bagno, A.; Marchesan, M.; De Rossi, G.; et al. A Comprehensive Comparison of Bovine and Porcine Decellularized Pericardia: New Insights for Surgical Applications. *Biomolecules* **2020**, *10*, 371. [CrossRef]
53. Sadat-Shojai, M.; Khorasani, M.-T.; Dinpanah-Khoshdargi, E.; Jamshidi, A. Synthesis methods for nanosized hydroxyapatite with diverse structures. *Acta Biomater.* **2013**, *9*, 7591–7621. [CrossRef]
54. Murugiah, K.; Zakaria, M.I.; Suhaimi, H.; Caesarendra, W.; Sambudi, N.S. Synthesis and Characterisation of Hydroxyapatite (HAp) from Asiatic Hard Clam (*Meretrix meretrix*) and Blood Cockle Clam (*Anadara granosa*) Using Wet Precipitation Process. In Proceedings of the 2021 IEEE National Biomedical Engineering Conference (NBEC), Kuala Lumpur, Malaysia, 9–10 November 2021; pp. 1–6. [CrossRef]
55. Mohd Pu'Ad, N.A.S.; Koshy, P.; Abdullah, H.Z.; Idris, M.I.; Lee, T.C. Syntheses of hydroxyapatite from natural sources. *Heliyon* **2019**, *5*, e01588. [CrossRef]
56. Karacan, I.; Ben-Nissan, B.; Sinutok, S. Marine-Based Calcium Phosphates from Hard Coral and Calcified Algae for Biomedical Applications. In *Marine-Derived Biomaterials for Tissue Engineering Applications*; Springer: Singapore, 2019; pp. 137–153.
57. Carluccio, D.; Demir, A.G.; Bermingham, M.J.; Dargusch, M.S. Challenges and Opportunities in the Selective Laser Melting of Biodegradable Metals for Load-Bearing Bone Scaffold Applications. *Met. Mater. Trans. A* **2020**, *51*, 3311–3334. [CrossRef]
58. Xie, Y.; Zhao, L.; Zhang, Z.; Wang, X.; Wang, R.; Cui, C. Fabrication and properties of porous Zn-Ag alloy scaffolds as biodegradable materials. *Mater. Chem. Phys.* **2018**, *219*, 433–443. [CrossRef]
59. Ku, S.H.; Lee, M.; Park, C.B. Carbon-Based Nanomaterials for Tissue Engineering. *Adv. Healthc. Mater.* **2013**, *2*, 244–260. [CrossRef]
60. Eivazzadeh-Keihan, R.; Maleki, A.; de la Guardia, M.; Bani, M.S.; Chenab, K.K.; Pashazadeh-Panahi, P.; Baradaran, B.; Mokhtarzadeh, A.; Hamblin, M.R. Carbon based nanomaterials for tissue engineering of bone: Building new bone on small black scaffolds: A review. *J. Adv. Res.* **2019**, *18*, 185–201. [CrossRef]
61. Girotti, A.; Gonzalez-Valdivieso, J.; Santos, M.; Martin, L.; Arias, F.J. Functional characterization of an enzymatically degradable multi-bioactive elastin-like recombinamer. *Int. J. Biol. Macromol.* **2020**, *164*, 1640–1648. [CrossRef]
62. Wen, Q.; Mithieux, S.M.; Weiss, A.S. Elastin Biomaterials in Dermal Repair. *Trends Biotechnol.* **2020**, *38*, 280–291. [CrossRef]
63. Gurumurthy, B.; Pal, P.; Griggs, J.A.; Janorkar, A.V. Optimization of collagen-elastin-like polypeptide-bioglass scaffold composition for osteogenic differentiation of adipose-derived stem cells. *Materialia* **2020**, *9*, 100572. [CrossRef]
64. Martín-Del-Campo, M.; Fernández-Villa, D.; Cabrera-Rueda, G.; Rojo, L. Antibacterial Bio-Based Polymers for Cranio-Maxillofacial Regeneration Applications. *Appl. Sci.* **2020**, *10*, 8371. [CrossRef]
65. Bedian, L.; Villalba-Rodríguez, A.M.; Hernández-Vargas, G.; Parra-Saldivar, R.; Iqbal, H.M. Bio-based materials with novel characteristics for tissue engineering applications—A review. *Int. J. Biol. Macromol.* **2017**, *98*, 837–846. [CrossRef]
66. Diekjürgen, D.; Grainger, D.W. Polysaccharide matrices used in 3D in vitro cell culture systems. *Biomaterials* **2017**, *141*, 96–115. [CrossRef] [PubMed]
67. Ehrlich, H.; Martinović, R.; Joksimović, D.; Petrenko, I.; Schiaparelli, S.; Wysokowski, M.; Tsurkan, D.; Stelling, A.L.; Springer, A.; Gelinsky, M.; et al. Conchixes: Organic scaffolds which resemble the size and shapes of mollusks shells, their isolation and potential multifunctional applications. *Appl. Phys. A* **2020**, *126*, 562. [CrossRef]
68. Liu, W.; Du, H.; Zhang, M.; Liu, K.; Liu, H.; Xie, H.; Zhang, X.; Si, C. Bacterial Cellulose-Based Composite Scaffolds for Biomedical Applications: A Review. *ACS Sustain. Chem. Eng.* **2020**, *8*, 7536–7562. [CrossRef]
69. Aki, D.; Ulag, S.; Unal, S.; Sengor, M.; Ekren, N.; Lin, C.-C.; Yılmazer, H.; Ustundag, C.B.; Kalaskar, D.M.; Gunduz, O. 3D printing of PVA/hexagonal boron nitride/bacterial cellulose composite scaffolds for bone tissue engineering. *Mater. Des.* **2020**, *196*, 109094. [CrossRef]
70. Namkaew, J.; Laowpanitchakorn, P.; Sawaddee, N.; Jirajessada, S.; Honsawek, S.; Yodmuang, S. Carboxymethyl Cellulose Entrapped in a Poly(vinyl) Alcohol Network: Plant-Based Scaffolds for Cartilage Tissue Engineering. *Molecules* **2021**, *26*, 578. [CrossRef]
71. Chen, P.-H.; Liao, H.-C.; Hsu, S.-H.; Chen, R.-S.; Wu, M.-C.; Yang, Y.-F.; Wu, C.-C.; Chen, M.-H.; Su, W.-F. A novel polyurethane/cellulose fibrous scaffold for cardiac tissue engineering. *RSC Adv.* **2015**, *5*, 6932–6939. [CrossRef]
72. Zha, F.; Chen, W.; Hao, L.; Wu, C.; Lu, M.; Zhang, L.; Yu, D. Electrospun cellulose-based conductive polymer nanofibrous mats: Composite scaffolds and their influence on cell behavior with electrical stimulation for nerve tissue engineering. *Soft Matter* **2020**, *16*, 6591–6598. [CrossRef]
73. Madub, K.; Goonoo, N.; Gimié, F.; Arsa, I.A.; Schönherr, H.; Bhaw-Luximon, A. Green seaweeds ulvan-cellulose scaffolds enhance in vitro cell growth and in vivo angiogenesis for skin tissue engineering. *Carbohydr. Polym.* **2021**, *251*, 117025. [CrossRef]
74. Maharjan, B.; Park, J.; Kaliannagounder, V.K.; Awasthi, G.P.; Joshi, M.K.; Park, C.H.; Kim, C.S. Regenerated cellulose nanofiber reinforced chitosan hydrogel scaffolds for bone tissue engineering. *Carbohydr. Polym.* **2021**, *251*, 117023. [CrossRef]
75. Pezeshki-Modaress, M.; Zandi, M.; Rajabi, S. Tailoring the gelatin/chitosan electrospun scaffold for application in skin tissue engineering: An in vitro study. *Prog. Biomater.* **2018**, *7*, 207–218. [CrossRef]
76. Saderi, N.; Rajabi, M.; Akbari, B.; Firouzi, M.; Hassannejad, Z. Fabrication and characterization of gold nanoparticle-doped electrospun PCL/chitosan nanofibrous scaffolds for nerve tissue engineering. *J. Mater. Sci. Mater. Med.* **2018**, *29*, 134. [CrossRef]
77. Sadeghianmaryan, A.; Naghieh, S.; Sardroud, H.A.; Yazdanpanah, Z.; Soltani, Y.A.; Sernaglia, J.; Chen, X. Extrusion-based printing of chitosan scaffolds and their in vitro characterization for cartilage tissue engineering. *Int. J. Biol. Macromol.* **2020**, *164*, 3179–3192. [CrossRef]

78. Ahmadi, P.; Nazeri, N.; Derakhshan, M.A.; Ghanbari, H. Preparation and characterization of polyurethane/chitosan/CNT nanofibrous scaffold for cardiac tissue engineering. *Int. J. Biol. Macromol.* **2021**, *180*, 590–598. [CrossRef]
79. Ghahremanzadeh, F.; Alihosseini, F.; Semnani, D. Investigation and comparison of new galactosylation methods on PCL/chitosan scaffolds for enhanced liver tissue engineering. *Int. J. Biol. Macromol.* **2021**, *174*, 278–288. [CrossRef]
80. Dutta, S.D.; Hexiu, J.; Patel, D.K.; Ganguly, K.; Lim, K.-T. 3D-printed bioactive and biodegradable hydrogel scaffolds of alginate/gelatin/cellulose nanocrystals for tissue engineering. *Int. J. Biol. Macromol.* **2020**, *167*, 644–658. [CrossRef]
81. Jadbabaei, S.; Kolahdoozan, M.; Naeimi, F.; Ebadi-Dehaghani, H. Preparation and characterization of sodium alginate–PVA polymeric scaffolds by electrospinning method for skin tissue engineering applications. *RSC Adv.* **2021**, *11*, 30674–30688. [CrossRef]
82. Shirehjini, L.M.; Sharifi, F.; Shojaei, S.; Irani, S. Poly-caprolactone nanofibrous coated with sol-gel alginate/ mesenchymal stem cells for cartilage tissue engineering. *J. Drug Deliv. Sci. Technol.* **2022**, *74*, 103488. [CrossRef]
83. Ghaderinejad, P.; Najmoddin, N.; Bagher, Z.; Saeed, M.; Karimi, S.; Simorgh, S.; Pezeshki-Modaress, M. An injectable anisotropic alginate hydrogel containing oriented fibers for nerve tissue engineering. *Chem. Eng. J.* **2021**, *420*, 130465. [CrossRef]
84. Ahmadi, M.; Mehdikhani, M.; Varshosaz, J.; Farsaei, S.; Torabi, H. Pharmaceutical evaluation of atorvastatin-loaded nanostructured lipid carriers incorporated into the gelatin/hyaluronic acid/polycaprolactone scaffold for the skin tissue engineering. *J. Biomater. Appl.* **2021**, *35*, 958–977. [CrossRef] [PubMed]
85. Entekhabi, E.; Nazarpak, M.H.; Moztarzadeh, F.; Sadeghi, A. Design and manufacture of neural tissue engineering scaffolds using hyaluronic acid and polycaprolactone nanofibers with controlled porosity. *Mater. Sci. Eng. C* **2016**, *69*, 380–387. [CrossRef]
86. Mirab, F.; Eslamian, M.; Bagheri, R. Fabrication and characterization of a starch-based nanocomposite scaffold with highly porous and gradient structure for bone tissue engineering. *Biomed. Phys. Eng. Express* **2018**, *4*, 055021. [CrossRef]
87. Yang, X.; Lu, Z.; Wu, H.; Li, W.; Zheng, L.; Zhao, J. Collagen-alginate as bioink for three-dimensional (3D) cell printing based cartilage tissue engineering. *Mater. Sci. Eng. C* **2018**, *83*, 195–201. [CrossRef]
88. Goodarzi, H.; Jadidi, K.; Pourmotabed, S.; Sharifi, E.; Aghamollaei, H. Preparation and in vitro characterization of cross-linked collagen–gelatin hydrogel using EDC/NHS for corneal tissue engineering applications. *Int. J. Biol. Macromol.* **2019**, *126*, 620–632. [CrossRef] [PubMed]
89. Zhang, W.; Wang, X.-C.; Li, X.-Y.; Zhang, L.-L.; Jiang, F. A 3D porous microsphere with multistage structure and component based on bacterial cellulose and collagen for bone tissue engineering. *Carbohydr. Polym.* **2020**, *236*, 116043. [CrossRef]
90. Vázquez, J.J.; Martínez, E.S.M. Collagen and elastin scaffold by electrospinning for skin tissue engineering applications. *J. Mater. Res.* **2019**, *34*, 2819–2827. [CrossRef]
91. Tabatabaei, F.; Moharamzadeh, K.; Tayebi, L. Fibroblast encapsulation in gelatin methacryloyl (GelMA) versus collagen hydrogel as substrates for oral mucosa tissue engineering. *J. Oral Biol. Craniofacial Res.* **2020**, *10*, 573–577. [CrossRef] [PubMed]
92. Wei, L.; Wu, S.; Kuss, M.; Jiang, X.; Sun, R.; Reid, P.; Qin, X.; Duan, B. 3D printing of silk fibroin-based hybrid scaffold treated with platelet rich plasma for bone tissue engineering. *Bioact. Mater.* **2019**, *4*, 256–260. [CrossRef] [PubMed]
93. Peifen, M.; Mengyun, L.; Jinglong, H.; Danqian, L.; Yan, T.; Liwei, X.; Han, Z.; Jianlong, D.; Lingyan, L.; Guanghui, Z.; et al. New skin tissue engineering scaffold with sulfated silk fibroin/chitosan/hydroxyapatite and its application. *Biochem. Biophys. Res. Commun.* **2023**, *640*, 117–124. [CrossRef]
94. Hong, H.; Seo, Y.B.; Kim, D.Y.; Lee, J.S.; Lee, Y.J.; Lee, H.; Ajiteru, O.; Sultan, T.; Lee, O.J.; Kim, S.H.; et al. Digital light processing 3D printed silk fibroin hydrogel for cartilage tissue engineering. *Biomaterials* **2020**, *232*, 119679. [CrossRef]
95. Forouzideh, N.; Nadri, S.; Fattahi, A.; Abdolahinia, E.D.; Habibizadeh, M.; Rostamizadeh, K.; Baradaran-Rafii, A.; Bakhshandeh, H. Epigallocatechin gallate loaded electrospun silk fibroin scaffold with anti-angiogenic properties for corneal tissue engineering. *J. Drug Deliv. Sci. Technol.* **2020**, *56*, 101498. [CrossRef]
96. Li, T.; Song, X.; Weng, C.; Wang, X.; Wu, J.; Sun, L.; Gong, X.; Zeng, W.-N.; Yang, L.; Chen, C. Enzymatically crosslinked and mechanically tunable silk fibroin/pullulan hydrogels for mesenchymal stem cells delivery. *Int. J. Biol. Macromol.* **2018**, *115*, 300–307. [CrossRef]
97. Wang, Y.; Yu, H.; Liu, H.; Fan, Y. Double coating of graphene oxide–polypyrrole on silk fibroin scaffolds for neural tissue engineering. *J. Bioact. Compat. Polym.* **2020**, *35*, 216–227. [CrossRef]
98. Sarrami, P.; Karbasi, S.; Farahbakhsh, Z.; Bigham, A.; Rafienia, M. Fabrication and characterization of novel polyhydroxybutyrate-keratin/nanohydroxyapatite electrospun fibers for bone tissue engineering applications. *Int J. Biol. Macromol.* **2022**, *220*, 1368–1389. [CrossRef]
99. Carvalho, C.R.; Costa, J.B.; Costa, L.; Silva-Correia, J.; Moay, Z.K.; Ng, K.W.; Reis, R.L.; Oliveira, J.M. Enhanced performance of chitosan/keratin membranes with potential application in peripheral nerve repair. *Biomater. Sci.* **2019**, *7*, 5451–5466. [CrossRef]
100. Dou, J.; Wang, Y.; Jin, X.; Li, P.; Wang, L.; Yuan, J.; Shen, J. PCL/sulfonated keratin mats for vascular tissue engineering scaffold with potential of catalytic nitric oxide generation. *Mater. Sci. Eng. C* **2020**, *107*, 110246. [CrossRef]
101. Ye, J.-P.; Gong, J.-S.; Su, C.; Liu, Y.-G.; Jiang, M.; Pan, H.; Li, R.-Y.; Geng, Y.; Xu, Z.-H.; Shi, J.-S. Fabrication and characterization of high molecular keratin based nanofibrous membranes for wound healing. *Colloids Surf. B Biointerfaces* **2020**, *194*, 111158. [CrossRef]
102. Lv, X.; Li, Z.; Chen, S.; Xie, M.; Huang, J.; Peng, X.; Yang, R.; Wang, H.; Xu, Y.; Feng, C. Structural and functional evaluation of oxygenating keratin/silk fibroin scaffold and initial assessment of their potential for urethral tissue engineering. *Biomaterials* **2016**, *84*, 99–110. [CrossRef]

103. Rosellini, E.; Madeddu, D.; Barbani, N.; Frati, C.; Graiani, G.; Falco, A.; Lagrasta, C.; Quaini, F.; Cascone, M.G. Development of Biomimetic Alginate/Gelatin/Elastin Sponges with Recognition Properties toward Bioactive Peptides for Cardiac Tissue Engineering. *Biomimetics* **2020**, *5*, 67. [CrossRef]
104. Kazemi, T.; Mohammadpour, A.A.; Matin, M.M.; Mahdavi-Shahri, N.; Dehghani, H.; Riabi, S.H.K. Decellularized bovine aorta as a promising 3D elastin scaffold for vascular tissue engineering applications. *Regen. Med.* **2021**, *16*, 1037–1050. [CrossRef]
105. Kuo, Y.-C.; Ku, H.-F.; Rajesh, R. Chitosan/γ-poly(glutamic acid) scaffolds with surface-modified albumin, elastin and poly- l -lysine for cartilage tissue engineering. *Mater. Sci. Eng. C* **2017**, *78*, 265–277. [CrossRef] [PubMed]
106. Liu, X.; Smith, L.A.; Hu, J.; Ma, P.X. Biomimetic nanofibrous gelatin/apatite composite scaffolds for bone tissue engineering. *Biomaterials* **2009**, *30*, 2252–2258. [CrossRef]
107. Han, F.; Dong, Y.; Su, Z.; Yin, R.; Song, A.; Li, S. Preparation, characteristics and assessment of a novel gelatin–chitosan sponge scaffold as skin tissue engineering material. *Int. J. Pharm.* **2014**, *476*, 124–133. [CrossRef]
108. Semitela, Â.; Girão, A.F.; Fernandes, C.; Ramalho, G.; Bdikin, I.; Completo, A.; Marques, P.A. Electrospinning of bioactive polycaprolactone-gelatin nanofibres with increased pore size for cartilage tissue engineering applications. *J. Biomater. Appl.* **2020**, *35*, 471–484. [CrossRef] [PubMed]
109. Zeinali, K.; Khorasani, M.T.; Rashidi, A.; Joupari, M.D. Preparation and characterization of graphene oxide aerogel/gelatin as a hybrid scaffold for application in nerve tissue engineering. *Int. J. Polym. Mater. Polym. Biomater.* **2021**, *70*, 674–683. [CrossRef]
110. Bonhome-Espinosa, A.B.; Campos, F.; Durand-Herrera, D.; Sánchez-López, J.D.; Schaub, S.; Durán, J.D.; Lopez-Lopez, M.T.; Carriel, V. In vitro characterization of a novel magnetic fibrin-agarose hydrogel for cartilage tissue engineering. *J. Mech. Behav. Biomed. Mater.* **2020**, *104*, 103619. [CrossRef]
111. Rajalekshmi, R.; Shaji, A.K.; Joseph, R.; Bhatt, A. Scaffold for liver tissue engineering: Exploring the potential of fibrin incorporated alginate dialdehyde–gelatin hydrogel. *Int. J. Biol. Macromol.* **2021**, *166*, 999–1008. [CrossRef] [PubMed]
112. Balagholi, S.; Kanavi, M.R.; Alizadeh, S.; Dabbaghi, R.; Karami, S.; Kheiri, B.; Daftarian, N. Effects of fibrin glue as a three-dimensional scaffold in cultivated adult human retinal pigment epithelial cells. *J. Biomater. Appl.* **2018**, *33*, 514–526. [CrossRef] [PubMed]
113. Hasanzadeh, E.; Ebrahimibarough, S.; Mirzaei, E.; Azami, M.; Tavangar, S.M.; Mahmoodi, N.; Basiri, A.; Ai, J. Preparation of fibrin gel scaffolds containing MWCNT/PU nanofibers for neural tissue engineering. *J. Biomed. Mater. Res. Part A* **2019**, *107*, 802–814. [CrossRef] [PubMed]
114. El Knidri, H.; Belaabed, R.; Addaou, A.; Laajeb, A.; Lahsini, A. Extraction, chemical modification and characterization of chitin and chitosan. *Int. J. Biol. Macromol.* **2018**, *120*, 1181–1189. [CrossRef]
115. Shamshina, J.L.; Berton, P.; Rogers, R.D. Advances in Functional Chitin Materials: A Review. *ACS Sustain. Chem. Eng.* **2019**, *7*, 6444–6457. [CrossRef]
116. Nezhad-Mokhtari, P.; Akrami-Hasan-Kohal, M.; Ghorbani, M. An injectable chitosan-based hydrogel scaffold containing gold nanoparticles for tissue engineering applications. *Int. J. Biol. Macromol.* **2020**, *154*, 198–205. [CrossRef]
117. Sahoo, D.R.; Biswal, T. Alginate and its application to tissue engineering. *SN Appl. Sci.* **2021**, *3*, 30. [CrossRef]
118. Farokhi, M.; Shariatzadeh, F.J.; Solouk, A.; Mirzadeh, H. Alginate Based Scaffolds for Cartilage Tissue Engineering: A Review. *Int. J. Polym. Mater. Polym. Biomater.* **2020**, *69*, 230–247. [CrossRef]
119. Kaczmarek-Pawelska, A. Alginate-Based Hydrogels in Regenerative Medicine. In *Alginate Recent Uses of This Natural Polymer*; Pereira, L., Ed.; IntechOpen: London, UK, 2020.
120. National Center for Biotechnology Information. PubChem Compound Summary for CID 14055602, Cellulose, Microcrystalline. Available online: https://pubchem.ncbi.nlm.nih.gov/compound/Cellulose_-microcrystalline (accessed on 15 January 2023).
121. National Center for Biotechnology Information. PubChem Compound Summary for CID 129662530, Chitosan. Available online: https://pubchem.ncbi.nlm.nih.gov/compound/129662530 (accessed on 15 January 2023).
122. National Center for Biotechnology Information. PubChem Compound Summary for CID 131704328, Alginate. Available online: https://pubchem.ncbi.nlm.nih.gov/compound/Alginate (accessed on 15 January 2023).
123. National Center for Biotechnology Information. PubChem Compound Summary for CID 24759, Hyaluronan. Available online: https://pubchem.ncbi.nlm.nih.gov/compound/Hyaluronan (accessed on 15 January 2023).
124. Mastalska-Popławska, J.; Sikora, M.; Izak, P.; Góral, Z. Applications of starch and its derivatives in bioceramics. *J. Biomater. Appl.* **2019**, *34*, 12–24. [CrossRef]
125. Waghmare, V.S.; Wadke, P.R.; Dyawanapelly, S.; Deshpande, A.; Jain, R.; Dandekar, P. Starch based nanofibrous scaffolds for wound healing applications. *Bioact. Mater.* **2018**, *3*, 255–266. [CrossRef] [PubMed]
126. National Center for Biotechnology Information. PubChem Compound Summary for CID 51003661, Starch Soluble. Available online: https://pubchem.ncbi.nlm.nih.gov/compound/Starch-soluble (accessed on 15 January 2023).
127. Lin, W.; Klein, J. Recent Progress in Cartilage Lubrication. *Adv. Mater.* **2021**, *33*, e2005513. [CrossRef] [PubMed]
128. Dovedytis, M.; Liu, Z.J.; Bartlett, S. Hyaluronic acid and its biomedical applications: A review. *Eng. Regen.* **2020**, *1*, 102–113. [CrossRef]
129. Zhai, P.; Peng, X.; Li, B.; Liu, Y.; Sun, H.; Li, X. The application of hyaluronic acid in bone regeneration. *Int. J. Biol. Macromol.* **2020**, *151*, 1224–1239. [CrossRef] [PubMed]
130. Mohammadi, F.; Samani, S.M.; Tanideh, N.; Ahmadi, F. Hybrid Scaffolds of Hyaluronic Acid and Collagen Loaded with Prednisolone: An Interesting System for Osteoarthritis. *Adv. Pharm. Bull.* **2018**, *8*, 11–19. [CrossRef]

131. Sieni, E.; Dettin, M.; De Robertis, M.; Bazzolo, B.; Conconi, M.T.; Zamuner, A.; Marino, R.; Keller, F.; Campana, L.G.; Signori, E. The Efficiency of Gene Electrotransfer in Breast-Cancer Cell Lines Cultured on a Novel Collagen-Free 3D Scaffold. *Cancers* **2020**, *12*, 1043. [CrossRef]
132. Zhang, X.; Wang, C.; Liao, M.; Dai, L.; Tang, Y.; Zhang, H.; Coates, P.; Sefat, F.; Zheng, L.; Song, J.; et al. Aligned electrospun cellulose scaffolds coated with rhBMP-2 for both in vitro and in vivo bone tissue engineering. *Carbohydr. Polym.* **2019**, *213*, 27–38. [CrossRef]
133. Hickey, R.J.; Pelling, A.E. Cellulose Biomaterials for Tissue Engineering. *Front. Bioeng. Biotechnol.* **2019**, *7*, 45. [CrossRef]
134. Kuzmenko, V.; Karabulut, E.; Pernevik, E.; Enoksson, P.; Gatenholm, P. Tailor-made conductive inks from cellulose nanofibrils for 3D printing of neural guidelines. *Carbohydr. Polym.* **2018**, *189*, 22–30. [CrossRef]
135. Luo, H.; Cha, R.; Li, J.; Hao, W.; Zhang, Y.; Zhou, F. Advances in tissue engineering of nanocellulose-based scaffolds: A review. *Carbohydr. Polym.* **2019**, *224*, 115144. [CrossRef]
136. Khan, S.; Ul-Islam, M.; Ikram, M.; Islam, S.U.; Ullah, M.W.; Israr, M.; Jang, J.H.; Yoon, S.; Park, J.K. Preparation and structural characterization of surface modified microporous bacterial cellulose scaffolds: A potential material for skin regeneration applications in vitro and in vivo. *Int. J. Biol. Macromol.* **2018**, *117*, 1200–1210. [CrossRef] [PubMed]
137. Fang, Y.; Zhang, T.; Song, Y.; Sun, W. Assessment of various crosslinking agents on collagen/chitosan scaffolds for myocardial tissue engineering. *Biomed. Mater.* **2020**, *15*, 045003. [CrossRef] [PubMed]
138. Maged, A.; Abdelkhalek, A.A.; Mahmoud, A.A.; Salah, S.; Ammar, M.M.; Ghorab, M.M. Mesenchymal stem cells associated with chitosan scaffolds loaded with rosuvastatin to improve wound healing. *Eur. J. Pharm. Sci.* **2019**, *127*, 185–198. [CrossRef] [PubMed]
139. Islam, M.; Shahruzzaman, M.; Biswas, S.; Sakib, N.; Rashid, T.U. Chitosan based bioactive materials in tissue engineering applications-A review. *Bioact. Mater.* **2020**, *5*, 164–183. [CrossRef]
140. Rodríguez-Vázquez, M.; Vega-Ruiz, B.; Ramos-Zúñiga, R.; Saldaña-Koppel, D.A.; Quiñones-Olvera, L.F. Chitosan and Its Potential Use as a Scaffold for Tissue Engineering in Regenerative Medicine. *BioMed Res. Int.* **2015**, *2015*, 821279. [CrossRef] [PubMed]
141. Shabunin, A.S.; Yudin, V.E.; Dobrovolskaya, I.P.; Zinovyev, E.V.; Zubov, V.; Ivan'kova, E.M.; Morganti, P. Composite Wound Dressing Based on Chitin/Chitosan Nanofibers: Processing and Biomedical Applications. *Cosmetics* **2019**, *6*, 16. [CrossRef]
142. Sadeghi, A.; Moztarzadeh, F.; Mohandesi, J.A. Investigating the effect of chitosan on hydrophilicity and bioactivity of conductive electrospun composite scaffold for neural tissue engineering. *Int. J. Biol. Macromol.* **2019**, *121*, 625–632. [CrossRef]
143. Niu, X.; Wei, Y.; Liu, Q.; Yang, B.; Ma, N.; Li, Z.; Zhao, L.; Chen, W.; Huang, D. Silver-loaded microspheres reinforced chitosan scaffolds for skin tissue engineering. *Eur. Polym. J.* **2020**, *134*, 109861. [CrossRef]
144. Naghieh, S.; Sarker, M.; Abelseth, E.; Chen, X. Indirect 3D bioprinting and characterization of alginate scaffolds for potential nerve tissue engineering applications. *J. Mech. Behav. Biomed. Mater.* **2019**, *93*, 183–193. [CrossRef]
145. Deepthi, S.; Jayakumar, R. Alginate nanobeads interspersed fibrin network as in situ forming hydrogel for soft tissue engineering. *Bioact. Mater.* **2018**, *3*, 194–200. [CrossRef]
146. Prasadh, S.; Wong, R.C.W. Unraveling the mechanical strength of biomaterials used as a bone scaffold in oral and maxillofacial defects. *Oral Sci. Int.* **2018**, *15*, 48–55. [CrossRef]
147. Hernández-González, A.C.; Téllez-Jurado, L.; Rodríguez-Lorenzo, L.M. Alginate hydrogels for bone tissue engineering, from injectables to bioprinting: A review. *Carbohydr. Polym.* **2019**, *229*, 115514. [CrossRef]
148. Sun, J.; Tan, H. Alginate-Based Biomaterials for Regenerative Medicine Applications. *Materials* **2013**, *6*, 1285–1309. [CrossRef]
149. Huang, S.; Wang, C.; Xu, J.; Ma, L.; Gao, C. In situ assembly of fibrinogen/hyaluronic acid hydrogel via knob-hole interaction for 3D cellular engineering. *Bioact. Mater.* **2017**, *2*, 253–259. [CrossRef] [PubMed]
150. Bacakova, L.; Pajorova, J.; Zikmundova, M.; Filova, E.; Mikes, P.; Jencova, V.; Kostakova, E.K.; Sinica, A. Nanofibrous Scaffolds for Skin Tissue Engineering and Wound Healing Based on Nature-Derived Polymers. In *Current and Future Aspects of Nanomedicine*; IntechOpen: London, UK, 2019; pp. 1–30.
151. Monteiro, I.P.; Shukla, A.; Marques, A.P.; Reis, R.L.; Hammond, P.T. Spray-assisted layer-by-layer assembly on hyaluronic acid scaffolds for skin tissue engineering. *J. Biomed. Mater. Res. Part A* **2015**, *103*, 330–340. [CrossRef] [PubMed]
152. Chircov, C.; Grumezescu, A.M.; Bejenaru, L.E. Hyaluronic acid-based scaffolds for tissue engineering. *Rom. J. Morphol. Embryol.* **2018**, *59*, 71–76. [PubMed]
153. Bejoy, J.; Wang, Z.; Bijonowski, B.; Yang, M.; Ma, T.; Sang, Q.-X.; Li, Y. Differential Effects of Heparin and Hyaluronic Acid on Neural Patterning of Human Induced Pluripotent Stem Cells. *ACS Biomater. Sci. Eng.* **2018**, *4*, 4354–4366. [CrossRef] [PubMed]
154. Movahedi, M.; Asefnejad, A.; Rafienia, M.; Khorasani, M.T. Potential of novel electrospun core-shell structured polyurethane/starch (hyaluronic acid) nanofibers for skin tissue engineering: In vitro and in vivo evaluation. *Int. J. Biol. Macromol.* **2020**, *146*, 627–637. [CrossRef]
155. Spearman, B.S.; Agrawal, N.K.; Rubiano, A.; Simmons, C.S.; Mobini, S.; Schmidt, C.E. Tunable methacrylated hyaluronic acid-based hydrogels as scaffolds for soft tissue engineering applications. *J. Biomed. Mater. Res. Part A* **2020**, *108*, 279–291. [CrossRef]
156. Thompson, R.E.; Pardieck, J.; Smith, L.; Kenny, P.; Crawford, L.; Shoichet, M.; Sakiyama-Elbert, S. Effect of hyaluronic acid hydrogels containing astrocyte-derived extracellular matrix and/or V2a interneurons on histologic outcomes following spinal cord injury. *Biomaterials* **2018**, *162*, 208–223. [CrossRef]

157. Roslan, M.R.; Nasir, N.F.M.; Cheng, E.M.; Amin, N.A.M. Tissue engineering scaffold based on starch: A review. In Proceedings of the 2016 International Conference on Electrical, Electronics, and Optimization Techniques (ICEEOT), Chennai, India, 3–5 March 2016; pp. 1857–1860.
158. Espigares, I.; Elvira, C.; Mano, J.; Vázquez, B.; Román, J.S.; Reis, R.L. New partially degradable and bioactive acrylic bone cements based on starch blends and ceramic fillers. *Biomaterials* **2002**, *23*, 1883–1895. [CrossRef]
159. Das, A.; Das, A.; Basu, A.; Datta, P.; Gupta, M.; Mukherjee, A. Newer guar gum ester/chicken feather keratin interact films for tissue engineering. *Int. J. Biol. Macromol.* **2021**, *180*, 339–354. [CrossRef]
160. Long, N.S.W.; Ahmad, M.; Hairom, N.H.H. Tensile and Thermogravimetry Analysis of Pullulan/Cellulose Films Incorporated with Carica Papaya Seeds Extract. In *Materials: Technology and Applications Series 1*; Ahmad, M.B., Hairom, N.H.H.B., Othman, S.A.B., Eds.; Penerbit UTHM Universiti Tun Hussein Onn Malaysia: Parit Raja, Malaysia, 2019; pp. 1–20.
161. Selvakumar, G.; Lonchin, S. Fabrication and characterization of collagen-oxidized pullulan scaffold for biomedical applications. *Int. J. Biol. Macromol.* **2020**, *164*, 1592–1599. [CrossRef]
162. Cavelier, S. New Strategies for Bone Graft Materials. Master's Thesis, McGill University, Montréal, QC, Canada, 2015.
163. Zarei, M.; Samimi, A.; Khorram, M.; Abdi, M.M.; Golestaneh, S.I. Fabrication and characterization of conductive polypyrrole/chitosan/collagen electrospun nanofiber scaffold for tissue engineering application. *Int. J. Biol. Macromol.* **2021**, *168*, 175–186. [CrossRef]
164. Grabska-Zielińska, S.; Sionkowska, A.; Carvalho, Â; Monteiro, F. Biomaterials with Potential Use in Bone Tissue Regeneration—Collagen/Chitosan/Silk Fibroin Scaffolds Cross-Linked by EDC/NHS. *Materials* **2021**, *14*, 1105. [CrossRef]
165. Szychlinska, M.A.; Calabrese, G.; Ravalli, S.; Dolcimascolo, A.; Castrogiovanni, P.; Fabbi, C.; Puglisi, C.; Lauretta, G.; Di Rosa, M.; Castorina, A.; et al. Evaluation of a Cell-Free Collagen Type I-Based Scaffold for Articular Cartilage Regeneration in an Orthotopic Rat Model. *Materials* **2020**, *13*, 2369. [CrossRef]
166. Jiang, J.P.; Liu, X.Y.; Zhao, F.; Zhu, X.; Li, X.-Y.; Niu, X.G.; Yao, Z.T.; Dai, C.; Xu, H.-Y.; Ma, K.; et al. Three-dimensional bioprinting collagen/silk fibroin scaffold combined with neural stem cells promotes nerve regeneration after spinal cord injury. *Neural Regen. Res.* **2020**, *15*, 959. [CrossRef]
167. Bayrak, E.; Huri, P.Y. Engineering Musculoskeletal Tissue Interfaces. *Front. Mater.* **2018**, *5*, 24. [CrossRef]
168. Ma, D.; Wang, Y.; Dai, W. Silk fibroin-based biomaterials for musculoskeletal tissue engineering. *Mater. Sci. Eng. C* **2018**, *89*, 456–469. [CrossRef]
169. Hadisi, Z.; Bakhsheshi-Rad, H.R.; Walsh, T.; Dehghan, M.M.; Farzad-Mohajeri, S.; Gholami, H.; Diyanoush, A.; Pagan, E.; Akbari, M. In vitro and in vivo evaluation of silk fibroin-hardystonite-gentamicin nanofibrous scaffold for tissue engineering applications. *Polym. Test.* **2020**, *91*, 106698. [CrossRef]
170. Zakeri-Siavashani, A.; Chamanara, M.; Nassireslami, E.; Shiri, M.; Hoseini-Ahmadabadi, M.; Paknejad, B. Three dimensional spongy fibroin scaffolds containing keratin/vanillin particles as an antibacterial skin tissue engineering scaffold. *Int. J. Polym. Mater. Polym. Biomater.* **2020**, *71*, 220–231. [CrossRef]
171. Feroz, S.; Muhammad, N.; Ratnayake, J.; Dias, G. Keratin-Based materials for biomedical applications. *Bioact. Mater.* **2020**, *5*, 496–509. [CrossRef]
172. Naderi, P.; Zarei, M.; Karbasi, S.; Salehi, H. Evaluation of the effects of keratin on physical, mechanical and biological properties of poly (3-hydroxybutyrate) electrospun scaffold: Potential application in bone tissue engineering. *Eur. Polym. J.* **2020**, *124*, 109502. [CrossRef]
173. Rojas-Martínez, L.; Flores-Hernandez, C.; López-Marín, L.; Martinez-Hernandez, A.; Thorat, S.; Vasquez, C.R.; Del Rio-Castillo, A.; Velasco-Santos, C. 3D printing of PLA composites scaffolds reinforced with keratin and chitosan: Effect of geometry and structure. *Eur. Polym. J.* **2020**, *141*, 110088. [CrossRef]
174. Wan, X.; Li, P.; Jin, X.; Su, F.; Shen, J.; Yuan, J. Poly(ε-caprolactone)/keratin/heparin/VEGF biocomposite mats for vascular tissue engineering. *J. Biomed. Mater. Res. Part A* **2020**, *108*, 292–300. [CrossRef]
175. National Center for Biotechnology Information. PubChem Compound Summary for CID 6913668, Collagen I, Alpha Chain (98–110). Available online: https://pubchem.ncbi.nlm.nih.gov/compound/Collagen-I_-alpha-chain-_98-110 (accessed on 15 January 2023).
176. National Center for Biotechnology Information. PubChem Compound Summary for CID 446715, Keratan. Available online: https://pubchem.ncbi.nlm.nih.gov/compound/Keratan (accessed on 15 January 2023).
177. National Center for Biotechnology Information. PubChem Compound Summary for CID 439199, Fibrin. Available online: https://pubchem.ncbi.nlm.nih.gov/compound/Fibrin (accessed on 15 January 2023).
178. National Center for Biotechnology Information. PubChem Compound Summary for CID 439221, Elastin. Available online: https://pubchem.ncbi.nlm.nih.gov/compound/Elastin (accessed on 15 January 2023).
179. Vazquez-Portalatin, N.; Alfonso-Garcia, A.; Liu, J.C.; Marcu, L.; Panitch, A. Physical, Biomechanical, and Optical Characterization of Collagen and Elastin Blend Hydrogels. *Ann. Biomed. Eng.* **2020**, *48*, 2924–2935. [CrossRef]
180. Wang, Z.; Liu, L.; Mithieux, S.M.; Weiss, A.S. Fabricating Organized Elastin in Vascular Grafts. *Trends Biotechnol.* **2021**, *39*, 505–518. [CrossRef]
181. Rodrigues, I.C.P.; Pereira, K.D.; Woigt, L.F.; Jardini, A.L.; Luchessi, A.D.; Lopes, É.S.N.; Webster, T.J.; Gabriel, L.P. A novel technique to produce tubular scaffolds based on collagen and elastin. *Artif. Organs* **2020**, *45*, E113–E122. [CrossRef]

182. Dubey, A.P. Carbon Nanofiber and Polymer Conjugate. In *Carbon Nanofibers: Fundamentals and Applications*; Wiley-Scrivener: Hoboken, NJ, USA, 2021; pp. 75–98.
183. Park, C.H.; Woo, K.M. Fibrin-Based Biomaterial Applications in Tissue Engineering and Regenerative Medicine. *Adv. Exp. Med. Biol.* **2018**, *1064*, 253–261. [CrossRef]
184. Noori, A.; Ashrafi, S.J.; Vaez-Ghaemi, R.; Hatamian-Zaremi, A.; Webster, T.J. A review of fibrin and fibrin composites for bone tissue engineering. *Int. J. Nanomed.* **2017**, *12*, 4937–4961. [CrossRef]
185. Purohit, S.D.; Singh, H.; Bhaskar, R.; Yadav, I.; Chou, C.-F.; Gupta, M.K.; Mishra, N.C. Gelatin—alginate—cerium oxide nanocomposite scaffold for bone regeneration. *Mater. Sci. Eng. C* **2020**, *116*, 111111. [CrossRef] [PubMed]
186. Abedinia, A.; Nafchi, A.M.; Sharifi, M.; Ghalambor, P.; Oladzadabbasabadi, N.; Ariffin, F.; Huda, N. Poultry gelatin: Characteristics, developments, challenges, and future outlooks as a sustainable alternative for mammalian gelatin. *Trends Food Sci. Technol.* **2020**, *104*, 14–26. [CrossRef]
187. Ashwin, B.; Abinaya, B.; Prasith, T.; Chandran, S.V.; Yadav, L.R.; Vairamani, M.; Patil, S.; Selvamurugan, N. 3D-poly (lactic acid) scaffolds coated with gelatin and mucic acid for bone tissue engineering. *Int. J. Biol. Macromol.* **2020**, *162*, 523–532. [CrossRef] [PubMed]
188. Kimura, A.; Yoshida, F.; Ueno, M.; Taguchi, M. Application of radiation crosslinking technique to development of gelatin scaffold for tissue engineering. *Radiat. Phys. Chem.* **2021**, *180*, 109287. [CrossRef]
189. Singh, S.; Dutt, D.; Kaur, P.; Singh, H.; Mishra, N.C. Microfibrous paper scaffold for tissue engineering application. *J. Biomater. Sci. Polym. Ed.* **2020**, *31*, 1091–1106. [CrossRef]
190. Goudarzi, Z.M.; Behzad, T.; Ghasemi-Mobarakeh, L.; Kharaziha, M. An investigation into influence of acetylated cellulose nanofibers on properties of PCL/Gelatin electrospun nanofibrous scaffold for soft tissue engineering. *Polymer* **2021**, *213*, 123313. [CrossRef]
191. Zhang, D.; Wu, X.; Chen, J.; Lin, K. The development of collagen based composite scaffolds for bone regeneration. *Bioact. Mater.* **2018**, *3*, 129–138. [CrossRef]
192. Lim, Y.-S.; Ok, Y.-J.; Hwang, S.-Y.; Kwak, J.-Y.; Yoon, S. Marine Collagen as A Promising Biomaterial for Biomedical Applications. *Mar. Drugs* **2019**, *17*, 467. [CrossRef]
193. Kim, H.; Jang, J.; Park, J.; Lee, K.P.; Lee, S.; Lee, D.M.; Kim, K.H.; Kim, H.K.; Cho, D.W. Shear-induced alignment of collagen fibrils using 3D cell printing for corneal stroma tissue engineering. *Biofabrication* **2019**, *11*, 3. [CrossRef]
194. Nabavi, M.H.; Salehi, M.; Ehterami, A.; Bastami, F.; Semyari, H.; Tehranchi, M.; Semyari, H. A collagen-based hydrogel containing tacrolimus for bone tissue engineering. *Drug Deliv. Transl. Res.* **2020**, *10*, 108–121. [CrossRef]
195. Chen, X.; Hao, W.; Li, X.; Xiao, Z.; Yao, Y.; Chu, Y.; Farkas, B.; Romano, I.; Brandi, F.; Dai, J. Functional Multichannel Poly(Propylene Fumarate)-Collagen Scaffold with Collagen-Binding Neurotrophic Factor 3 Promotes Neural Regeneration After Transected Spinal Cord Injury. *Adv. Heal. Mater.* **2018**, *7*, e1800315. [CrossRef]
196. Chen, Z.; Zhang, Q.; Li, H.; Wei, Q.; Zhao, X.; Chen, F. Elastin-like polypeptide modified silk fibroin porous scaffold promotes osteochondral repair. *Bioact. Mater.* **2021**, *6*, 589–601. [CrossRef]
197. Gholipourmalekabadi, M.; Sapru, S.; Samadikuchaksaraei, A.; Reis, R.L.; Kaplan, D.L.; Kundu, S.C. Silk fibroin for skin injury repair: Where do things stand? *Adv. Drug Deliv. Rev.* **2020**, *153*, 28–53. [CrossRef] [PubMed]
198. Kundu, B.; Rajkhowa, R.; Kundu, S.C.; Wang, X. Silk fibroin biomaterials for tissue regenerations. *Adv. Drug Deliv. Rev.* **2013**, *65*, 457–470. [CrossRef] [PubMed]
199. Keirouz, A.; Zakharova, M.; Kwon, J.; Robert, C.; Koutsos, V.; Callanan, A.; Chen, X.; Fortunato, G.; Radacsi, N. High-throughput production of silk fibroin-based electrospun fibers as biomaterial for skin tissue engineering applications. *Mater. Sci. Eng. C* **2020**, *112*, 110939. [CrossRef] [PubMed]
200. Gupta, P.; Lorentz, K.L.; Haskett, D.G.; Cunnane, E.M.; Ramaswamy, A.K.; Weinbaum, J.S.; Vorp, D.A.; Mandal, B.B. Bioresorbable silk grafts for small diameter vascular tissue engineering applications: In vitro and in vivo functional analysis. *Acta Biomater.* **2020**, *105*, 146–158. [CrossRef]
201. Atrian, M.; Kharaziha, M.; Emadi, R.; Alihosseini, F. Silk-Laponite®fibrous membranes for bone tissue engineering. *Appl. Clay Sci.* **2019**, *174*, 90–99. [CrossRef]
202. De Torre, I.G.; Alonso, M.; Rodriguez-Cabello, J.-C. Elastin-Based Materials: Promising Candidates for Cardiac Tissue Regeneration. *Front. Bioeng. Biotechnol.* **2020**, *8*, 657. [CrossRef]
203. Miranda-Nieves, D.; Chaikof, E.L. Collagen and Elastin Biomaterials for the Fabrication of Engineered Living Tissues. *ACS Biomater. Sci. Eng.* **2017**, *3*, 694–711. [CrossRef]
204. Nguyen, T.-U.; Shojaee, M.; Bashur, C.; Kishore, V. Electrochemical fabrication of a biomimetic elastin-containing bi-layered scaffold for vascular tissue engineering. *Biofabrication* **2019**, *11*, 015007. [CrossRef]
205. Wang, X.; Ali, M.S.; Lacerda, C.M.R. A Three-Dimensional Collagen-Elastin Scaffold for Heart Valve Tissue Engineering. *Bioengineering* **2018**, *5*, 69. [CrossRef] [PubMed]
206. Silva, R.; Singh, R.; Sarker, B.; Papageorgiou, D.G.; Juhasz-Bortuzzo, J.A.; Roether, J.A.; Cicha, I.; Kaschta, J.; Schubert, D.W.; Chrissafis, K.; et al. Hydrogel matrices based on elastin and alginate for tissue engineering applications. *Int. J. Biol. Macromol.* **2018**, *114*, 614–625. [CrossRef] [PubMed]
207. Khalili, S.; Khorasani, S.N.; Razavi, S.M.; Hashemibeni, B.; Tamayol, A. Nanofibrous Scaffolds with Biomimetic Composition for Skin Regeneration. *Appl. Biochem. Biotechnol.* **2019**, *187*, 1193–1203. [CrossRef] [PubMed]

208. Tian, L.; Prabhakaran, M.P.; Ramakrishna, S. Strategies for regeneration of components of nervous system: Scaffolds, cells and biomolecules. *Regen. Biomater.* **2015**, *2*, 31–45. [CrossRef] [PubMed]
209. Afewerki, S.; Sheikhi, A.; Kannan, S.; Ahadian, S.; Khademhosseini, A. Gelatin-polysaccharide composite scaffolds for 3D cell culture and tissue engineering: Towards natural therapeutics. *Bioeng. Transl. Med.* **2019**, *4*, 96–115. [CrossRef] [PubMed]
210. Tytgat, L.; Van Damme, L.; Van Hoorick, J.; Declercq, H.; Thienpont, H.; Ottevaere, H.; Blondeel, P.; Dubruel, P.; Van Vlierberghe, S. Additive manufacturing of photo-crosslinked gelatin scaffolds for adipose tissue engineering. *Acta Biomater.* **2019**, *94*, 340–350. [CrossRef] [PubMed]
211. Ye, W.; Li, H.; Yu, K.; Xie, C.; Wang, P.; Zheng, Y.; Zhang, P.; Xiu, J.; Yang, Y.; He, Y.; et al. 3D printing of gelatin methacrylate-based nerve guidance conduits with multiple channels. *Mater. Des.* **2020**, *192*, 108757. [CrossRef]
212. Conrad, B.; Han, L.-H.; Yang, F. Gelatin-Based Microribbon Hydrogels Accelerate Cartilage Formation by Mesenchymal Stem Cells in Three Dimensions. *Tissue Eng. Part A* **2018**, *24*, 1631–1640. [CrossRef]
213. Celikkin, N.; Mastrogiacomo, S.; Jaroszewicz, J.; Walboomers, X.F.; Swieszkowski, W. Gelatin methacrylate scaffold for bone tissue engineering: The influence of polymer concentration. *J. Biomed. Mater. Res. Part A* **2018**, *106*, 201–209. [CrossRef]
214. Rezaeeyazdi, M.; Colombani, T.; Memic, A.; Bencherif, S.A. Injectable Hyaluronic Acid-co-Gelatin Cryogels for Tissue-Engineering Applications. *Materials* **2018**, *11*, 1374. [CrossRef]
215. Song, H.-H.G.; Rumma, R.T.; Ozaki, C.K.; Edelman, E.R.; Chen, C.S. Vascular Tissue Engineering: Progress, Challenges, and Clinical Promise. *Cell Stem Cell* **2018**, *22*, 340–354. [CrossRef] [PubMed]
216. Abelseth, E.; Abelseth, L.; De la Vega, L.; Beyer, S.T.; Wadsworth, S.J.; Willerth, S.M. 3D Printing of Neural Tissues Derived from Human Induced Pluripotent Stem Cells Using a Fibrin-Based Bioink. *ACS Biomater. Sci. Eng.* **2019**, *5*, 234–243. [CrossRef] [PubMed]
217. Bachmann, B.; Spitz, S.; Rothbauer, M.; Jordan, C.; Purtscher, M.; Zirath, H.; Schuller, P.; Eilenberger, C.; Ali, S.F.; Mühleder, S.; et al. Engineering of three-dimensional pre-vascular networks within fibrin hydrogel constructs by microfluidic control over reciprocal cell signaling. *Biomicrofluidics* **2018**, *12*, 042216. [CrossRef]
218. Soleimannejad, M.; Ebrahimibarough, S.; Soleimani, M.; Nadri, S.; Tavangar, S.M.; Roohipoor, R.; Yazdankhah, M.; Bayat, N.; Riazi-Esfahani, M.; Ai, J. Fibrin gel as a scaffold for photoreceptor cells differentiation from conjunctiva mesenchymal stem cells in retina tissue engineering. *Artif. Cells Nanomed. Biotechnol.* **2018**, *46*, 805–814. [CrossRef] [PubMed]
219. Wang, X.; Liu, C. Fibrin Hydrogels for Endothelialized Liver Tissue Engineering with a Predesigned Vascular Network. *Polymers* **2018**, *10*, 1048. [CrossRef] [PubMed]
220. Zhao, P.; Gu, H.; Mi, H.; Rao, C.; Fu, J.; Turng, L.-S. Fabrication of scaffolds in tissue engineering: A review. *Front. Mech. Eng.* **2018**, *13*, 107–119. [CrossRef]
221. Chia, H.N.; Wu, B.M. Recent advances in 3D printing of biomaterials. *J. Biol. Eng.* **2015**, *9*, 4. [CrossRef] [PubMed]
222. Touri, M.; Kabirian, F.; Saadati, M.; Ramakrishna, S.; Mozafari, M. Additive Manufacturing of Biomaterials—The Evolution of Rapid Prototyping. *Adv. Eng. Mater.* **2019**, *21*, 1800511. [CrossRef]
223. Yuan, B.; Zhou, S.Y.; Chen, X.S. Rapid prototyping technology and its application in bone tissue engineering. *J. Zhejiang Univ. Sci. B* **2017**, *18*, 303–315. [CrossRef]
224. Fereshteh, Z. Freeze-drying technologies for 3D scaffold engineering. In *Functional 3D Tissue Engineering Scaffolds—Materials, Technologies and Applications*; Elsevier: Amsterdam, The Netherlands, 2018; pp. 151–174.
225. Brougham, C.M.; Levingstone, T.J.; Shen, N.; Cooney, G.M.; Jockenhoevel, S.; Flanagan, T.C.; O'Brien, F.J. Freeze-Drying as a Novel Biofabrication Method for Achieving a Controlled Microarchitecture within Large, Complex Natural Biomaterial Scaffolds. *Adv. Heal. Mater.* **2017**, *6*, 1700598. [CrossRef]
226. Aghmiuni, A.I.; Keshel, S.H.; Sefat, F.; AkbarzadehKhiyavi, A. Fabrication of 3D hybrid scaffold by combination technique of electrospinning-like and freeze-drying to create mechanotransduction signals and mimic extracellular matrix function of skin. *Mater. Sci. Eng. C* **2021**, *120*, 111752. [CrossRef]
227. Mikos, A.G.; Sarakinos, G.; Leite, S.M.; Vacant, J.P.; Langer, R. Laminated three-dimensional biodegradable foams for use in tissue engineering. *Biomaterials* **1993**, *14*, 323–330. [CrossRef] [PubMed]
228. Thadavirul, N.; Pavasant, P.; Supaphol, P. Development of polycaprolactone porous scaffolds by combining solvent casting, particulate leaching, and polymer leaching techniques for bone tissue engineering. *J. Biomed. Mater. Res. Part A* **2014**, *102*, 3379–3392. [CrossRef]
229. Prasad, A.; Sankar, M.; Katiyar, V. State of Art on Solvent Casting Particulate Leaching Method for Orthopedic ScaffoldsFabrication. *Mater. Today Proc.* **2017**, *4*, 898–907. [CrossRef]
230. Harris, L.D.; Kim, B.; Mooney, D.J. Open pore biodegradable matrices formed with gas foaming. *J. Biomed. Mater. Res.* **1998**, *42*, 396–402. [CrossRef]
231. Kishan, A.P.; Cosgriff-Hernandez, E.M. Recent advancements in electrospinning design for tissue engineering applications: A review. *J. Biomed. Mater. Res. Part A* **2017**, *105*, 2892–2905. [CrossRef] [PubMed]
232. Friend, D.F.L.; González, M.E.L.; Caraballo, M.M.; de Queiroz, A.A.A. Biological properties of electrospun cellulose scaffolds from biomass. *J. Biomater. Sci. Polym. Ed.* **2019**, *30*, 1399–1414. [CrossRef] [PubMed]
233. Nam, Y.S.; Park, T.G. Biodegradable polymeric microcellular foams by modified thermally induced phase separation method. *Biomaterials* **1999**, *20*, 1783–1790. [CrossRef] [PubMed]

234. Melchels, F.P.W.; Feijen, J.; Grijpma, D.W. A review on stereolithography and its applications in biomedical engineering. *Biomaterials* **2010**, *31*, 6121–6130. [CrossRef]
235. Kamboj, N.; Ressler, A.; Hussainova, I. Bioactive Ceramic Scaffolds for Bone Tissue Engineering by Powder Bed Selective Laser Processing: A Review. *Materials* **2021**, *14*, 5338. [CrossRef]
236. Xia, X.; Xu, X.; Lin, C.; Yang, Y.; Zeng, L.; Zheng, Y.; Wu, X.; Li, W.; Xiao, L.; Qian, Q.; et al. Microalgal-Immobilized Biocomposite Scaffold Fabricated by Fused Deposition Modeling 3D Printing Technology for Dyes Removal. *ES Mater. Manuf.* **2020**, *7*, 40–50. [CrossRef]
237. Zhang, B.; Cristescu, R.; Chrisey, D.B.; Narayan, R.J. Solvent-based Extrusion 3D Printing for the Fabrication of Tissue Engineering Scaffolds. *Int. J. Bioprint.* **2019**, *6*, 19. [CrossRef] [PubMed]
238. Huang, Y.; Zhang, X.-F.; Gao, G.; Yonezawa, T.; Cui, X. 3D bioprinting and the current applications in tissue engineering. *Biotechnol. J.* **2017**, *12*, 8. [CrossRef]
239. Ćatić, N.; Wells, L.; Al Nahas, K.; Smith, M.; Jing, Q.; Keyser, U.F.; Cama, J.; Kar-Narayan, S. Aerosol-jet printing facilitates the rapid prototyping of microfluidic devices with versatile geometries and precise channel functionalization. *Appl. Mater. Today* **2020**, *19*, 100618. [CrossRef]
240. Salmoria, G.V.; Klauss, P.; Paggi, R.A.; Kanis, L.A.; Lago, A. Structure and mechanical properties of cellulose based scaffolds fabricated by selective laser sintering. *Polym. Test.* **2009**, *28*, 648–652. [CrossRef]
241. Sharmila, G.; Muthukumaran, C.; Kirthika, S.; Keerthana, S.; Kumar, N.M.; Jeyanthi, J. Fabrication and characterization of Spinacia oleracea extract incorporated alginate/carboxymethyl cellulose microporous scaffold for bone tissue engineering. *Int. J. Biol. Macromol.* **2020**, *156*, 430–437. [CrossRef] [PubMed]
242. Deepthi, S.; Viha, C.V.S.; Thitirat, C.; Furuike, T.; Tamura, H.; Jayakumar, R. Fabrication of Chitin/Poly(butylene succinate)/Chondroitin Sulfate Nanoparticles Ternary Composite Hydrogel Scaffold for Skin Tissue Engineering. *Polymers* **2014**, *6*, 2974–2984. [CrossRef]
243. Entekhabi, E.; Nazarpak, M.H.; Shafieian, M.; Mohammadi, H.; Firouzi, M.; Hassannejad, Z. Fabrication and in vitro evaluation of 3D composite scaffold based on collagen/hyaluronic acid sponge and electrospun polycaprolactone nanofibers for peripheral nerve regeneration. *J. Biomed. Mater. Res. Part A* **2021**, *109*, 300–312. [CrossRef]
244. Kitsara, M.; Joanne, P.; Boitard, S.E.; Ben Dhiab, I.; Poinard, B.; Menasché, P.; Gagnieu, C.; Forest, P.; Agbulut, O.; Chen, Y. Fabrication of cardiac patch by using electrospun collagen fibers. *Microelectron. Eng.* **2015**, *144*, 46–50. [CrossRef]
245. Morris, V.B.; Nimbalkar, S.; Younesi, M.; McClellan, P.; Akkus, O. Mechanical Properties, Cytocompatibility and Manufacturability of Chitosan:PEGDA Hybrid-Gel Scaffolds by Stereolithography. *Ann. Biomed. Eng.* **2017**, *45*, 286–296. [CrossRef]
246. Rastegar, A.; Mahmoodi, M.; Mirjalili, M.; Nasirizadeh, N. Platelet-Rich Fibrin-Loaded PCL/Chitosan Core-Shell fibers Scaffold for Enhanced Osteogenic Differentiation of Mesenchymal Stem Cells. *Carbohydr. Polym.* **2021**, *269*, 118351. [CrossRef] [PubMed]
247. Zhong, N.; Dong, T.; Chen, Z.; Guo, Y.; Shao, Z.; Zhao, X. A novel 3D-printed silk fibroin-based scaffold facilitates tracheal epithelium proliferation in vitro. *J. Biomater. Appl.* **2019**, *34*, 3–11. [CrossRef] [PubMed]
248. Kim, H.; Yang, G.H.; Choi, C.H.; Cho, Y.S.; Kim, G. Gelatin/PVA scaffolds fabricated using a 3D-printing process employed with a low-temperature plate for hard tissue regeneration: Fabrication and characterizations. *Int. J. Biol. Macromol.* **2018**, *120*, 119–127. [CrossRef] [PubMed]
249. Chen, S.; Zhao, X.; Du, C. Macroporous poly (l-lactic acid)/chitosan nanofibrous scaffolds through cloud point thermally induced phase separation for enhanced bone regeneration. *Eur. Polym. J.* **2018**, *109*, 303–316. [CrossRef]
250. Beh, C.Y.; Cheng, E.M.; Nasir, N.F.M.; Majid, M.S.A.; Roslan, M.R.M.; You, K.Y.; Khor, S.F.; Ridzuan, M.J.M. Fabrication and characterization of three-dimensional porous cornstarch/n-HAp biocomposite scaffold. *Bull. Mater. Sci.* **2020**, *43*, 249. [CrossRef]
251. Li, P.; Wang, Y.; Jin, X.; Dou, J.; Han, X.; Wan, X.; Yuan, J.; Shen, J. Fabrication of PCL/keratin composite scaffolds for vascular tissue engineering with catalytic generation of nitric oxide potential. *J. Mater. Chem. B* **2020**, *8*, 6092–6099. [CrossRef]
252. Wang, Q.; Zhou, S.; Wang, L.; You, R.; Yan, S.; Zhang, Q.; Li, M. Bioactive silk fibroin scaffold with nanoarchitecture for wound healing. *Compos. Part B Eng.* **2021**, *224*, 109165. [CrossRef]
253. Hejazi, F.; Ebrahimi, V.; Asgary, M.; Piryaei, A.; Fridoni, M.J.; Kermani, A.A.; Zare, F.; Abdollahifar, M.-A. Improved healing of critical-size femoral defect in osteoporosis rat models using 3D elastin/polycaprolactone/nHA scaffold in combination with mesenchymal stem cells. *J. Mater. Sci. Mater. Med.* **2021**, *32*, 27. [CrossRef]

gels

Review

Cryopreservation of Cell Sheets for Regenerative Therapy: Application of Vitrified Hydrogel Membranes

Yoshitaka Miyamoto [1,2,3,4]

1 Department of Reproductive Biology, National Research Institute for Child Health and Development, Setagaya-ku, Tokyo 157-8535, Japan; miyamoto-ys@ncchd.go.jp or myoshi1230@gmail.com; Tel.: +81-3-3416-0181
2 Department of Maternal-Fetal Biology, National Research Institute for Child Health and Development, Setagaya-ku, Tokyo 157-8535, Japan
3 Graduate School of BASE, Tokyo University of Agriculture and Technology, Koganei, Tokyo 184-8588, Japan
4 Department of Mechanical Engineering, Tokyo Institute of Technology, Meguro-ku, Tokyo 152-8552, Japan

Abstract: Organ transplantation is the first and most effective treatment for missing or damaged tissues or organs. However, there is a need to establish an alternative treatment method for organ transplantation due to the shortage of donors and viral infections. Rheinwald and Green et al. established epidermal cell culture technology and successfully transplanted human-cultured skin into severely diseased patients. Eventually, artificial cell sheets of cultured skin were created, targeting various tissues and organs, including epithelial sheets, chondrocyte sheets, and myoblast cell sheets. These sheets have been successfully used for clinical applications. Extracellular matrix hydrogels (collagen, elastin, fibronectin, and laminin), thermoresponsive polymers, and vitrified hydrogel membranes have been used as scaffold materials to prepare cell sheets. Collagen is a major structural component of basement membranes and tissue scaffold proteins. Collagen hydrogel membranes (collagen vitrigel), created from collagen hydrogels through a vitrification process, are composed of high-density collagen fibers and are expected to be used as carriers for transplantation. In this review, the essential technologies for cell sheet implantation are described, including cell sheets, vitrified hydrogel membranes, and their cryopreservation applications in regenerative medicine.

Keywords: cell sheet; cryopreservation; vitrified hydrogel membrane; collagen; regenerative therapy; cell therapy

Citation: Miyamoto, Y. Cryopreservation of Cell Sheets for Regenerative Therapy: Application of Vitrified Hydrogel Membranes. *Gels* **2023**, *9*, 321. https://doi.org/10.3390/gels9040321

Academic Editors: Arish Dasan, Ashokraja Chandrasekar and Nupur Kohli

Received: 10 February 2023
Revised: 7 April 2023
Accepted: 7 April 2023
Published: 10 April 2023

1. Introduction

Organ transplantation is the first and most effective treatment when a tissue or organ is missing or damaged. However, there is a need to establish an alternative treatment method for organ transplantation due to the shortage of donors and viral infections. As an alternative, Rheinwald and Green et al. established epidermal cell culture technology [1,2] and successfully transplanted human-cultured skin into severely diseased patients [3]. In other words, this was the beginning of regenerative medicine. Then, combining the fields of medicine and engineering, Langer, R., and Vacanti, J.P. proposed "tissue engineering" to regenerate organs by drawing out the regenerative ability of cells [4]. In "tissue engineering," cells, scaffolds, and growth factors play significant roles [5], and cell sheets such as cultured skin can be created using these key factors. For example, cell monolayer sheets have been successfully fabricated using thermoresponsive polymers [6] and applied clinically to myocardial tissues [7]. In addition, focusing on scaffolds, the extracellular matrix (ECM) is a non-cellular constituent of all tissues and organs [8,9]. The major fibrous proteins were collagen, elastin, fibronectin, and laminin. Collagen is a major structural component of basement membranes and tissue scaffold proteins. Therefore, collagen is widely used in tissue engineering and regenerative medicine to process hydrogels and membrane structures. In particular, collagen hydrogel membranes (collagen vitrigel), created from collagen hydrogels through a vitrification process, are composed of high-density collagen fibers [10] and are expected to be used as carriers for transplantation. Furthermore, it is necessary to provide large-volume quality-controlled cell sheets for widespread medical applications. Here, I introduce the advances in cryopreservation that can provide a stable supply of cell sheets for regenerative therapy (Figure 1): 1. Targeted autologous or allogeneic cells are isolated to treat diseases. 2. Cell sheets are prepared and cryopreserved using a hydrogel membrane. 3. The frozen cell sheets are thawed and transplanted into diseased patients.

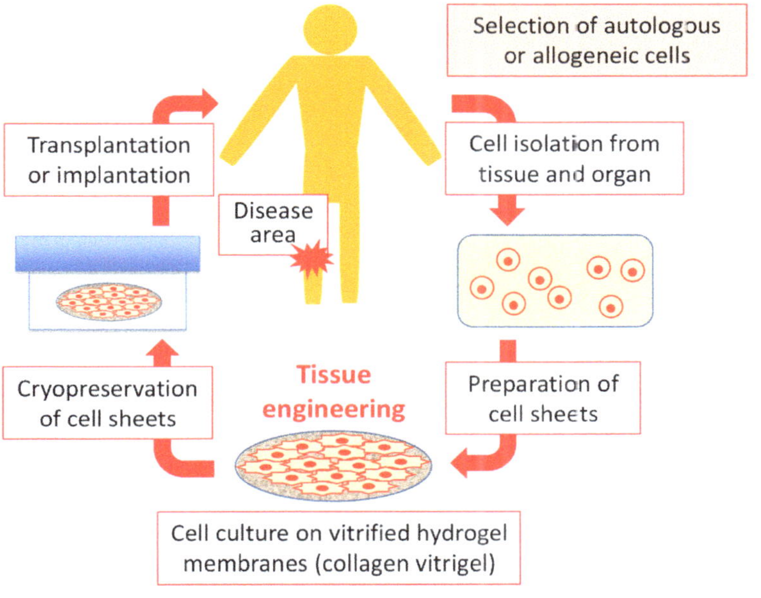

Figure 1. Regenerative medicine concept of cell sheets.

In this review, the essential technologies for cell sheet implantation are described, including cell sheets, vitrified hydrogel membranes, and their cryopreservation applications in regenerative medicine.

2. The History of Tissue Culture and Tissue Engineering

Harrison, R.G. (1907) observed the growth of nerve fibers in the embryonic tissue fragments of frogs removed from the body [11]. This was the beginning of the proof of the concept of "tissue culture" from animal tissue culture. Tissue culture is used to maintain, culture, and revive (regenerate) tissues, organs, or cells. The concept of "cell culture," which is widely used in the world today, is thought to have started when Rous and Jones (1916) succeeded in growing and culturing free cells by trypsin treatment of animal tissues [12]. The world's oldest mouse-derived L cell line [13,14] and human-derived HeLa cell line [15] were established and cell culture technology has rapidly developed. The tissue culture method of Enders et al. (1949) proved that polioviruses can be grown in cells derived from human tissues, including human connective tissue, intestine, liver, and kidneys [16]. Enders, J.F., Weller, T.H., and Robbins, F.C. won the Nobel Prize in Physiology in 1954 for their "Tissue Culture of Poliovirus". With the development and production of poliovirus vaccines [17–19], synthetic culture media such as 199 medium [20], Eagle's medium [21], and Dulbecco's medium [22] have been developed, and stable two-dimensional cultures have become possible.

In contrast, three-dimensional culture began with van Wezel's (1967) technique of agitating cells by attaching them to microcarrier supports [23] and Knazek et al.'s (1972) technique of perfusion culture of cells in hollow fiber supports [24], which aimed at high-density culture. Eventually, not only a three-dimensional environment but also an environment that more closely mimics the in vivo environment was created by utilizing the ECM. For example, Kleinman et al. (1986) used a mouse tumor tissue extract (Matrigel) to create basement membrane-like structures [25]. Bell et al. (1979) used collagen gels to create artificial dermal tissue [26] and artificial skin tissue [27]. From 1970 to 1990, culture techniques for tissue and organ regeneration using biomaterials, such as biological materials and synthetic polymers, spread rapidly. Eventually, Langer and Vacanti (1993) proposed "tissue engineering" to develop organ and tissue substitutes that enable the regeneration, maintenance, and repair of vital functions [4].

3. Regenerative Medicine and Cultured Human Skin

Conventional medicine promotes the healing of damaged tissues and organs through pharmaceutical and surgical treatments.

Regenerative medicine involves the repair and regeneration of lost tissue and organ functions using patient cells, other cells, or artificial tissue. Specifically, this refers to medical treatments that regenerate lost functions:

1. Stem cells and other cells are artificially cultured outside the patient's body.
2. Tissue is artificially constructed from stem cells and other cells outside the patient's body.
3. Devices incorporating living cells activate and differentiate intrinsic stem cells with cell growth differentiation factors.

Regenerative medicine has the potential to provide novel treatment options for previously untreatable diseases. Cells used in regenerative medicine include somatic cells (skin cells, muscle cells, etc.), which comprise the body, somatic stem cells, embryonic stem cells (ES), and induced pluripotent stem (iPS) cells. In the 1970s, epidermal cell culture technology was established by Rheinwald and Green et al. [1,2]. In 1981, they succeeded in cultured human skin transplantation for patients with severe diseases, which marked the beginning of regenerative medicine [3].

The cultured human skin was then used as a cell sheet in which human epidermal cells were cultured and grown in a medium containing fetal bovine serum, using mouse 3T3 cells (fibroblasts) as a support cell layer (feeder cells). The key cell culture technology involves the use of feeder cells. The technique is as follows: (1) a co-culture of human epidermal cells and irradiated mouse 3T3 cells to selectively proliferate human epidermal cells; (2) irradiation causes 3T3 cells to lose proliferative ability but maintains cell adhesion and promotes the proliferation of human epidermal cells; and (3) the presence of 3T3 cells suppresses the proliferation of human fibroblasts, which are mixed with each cell. In the United States, the cultured epidermal autografts (Epicel®, Genzyme Corporation: Boston, MA, USA, 1987) and the autologous cultured chondrocytes (Carticel®, Genzyme Corporation: Boston, MA, USA, 1997) received authorization from the Food and Drug Administration (FDA). About 20 years later, the cultured epidermal autograft (JACE®, Japan Tissue Engineering Co., Ltd.: Aichi, Japan, 2007) received authorization for Japan's first regenerative medical products (cellular and tissue-based products). Currently, tissue-engineering technology for preparing cells and tissues for transplantation is essential for cellular and regenerative medicine.

4. Hydrogel and Artificial Cell Sheets

ECM hydrogels (fibrous proteins such as collagen, elastin, fibronectin, and laminin, and polysaccharides such as hyaluronic acid and proteoglycans), synthetic polymer hydrogels, and rigid polymer materials (polystyrene) have been used as scaffold materials for the preparation of cell sheets such as cultured epidermal autografts [10,25–28]. Okano et al. [6] used cell sheets cultured on a temperature-responsive polymer that detached in sheet form upon temperature changes as carriers for transplantation. Medical treatments using this technology have been developed for several tissues and organs, including cardiomyocytes [7]. Takezawa et al. developed a cell culture carrier (collagen vitrigel) that closely resembled the density of collagen fibers in vivo [10] and applied it to drug discovery research and regenerative medicine [29,30].

4.1. Extracellular Matrix (ECM)

The ECM is a non-cellular constituent of all tissues and organs [8,9]. Major fibrous proteins such as collagen, elastin, fibronectin, and laminin, polysaccharides such as hyaluronic acid, and proteoglycans are known components of the ECM. These biopolymers form fibrous or net-like structures, or hydrogels containing many water molecules. ECM serves as a physical scaffold for cells and provides the signals required for tissue morphogenesis, differentiation, and homeostasis [9].

4.1.1. Collagen and Gelatin

Collagen is the major protein component of connective tissue and basement membranes and exists in numerous forms (Types I–XVIII) with varying tensile strengths and tissue distributions [31]. Stiffness, flexibility, and structural changes in many body tissues are caused by changes in collagen composition, cell restriction, and compartmentalization. Collagen in vivo consists of rigid triple helical structures of three molecular collagen chains that aggregate to form nanometer-scale collagen fibrils, forming hierarchically ordered higher-order structures [32]. Collagen extracted from living organisms is insoluble in water due to the presence of hydrophobic amino acid residues outside the collagen fibrils. Atelocollagen, prepared by hydrolyzing collagen by acid treatment, dissolves in an acidic aqueous solution as the collagen fibrils dissociate, while maintaining the triple-helical structure. When an acidic solution of atelocollagen is neutralized and maintained at 37 °C, the collagen fibrils aggregate into a network due to hydrophobic interactions to form a hydrogel. The resulting collagen hydrogel is widely used as a base material for cell adhesion and implantation owing to its high biocompatibility and physiological activity [33–36].

Gelatin is a denatured form of collagen, the main component of connective tissues such as the skin, bones, and tendons. The main chemical component is a linear amino acid polymer. Gelatin is denatured by (1) acid and heat treatment and (2) alkali treatment. The alkali treatment method produces more carboxyl groups than the acid and heat treatment method [33]. When heated, gelatin exists in a sol state (randomly coiled molecular structure). When the solution is cooled, some gelatin molecules change to their original collagen-like helical structure (triple-helical structure), forming a network that eventually loses fluidity and becomes a gel. This gel undergoes a reversible sol-gel structural change upon heating and cooling. Gelatin hydrogels have high biocompatibility and oxygen permeability. Nutrients are transported through water diffusion via the hydrogel. Additionally, cells can be incorporated into gelatin hydrogels. However, owing to their low mechanical strength, researchers have attempted to improve the strength of these materials [37]. The resulting gelatin hydrogel, similar to the collagen hydrogel described above, is widely used as a base material for cell adhesion and regenerative medicine [38].

4.1.2. Elastin

Elastin is a protein of the ECM involved in the elastic recoiling ability. Elasticity is the property of "elastic fibers" that return to their original state when force is removed. In addition to its elastic recoil, elastin is biocompatible because it is chemically inert and can be used as a hydrogel [39,40]. It mainly forms fibrous elastic tissues in the skin, blood vessels, and ligaments and provides elasticity to tissues [41]. However, biogenic elastin is highly crosslinked and insoluble. Therefore, it is difficult to obtain elastin as a homogeneous and easily handled material. Specific amino acid repeat sequences are evident in the properties of elastin, with Val-Pro-Gly-Val-Gly (VPGVG) being the most abundant in elastin [42–44]. Elastin-like peptides (ELPs) based on this repeating sequence exhibit a reversible phase transition called inverse temperature transition (ITT) in water. Monomer ELP genes are synthesized by ligating double stranded oligonucleotide cassettes or in pUC19 [45]. As the temperature changes, the ELP exhibits hydrophobic properties above the phase transition temperature (Tt) and changes to hydrophilic properties below Tt. Because of their excellent self-assembly and high biocompatibility, ELPs are expected to be used in medical engineering applications, such as DDS and ECM scaffold materials. Sugawara-Narutaki et al. created a novel block ELP by combining two types of sequence motifs derived from elastin and reported that it self-assembled upon temperature stimulation in water to form a fiber structure (nanofibers) similarly to elastin derived from organisms [46]. Furthermore, hydrogels were obtained with 0.3 wt % ELP, and uniform nanofibers were successfully formed. Because elastin is highly hydrophobic, heterogeneous, and prone to aggregation, it is difficult to create uniform gels, and there are few reported cases [47–49]. In the future, elastin-like hydrogels are expected to be used as carriers for implantation in cellular and regenerative medicine.

4.1.3. Fibronectin

Fibronectin (FN) is a large glycoprotein and cell-adhesion molecule [50]. Cellular fibronectin is present in many tissues, including the spleen, lymph nodes, blood vessel walls, liver, kidney, muscle, skin, brain, and peripheral nerves. The interaction between FN and cell surface receptors such as integrins promote cell adhesion, shape, migration, growth, and differentiation in vitro [51]. They have a variety of functions in vivo, including cell adhesion to the ECM and connective tissue formation and retention. In addition, the amino acid sequences of FN and vitronectin contain many portions in which arginine (Arg), glycine (Gly), and aspartic acid (Asp) form a continuous motif that binds to integrin molecules on various cell membranes and promotes cell adhesion and survival.

Recently, Trujilloa et al. reported the development of FN-based 3D hydrogels of controlled stiffness and degradability [52]. By incorporating vascular endothelial growth factor (VEGF) into FN-based 3D hydrogels, it is expected to be a useful implantation carrier in tissue engineering and regenerative medicine.

4.1.4. Laminin

Laminin is a large protein that constitutes the basement membrane of the ECM [50]. It promotes the establishment and maintenance of multicellular systems and tissues, as well as cell adhesion, migration, and proliferation [51]. Laminin plays an essential role in the formation of the basement membrane and conferring cell adhesion. The cell adhesion activity of laminin, mediated by cell surface integrins, is extremely strong compared with that of other cell adhesion proteins. Therefore, laminin is attracting attention as a feeder-free culture substrate for human embryonic stem cells (hESCs) and induced pluripotent stem cells (hiPSCs) [53]. Human laminin-511 supports the stable culture of hESCs and hiPSCs [54]. Laminin is an important constituent in neuronal tissue and brain [55]. To create three-dimensional neuronal models with neurons that are similar to those of living organisms, it is useful to incorporate laminin into hydrogels. Azide-modified laminin is conjugated to hyaluronan–poly(ethylene glycol) (HA:PEG) hybrid hydrogels. Encapsulated human neuronal cells demonstrate high viability and grow into cross-linked hyaluronan–laminin hydrogels.

4.1.5. Hyaluronan (Hyaluronic Acid)

Hyaluronan (HA) is a polysaccharide consisting of N-acetylglucosamine and D-glucuronic acid (GlcNAcβ1-4GlcAβ1-3) linked on a linear chain [56]. It is widely distributed in vivo and plays important roles in the skin, cartilage, and eyes. HA mediates its activity in cellular signaling, wound repair, morphogenesis, and matrix organization [57–59]. HA has an extremely high molecular weight, ranging from 100,000 Da in the serum to 8,000,000 Da in the vitreous of the eye [60], with a minimum molecular weight of 411. As hyaluronan contains carboxyl and hydroxyl groups, it is easily chemically modified and can be used to create hydrogels with gelators. As an example, thiol, haloacetate, dihydrazide, aldehyde, tyramine, and Huisgen cycloaddition are chemically modified with hyaluronic acid. These modified HA monomers are used to create HA hydrogels by photopolymerization and electropolymerization reactions. Radical polymerization has been applied to the formation of HA hydrogels as an example of photopolymerization. In addition, HA hydrogels are dissolved by hyaluronidase, which is present in vivo and widely used as a culture substrate, 3D tissues, and carrier for transplantation [61].

4.1.6. Alginic Acid

Alginic acid is a polysaccharide found in brown algae and other algae and is composed of β-D-mannuronate and α-L-guluronate in a β-1,4 bonded structure [62]. The characteristic feature of alginic acid is that when a divalent metal cation is added to an aqueous solution, the alginic acid molecules form an egg-box structure [63], and a hydrogel can be created [64]. Alginic acid is insoluble in water, but is extracted as a soluble salt, such as sodium alginate, and is used as a food additive. Alginate is able to form gels independently of temperature changes [62]. The formation of alginate gels can be achieved by ionic bonding with cations or acid deposition [65]. In other medical applications, alginate is used in fibrous gels (surgical threads), alginate salts (wound dressings), drug delivery, and tissue engineering [66]. Islet transplantation is an effective therapeutic modality to stabilize glycemic control in type 1 diabetes patients [67]. Islet encapsulation in an alginate hydrogel can immunosuppress, and maintain long-term patient survival.

4.1.7. Synthetic Polymer

Polyethylene glycol (PEG) is a hydrophilic polymer with high biocompatibility. PEG has been used as a medical material for a long time. PEG-based materials are the most well-known DDS carriers made of PEG [68–70]. They can improve blood retention and enhance drug efficacy, but are not biodegradable or cell-adhesive. PEG is also expected to have potential as a carrier for transplantation, because it can incorporate bioactive molecules and other substances into its network to create functionalized hydrogels [71].

Polyvinyl alcohol (PVA) is a polymer obtained from polyvinyl acetate via alcoholysis, hydrolysis, and aminolysis [33]. An aqueous PVA solution was subjected to repeated freeze-thaw cycles to crystallize the molecular chains and form hydrogels [72]. PVA hydrogels are used as biomaterials in artificial cartilage and joints [73].

Polylactic acid (PLA) is a polyester that is used as a medical material for surgical sutures. PLA has D isomer (poly-D-lactic acid, PDLA), L isomer (poly-L-lactic acid, PLLA), and racemic forms depending on the structure of the monomer unit, which has a chiral carbon [74]. PLA synthesis requires control of conditions (temperature, pressure, and pH), and the use of catalysts [75]. PDLA and PLLA form a stereocomplex and aggregated to form a gel. PLA is a very safe material for living organisms because it is easily degraded by hydrolysis, and its degradation product is lactic acid. Therefore, PLA-based materials have been applied in orthopaedic regenerative engineering [76].

4.2. *Thermoresponsive Polymer*

Poly-N-isopropyl acrylamide (PNIPAm) is a well-known temperature-responsive polymer. PNIPAm undergoes a hydration/dehydration transition with a phase transition temperature of 32 °C as a boundary. To utilize the structural change in PNIPAm, a polymer layer is synthesized by graft polymerization of the N-isopropyl acrylamide monomer (NIPAm) on the surface of the culture substrate. The PNIPAm, which is strictly controlled on the surface of the culture substrate, shows hydrophobic properties at 37 °C and changes to hydrophilic properties at 32 °C. Cells adhere and proliferate on the surface of the culture substrate at 37 °C, but cells detach from the surface of the culture substrate and can be recovered at temperatures below 32 °C [6]. Tissue engineering using temperature-responsive polymers can be used to create and recover various cell sheets, including fibroblasts, epithelial cells, and cardiomyocytes. For example, cell sheets created from cardiomyocytes beat independently. When cardiomyocyte sheets are stacked on a fibrin gel, beating can be observed with the naked eye. Furthermore, thicker and more functional myocardial tissue was created by adding a vascular network to the cardiomyocyte sheet and increasing the number of stacked sheets [77]. According to previous reports, the clinical application of cardiomyocyte sheets has been successful. Cartilage is exhausted owing to trauma and aging, and many patients experience knee osteoarthritis. To treat these diseases, laminated autologous cartilage sheets have been created and various cytokines are expressed on these sheets to support tissue repair and regeneration. Other artificial cell sheets have been created using temperature-responsive polymers, such as corneal epithelial sheets [78], oral mucosal epithelial cell sheets [79,80], and periodontal tissue sheets [81], which are expected to have clinical applications.

4.3. Vitrified Hydrogel Membranes

Vitrification gradually removes free and bound water by drying, thereby converting the material into one with glass-like properties with excellent strength and transparency [82]. Takezawa et al. produced vitrified collagen hydrogel membranes (collagen vitrigels) composed of high-density collagen fibers via gelation, vitrification, and rehydration from a collagen hydrogel composed of conventional low-density collagen fibers [10]. The collagen vitrigel is a thin sheet with superior strength and transparency ranging from 10–100 μm in thickness, similar to the connective tissue in vivo. In addition to the ability to culture heterologous cells on both sides of the collagen vitrigel, they are permeable to proteins and drugs [29,30,83]. Numerous applications have been reported, including a corneal microtissue patch test [84]. Carriers such as collagen vitrigels induce bone regeneration by local and sustained delivery of bone morphogenetic protein-2 [85]. In addition, researchers have reported its use as an artificial corneal endothelial graft as a cell scaffold [86] and as an implantable material for the reconstruction of the cornea [87] and trachea [88].

Vitrification and rehydration can convert other vitrified hydrogel membranes into new stable physical states. However, there are only a few reports on gelatin, a chondroitin sulfate-polyethylene glycol (CS-PEG) adhesive, a collagen-based membrane (collagen vitrigel, CV) combination [89], and a bi-layered carboxymethyl cellulose-collagen vitrigel dual-surface adhesion-prevention membrane [90]. In addition, collagen-elastin (CE) membranes have been fabricated on polydimethylsiloxane (PDMS), and their Young's modulus has been evaluated using atomic force microscopy (AFM) to design precise hydrogel membranes that mimic the ECM [91]. Precisely designed vitrified hydrogel membranes have new possibilities and have been developed as carriers for implantation.

5. Artificial Cell Sheets and Cryopreservation

5.1. Cryopreservation of Cellular and Tissue-Based Products and Simple Cell Suspensions

To generalize regenerative and cell medicine, it is necessary to provide therapeutic cells in large quantities with quality control [92]. Therefore, it is essential to establish a cryopreservation system that provides a stable cell supply. However, cellular damage can be triggered by ice crystal formation and dehydration within cells during freezing and thawing [93–96]. Therefore, to obtain high-quality cells, it is necessary to consider the freezing technique, composition of the cryopreservation solution, and thawing method [95,97,98]. In 1949, Polge et al. discovered that glycerol exhibits cryoprotective effects on human and avian sperm [99]. Ten years later, Lovelock et al. found that dimethyl sulfoxide (DMSO) was superior to glycerol in cell penetration and removal after thawing [100]. Therefore, it is now widely used as a mainstream cryopreservation solution [101].

Since 1998, the Food and Drug Administration (FDA: Silver Spring, MD, USA) has updated its Cellular & Gene Therapy Products guidelines [102]. In addition, the Pharmaceuticals and Medical Devices Agency (PMDA: Tokyo, Japan) [103] and European Medicines Agency (EMA, European Union) [104] updated their guidelines in the 2000s. Many mesenchymal stem cell (MSCs) products are available at clinicaltrials.gov [105]. Kymriah (tisagenlecleucel) and TEMCELL HS Inj. (MSCs) containing 10% to 7.5 vol% DMSO are used in clinical practice [80]. As these products are prone to side effects when administered as-is after freeze-thawing, the DMSO concentration should be reduced to 3–4%. Thus, the cytotoxicity of DMSO reduces cell survival due to mitochondrial swelling, membrane potential damage, and the production of reactive oxygen species [106].

In recent years, the shortage of donors in transplantation medicine has become increasingly serious, and it is essential to supply therapeutic cells to patients. Miyamoto et al. developed a cryopreservation solution for the transplantation of hepatocytes [107,108] and stem cells [109]. Similar to the cellular- and tissue-based products described above, we used a cryopreservation solution containing 10% DMSO. Furthermore, cell damage was reduced by combining it with other cryoprotectants (sericin and maltose). Yamatoya and Miyamoto et al. also reported that this cryopreservation solution was effective for differentiated neuronal cells [110].

5.2. Cryopreservation of Tissues and Cell Sheets

Cryopreservation of tissues results in significantly different cooling, warming, and dehydration responses compared with simple cell suspensions. These differences are due to differences in tissue structure and significant differences in the freezing and thawing methods [111]. Medical treatments such as bone and tendon treatments can be successfully performed without living cell components after freezing [112]. However, many medical treatments, such as those for the heart, require the maintenance of a live cell component after freezing. Therefore, freezing and thawing conditions and cryopreservation solutions must be designed to prevent ice formation inside and outside various tissues. Thus, vitrification is an effective ice-free cryopreservation technique [113].

A few therapeutic frozen cell sheets have been reported as cellular- and tissue-based products. To be approved as a cellular- and tissue-based product, frozen cell sheets must be scientifically evaluated in terms of chemistry, biophysics, toxicology, and cryobiology [93,111–114]. The critical evaluation criteria were cell viability within the cellular tissue sheet and the in vivo tissue function. The vitrification of various tissues has been reported [115–118], with articular cartilage being a good example [112]. However, in slow freezing, most chondrocytes die by ice formation and the ECM is damaged, which is a significant obstacle to clinical application [119,120]. Nevertheless, no difference was observed between fresh and vitrified cartilage during transplantation, and approximately 85% of the cellular metabolic activity was maintained [112,121]. Vitrification has also been reported to be effective in many small tissue structures such as spheroids [122], organoids [123], and encapsulated cells [124]. Cryopreservation of skin grafts has been used for several years. Each skin graft is banked in a manner desirable for long-term preservation, improving graft performance and safety, and reducing risks to the recipient [125].

The preparation and cryopreservation of cell sheets covers a variety of tissues, including epithelial sheets [126–130], chondrocyte sheets [131,132], myoblast [133], stem cell [134–136] and other cell sheets [137–139] (Tables 1 and 2). In cryopreservation of cell sheets, the basic composition is a freezing solution containing DMSO or EG [129–139]. Rewarming solutions containing sucrose are often used to thaw cryopreserved cell sheets. Oliva et al. examined vitrification and storage against oral mucosa epithelial cell sheets (CAOMECS) [129]. CAOMECS were cryopreserved by Vitrification procedures 1 and 2 in solutions as reported by Vitrification procedure 1 (Sheikhi et al.) [140], and Vitrification procedure 2 (Marco-Jimenez et al.) [141]. In contrast, Vitrification procedure 3 (Li et al.) [142] was cryopreserved CAOMECS in bulk. CAOMECS were prepared in transwells and placed in plastic containers. The plastic containers were placed in a liquid nitrogen freezer. The CAOMECS were preserved with the vitrification solution in the liquid nitrogen using Vitrification procedure 1 and 2. The cell sheets broke during the thawing process. In contrast, Vitrification procedure 3 was effective dry, in the absence of the vitrification solution, because the cell sheets were not damaged by the thawing process. Cryopreservation under liquid nitrogen using the vitrification solution were effective for the chondrocytes [132] and the myoblast sheets [133]. These cell sheets were placed in a bag held 1 cm above the surface of liquid nitrogen vapor (approximately −150 °C). The vitrified cell sheets can be stored for a long time and have potential for clinical application. Without vitrification or a controlled-rate freezer, MSCs [136] and the fibroblasts sheets [139] can maintain sheet morphology and function by slow freezing. Miyamoto et al. also succeeded in cryopreserving cell sheets such as primary rat hepatocytes, mouse embryonic fibroblasts (MEF), and mouse embryonic stem cells (ESCs) cultured on collagen vitrigel membranes [143]. These cells were used because cryopreservation of hepatocytes is difficult. Therefore, the recovery rate of adherent cells is significantly compromised by collection, freezing, thawing, and reattachment. Similar to the cellular- and tissue-based products described above, we used a cryopreservation solution containing 10% DMSO. The freezing procedure was performed in a controlled-rate freezer [108], and the storage temperature was maintained in liquid nitrogen. Compared with these cell sheets, the cryopreservation of collagen vitrigel adherent to MEF and mouse ESCs was more effective than that of primary rat hepatocytes.

Table 1. Preparation of cell sheets.

Cells	Origin	Species	Sheet Preparation	In Vivo Test	Year of Publication	Reference
Epithelial cells	Oral mucosa	Rabbit	Oral mucosal epithelium cells were cultured with MMC-treated NIH/3T3 feeder cells.	YES	2019	[129]
Epithelial cells	Foreskin keratinocytes	Human	Human epithelial cell sheets (ECSs) were cultured on plastic dishes.	YES	2021	[130]
Chondrocytes	Knee cartilage	Rabbit	Primary cultured cells derived from the knee cartilage were plated onto temperature-responsive culture dishes (UpCell®, CellSeed).		2020	[132]
Myoblast cells	Vastus medialis muscle	Human	Cell sheets consisting of myoblast cells were prepared using temperature-responsive culture dishes (UpCell®, CellSeed).	YES	2018	[133]
Mesenchymal stem cells (MSCs)	The bone marrow of femurs	Rat	Cell sheets consisting of mesenchymal stem cells were prepared using temperature-responsive culture dishes.	YES	2018	[136]
Fibroblasts	Tail skin	Mouse	Primary fibroblasts were inoculated on plastic dishes	YES	2022	[139]
Hepatocytes	Liver	Rat	Primary hepatocytes were inoculated onto collagen vitrigel membranes.		2009	[143]
Embryonic stem cells (ESCs)	Fertilized egg	Mouse	1st step: Primary mouse embryonic fibroblasts (MEF feeder cells) were inoculated onto UV-irradiated collagen vitrigels. 2nd step: A mouse ES cell culture was performed with mitomycin C-treated MEF feeder cell layers.		2009	[143]
Embryonic fibroblasts	Fetal tissues	Mouse	Primary mouse embryonic fibroblasts (MEF feeder cells) were inoculated onto collagen vitrigel membranes.		2009	[143]

Table 2. Freezing and thawing procedure of cell sheets.

Cells	Freezing Procedure	Thawing Procedure	Reference
Epithelial cells	**Vitrification of epithelial cell sheets** Vitrification Procedures (1, and 2) in solutions Vitrification Procedures (3) in bulk The plastic containers were placed in the liquid nitrogen freezer	**Stepwise** The cell sheets were thawed in four steps using the solutions (culture media, culture media supplemented with 0.2, and 0.1 M sucrose) at 37 °C.	[129]
Epithelial cells	**Programmed freezing** Epithelial cells were cryopreserved for storage in KGM and 10 µM Y27632 at 4 °C. Or cryochamber (Planer KRYO 10 Series III Freezer, UK) was run by the pre-set cooling program.	**Rapid thawing** Thawing was carried out rapidly by holding in air for 1 min to boil off any liquid nitrogen and swirling in a water bath at 40 °C.	[130]

Table 2. *Cont.*

Cells	Freezing Procedure	Thawing Procedure	Reference
Chondrocytes	**Vitrification of the chondrocyte sheets** The chondrocyte sheets in the circulating vitrification bag were held 1 cm above the surface of liquid nitrogen vapor (approximately −150 °C).	**Rapid thawing** The circulating vitrification bag was placed on a heating plate at 45 °C. The gel was placed on top of the cell sheet to thaw it rapidly.	[132]
Myoblast cells	**Vitrification of the myoblast sheets** The myoblast sheets were detached from the dish at 22 °C. The cell sheets were placed in thin polyethylene films and held 1 cm above the surface of liquid nitrogen vapor.	**Rapid thawing** Thin polyethylene films was placed on a heating plate at 37 °C. The gel was placed on top of the cell sheet to thaw it rapidly.	[133]
Mesenchymal stem cells (MSCs)	**Slow freezing** MSCs sheets were placed directly into a deep-freezer set at −80 °C.	**Rapid thawing** MSCs sheets were placed in a 37 °C water bath for rapid thawing until almost no ice was detectable.	[136]
Fibroblasts	**Stepwise** The fibroblasts sheets were placed into a 3D freezer (Koga Sangyo Co., Ltd.) at −35 °C for 20 min to freeze the cells and then transferred to a −80 °C deep freezer.	**Rapid thawing** The fibroblasts sheets were placed in a 37 °C water bath for rapid thawing until almost no ice was detectable.	[139]
Hepatocytes	**Programmed freezing** The hepatocytes sheets were placed in a controlled rate freezer (Kryo10, Planer) and frozen at a rate of 1 °C/min [108].	**Rapid thawing** To thaw the cells, 2 mL of culture medium warmed to 37 °C was added.	[143]
Embryonic stem cells (ESCs)	**Programmed freezing** The ESCs sheets were placed in a controlled rate freezer (Kryo10, Planer) and frozen at a rate of 1 °C/min [108].	**Rapid thawing** To thaw the cells, 2 mL of culture medium warmed to 37 °C was added.	[143]
Embryonic fibroblast	**Programmed freezing** The embryonic fibroblast sheets were placed in a controlled rate freezer (Kryo10, Planer) and frozen at a rate of 1 °C/min [108].	**Rapid thawing** To thaw the cells, 2 mL of culture medium warmed to 37 °C was added.	[143]

6. Conclusions and Future Perspectives

Applications of various hydrogels will be considered in regenerative and cellular medicine [144,145]. Various cells have been encapsulated into hydrogels to construct desired tissues and organs. However, it is important to develop cryopreservation technology to utilize the created tissues and organs more effectively. 3D hydrogel cryopreservation, using alginic acid, synthetic polymers, and supramolecular gels, has been reported as effective [146]. Manferdini et al. focused on hydrogels for embedding the MSCs in vitro/ex vivo studies and reported a systematic review in an in vivo osteoarthritis (OA) model. [147]. The objective is cartilage regeneration in OA using hydrogels. In conclusion, the basic study results are positive, but cartilage strength and function need to be improved for clinical application. Next, we will discuss cell sheets from 3D hydrogel.

This review describes cell sheets, vitrified hydrogel membranes, and their cryopreservation applications in regenerative and cellular medicine. Synthetic polymers, thermoresponsive polymers, and ECMs such as collagen and gelatin, have been used to fabricate cell sheets. Numerous regenerative medical products for single-cell transplantation have been cryopreserved using DMSO. Similarly, cryopreservation of cell sheets is expected to further promote the medical industry. When using cell-sheet products, the highest priority should be the therapeutic effect on patients. If there is no difference in quality between frozen and unfrozen cellular products, such as artificial cell sheets, frozen cellular products may be chosen. In fact, ice-free vitrification cryopreservation methods were effective for the epithelial constructs (EpiDerm) from MatTek (www.mattek.com, accessed on 6 December 2022) for the testing of new chemicals and drug screening [148]. The safety and efficacy of cellular- and tissue-based products, as well as the manufacturing time, process, and cost, must be considered and judged as a comprehensive medical system. Because various cellular and tissue-based products (living and non-living cell products) have different uses, selecting the appropriate manufacturing and storage processes for cellular- and tissue-based products is essential, considering the adequacy of the facility's equipment and medical system.

Funding: This work was supported in part by The Grant of National Center for Child Health and Development (2022C-29, and 2023C-25); JSPS KAKENHI Grant Numbers JP16KK0192. The APC was funded by The Grant of National Center for Child Health and Development (2022C-29).

Institutional Review Board Statement: Not applicable.

Informed Consent Statement: Not applicable.

Data Availability Statement: Not applicable.

Conflicts of Interest: The author declares no conflict of interest.

References

1. Rheinwald, J.G.; Green, H. Serial cultivation of strains of human epidermal keratinocytes: The formation of keratinizing colonies from single cells. *Cell* **1975**, *6*, 331–343. [CrossRef] [PubMed]
2. Green, H.; Kehinde, O.; Thomas, J. Growth of cultured human epidermal cells into multiple epithelia suitable for grafting. *Proc. Natl. Acad. Sci. USA* **1979**, *76*, 5665–5668. [CrossRef] [PubMed]
3. O'Connor, N.E.; Mulliken, J.B.; Banks-Schlegel, S.; Kehinde, O.; Green, H. Grafting of burns with cultured epithelium prepared from autologous epidermal cells. *Lancet* **1981**, *1*, 75–78. [CrossRef]
4. Langer, R.; Vacanti, J.P. Tissue engineering. *Science* **1993**, *260*, 920–926. [CrossRef]
5. Vacanti, C.A.; Vacanti, J.P. The history and scope of tissue engineering. In *Principles of Tissue Engineering*, 3rd ed.; Lanza, R., Langer, R., Vacanti, J.P., Eds.; Academic Press, Elsevier: Waltham, MA, USA, 2007; pp. 3–6.
6. Yamada, N.; Okano, T.; Sakai, H.; Karikusa, F.; Sawasaki, Y.; Sakurai, Y. Thermo-responsive polymeric surfaces; control of attachment and detachment of cultured cells. *Makromol. Chem. Rapid Commun.* **1990**, *11*, 571–576. [CrossRef]
7. Sawa, Y.; Miyagawa, S.; Sakaguchi, T.; Fujita, T.; Matsuyama, A.; Saito, A.; Shimizu, T.; Okano, T. Tissue engineered myoblast sheets improved cardiac function sufficiently to discontinue LVAS in a patient with DCM: Report of a case. *Surg. Today* **2012**, *42*, 181–184. [CrossRef]
8. Järveläinen, H.; Sainio, A.; Koulu, M.; Wight, T.N.; Penttinen, R. Extracellular matrix molecules: Potential targets in pharmacotherapy. *Pharmacol. Rev.* **2009**, *61*, 198–223. [CrossRef]
9. Frantz, C.; Stewart, K.M.; Weaver, V.M. The extracellular matrix at a glance. *J. Cell Sci.* **2010**, *123 Pt 24*, 4195–4200. [CrossRef]
10. Takezawa, T.; Ozaki, K.; Nitani, A.; Takabayashi, C.; Shimo-Oka, T. Collagen vitrigel: A novel scaffold that can facilitate a three-dimensional culture for reconstructing organoids. *Cell Transplant.* **2004**, *13*, 463–473. [CrossRef]
11. Harrison, R.G. Observations on the living developing nerve fiber. *Proc. Soc. Exp. Biol. Med.* **1907**, *4*, 140–143. [CrossRef]
12. Rous, P.; Jones, F.S. A method for obtaining suspensions of living cells from the fixed tissues, and for the plating out of individual cells. *J. Exp. Med.* **1916**, *23*, 549–555. [CrossRef] [PubMed]
13. Earle, W.R. Changes Induced in a Strain of Fibro blasts From a Strain C3H Mouse by the Action of 20-Methylcholanthrene (Preliminary Report). *J. Natl. Cancer Inst.* **1943**, *3*, 555–558.
14. Earle, W.R.; Schilling, E.L.; Stark, T.H.; Straus, N.P.; Brown, M.F.; Shelton, E. Production of Malignancy in Vitro. IV. The Mouse Fibroblast Cultures and Changes Seen in the Living Cells. *J. Natl. Cancer Inst.* **1943**, *4*, 165–212.
15. Gey, G.O.; Coffman, W.D.; Kubicek, M.T. Tissue culture studies of the proliferative capacity of cervical carcinoma and normal epithelium. *Cancer Res.* **1952**, *12*, 264–265.
16. Enders, J.F.; Weller, T.H.; Robbins, F.C. Cultivation of the Lansing Strain of Poliomyelitis Virus in Cultures of Various Human Embryonic Tissues. *Science* **1949**, *109*, 85–87. [CrossRef]
17. Salk, J.E.; Youngner, J.S.; Ward, E.N. Use of color change of phenol red as the indicator in titrating poliomyelitis virus or its antibody in a tissue-culture system. *Am. J. Hyg.* **1954**, *60*, 214–230. [CrossRef]
18. Salk, J.E.; Krech, U.; Youngner, J.S.; Bennett, B.L.; Lewis, L.J.; Bazeley, P.L. Formaldehyde treatment and safety testing of experimental poliomyelitis vaccines. *Am. J. Public Health Nations Health* **1954**, *44*, 563–570. [CrossRef]
19. Salk, J.E.; Bazeley, P.L.; Bennett, B.L.; Krech, U.; Lewis, L.J.; Ward, E.N.; Youngner, J.S. Studies in human subjects on active immunization against poliomyelitis. II. A practical means for inducing and maintaining antibody formation. *Am. J. Public Health Nations Health* **1954**, *44*, 994–1009. [CrossRef]
20. Morgan, J.F.; Morton, H.J.; Parker, R.C. Nutrition of animal cells in tissue culture; initial studies on a synthetic medium. *Proc. Soc. Exp. Biol. Med.* **1950**, *73*, 1–8. [CrossRef]
21. Eagle, H. Amino acid metabolism in mammalian cell cultures. *Science* **1959**, *130*, 432–437. [CrossRef]
22. Dulbecco, R.; Freeman, G. Plaque production by the polyoma virus. *Virology* **1959**, *8*, 396–397. [CrossRef] [PubMed]
23. Van Wezel, A.L. Growth of cell-strains and primary cells on micro-carriers in homogeneous culture. *Nature* **1967**, *216*, 64–65. [CrossRef] [PubMed]
24. Knazek, R.A.; Gullino, P.M.; Kohler, P.O.; Dedrick, R.L. Cell culture on artificial capillaries: An approach to tissue growth in vitro. *Science* **1972**, *178*, 65–66. [CrossRef] [PubMed]
25. Kleinman, H.K.; McGarvey, M.L.; Hassell, J.R.; Star, V.L.; Cannon, F.B.; Laurie, G.W.; Martin, G.R. Basement membrane complexes with biological activity. *Biochemistry* **1986**, *25*, 312–318. [CrossRef]

26. Bell, E.; Ivarsson, B.; Merrill, C. Production of a tissue-like structure by contraction of collagen lattices by human fibroblasts of different proliferative potential in vitro. *Proc. Natl. Acad. Sci. USA* **1979**, *76*, 1274–1278. [CrossRef]
27. Bell, E.; Ehrlich, H.P.; Buttle, D.J.; Nakatsuji, T. Living tissue formed in vitro and accepted as skin-equivalent tissue of full thickness. *Science* **1981**, *211*, 1052–1054. [CrossRef]
28. Berthiaume, F.; Moghe, P.V.; Toner, M.; Yarmush, M.L. Effect of extracellular matrix topology on cell structure, function, and physiological responsiveness: Hepatocytes cultured in a sandwich configuration. *FASEB J.* **1996**, *10*, 1471–1484. [CrossRef]
29. Takezawa, T.; Takeuchi, T.; Nitani, A.; Takayama, Y.; Kino-Oka, M.; Taya, M.; Enosawa, S. Collagen vitrigel membrane useful for paracrine assays in vitro and drug delivery systems in vivo. *J. Biotechnol.* **2007**, *131*, 76–83. [CrossRef]
30. Takezawa, T.; Nishikawa, K.; Wang, P.C. Development of a human corneal epithelium model utilizing a collagen vitrigel membrane and the changes of its barrier function induced by exposing eye irritant chemicals. *Toxicol. In Vitro* **2011**, *25*, 1237–1241. [CrossRef]
31. Kadler, K.E.; Baldock, C.; Bella, J.; Boot-Handford, R.P. Collagens at a glance. *J. Cell Sci.* **2007**, *120 Pt 12*, 1955–1958. [CrossRef]
32. Kotch, F.W.; Raines, R.T. Self-assembly of synthetic collagen triple helices. *Proc. Natl. Acad. Sci. USA* **2006**, *103*, 3028–3033. [CrossRef]
33. Lee, K.Y.; Mooney, D.J. Hydrogels for Tissue Engineering. *Chem. Rev.* **2001**, *101*, 1869–1880. [CrossRef]
34. Sarrigiannidis, S.O.; Rey, J.M.; Dobre, O.; González-García, C.; Dalby, M.J.; Salmeron-Sanchez, M. A tough act to follow: Collagen hydrogel modifications to improve mechanical and growth factor loading capabilities. *Mater. Today Bio* **2021**, *10*, 100098. [CrossRef] [PubMed]
35. Zhang, Y.; Wang, Y.; Li, Y.; Yang, Y.; Jin, M.; Lin, X.; Zhuang, Z.; Guo, K.; Zhang, T.; Tan, W. Application of Collagen-Based Hydrogel in Skin Wound Healing. *Gels* **2023**, *9*, 185. [CrossRef]
36. Schussler, O.; Falcoz, P.E.; Chachques, J.C.; Alifano, M.; Lecarpentier, Y. Possible Treatment of Myocardial Infarct Based on Tissue Engineering Using a Cellularized Solid Collagen Scaffold Functionalized with Arg-Glyc-Asp (RGD) Peptide. *Int. J. Mol. Sci.* **2021**, *22*, 12563. [CrossRef]
37. Saotome, T.; Shimada, N.; Matsuno, K.; Nakamura, K.; Tabata, Y. Gelatin hydrogel nonwoven fabrics of a cell culture scaffold to formulate 3-dimensional cell constructs. *Regen. Ther.* **2021**, *18*, 418–429. [CrossRef]
38. Barreto, M.E.V.; Medeiros, R.P.; Shearer, A.; Fook, M.V.L.; Montazerian, M.; Mauro, J.C. Gelatin and Bioactive Glass Composites for Tissue Engineering: A Review. *J. Funct. Biomater.* **2022**, *14*, 23. [CrossRef]
39. Mahmood, A.; Patel, D.; Hickson, B.; DesRochers, J.; Hu, X. Recent Progress in Biopolymer-Based Hydrogel Materials for Biomedical Applications. *Int. J. Mol. Sci.* **2022**, *23*, 1415. [CrossRef]
40. Rodgers, U.R.; Weiss, A.S. Cellular interactions with elastin. *Pathol. Biol.* **2005**, *53*, 390–398. [CrossRef] [PubMed]
41. Mithieux, S.M.; Weiss, A.S. Elastin. *Adv. Protein Chem.* **2005**, *70*, 437–461. [PubMed]
42. Urry, D.W. Free energy transduction in polypeptides and proteins based on inverse temperature transitions. *Prog. Biophys. Mol. Biol.* **1992**, *57*, 23–57. [CrossRef] [PubMed]
43. Urry, D.W. Molecular Machines: How Motion and other Functions of Living Organisms can Result from Reversible Chemical Changes. *Angew. Chem. Int. Ed.* **1993**, *32*, 819–841. [CrossRef]
44. Urry, D.W. Physical Chemistry of Biological Free Energy Transduction as Demonstrated by Elastic Protein-Based Polymers. *J. Phys. Chem. B* **1997**, *101*, 11007–11028. [CrossRef]
45. Lim, D.W.; Trabbic-Carlson, K.; Mackay, J.A.; Chilkoti, A. Improved non-chromatographic purification of a recombinant protein by cationic elastin-like polypeptides. *Biomacromolecules* **2007**, *8*, 1417–1424. [CrossRef]
46. Le, D.H.; Hanamura, R.; Pham, D.H.; Kato, M.; Tirrell, D.A.; Okubo, T.; Sugawara-Narutaki, A. Self-assembly of elastin-mimetic double hydrophobic polypeptides. *Biomacromolecules* **2013**, *14*, 1028–1034. [CrossRef]
47. Lim, D.W.; Nettles, D.L.; Setton, L.A.; Chilkoti, A. Rapid cross-linking of elastin-like polypeptides with (hydroxymethyl)phosphines in aqueous solution. *Biomacromolecules* **2007**, *8*, 1463–1470. [CrossRef] [PubMed]
48. Trabbic-Carlson, K.; Setton, L.A.; Chilkoti, A. Swelling and mechanical behaviors of chemically cross-linked hydrogels of elastin-like polypeptides. *Biomacromolecules* **2003**, *4*, 572–580. [CrossRef]
49. Zhang, Y.N.; Avery, R.K.; Vallmajo-Martin, Q.; Assmann, A.; Vegh, A.; Memic, A.; Olsen, B.D.; Annabi, N.; Khademhosseini, A. A Highly Elastic and Rapidly Crosslinkable Elastin-Like Polypeptide-Based Hydrogel for Biomedical Applications. *Adv. Funct. Mater.* **2015**, *25*, 4814–4826. [CrossRef] [PubMed]
50. Saiki, I. Cell adhesion molecules and cancer metastasis. *Jpn. J. Pharmacol.* **1997**, *75*, 215–242. [CrossRef]
51. Kaur, J.; Reinhardt, D.P. Extracellular Matrix (ECM) Molecules. In *Stem Cell Biology and Tissue Engineering in Dental Sciences*; Vishwakarma, A., Sharpe, P., Shi, S., Ramalingam, M., Eds.; Academic Press, Elsevier Inc.: Waltham, MA, USA, 2015; pp. 25–45.
52. Trujillo, S.; Gonzalez-Garcia, C.; Rico, P.; Reid, A.; Windmill, J.; Dalby, M.J.; Salmeron-Sanchez, M. Engineered 3D hydrogels with full-length fibronectin that sequester and present growth factors. *Biomaterials* **2020**, *252*, 120104. [CrossRef]
53. Nakagawa, M.; Taniguchi, Y.; Senda, S.; Takizawa, N.; Ichisaka, T.; Asano, K.; Morizane, A.; Doi, D.; Takahashi, J.; Nishizawa, M.; et al. A novel efficient feeder-free culture system for the derivation of human induced pluripotent stem cells. *Sci. Rep.* **2014**, *4*, 3594. [CrossRef] [PubMed]
54. Rodin, S.; Domogatskaya, A.; Ström, S.; Hansson, E.M.; Chien, K.R.; Inzunza, J.; Hovatta, O.; Tryggvason, K. Long-term self-renewal of human pluripotent stem cells on human recombinant laminin-511. *Nat. Biotechnol.* **2010**, *28*, 611–615. [CrossRef]

55. Jury, M.; Matthiesen, I.; Rasti Boroojeni, F.; Ludwig, S.L.; Civitelli, L.; Winkler, T.E.; Selegård, R.; Herland, A.; Aili, D. Bioorthogonally Cross-Linked Hyaluronan-Laminin Hydrogels for 3D Neuronal Cell Culture and Biofabrication. *Adv. Healthc. Mater.* **2022**, *11*, e2102097. [CrossRef]
56. Drury, J.L.; Mooney, D.J. Hydrogels for tissue engineering: Scaffold design variables and applications. *Biomaterials* **2003**, *24*, 4337–4351. [CrossRef]
57. Toole, B.P. Hyaluronan in morphogenesis. *J. Intern. Med.* **1997**, *242*, 35–40. [CrossRef]
58. Toole, B.P. Hyaluronan: From extracellular glue to pericellular cue. *Nat. Rev. Cancer* **2004**, *4*, 528–539. [CrossRef] [PubMed]
59. Sahoo, S.; Chung, C.; Khetan, S.; Burdick, J.A. Hydrolytically degradable hyaluronic acid hydrogels with controlled temporal structures. *Biomacromolecules* **2008**, *9*, 1088–1092. [CrossRef] [PubMed]
60. Burdick, J.A.; Prestwich, G.D. Hyaluronic acid hydrogels for biomedical applications. *Adv. Mater.* **2011**, *23*, H41–H56. [CrossRef] [PubMed]
61. Horder, H.; Guaza Lasheras, M.; Grummel, N.; Nadernezhad, A.; Herbig, J.; Ergün, S.; Teßmar, J.; Groll, J.; Fabry, B.; Bauer-Kreisel, P.; et al. Bioprinting and Differentiation of Adipose-Derived Stromal Cell Spheroids for a 3D Breast Cancer-Adipose Tissue Model. *Cells.* **2021**, *10*, 803. [CrossRef]
62. Abka-Khajouei, R.; Tounsi, L.; Shahabi, N.; Patel, A.K.; Abdelkafi, S.; Michaud, P. Structures, Properties and Applications of Alginates. *Mar. Drugs* **2022**, *20*, 364. [CrossRef]
63. Mørch, Y.A.; Donati, I.; Strand, B.L.; Skjåk-Braek, G. Effect of Ca^{2+}, Ba^{2+}, and Sr^{2+} on alginate microbeads. *Biomacromolecules* **2006**, *7*, 1471–1480. [CrossRef] [PubMed]
64. Draget, K.I.; Strand, B.; Hartmann, M.; Valla, S.; Smidsrød, O.; Skjåk-Braek, G. Ionic and Acid Gel Formation of Epimerised Alginates; the Effect of Alge4. *Int. J. Biol. Macromol.* **2000**, *27*, 117–122. [CrossRef] [PubMed]
65. Ching, S.H.; Bansal, N.; Bhandari, B. Alginate Gel Particles—A Review of Production Techniques and Physical Properties. *Crit. Rev. Food Sci. Nutr.* **2017**, *57*, 1133–1152. [CrossRef] [PubMed]
66. He, Q.; Tong, T.; Yu, C.; Wang, Q. Advances in Algin and Alginate-Hybrid Materials for Drug Delivery and Tissue Engineering. *Mar. Drugs* **2022**, *21*, 14. [CrossRef]
67. Zhang, Q.; Gonelle-Gispert, C.; Li, Y.; Geng, Z.; Gerber-Lemaire, S.; Wang, Y.; Buhler, L. Islet Encapsulation: New Developments for the Treatment of Type 1 Diabetes. *Front. Immunol.* **2022**, *13*, 869984. [CrossRef]
68. Sun, S.; Cui, Y.; Yuan, B.; Dou, M.; Wang, G.; Xu, H.; Wang, J.; Yin, W.; Wu, D.; Peng, C. Drug delivery systems based on polyethylene glycol hydrogels for enhanced bone regeneration. *Front. Bioeng. Biotechnol.* **2023**, *11*, 1117647. [CrossRef]
69. Yin, L.; Pang, Y.; Shan, L.; Gu, J. The In Vivo Pharmacokinetics of Block Copolymers Containing Polyethylene Glycol Used in Nanocarrier Drug Delivery Systems. *Drug Metab. Dispos.* **2022**, *50*, 827–836. [CrossRef]
70. Ju, H.J.; Ji, Y.B.; Kim, S.; Yun, H.W.; Kim, J.H.; Min, B.H.; Kim, M.S. Development and In Vivo Assessment of an Injectable Crosslinked Cartilage Acellular Matrix-PEG Hydrogel Scaffold Derived from Porcine Cartilage for Tissue Engineering. *Macromol. Biosci.* **2023**, e2300029. [CrossRef]
71. Friend, N.E.; McCoy, A.J.; Stegemann, J.P.; Putnam, A.J. A combination of matrix stiffness and degradability dictate microvascular network assembly and remodeling in cell-laden poly(ethylene glycol) hydrogels. *Biomaterials* **2023**, *295*, 122050. [CrossRef]
72. Mongia, N.K.; Anseth, K.S.; Peppas, N.A. Mucoadhesive poly(vinyl alcohol) hydrogels produced by freezing/thawing processes: Applications in the development of wound healing systems. *J. Biomater. Sci. Polym. Ed.* **1996**, *7*, 1055–1064. [CrossRef]
73. Zhu, C.; Huang, C.; Zhang, W.; Ding, X.; Yang, Y. Biodegradable-Glass-Fiber Reinforced Hydrogel Composite with Enhanced Mechanical Performance and Cell Proliferation for Potential Cartilage Repair. *Int. J. Mol. Sci.* **2022**, *23*, 8717. [CrossRef]
74. Singhvi, M.S.; Zinjarde, S.S.; Gokhale, D.V. Polylactic acid: Synthesis and biomedical applications. *J. Appl. Microbiol.* **2019**, *127*, 1612–1626. [CrossRef]
75. Savioli Lopes, M.; Jardini, A.L.; Maciel Filho, R. Poly (Lactic Acid) Production for Tissue Engineering Applications. *Procedia Eng.* **2012**, *42*, 1402–1413. [CrossRef]
76. Narayanan, G.; Vernekar, V.N.; Kuyinu, E.L.; Laurencin, C.T. Poly (lactic acid)-based biomaterials for orthopaedic regenerative engineering. *Adv. Drug Deliv. Rev.* **2016**, *107*, 247–276. [CrossRef] [PubMed]
77. Sekine, H.; Shimizu, T.; Sakaguchi, K.; Dobashi, I.; Wada, M.; Yamato, M.; Kobayashi, E.; Umezu, M.; Okano, T. In vitro fabrication of functional three-dimensional tissues with perfusable blood vessels. *Nat. Commun.* **2013**, *4*, 1399. [CrossRef] [PubMed]
78. Nishida, K.; Yamato, M.; Hayashida, Y.; Watanabe, K.; Yamamoto, K.; Adachi, E.; Nagai, S.; Kikuchi, A.; Maeda, N.; Watanabe, H.; et al. Corneal reconstruction with tissue-engineered cell sheets composed of autologous oral mucosal epithelium. *N. Engl. J. Med.* **2004**, *351*, 1187–1196. [CrossRef]
79. Ohki, T.; Yamato, M.; Murakami, D.; Takagi, R.; Yang, J.; Namiki, H.; Okano, T.; Takasaki, K. Treatment of oesophageal ulcerations using endoscopic transplantation of tissue-engineered autologous oral mucosal epithelial cell sheets in a canine model. *Gut* **2006**, *55*, 1704–1710. [CrossRef]
80. Ohki, T.; Yamato, M.; Ota, M.; Takagi, R.; Murakami, D.; Kondo, M.; Sasaki, R.; Namiki, H.; Okano, T.; Yamamoto, M. Prevention of esophageal stricture after endoscopic submucosal dissection using tissue-engineered cell sheets. *Gastroenterology* **2012**, *143*, 582–588.e2. [CrossRef] [PubMed]
81. Iwata, T.; Yamato, M.; Tsuchioka, H.; Takagi, R.; Mukobata, S.; Washio, K.; Okano, T.; Ishikawa, I. Periodontal regeneration with multi-layered periodontal ligament-derived cell sheets in a canine model. *Biomaterials* **2009**, *30*, 2716–2723. [CrossRef] [PubMed]
82. Takushi, E.; Asato, L.; Nakada, T. Edible eyeballs from fish. *Nature* **1990**, *345*, 298–299. [CrossRef]

83. Takezawa, T.; Nitani, A.; Shimo-Oka, T.; Takayama, Y. A protein-permeable scaffold of a collagen vitrigel membrane useful for reconstructing crosstalk models between two different cell types. *Cells Tissues Organs* **2007**, *185*, 237–241. [CrossRef] [PubMed]
84. Puleo, C.M.; Ambrose, W.M.; Takezawa, T.; Elisseeff, J.; Wang, T.H. Integration and application of vitrified collagen in multilayered microfluidic devices for corneal microtissue culture. *Lab Chip* **2009**, *9*, 3221–3227. [CrossRef]
85. Zhao, J.; Shinkai, M.; Takezawa, T.; Ohba, S.; Chung, U.I.; Nagamune, T. Bone regeneration using collagen type I vitrigel with bone morphogenetic protein-2. *J. Biosci. Bioeng.* **2009**, *107*, 318–323. [CrossRef] [PubMed]
86. Yoshida, J.; Oshikata-Miyazaki, A.; Yokoo, S.; Yamagami, S.; Takezawa, T.; Amano, S. Development and evaluation of porcine atelocollagen vitrigel membrane with a spherical curve and transplantable artificial corneal endothelial grafts. *Investig. Ophthalmol. Vis. Sci.* **2014**, *55*, 4975–4981. [CrossRef] [PubMed]
87. Chae, J.J.; Ambrose, W.M.; Espinoza, F.A.; Mulreany, D.G.; Ng, S.; Takezawa, T.; Trexler, M.M.; Schein, O.D.; Chuck, R.S.; Elisseeff, J.H. Regeneration of corneal epithelium utilizing a collagen vitrigel membrane in rabbit models for corneal stromal wound and limbal stem cell deficiency. *Acta Ophthalmol.* **2015**, *93*, e57–e66. [CrossRef] [PubMed]
88. Okuyama, H.; Ohnishi, H.; Nakamura, R.; Yamashita, M.; Kishimoto, Y.; Tateya, I.; Suehiro, A.; Gotoh, S.; Takezawa, T.; Nakamura, T.; et al. Transplantation of multiciliated airway cells derived from human iPS cells using an artificial tracheal patch into rat trachea. *J. Tissue Eng. Regen. Med.* **2019**, *13*, 1019–1030. [CrossRef]
89. Chae, J.J.; Mulreany, D.G.; Guo, Q.; Lu, Q.; Choi, J.S.; Strehin, I.; Espinoza, F.A.; Schein, O.; Trexler, M.M.; Bower, K.S.; et al. Application of a collagen-based membrane and chondroitin sulfate-based hydrogel adhesive for the potential repair of severe ocular surface injuries. *Mil. Med.* **2014**, *179*, 686–694. [CrossRef] [PubMed]
90. Wang, Y.; Kanie, K.; Takezawa, T.; Horikawa, M.; Kaneko, K.; Sugimoto, A.; Yamawaki-Ogata, A.; Narita, Y.; Kato, R. Bi-layered carboxymethyl cellulose-collagen vitrigel dual-surface adhesion-prevention membrane. *Carbohydr. Polym.* **2022**, *285*, 119223. [CrossRef] [PubMed]
91. Zamprogno, P.; Thoma, G.; Cencen, V.; Ferrari, D.; Putz, B.; Michler, J.; Fantner, G.E.; Guenat, O.T. Mechanical Properties of Soft Biological Membranes for Organ-on-a-Chip Assessed by Bulge Test and AFM. *ACS Biomater. Sci. Eng.* **2021**, *7*, 2990–2997. [CrossRef]
92. Stacey, G.N.; Day, J.G. Long-term Ex Situ conservation of biological resources and the role of biological resource centers. In *Cryopreservation and Freeze-Drying Protocols*, 2nd ed.; Day, J.G., Stacey, G.N., Eds.; Humana Press: Totowa, NJ, USA, 2007; pp. 1–14.
93. Mazur, P. Kinetics of Water Loss from Cells at Subzero Temperatures and the Likelihood of Intracellular Freezing. *J. Gen. Physiol.* **1963**, *47*, 347–369. [CrossRef]
94. Mazur, P.; Leibo, S.P.; Chu, E.H. A two-factor hypothesis of freezing injury. Evidence from Chinese hamster tissue-culture cells. *Exp. Cell Res.* **1972**, *71*, 345–355. [CrossRef] [PubMed]
95. Pegg, D.E. Principles of cryopreservation. In *Cryopreservation and Freeze-Drying Protocols*, 2nd ed.; Day, J.G., Stacey, G.N., Eds.; Humana Press: Totowa, NJ, USA, 2007; pp. 39–57.
96. Taylor, M.J. Biology of cell survival in the cold: The basis for biopreservation of tissues and organs. In *Advances in Biopreservation*; Baust, J.G., Baust, J.M., Eds.; CRC Press: New York, NY, USA; Taylor and Francis: New York, NY, USA, 2007; pp. 15–62.
97. Baust, J.G. Concepts in biopreservation. In *Advances in Biopreservation*; Baust, J.G., Baust, J.M., Eds.; CRC Press: New York, NY, USA; Taylor and Francis: New York, NY, USA, 2007; pp. 1–14.
98. Acker, J.P. The use of intracellular protectants in cell biopreservation. In *Advances in Biopreservation*; Baust, J.G., Baust, J.M., Eds.; CRC Press: New York, NY, USA; Taylor and Francis: New York, NY, USA, 2007; pp. 299–320.
99. Polge, C.; Smith, A.U.; Parkes, A.S. Revival of spermatozoa after vitrification and dehydration at low temperatures. *Nature* **1949**, *164*, 666. [CrossRef]
100. Lovelock, J.E.; Bishop, M.W. Prevention of freezing damage to living cells by dimethyl sulphoxide. *Nature* **1959**, *183*, 1394–1395. [CrossRef]
101. Morris, C.B. Cryopreservation of animal and human cell lines. In *Cryopreservation and Freeze-Drying Protocols*, 2nd ed.; Day, J.G., Stacey, G.N., Eds.; Humana Press: Totowa, NJ, USA, 2007; pp. 227–236.
102. U.S. Food and Drug Administration (FDA). Cellular & Gene Therapy Guidances. Available online: https://www.fda.gov/vaccines-blood-biologics/biologics-guidances/cellular-gene-therapy-guidances (accessed on 6 December 2022).
103. The Pharmaceuticals and Medical Devices Agency (PMDA). Review Reports: Regenerative Medical Products. Available online: https://www.pmda.go.jp/english/review-services/reviews/approved-information/0004.html (accessed on 6 December 2022).
104. The European Medicines Agency (EMA). Human Cell-Based Medicinal Products—Scientific Guideline. Available online: https://www.ema.europa.eu/en/human-cell-based-medicinal-products-scientific-guideline (accessed on 6 December 2022).
105. ClinicalTrials.gov. Available online: https://clinicaltrials.gov/ (accessed on 6 December 2022).
106. Yuan, C.; Gao, J.; Guo, J.; Bai, L.; Marshall, C.; Cai, Z.; Wang, L.; Xiao, M. Dimethyl sulfoxide damages mitochondrial integrity and membrane potential in cultured astrocytes. *PLoS ONE* **2014**, *9*, e107447. [CrossRef]
107. Miyamoto, Y.; Teramoto, N.; Hayashi, S.; Enosawa, S. An improvement in the attaching capability of cryopreserved human hepatocytes by a proteinaceous high molecule, sericin, in the serum-free solution. *Cell Transplant.* **2010**, *19*, 701–706. [CrossRef] [PubMed]
108. Miyamoto, Y.; Suzuki, S.; Nomura, K.; Enosawa, S. Improvement of hepatocyte viability after cryopreservation by supplementation of long-chain oligosaccharide in the freezing medium in rats and humans. *Cell Transplant.* **2006**, *15*, 911–919. [CrossRef] [PubMed]

109. Miyamoto, Y.; Oishi, K.; Yukawa, H.; Noguchi, H.; Sasaki, M.; Iwata, H.; Hayashi, S. Cryopreservation of human adipose tissue-derived stem/progenitor cells using the silk protein sericin. *Cell Transplant.* **2012**, *21*, 617–622. [CrossRef]
110. Yamatoya, K.; Nagai, Y.; Teramoto, N.; Kang, W.; Miyado, K.; Nakata, K.; Yagi, T.; Miyamoto, Y. Cryopreservation of undifferentiated and differentiated human neuronal cells. *Regen. Ther.* **2022**, *19*, 58–68. [CrossRef] [PubMed]
111. Taylor, M.J. Sub-zero preservation and the prospect of long-term storage of multicellular tissues and organs. In *Transplantation Immunology: Clinical and Experimental*; Calne, R.Y., Ed.; Oxford University Press: Oxford, UK, 1984; pp. 360–390.
112. Brockbank, K.G.M.; Taylor, M.J. Tissue preservation. In *Advances in Biopreservation*; Baust, J.G., Baust, J.M., Eds.; CRC Press: New York, NY, USA; Taylor and Francis: New York, NY, USA, 2007; pp. 157–196.
113. Taylor, M.J.; Song, Y.C.; Brockbank, K.G.M. Vitrification in tissue preservation: New developments. In *Life in the Frozen State*; Fuller, B.J., Lane, N., Benson, E.E., Eds.; CRC Press: London, UK; Taylor and Francis: London, UK, 2004; pp. 603–641.
114. Karlsson, J.O.; Toner, M. Long-term storage of tissues by cryopreservation: Critical issues. *Biomaterials* **1996**, *17*, 243–256. [CrossRef]
115. Takahashi, T.; Hirsh, A.; Erbe, E.F.; Bross, J.B.; Steere, R.L.; Williams, R.J. Vitrification of human monocytes. *Cryobiology* **1986**, *23*, 103–115. [CrossRef]
116. Jutte, N.H.; Heyse, P.; Jansen, H.G.; Bruining, G.J.; Zeilmaker, G.H. Vitrification of mouse islets of Langerhans: Comparison with a more conventional freezing method. *Cryobiology* **1987**, *24*, 292–302. [CrossRef] [PubMed]
117. Jutte, N.H.; Heyse, P.; Jansen, H.G.; Bruining, G.J.; Zeilmaker, G.H. Vitrification of human islets of Langerhans. *Cryobiology* **1987**, *24*, 403–411. [CrossRef] [PubMed]
118. Van Wagtendonk-De Leeuw, A.M.; Den Daas, J.H.; Kruip, T.A.; Rall, W.F. Comparison of the efficacy of conventional slow freezing and rapid cryopreservation methods for bovine embryos. *Cryobiology* **1995**, *32*, 157–167. [CrossRef] [PubMed]
119. Ohlendorf, C.; Tomford, W.W.; Mankin, H.J. Chondrocyte survival in cryopreserved osteochondral articular cartilage. *J. Orthop. Res.* **1996**, *14*, 413–416. [CrossRef] [PubMed]
120. Tomford, W.W.; Fredericks, G.R.; Mankin, H.J. Studies on cryopreservation of articular cartilage chondrocytes. *J. Bone Jt. Surg. Am.* **1984**, *66*, 253–259. [CrossRef]
121. Song, Y.C.; An, Y.H.; Kang, Q.K.; Li, C.; Boggs, J.M.; Chen, Z.; Taylor, M.J.; Brockbank, K.G. Vitreous preservation of articular cartilage grafts. *J. Investig. Surg.* **2004**, *17*, 65–70. [CrossRef]
122. Chong, Y.K.; Toh, T.B.; Zaiden, N.; Poonepalli, A.; Leong, S.H.; Ong, C.E.; Yu, Y.; Tan, P.B.; See, S.J.; Ng, W.H.; et al. Cryopreservation of neurospheres derived from human glioblastoma multiforme. *Stem Cells* **2009**, *27*, 29–39. [CrossRef]
123. Spurrier, R.G.; Speer, A.L.; Grant, C.N.; Levin, D.E.; Grikscheit, T.C. Vitrification preserves murine and human donor cells for generation of tissue-engineered intestine. *J. Surg. Res.* **2014**, *190*, 399–406. [CrossRef]
124. Mukherjee, N.; Chen, Z.; Sambanis, A.; Song, Y. Effects of cryopreservation on cell viability and insulin secretion in a model tissue-engineered pancreatic substitute (TEPS). *Cell Transplant.* **2005**, *14*, 449–456. [CrossRef]
125. Kearney, J.N. Guidelines on processing and clinical use of skin allografts. *Clin. Dermatol.* **2005**, *23*, 357–364. [CrossRef]
126. Kinoshita, S.; Nakamura, T. Development of cultivated mucosal epithelial sheet transplantation for ocular surface reconstruction. *Artif. Organs* **2004**, *28*, 22–27. [CrossRef] [PubMed]
127. Kito, K.; Kagami, H.; Kobayashi, C.; Ueda, M.; Terasaki, H. Effects of cryopreservation on histology and viability of cultured corneal epithelial cell sheets in rabbit. *Cornea* **2005**, *24*, 735–741. [CrossRef]
128. Chen, F.; Zhang, W.; Wu, W.; Jin, Y.; Cen, L.; Kretlow, J.D.; Gao, W.; Dai, Z.; Wang, J.; Zhou, G.; et al. Cryopreservation of tissue-engineered epithelial sheets in trehalose. *Biomaterials* **2011**, *32*, 8426–8435. [CrossRef] [PubMed]
129. Oliva, J.; Florentino, A.; Bardag-Gorce, F.; Niihara, Y. Vitrification and storage of oral mucosa epithelial cell sheets. *J. Tissue Eng. Regen. Med.* **2019**, *13*, 1153–1163. [CrossRef]
130. Zhang, D.; Shao, J.; Zhuang, J.; Zhou, S.; Yin, S.; Wu, F.; Hou, J.; Wang, X. Biobanked human foreskin epithelial cell sheets reduce inflammation and promote wound healing in a nude mouse model. *BMC Biotechnol.* **2021**, *21*, 11. [CrossRef]
131. Maehara, M.; Sato, M.; Watanabe, M.; Matsunari, H.; Kokubo, M.; Kanai, T.; Sato, M.; Matsumura, K.; Hyon, S.H.; Yokoyama, M.; et al. Development of a novel vitrification method for chondrocyte sheets. *BMC Biotechnol.* **2013**, *13*, 58. [CrossRef] [PubMed]
132. Hayashi, A.; Maehara, M.; Uchikura, A.; Matsunari, H.; Matsumura, K.; Hyon, S.H.; Sato, M.; Nagashima, H. Development of an efficient vitrification method for chondrocyte sheets for clinical application. *Regen. Ther.* **2020**, *14*, 215–221. [CrossRef] [PubMed]
133. Ohkawara, H.; Miyagawa, S.; Fukushima, S.; Yajima, S.; Saito, A.; Nagashima, H.; Sawa, Y. Development of a vitrification method for preserving human myoblast cell sheets for myocardial regeneration therapy. *BMC Biotechnol.* **2018**, *18*, 56. [CrossRef] [PubMed]
134. Ochiai, J.; Niihara, Y.; Oliva, J. Measurement of the Adipose Stem Cells Cell Sheets Transmittance. *Bioengineering* **2021**, *8*, 93. [CrossRef]
135. Li, M.; Feng, C.; Gu, X.; He, Q.; Wei, F. Effect of cryopreservation on proliferation and differentiation of periodontal ligament stem cell sheets. *Stem Cell Res. Ther.* **2017**, *8*, 77. [CrossRef]
136. Motoike, S.; Kajiya, M.; Komatsu, N.; Takewaki, M.; Horikoshi, S.; Matsuda, S.; Ouhara, K.; Iwata, T.; Takeda, K.; Fujita, T.; et al. Cryopreserved clumps of mesenchymal stem cell/extracellular matrix complexes retain osteogenic capacity and induce bone regeneration. *Stem Cell Res. Ther.* **2018**, *9*, 73. [CrossRef] [PubMed]

137. Ohashi, K.; Mukobata, S.; Utoh, R.; Yamashita, S.; Masuda, T.; Sakai, H.; Okano, T. Production of islet cell sheets using cryopreserved islet cells. *Transplant. Proc.* **2011**, *43*, 3188–3191. [CrossRef] [PubMed]
138. Shimizu, T.; Akahane, M.; Ueha, T.; Kido, A.; Omokawa, S.; Kobata, Y.; Murata, K.; Kawate, K.; Tanaka, Y. Osteogenesis of cryopreserved osteogenic matrix cell sheets. *Cryobiology* **2013**, *66*, 326–332. [CrossRef] [PubMed]
139. Ike, S.; Ueno, K.; Yanagihara, M.; Mizoguchi, T.; Harada, T.; Suehiro, K.; Kurazumi, H.; Suzuki, R.; Kondo, T.; Murata, T.; et al. Cryopreserved allogenic fibroblast sheets: Development of a promising treatment for refractory skin ulcers. *Am. J. Transl. Res.* **2022**, *14*, 3879–3892.
140. Sheikhi, M.; Hultenby, K.; Niklasson, B.; Lundqvist, M.; Hovatta, O. Clinical grade vitrification of human ovarian tissue: An ultrastructural analysis of follicles and stroma in vitrified tissue. *Hum. Reprod.* **2011**, *26*, 594–603. [CrossRef]
141. Marco-Jimenez, F.; Garcia-Dominguez, X.; Jimenez-Trigos, E.; Vera-Donoso, C.D.; Vicente, J.S. Vitrification of kidney precursors as a new source for organ transplantation. *Cryobiology* **2015**, *70*, 278–282. [CrossRef]
142. Li, T.; Zhou, C.; Liu, C.; Mai, Q.; Zhuang, G. Bulk vitrification of human embryonic stem cells. *Hum. Reprod.* **2008**, *23*, 358–364. [CrossRef]
143. Miyamoto, Y.; Enosawa, S.; Takeuchi, T.; Takezawa, T. Cryopreservation in situ of cell monolayers on collagen vitrigel membrane culture substrata: Ready-to-use preparation of primary hepatocytes and ES cells. *Cell Transplant.* **2009**, *18*, 619–626. [CrossRef]
144. Martinez-Garcia, F.D.; Fischer, T.; Hayn, A.; Mierke, C.T.; Burgess, J.K.; Harmsen, M.C. A Beginner's Guide to the Characterization of Hydrogel Microarchitecture for Cellular Applications. *Gels* **2022**, *8*, 535. [CrossRef]
145. Revete, A.; Aparicio, A.; Cisterna, B.A.; Revete, J.; Luis, L.; Ibarra, E.; Segura González, E.A.; Molino, J.; Reginensi, D. Advancements in the Use of Hydrogels for Regenerative Medicine: Properties and Biomedical Applications. *Int. J. Biomater.* **2022**, *2022*, 3606765. [CrossRef]
146. Zhang, C.; Zhou, Y.; Zhang, L.; Wu, L.; Chen, Y.; Xie, D.; Chen, W. Hydrogel Cryopreservation System: An Effective Method for Cell Storage. *Int. J. Mol. Sci.* **2018**, *19*, 3330. [CrossRef] [PubMed]
147. Manferdini, C.; Gabusi, E.; Saleh, Y.; Lenzi, E.; D'Atri, G.; Ricotti, L.; Lisignoli, G. Mesenchymal Stromal Cells Laden in Hydrogels for Osteoarthritis Cartilage Regeneration: A Systematic Review from In Vitro Studies to Clinical Applications. *Cells* **2022**, *11*, 3969. [CrossRef] [PubMed]
148. Campbell, L.H.; Brockbank, K.G.M. Development of a Vitrification Preservation Process for Bioengineered Epithelial Constructs. *Cells* **2022**, *11*, 1115. [CrossRef] [PubMed]

gels

Article

Curative Effects of Copper Iodide Embedded on Gallic Acid Incorporated in a Poly(vinyl alcohol) (PVA) Liquid Bandage

Putita Phetcharat [1], Pakakrong Sangsanoh [1,2], Chasuda Choipang [1,2], Sonthaya Chaiarwut [1,2], Orawan Suwantong [3,4], Piyachat Chuysinuan [5] and Pitt Supaphol [1,2,*]

1 The Petroleum and Petrochemical College, Chulalongkorn University, Bangkok 10330, Thailand
2 Research Unit on Herbal Extracts-Infused Advanced Wound Dressing, Chulalongkorn University, Bangkok 10330, Thailand
3 School of Science, Mae Fah Luang University, Chiang Rai 57100, Thailand
4 Center of Chemical Innovation for Sustainability (CIS), Mae Fah Luang University, Chiang Rai 57100, Thailand
5 Laboratory of Organic Synthesis, Chulabhorn Research Institute, Bangkok 10210, Thailand
* Correspondence: pitt.s@chula.ac.th; Tel.: +66-2-21-84117

Abstract: In daily life, people are often receiving minor cuts due to carelessness, leaving wounds on the skin. If wound healing is interrupted and the healing process does not finish, pathogens can easily enter wounds and cause infection. Liquid bandages are a fast and convenient way to help stop the bleeding of superficial wounds. Moreover, antibacterial agents in liquid bandages can promote wound restoration and fight bacteria. Herein, a poly(vinyl alcohol) (PVA) liquid bandage incorporating copper iodide nanoparticles (CuI NPs) was developed. CuI NPs were synthesized through green synthesis using gallic acid (GA) as a reducing and capping agent. The sizes of the CuI NPs, which were dependent on the concentration of GA, were 41.45, 43.51 and 49.71 nm, with the concentrations of gallic acid being 0, 2.5 mM and 5.0 mM, respectively. CuI NPs were analyzed using FTIR, XRD and SEM and tested for peroxidase-like properties and antibacterial activity. Then, PVA liquid bandages were formulated with different concentrations of stock CuI suspension. The results revealed that PVA liquid bandages incorporating 0.190% CuI synthesized with 5.0 mM of GA can kill bacteria within 24 h and have no harmful effects on human fibroblast cells.

Keywords: poly(vinyl alcohol) (PVA); cuprous iodide (CuI); green synthesis; liquid bandage

Citation: Phetcharat, P.; Sangsanoh, P.; Choipang, C.; Chaiarwut, S.; Suwantong, O.; Chuysinuan, P.; Supaphol, P. Curative Effects of Copper Iodide Embedded on Gallic Acid Incorporated in a Poly(vinyl alcohol) (PVA) Liquid Bandage. *Gels* **2023**, *9*, 53. https://doi.org/10.3390/gels9010053

Academic Editors: Arish Dasan, Ashokraja Chandrasekar and Nupur Kohli

Received: 20 October 2022
Revised: 23 December 2022
Accepted: 25 December 2022
Published: 8 January 2023

1. Introduction

Damage to the skin makes the human body vulnerable to pathogenic invasion. Pathogens colonize their hosts and penetrate deeper tissues. As a result, wound healing can be disrupted and delayed. Therefore, methodical wound dressing to cover wound sites is a traditional approach to prevent bacterial wound infection. Currently, various cyto-compatible polymers, especially synthetic polymers, are widely applied for wound bandage formulation and several biomedical applications. Examples of synthetic polymers used for wound dressing are poly(vinyl alcohol) (PVA), poly(lactic-co-glycolic acid) (PLGA), polylactide (PLA), polyurethanes (PUs), poly(ethylene oxide) (PEO)/poly(ethylene glycol) (PEG), poly(hydroxyethyl methacrylate) (PHEMA), and poly(vinyl pyrrolidone) (PVP), due to their excellent biocompatibility, which makes them non-toxic to human cells. Additionally, most of these have been synthesized in 3D network hydrogels, which are a main focus of biomedical research [1,2].

Ideal wound dressings must have an appropriate water vapor transmission rate (WVTR), capability to provide moisture, suitable temperature and humidity to enhance wound healing, gaseous permeation, excellent antimicrobial properties, strong mechanical performance and the capacity to deliver active agents [3]. Poly(vinyl alcohol), a conventional synthetic polymer, has long been used for biomedical applications due to its non-toxicity to mammalian cells, biodegradability and biocompatibility. PVA has a large amount of hydroxyl groups in its structure, which provide a hydrophilic structure and sufficient swelling capacity, allowing it to absorb exudation to stop bleeding. Poly(vinyl alcohol) wound dressings, which are also popularly used in commercial products, are liquid gel bandages with quick film formation to prevent infection. In addition, PVA thin film is breathable to allow the exchange of gas and moisture in the wound, providing an optimal environment for wound healing [4,5].

Despite the fact that PVA is widely used for liquid bandages, it does not have inherent antibacterial activity because of its chemical structure. Therefore, active bacteria-killing agents play a significant role in PVA wound dressings in the production of alternative antibacterial bandages. Presently, there is various research reporting on polymeric materials containing metal nanoparticles to promote the wound healing process. It is widely known that metal NPs have significant potential as antibiotics [6–8]. Some reports have shown the strong ability of cyto-compatible polymers to kill bacteria when fabricated with silver nanoparticles or gold nanoparticles, due to their small surface-area-to-volume ratio [2,9,10]. Copper nanoparticles and copper-based compounds have been employed as practical materials in human tissues for centuries and have been used in biomedical and pharmaceutical applications since early 2005 [11]. They have been proven to be toxic to several bacterial strains through various mechanisms, including ROS production, metal ions released from the metal surface and denaturation of the biomolecules in bacteria [12–14]. Copper (I) iodide or cuprous iodide has been used as an active agent in wound dressings for its significant antimicrobial activity compared to other copper (II) or cupric compounds [15]. The reaction kinetics of copper compounds kill bacteria in a manner similar to that of natural enzymes in the human body, making them more attractive for the development of nano-enzyme antibacterial systems [16]. However, suitable content and crystallite size of metal nanoparticles should be considered for these materials to be safely applied to wound sites. Otherwise, they may be harmful to cell tissue.

Various research has reported on diverse methods of controlling the crystallite size of synthesized copper iodide, such as the chemical route, co-precipitation and the microwave-hydrothermal assisted method. However, these syntheses require hazardous raw materials, complicated synthetic steps and high temperature, and are time consuming, resulting in high cost and biological harm. Consequently, less toxic processes are essential for copper iodide synthesis [17]. Green synthesis or biosynthesis is an eco-friendly method proposed to eliminate toxic and polluting chemical agents, consume less energy and allow the use of green solvents (water, ethanol, ethyl acetate, etc.) [17]. In addition, in green synthesis of copper iodide, the size of metal nanoparticles can be controlled by varying the concentrations of biochemical substances [18]. The usage of green chemical substances to synthesize copper iodide nanoparticles has become an extensive field, since they provide much greener and ecologically friendly routes of synthesis. Gallic acid (GA) is a well-known natural antioxidant derived from several plants, such as berries, fruit and tea. Gallic acid is a type of phenolic compound that reduces inflammation and oxidative stress damage. There are several reports on the use of gallic acid as a bio-stabilizer and bio-reducer in the synthesis of functional nano-enzyme materials [19]. By varying the concentration of gallic acid, one can characteristically produce a wide range of nanoparticles and control their size.

In this work, we study the therapeutic effects of antibacterial liquid bandages produced with gallic acid–copper iodide nanoparticles (GA-CuI NPs) incorporated in poly(vinyl alcohol) through ultrasonic homogenization. The appropriate content of GA-CuI NPs to load into PVA liquid bandages was investigated. The physical and chemical properties of PVA liquid bandages with different concentrations of CuI suspension were characterized by various methodologies, including FTIR, XRD, FESEM, drying time and water vapor transmission rate. In addition, the biological properties of synthesized liquid bandages were tested in vitro for cytotoxicity to human dermal fibroblast cells and antibacterial activity against *E. coli* (Gram-negative) and MRSA (Gram-positive) to confirm their practicality for use in biomedical applications.

2. Results and Discussion

2.1. Encapsulation and Stabilization of Gallic Acid–Cuprous Iodide Nanoparticles

The synthesis of cuprous iodide (CuI) nanoparticles that is stabilized and reduced by gallic acid is a simple and convenient approach of green chemistry. Green synthesis of metallic compounds is rapid and highly effective when using phytochemicals, which are bioactive molecules from plants (such as gallic acid (GA)), as reducing and stabilizing agents, because they have abundant hydroxyl groups, phenolic groups and carboxylic groups, which are beneficial for capping metallic components. Several studies have shown that gallic acid concentration affects the size of metallic nanoparticles due to its bifunctional nature, i.e., its enol form and keto form [20–22]. Gallic acid is subjected to a two-electron oxidation process with two hydroxyl groups, which converts the enol form to the keto form. This formation of keto(quinone) enhances the interaction with the cuprous iodide nanoparticles and stabilizes the negative carboxyl group.

2.1.1. X-ray Diffraction (XRD) Analysis

The crystalline structure and phase transition of synthesized CuI NPs and GA-CuI NPs were investigated using X-ray diffraction (XRD) analysis. Figure 1a shows the XRD pattern of CuI NPs synthesized without gallic acid. The crystal lattice of CuI is reported to crystallize in a face-centered cubic (FCC) lattice; the γ phase is shown because it is the most stable structure at room temperature. The XRD pattern of CuI powder was found to relate to the database card numbers 01-089-7072 and 01-077-9397. Detailed peaks were noticed at the 2θ values 25.43°, 29.42°, 49.80°, 52.30°, 61.20°, 67.34° and 77.12°, correlating to the (111), (200), (220), (311), (222), (400), (331), (420) and (422) planes, respectively. These XRD peaks correspond to CuI synthesized using black soybean [23]. The crystallite size was calculated using Scherrer's equation; the average crystallite size was found to be 41.45 nm, which is in the nano-size range.

$$t = \frac{K\lambda}{\beta \cos \theta}$$

t = crystallite size;
K = shape factor (0.9);
λ = wavelength of x-ray (0.154 nm);
β = full width at half maximum (FWHM);
θ = diffraction angle.

The crystal structure of cuprous iodide (CuI) synthesized using different concentrations of gallic acid as a stabilizing and reducing agent was also studied by X-ray diffraction (XRD). The XRD patterns of CuI stabilized with 2.5 and 5.0 mM of gallic acid are illustrated in Figure 1b,c. The well-defined peaks of both 2.5 GA-CuI and 5.0 GA-CuI correspond to JCPDS card number 82-2111, space group: F-43 m [24] and database card numbers 01-082-2111, 01-083-1106 and 01-075-0832, presenting the FCC unit cell. In the case of 2.5 GA-CuI NPs, peaks were observed at the 2θ values 25.46°, 29.46°, 42.20°, 49.8°, 61.20°, 67.38° and 77.08°, corresponding with the (111), (200), (220), (311), (222), (400), (331), (420) and (422) planes, respectively. The average size of 2.5 GA-CuI NPs is 43.51 nm. XRD peaks of 5.0 GA-CuI NPs were detected at 2θ values 25.42°, 29.44°, 42.25°, 49.83°, 61.10°, 67.22° and 77.03°. These peaks were matched with the (111), (200), (220), (311), (222), (400), (331), (420) and (422) planes, similar to the previous CuI NPs. The mean crystallite size of 5.0 GA-CuI is 49.71 nm. In addition, some lattice planes of CuI synthesized with different concentration of GA, especially (220), (311), (222) and (422) planes, demonstrate an insignificant shift. The 5.0 GA-CuI represented by the blue line shows a minor shift to a lower angle compared to CuI and 2.5 GA-CuI. This result is attributable to the larger crystallite size of 5.0 GA-CuI, which broadens the lattice planes and causes the lower angle shift. This result matches earlier research on the different sizes of synthesized gold nanoparticles [25].

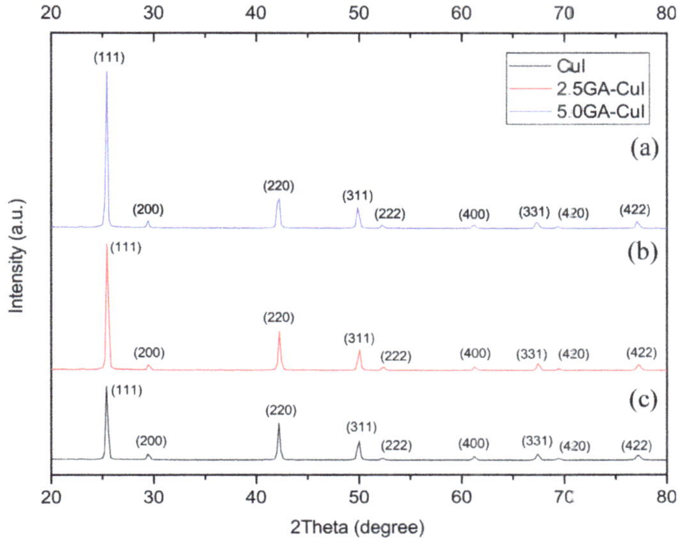

Figure 1. XRD patterns of synthesized nanoparticles (**a**) 5.0GA-CuI NPs (**b**) 2.5GA-CuI NPs and (**c**) CuI NPs.

According to the XRD patterns and different crystallite sizes of synthesized CuI NPs due to varying gallic acid concentration, it is clear that the gallic acid concentration is essential for controlling the size of these CuI NPs under normal pH conditions (pH~7). The size of CuI NPs increases with greater concentrations of gallic acid [20,26]. This result may be attributed to the higher negative charge of gallic acid, which remains on the copper surface and creates a repulsive force in the copper complex. The presence of this negative charge is due to the carboxyl group of gallic acid attached to the copper nanoparticle surface, and it provides stability through the electrostatic repulsion of CuI NPs in the solution.

2.1.2. X-ray Photoelectron Spectroscopy (XPS) Study

The chemical state of synthesized CuI NPs was characterized by XPS analysis, presented in XPS spectra (Figure 2). The XPS spectra were scanned using adventitious carbon (C 1 s–284.6 eV) as the reference and used to investigate the exact oxidation state of the elements composing CuI NPs. Figure 2a shows the XPS peaks of the Cu 2p orbital. It can be noted that the binding energies around 932 eV and 952.4 eV belong to Cu $2p_{3/2}$ and Cu $2p_{1/2}$, respectively. Figure 2b illustrates the binding energies at 619.4 eV and 631 eV, which correspond to the I $3d_{5/2}$ and I $3d_{3/2}$ orbitals, clearly confirming that I^- is a component in synthesized CuI, 2.5 GA-CuI and 5.0 GA-CuI NPS. The high resolution of the Cu 2p and I 3d spectra show the creation of CuI synthesized both using GA and without using GA, as well as only one oxidation for both Cu and I. Additionally, Figure 2c shows core level spectra for O 1 s orbital peaks in all synthesized CuI NPs, positioned around the 532 eV binding energy, which is assigned to absorbed H_2O and gallic acid [26].

2.1.3. Field Emission Scanning Electron Microscope (FESEM) Analysis of CuI

The morphology of the synthesized CuI crystals was studied using field emission scanning electron microscopy (FESEM) with a voltage of 2.0 kV. FESEM was performed at 1000× and 10,000× magnification to observe the crystallite shape in various synthesized condition. The results shown in Figure 3 reveal that when gallic acid is not used as a biostabilizer, a triangular shape of copper iodide nanostructures is typically produced. Using 2.5 mM and 5.0 mM of gallic acid produced γ-CuI nanosheets instead of nanoparticles [27].

Triangle-like nanosheets of γ-CuI were grown on each other and generated multi-layer stacks; after gallic acid was added into copper sulfate solution, the copper nanoparticles were generated as a substrate [28]. When potassium iodide was added to the mixture, copper iodide nanosheets were grown on copper substrate. Therefore, using gallic acid as a reducing and stabilizing agent produced triangle-shaped nanosheets. Additionally, focusing on the 1000× magnification, biosynthesis of copper iodide without gallic acid provided more agglomeration of particles than when using gallic acid. This result reveals the important role of gallic acid in particle distribution.

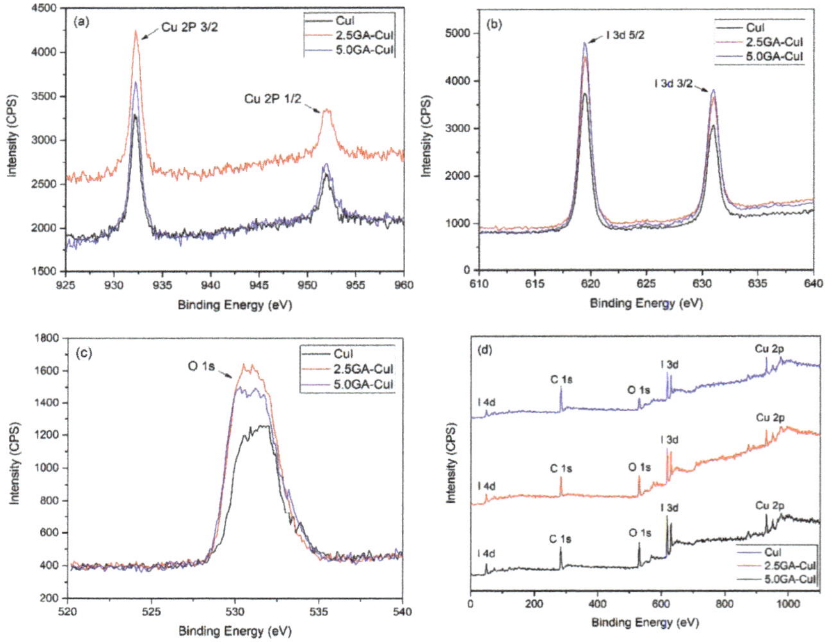

Figure 2. XPS spectra of (**a**) Cu 2p, (**b**) I 3d, (**c**) O 2 s and (**d**) survey spectra of CuI NPs.

2.1.4. Fourier Transform Infrared Spectrum (FTIR)

Figure 4 compares the FTIR spectra of copper iodide nanoparticles (CuI NPs) synthesized with different contents of gallic acid. There was only one interesting peak at wave number 553.76 cm^{-1} in the spectrum of CuI, related to Cu-I stretching vibrations. The FTIR spectrum of 2.5 GA-CuI only shows important peaks of CuI formation. In the final synthesis, the FTIR spectrum of 5.0 GA-CuI shows a broad band around 3322.94 cm^{-1}, assigned to the inter- or intramolecular hydrogen bond (-OH) stretching vibration in gallic acid, which serves as a stabilizer and capping agent of 2.5 GA-CuI. The FTIR peaks at 2921.44 and 2852.29 cm^{-1} represent -CH stretching in gallic acid, while the Cu-I stretching vibration appears in the FTIR peak at a wave number of 544.78 cm^{-1}.

After comparing the FTIR spectra of all synthesized CuI NPs, it can be seen from the spectra that gallic acid serves as a biostabilizer and bioreducer in CuI NP synthesis. Gallic acid at a concentration of 5.0 mM has a greater ability to stabilize CuI than at 2.5 mM, because the higher concentration of gallic acid disassociate in water. In addition, the FTIR peak of 5.0 GA-CuI shows a broad band and a high percent of transmittance at the -OH group. However, 2.5 GA-CuI and CuI NPs did not show any -OH groups in their FTIR spectra. This result is due to the low concentration of GA, which was used for reducing copper ions. GA did not play a biostabilizing role in 2.5 GA-CuI but did play a bioreducing role. Therefore, GA did not attach to CuI and 2.5 GA-CuI, resulting in no -OH groups in their FTIR spectra. The FTIR spectra of all CuI NPs were quite different from those of Ann Candice Fernandez et al. [29], who prepared copper iodide nanoparticles by using cyanidin-3-diglucoside-5-glucoside from red cabbage extract. Their FTIR results for CuI showed an obvious peak of cyanidin-3-fdiglucoside-5-glucoside, which was used in CuI stabilization because cyanidin-3-diglucoside-5-glucoside has a strong hydroxyl group, allowing it to act as a biostabilizer.

Figure 3. SEM images of CuI NPs with magnifications of (**a**) 1000× (CuI), (**b**) 10,000× (CuI), (**c**) 1000× (2.5 GA-CuI), (**d**) 10,000× (2.5 GA-CuI), (**e**) 1000× (5.0 GA-CuI) and (**f**) 10,000× (5.0 GA-CuI).

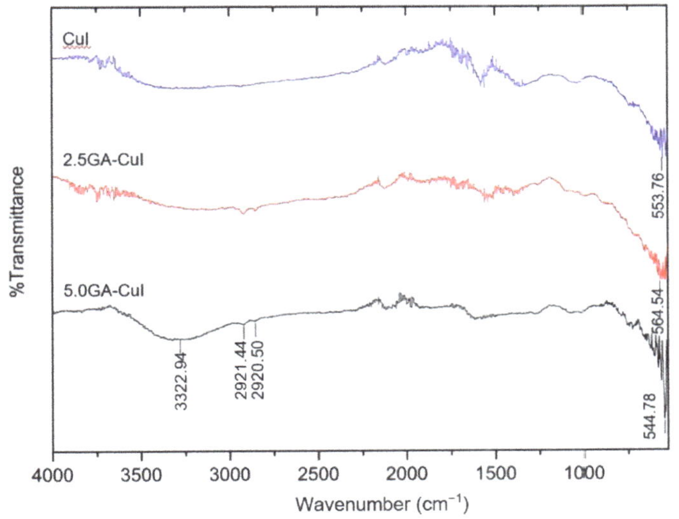

Figure 4. FTIR spectra of green synthesized CuI NPs using different gallic acid (GA) concentrations of 0, 2.5 and 5.0 mM.

2.1.5. Zeta Potential Analysis

The physical properties of synthesized nanoparticles in colloidal systems are generally measured in terms of zeta potential. Zeta potential indicates the dispersion stability and surface properties of molecules or nanoparticles in a liquid medium. Principally, if the magnitude of absolute zeta potential (±ve) is high, molecules or nanoparticles will be stable in a colloidal system because they have high repulsion force, preventing nanoparticles from aggregating [30]. The zeta potentials of the synthesized CuI NPs in different conditions are shown in Table 1. CuI NPs prepared using 0, 2.5 and 5.0 mM gallic acid have average zeta potential values of −13.47, −18.73 and −20.47 mV, respectively. The results show that synthesizing CuI NPs using a greater content of gallic acid leads to greater zeta potential due to the bioreducing properties of gallic acid; thus, these CuI NPs have greater dispersion stability.

Table 1. Zeta potential of synthesized CuI NPs.

Sample	Gallic Acid (GA) Contents (mM)	Zeta Potential (mV)	Average Value (mV)
CuI	0	−13.3	−13.47
		−13.6	
		−13.5	
2.5 GA-CuI	2.5	−18.8	−18.73
		−18.6	
		−18.8	
5.0 GA-CuI	5	−20.7	−20.47
		−19.8	
		−20.9	

2.1.6. Nanosizer/Dynamic Light Scattering

The nano-size of synthesized CuI NPs was measured by using a nanosizer/dynamic light scattering, which illustrates the relationship between intensity and size distribution (nm), as shown in Figure 5. It can be seen from Figure 5a, that the diameter was about 869 nm for CuI NPs synthesized without a biostabilizer. When gallic acid was used to reduce and stabilize CuI NPs, the size distributions of 2.5 GA-CuI and 5.0 GA-CuI were 2473 and 3062 nm, accordingly, as presented in Figure 5b,c. Both 2.5 GA-CuI and 5.0 GA-CuI had greater diameters than CuI NPs. This result reveals that green synthesis of CuI NPs provides electrostatic repulsion from gallic acid molecules, causing larger size distribution [20].

2.1.7. Peroxidase-like Properties of GA-CuI NPs

The therapeutic use of synthesized GA-CuI NPs has been investigated for their enzyme-like properties, which include peroxidase-like activity. This activity is important for the wound healing ability of GA-CuI, because copper nanoparticles function as redox enzymes due to the increased electron exchange of their atoms [31]. The improved antibacterial activity of metal nanoparticles were analyzed in terms of their oxidase-like and peroxidase-like properties, as seen in Equations (1) and (2). When GA-CuI NPs come in contact with bacteria in the presence of oxygen (O_2), antioxidant chemical (AH_2) is oxidized and produces hydrogen peroxide (H_2O_2). Then, H_2O_2 is further catalyzed by the peroxidase-like enzymes of GA-CuI NPs, generating hydroxyl radicals (•OH). Hydroxyl radicals (•OH) are important molecules for killing bacteria by interacting with the bacterial cell wall and oxidizing biomolecule. Therefore, the oxidase- and peroxidase-like activity of GA-CuI NPs gives them a strong antibacterial capacity through the mechanisms of hydrogen peroxide and hydroxyl radical production.

$$AH_2 + O_2 \rightarrow A + H_2O_2 \quad (1)$$

$$H_2O_2 \xrightarrow{GA - CuI\ NPs} 2{\cdot}OH \quad (2)$$

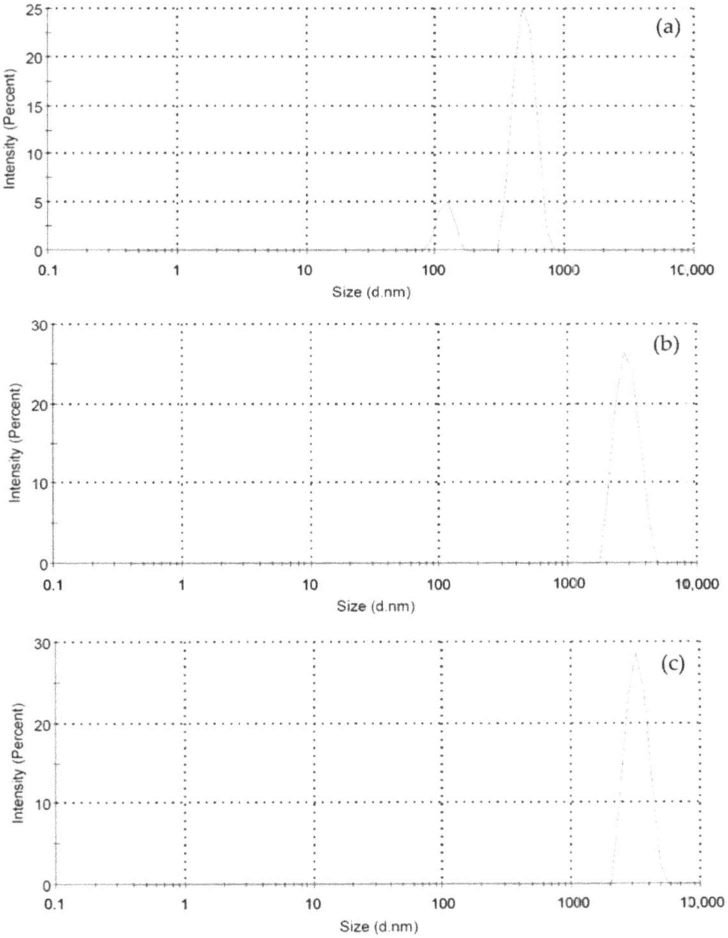

Figure 5. Size distribution of (**a**) CuI NPs, (**b**) 2.5 GA-CuI NPs, and (**c**) 5.0 GA-CuI NPs determined using a nanosizer.

In the investigation of peroxidase-like activity, 3,3′,5,5′-tetramethyl-benzidine (TMB) was used as a substrate for peroxidase. Notice the color change after treatment with CuI, 2.5 GA-CuI and 5.0 GA-CuI NPs. A UV–Vis spectrometer was employed to observe the absorbance peaks of oxidized TMB (TMB^+). It can be seen from Figure 6 that the absorbance peaks of TMB treated with CuI, 2.5 GA-CuI and 5.0 GA-CuI were detected around 650 nm. TMB oxidation peaks appeared after treatment with CuI and 2.5 GA-CuI, while 5.0 GA-CuI exhibited a weak oxidizing effect on TMB. This may be due to the decreased amount of CuI when concentrations of gallic acid are increased. Consequently, the capacity of 5.0 GA-CuI to oxidize TMB is poor. However, the weight loss of CuI, 2.5 GA-CuI and 5.0 GA-CuI must be examined to determine the amount of each chemical in the compound.

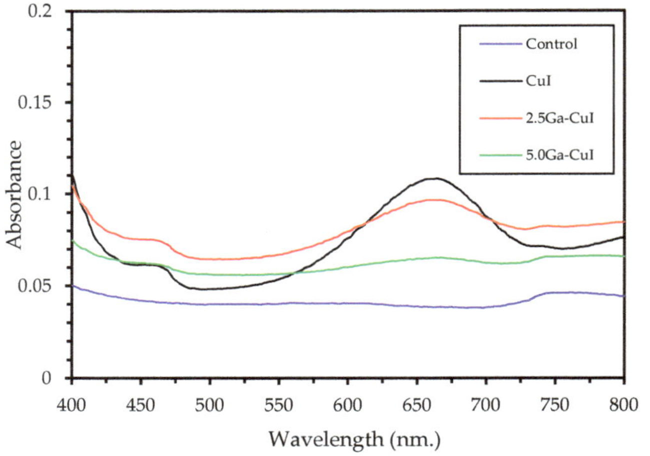

Figure 6. UV–Vis absorption spectra of 3,3′,5,5′-tetramethyl-benzidine (TMB) oxidation, treated with acetate buffer (light blue), CuI (black), 2.5 GA-CuI (red) and 5.0 GA-CuI (green).

2.1.8. Evaluation of Minimum Inhibitory Concentration (MIC) and Minimum Bactericidal Concentration (MBC) of CuI NPs

The antibacterial activity of synthesized CuI NPs was tested against *E. coli* (Gram-negative bacteria) and MRSA (Gram-positive bacteria), as these are the most common bacterial infections in wounds and the bloodstream. The inhibitory effect of CuI NPs on bacterial growth was studied with various concentration of CuI NPs, using the broth serial dilution method to evaluate minimum inhibitory concentration (MIC) and minimum bactericidal concentration (MBC). Figure 7 illustrates the results of bacterial growth on agar plates; these *E. coli* and MRSA were treated with different concentrations of synthesized CuI NPs. The CuI NPs were diluted to 256, 128, 64, 32, 16, 8, 4, 2, 1 and 0.5 mg/mL, represented by numbers 1, 2, 3, 4, 5, 6, 7, 8, 9 and 10, respectively. In addition, Figure 7a,b show bacterial growth after treatment with CuI synthesized without gallic acid. Figure 7c,d present bacterial growth after treatment with 2.5 GA-CuI. Bacterial growth after treatment with 5.0 GA-CuI is shown in Figure 7e,f. It can be seen from Table 2, which shows the MIC and MBC values of different CuI NPs for *E. coli* and MRSA strains, that CuI at 32 mg/mL completely inhibited both Gram-negative (*E. coli*) and Gram-positive (MRSA) bacteria, whereas 2.5 GA-CuI and 5.0 GA-CuI eliminated Gram-negative bacteria (*E. coli*) at a higher concentration of 64 mg/mL. Therefore, it can be concluded that CuI has a greater antibacterial effect than 2.5 GA-CuI and 5.0 GA-CuI, because 2.5 GA-CuI and 5.0 GA-CuI were composed of gallic acid and copper iodide. The antibacterial mechanism of CuI occurs through the denaturation of solid-state compounds, which deforms the bacterial cell wall. Synthesized CuI NPs have a high potential to bind with the thiol group of proteins in the cell wall, forming peptide/disulfide complexes. This activity can cause cell death, as illustrated in Figure 8 [32].

The proportion of pure CuI NPs in 2.5 GA-CuI and 5.0 GA-CuI is lower than in CuI, and CuI is the main bactericidal factor. Additionally, CuI synthesized without GA is smaller in size than with GA. The bacteria-killing ability of CuI NPs is also dependent on particle size, as CuI NPs of smaller size have more potential to damage bacterial cells than those of larger size. This is because smaller CuI NPs have greater ability to enter the cell wall and create physical attachments deforming it. Therefore, CuI NPs are more efficient at killing bacteria [16].

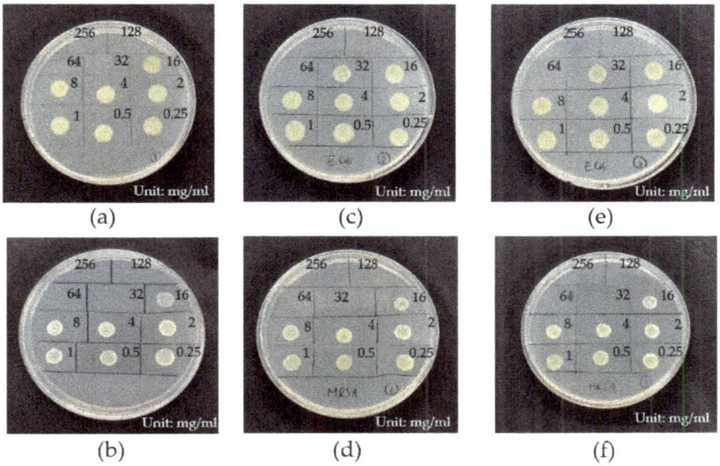

Figure 7. Bacterial growth after treatment with different synthesized CuI NPs for each bacterial strains: (**a**) *E. coli* treated with CuI; (**b**) MRSA treated with CuI; (**c**) *E. coli* treated with 2.5 GA-CuI; (**d**) MRSA treated with 2.5 GA-CuI; (**e**) *E. coli* treated with 5.0 GA-CuI; (**f**) MRSA treated with 5.0 GA-CuI.

Table 2. Minimum inhibitory concentration (MIC), minimum bactericidal concentration (MBC) and ratio of CuI NPs for each bacterial strain.

CuI Sample	Gallic Acid Concentration (mM)	MIC (mg/mL)		MBC (mg/mL)	
		E. coli (−)	MRSA (+)	*E. coli* (−)	MRSA (+)
CuI	0	8	16	32	32
2.5 GA-CuI	2.5	16	32	64	32
5.0 GA-CuI	5.0	16	32	64	32

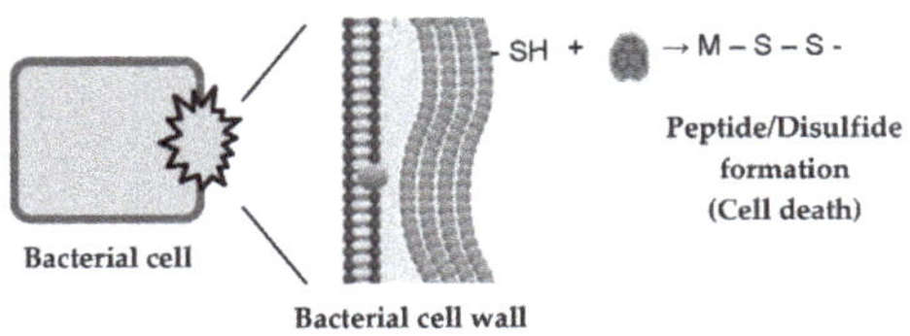

Figure 8. Antibacterial mechanism of CuI NPs through denaturation.

In addition, Gram-negative bacteria (*E. coli*) are more resistant to CuI NPs than Gram-positive bacteria (MRSA). This could be because *E. coli* has a more complex structure and because the outer membrane of Gram-negative bacteria acts as a barrier, providing additional protection against CuI NPs.

2.2. PVA/CuI Liquid Bandage Formulation

2.2.1. Fourier Transform Infrared Spectroscopy (FTIR)

Figure 9 illustrates the FTIR spectra of pure PVA and PVA with different concentrations of CuI suspension (0.095% and 0.190%). The characteristic peaks of PVA were observed at 3268.62, 2940.72, 2911.56, 1653.96, 1414.07, 1328.07, 1236.24, 1090.47, 1039.52 and 842.91 cm^{-1}. These peaks were designated as -OH stretching, asymmetric stretching of CH_2, symmetric stretching of CH_2, C=O carbonyl stretching, -CH_2 bending, C-H wagging,

C-OH stretching in alcohol, C-O stretching, C=O stretching and bending of -OH and C-C stretching, respectively. In the PVA solution with 0.095% CuI, peaks were noticed at wave numbers 3328.42, 2940.72, 2911.56, 1703.27, 1447.77, 1317.24, 1239.70, 1088.07, 1042.26 and 796.64 cm^{-1}. These peaks were assigned to characteristic PVA peaks with a minor shift, comparable with pure PVA [33]. Similarly, important peaks of PVA were observed in PVA with 0.190% CuI. The vibration spectra of the materials are presented in Table 3. The C–H scissoring vibration mode is predicted at about 1440 cm^{-1} for pure PVA. In this experiment, the C–H scissoring vibration was clearly observed only after the addition of CuI NPs to the PVA solution, which causes a structural deformation of the PVA backbone. In addition, for pure PVA, the vibration pattern of carbonyl stretching is expected at about 1650 cm^{-1}, which is the result of residual acetate in the polymerization. The carbonyl (C=O) stretching vibration was found at 1703.27 cm^{-1} and 1698.17 cm^{-1} for PVA/CuI (0.095%) and PVA/CuI (0.190%), respectively. It is clear that the transmittance of the carbonyl stretching was increased in PVA with CuI NPs, as they increase the presence of carbonyl moieties. This is due to the addition of CuI complexes, while polymerizing PVA can cause the breaking of –H and –OH bonds and, as a result, the formation of more carbonyl double bonds (C=O). Accordingly, the intensity of the C=O stretching peaks was increased and shifted to higher wavelengths. Furthermore, the characteristic peak of CuI NPs was not present in the FTIR spectra of PVA/CuI (0.095%) and PVA/CuI (0.190%) because this characteristic peak is lower than 525 cm^{-1}, while 525 cm^{-1} is the minimum wave number for this FTIR spectroscopy. However, PVA with CuI NPs in its structure can be confirmed from intensity and shift of the wave numbers.

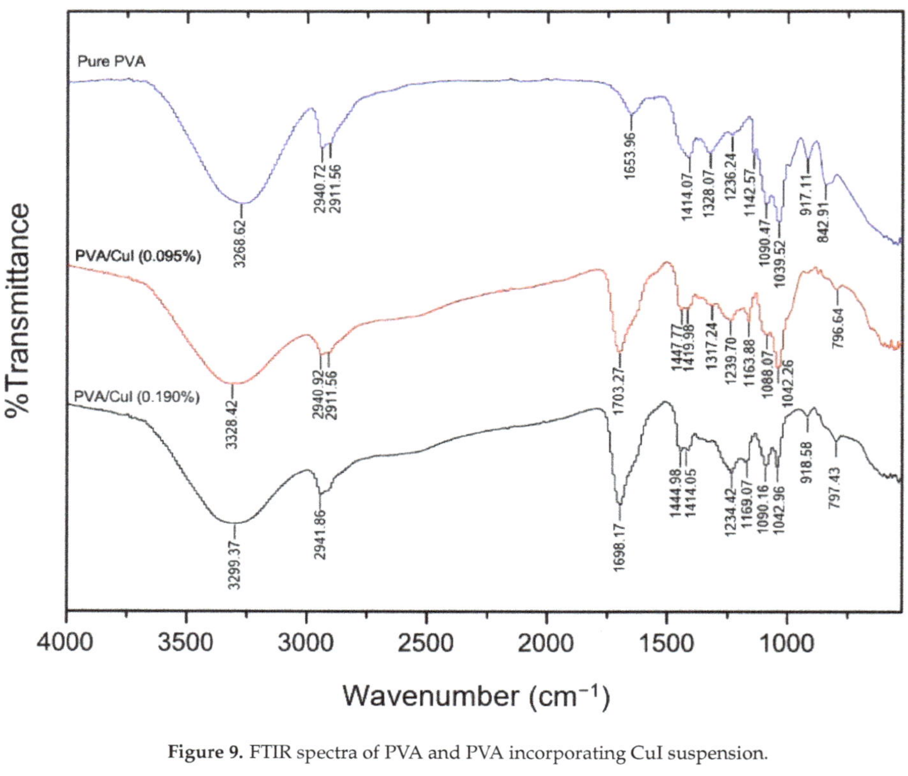

Figure 9. FTIR spectra of PVA and PVA incorporating CuI suspension.

Table 3. Wave numbers of FTIR results for PVA and PVA loaded with CuI suspension.

Assignment	Observed Wavelength (cm^{-1})			Reference
	Pure PVA	PVA/CuI (0.095%)	PVA/CuI (0.190%)	
OH stretching	3268.62	3328.42	3299.37	[34]
Asymmetric stretching of CH_2	2940.72	2940.72	2941.86	[34]
Symmetric stretching of CH_2	2911.56	2911.56	-	[34]
C=O carbonyl stretching	1653.96	1703.27	1698.17	[34]
CH_2 scissoring	-	1447.77	1444.98	[34]
CH_2 bending	1414.07	1419.38	1414.05	[34]
CH_2 wagging	1328.07	1317.24	-	[34]
C-OH stretching in alcohol	1236.24	1239.7	1234.42	[34]
C-O stretching	1090.47	1088.07	1090.16	[34]
Stretching of C=O and bending of OH	1039.52	1042.26	1042.96	[34]
C-C stretching	842.91	796.64	797.43	[34]

2.2.2. Surface Morphology Analysis of Cast PVA Film Using Field Emission Scanning Electron Microscopy (FESEM)

Figure 10 presents SEM images of the poly(vinyl alcohol) nanocomposite material after the incorporation of copper iodide nanoparticles (5.0 GA-CuI) in the polymer matrix prepared by cast film method. The magnification used to study the morphological surface was 1000×. The surface morphology of PVA film with 0, 0.095% and 0.190% CuI suspension is shown in Figure 10. Figure 10a shows the smooth surface observed on pure PVA film. After 0.095% copper iodide nanosheets were loaded onto PVA film, there were multi-sized granular particles on the surface of the composite. This shows the good distribution of smaller non-agglomerated particles. Then, more CuI suspension (0.190%) was added to the PVA solution. The morphology of the surface demonstrated that CuI NPs were excellently distributed in PVA film. In spite of the high concentration of copper iodide particles, excellent distribution was achieved due to GA-CuI stability.

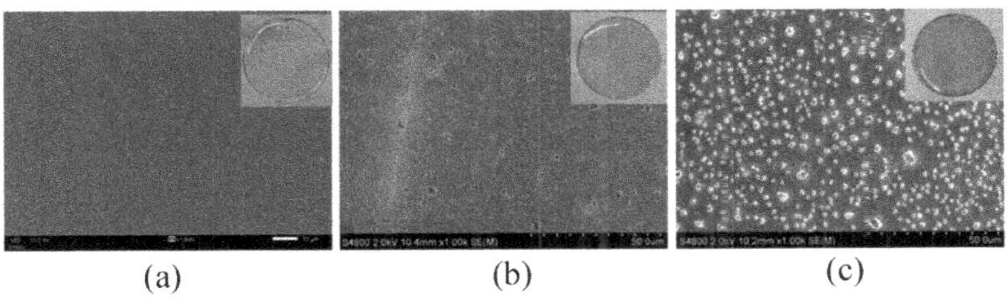

Figure 10. SEM image of cast PVA film: (**a**) pure PVA, (**b**) PVA/CuI (0.095%) and (**c**) PVA/CuI (0.190%).

2.2.3. Swelling Behavior of Pure PVA and PVA/CuI Film

The swelling ratio is used to determine the resistance of polymer film when exposed to high humidity or water in the bloodstream or at the wound site [34]. Tests of the swelling behavior of PVA film were conducted using gravimetric measurement [35]. The average thicknesses of pure PVA, PVA/CuI (0.095%) and PVA/CuI (0.190%) were 0.53, 0.59 and 0.60 mm, respectively. All PVA films were immersed in PBS buffer solution at 37 °C to provide a constant pH of 7.2 ± 0.1. After the specific swelling time, each film was removed from the PBS solution, and surface water was rapidly cleaned using Kimwipes paper. Then, each sample was weighed and recorded with an electronic balance. The swelling ratio of PVA films was calculated.

Figure 11 illustrates the swelling behavior of pure PVA, PVA/CuI (0.095%) and PVA/CuI (0.190%), shown by the black, red and blue lines, respectively. An increase in the swelling ratio of all PVA films over time was noticed; the swelling rate considerably increased at the initial time and then leveled off. The red and blue lines show the incorporation of CuI NPs in PVA films, which significantly reduced the swelling behavior. The swelling ratio decreased with the increasing load of CuI NPs. This is due to the hydrophobic nature of CuI NPs. The more CuI NPs are loaded in PVA films, the greater the repulsion of water molecules. Therefore, CuI NPs loaded in PVA liquid bandages enhance resistance to humidity and blood after they are cast to thin film.

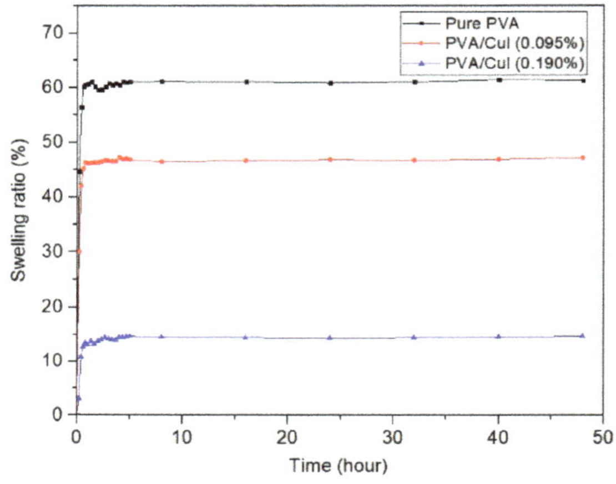

Figure 11. The swelling behavior of pure PVA, PVA/CuI (0.095%) and PVA/CuI (0.190%).

2.2.4. Drying Time of PVA Liquid Bandages

The drying time of liquid bandage formulations synthesized with different contents of poly(vinyl alcohol) was tested by applying 0.2 mL of 12%, 13% and 14% PVA liquid solution on the inside forearm of a volunteer and letting it dry. The drying time and physical properties of each PVA concentration were recorded after there was no visible PVA liquid on the skin. The drying time and physical properties of the liquid bandages are shown in Figure 12. The drying time of 12%, 13% and 14% PVA liquid bandages were 2.5, 4.0 and 6.0 min, respectively. Even though 12% PVA provided the minimum drying time, the PVA content was not enough for it to spread as a matrix, as shown in Figure 12a. Additionally, 14% PVA content showed significantly increased drying time due to the greater amount of PVA, which has high viscosity and requires more time for particles to coalesce and form a film. Therefore, the most suitable concentration of PVA is 13%.

After the appropriate PVA content was determined, the drying times of different PVA liquid bandages loaded with CuI suspension were subsequently examined. PVA solution incorporating different concentrations of CuI suspension (0, 0.095% and 0.190%) were prepared and drying time was tested. It can be seen from Figure 12b that the drying time increased with increasing concentrations of CuI suspension, showing that higher contents of CuI result in longer drying times, because a large amount of CuI NPs provides high viscosity and a difficult pathway for evaporation of the cosolvent [36].

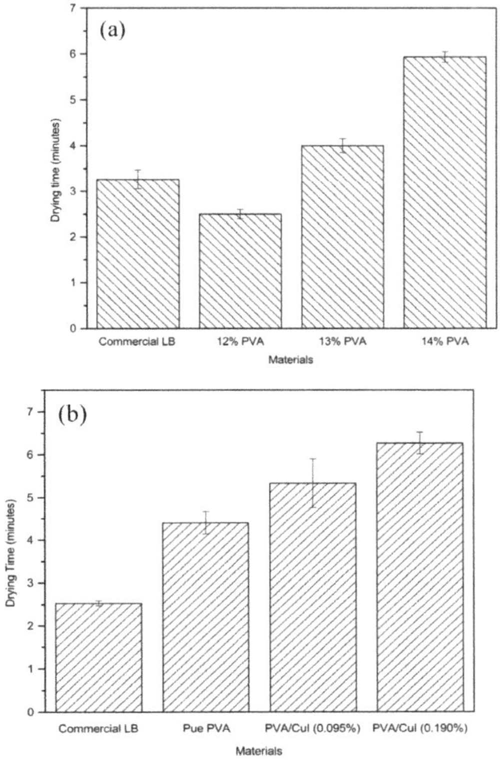

Figure 12. Drying time of film compared with a commercial liquid bandage (LB): (**a**) content of PVA; (**b**) PVA composed of CuI suspension.

2.2.5. Water Vapor Transmission Rate (WVTR)

It is necessary to study the water vapor transmission rate because this property helps promotes wound healing and contraction of the wound. Bleeding will occur from superficial wounds. Water is the most important component of blood (90%); therefore, a good water vapor transmission rate allows blood to exit the wound dressing and enhances wound drying. The water vapor transmission rate (WVTR) of the PVA samples was tested under virtual conditions, compared with a centrifuged tube without any cover as a control (+), as shown in Figure 13. The WVTRs of pure PVA, PVA/CuI (0.095%), PVA/CuI (0.190%) and a commercial liquid bandage (LB) were 384.33, 333.64, 278.23 and 81.35 g/m^2 • day, respectively. It can be seen that the water vapor transmission rate (WVTR) decreases with increasing concentrations of CuI due to the decreased diffusion rate of water vapor through the PVA composite film [37]. Compared with commercial liquid bandage, PVA film has a higher WVTR. Therefore, PVA film has good ability to allow water to evaporate from wounds, promoting rapid wound healing [38].

2.2.6. In Vitro Cytotoxicity and Biocompatibility of Liquid Bandage

In vitro cytotoxicity was tested to determine the cytotoxicity of various concentrations of CuI suspension in PVA liquid bandages. In this method, the ratio of cell viability of human dermal fibroblasts (HDFa) after 24 h incubation in sample extractions at different concentrations (2.5, 5 and 10 mg/mL) is measured, and exposure to the various liquid bandage was used as an indicator of whether these substances can cause damage to human cells [39]. Figure 14 shows the results, revealing that after HDFa was incubated in sample extractions for 24 h, cell viabilities with 2.5, 5 and 10 mg/mL pure PVA were 108.363, 99.7611 and 98.447%, respectively. HDFa incubated in 2.5, 5 and 10 mg/mL PVA/CuI (0.095%) showed cell viabilities of 98.685, 94.624 and 96.177%, respectively. HDFa incubated in 2.5, 5 and 10 mg/mL PVA/CuI (0.190%) showed cell viabilities of 94.982, 87.814 and 91.517%. The results reveal that total PVA liquid bandages were non-cytotoxic to HDFa at the concentrations tested. Nevertheless, the number of dead cells increased with increasing content of CuI suspension. The increase in cytotoxicity may show that high concentrations of CuI metallic complexes can cause cell damage. However, PVA liquid bandages with CuI suspension demonstrated biocompatibility, allowing them to be used for antibacterial wound dressing.

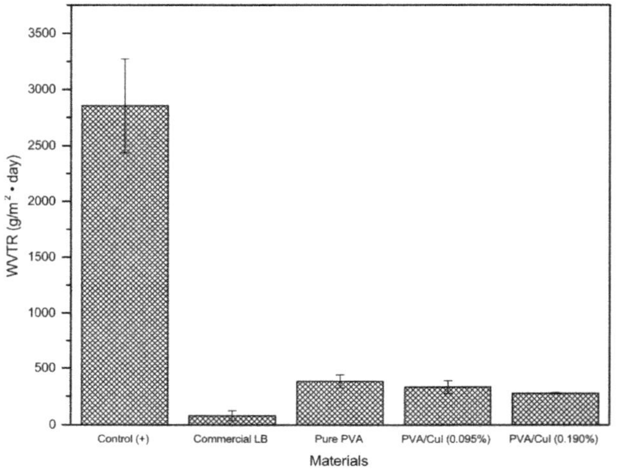

Figure 13. Water vapor transmission rate (WVTR) for each synthesized liquid bandage.

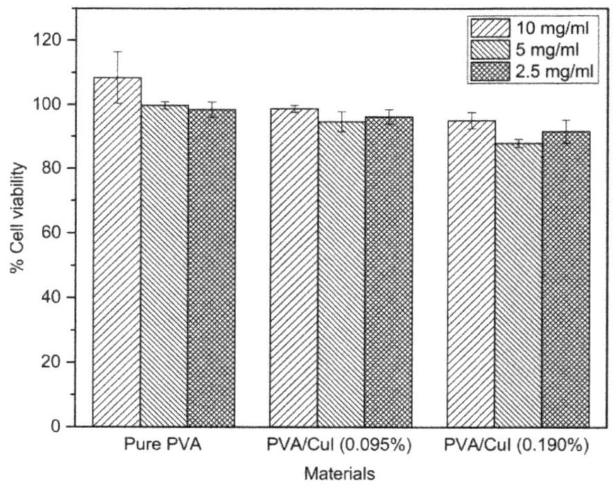

Figure 14. Surviving cells after treatment with each sample extraction.

2.2.7. Antibacterial Efficacy Testing by Time–Kill Assay

The time–kill assay was employed to test antibacterial efficacy, exploring the rate of bacterial reduction after being incubated in different concentrations of materials for

1, 3, 6, 12 and 24 h. In vitro antibacterial tests were conducted against Escherichia coli (*E. coli* ATCC25922) and methicillin-resistant *Staphylococcus aureus* (MRSA), which are Gram-positive and Gram-negative pathogens, respectively. Figure 15 shows the rate of bacterial reduction, with the concentration-dependent and time-dependent bactericidal activities of pure PVA, PVA/CuI (0.095%) and PVA/CuI (0.190%) against *E. coli* (a) and MRSA (b).

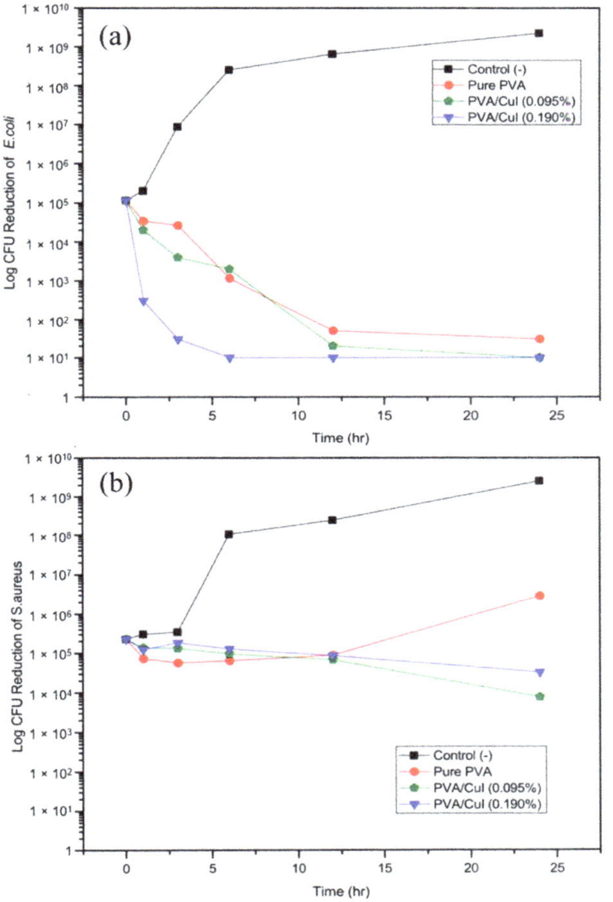

Figure 15. Bacterial reduction of *E. coli* (**a**) and MRSA (**b**) after incubation in liquid bandage.

Regarding the ability of the materials to kill Gram-negative bacteria (*E. coli* ATCC2592) measured in the rate of log CFU reduction and surviving bacteria counted, it can be observed from Figure 15a that there was a decline in the number of bacterial cells after treatment with pure PVA, PVA/CuI (0.095%) and PVA/CuI (0.190%). Focusing on the bactericidal properties of the pure PVA bandage solution, even though there was no CuI suspension in the PVA solution, a reduction of *E. coli* was noticed because ethanol was added into the mixture as a cosolvent with water. Therefore, ethanol can play an important role in killing bacteria. PVA/CuI (0.095%) and PVA/CuI (0.190%) were able to effectively kill the bacteria. PVA/CuI (0.190%), in particular, was toxic to *E. coli* after only 6 h of incubation due to its high concentration of CuI. PVA/CuI (0.095%) eliminated *E. coli* after treatment for 24 h. The bacterial reduction corresponded to the amount of CuI suspension in the PVA solution.

Regarding bactericidal properties against MRSA (Figure 15b), the direction of the log CFU reduction for MRSA was completely different from that for *E. coli*. PVA solution without CuI suspension when used in bandages has relatively low capability to inhibit bacterial growth. However, bacterial cell count was reduced when compared with the control (-) due to ethanol. When 0.095% and 0.190% CuI was included in the PVA solution, the trend of cell reduction marginally declined; thus, these materials were not toxic to MRSA. The difference in bacterial reduction for *E. coli* and MRSA is related to the differing bacterial cell structures of Gram-positive and Gram-negative bacteria. The presence of CuI NPs did not change the viability of MRSA, most likely because of its complex structure. These bacteria are strongly resistant to antibiotic drugs. Some strains can develop high-level resistance and pose a serious threat.

The overall results reveal that PVA liquid bandages incorporating CuI suspension can damage both Gram-positive and Gram-negative bacteria, as can be seen from decline of log CFU reduction. However, the antibacterial activities of the materials against *E. coli* were greater than against MRSA.

3. Conclusions

In conclusion, antibacterial liquid bandages combining the biocompatible polymer poly(vinyl alcohol) and the antibacterial agent copper iodide, synthesized using gallic acid as a bioreducer and biostabilizer, were effectively designed and characterized. The XRD results of CuI synthesized with gallic acid confirm the crystallinity of synthesized CuI in the γ-CuI phase. As calculated with Scherrer's equation, the nanoparticle size of copper iodide increased with increasing concentrations of gallic acid. The stability of synthesized copper iodide nanoparticles (CuI NPs) was confirmed by zeta potential analysis. This result agreed with the FTIR spectra of synthesized CuI NPs, which showed that more gallic acid creates more chelation in CuI NPs, seen in the -OH stretching peak, thus leading to high particle stability. The curative effects of synthesized CuI loaded in PVA solution to formulate a liquid bandage were characterized in terms of antibacterial activities against *E. coli* and MRSA, and cytotoxicity to human dermal fibroblasts (HDFa). The antibacterial test (time–kill assay) showed that the synthesized liquid bandages can successfully kill *E. coli* within 6 h and cause a reduction in MRSA. In addition, PVA liquid bandages are non-toxic to HDFa cells with CuI loads of 0.095% and 0.190% w/v.

4. Materials and Methods

4.1. Materials

Gallic acid (GA, 97.5–102.5% titration) with molecular weight of 170.12 g/mol was purchased from Sigma-Aldrich, St. Louis, MO, USA. PVA with a molecular weight of 89,000–90,000 g/moL, copper sulfate tetrahydrate ($CuSO_4 \bullet 5H_2O$) and potassium iodide (KI) were purchased from Sigma-Aldrich, USA.

4.2. Preparation of Gallic Acid–Cuprous Iodide Nanoparticles

The cuprous iodide nanoparticles stabilized by gallic acid (GA-CuI NPs) were prepared using different concentrations of gallic acid (0, 2.5 and 5.0 mM). In the case of cuprous iodide without gallic acid, 1.2400 g. of $CuSO_4 \bullet 5H_2O$ was dissolved in 50 mL deionized water in a 100 mL volumetric flask to make 50 mM of copper sulfate solution. Then, 50 mM of potassium iodide was prepared. Potassium iodide (KI) in the amount of 0.8300 g was dissolved in 50 mL deionized water in a 100 mL volumetric flask, and the concentration was adjusted. After, potassium iodide was added drop-wise to the copper sulfate solution with continuous stirring for 45 min. Then, the solution was centrifuged at 6000 rpm for 20 min, washed with deionized water to remove unreacted substances and centrifuged again. Finally, the precipitation was collected and dried in an oven at 60 °C for 6 h.

For the varying gallic acid concentrations, 0.0123 and 0.0245 g of gallic acid was dissolved in 30 mL deionized water to prepare 2.5 mM and 5.0 mM of gallic acid, respectively. Then, a constant amount of copper sulfate (50 mM) and potassium iodide (50 mM) were prepared similarly to the previous step. After, a varying amount of gallic acid (2.5 mM and 5.0 mM) was added drop-wise to the copper sulfate solution in 250 mL beakers with constant stirring for 15 min, allowing the gallic acid to stabilize and reduce the Cu^{2+} to Cu^{+}. A color change was observed. Each beaker of potassium iodide was added drop-wise to the mixture and stirred for 45 min. The solution was centrifuged at 6000 rpm for 20 min, washed by ethanol and deionized water to remove unreacted substances and centrifuged again. Finally, the precipitation was collected and dried in an oven at 60 °C for 6 h.

4.3. Preparation of PVA Antibacterial Liquid Dressing with Additional Gallic Acid–Cuprous Iodide

After the most suitable CuI NP solution was selected, copper iodide was loaded into the adhesive bandage formulation, in which PVA was used as a film former and glycerol as a softening agent. This creates a kind of liquid adhesive bandage. Calculated by weight percentage, 13% w/v of PVA was dissolved in 40% ethanol in water. Then, the mixture was heated to 75 °C, allowing the film former to fully dissolve. After, 6% w/v of glycerol softening agent was added to the PVA solution under continuous stirring, 0, 0.095% and 0.190% w/v of selected CuI NP antibacterial agent was separately added to the PVA solution and stirred for 15 min. Finally, the antibacterial PVA liquid bandage solutions were obtained.

4.4. Crystal Structure of CuI NPs

The crystal structure of the synthesized CuI, 2.5 GA-CuI and 5.0 GA-CuI NPs was characterized by X-ray diffraction (Rigaku SmartLab X-ray diffractometer), using standard XRD measurements with a wavelength of 1.54 Å Cu Kα radiation. The X-ray diffractometer was applied at a voltage of 40 kV and a current of 30 mA. All samples were scanned with a diffraction angle of 2θ from 20 to 80 degrees, with a scanning rate of 0.02 degrees per second.

4.5. Characterization of Microstructure and Morphology of Synthesized CuI NPs and PVA Liquid Bandage

Scanning electron microscopy (SEM; JEOL JSM-6610LV, Oxford X-Max 50, Tokyo, Japan) was used to investigate the microstructure, morphology and average size of gallic acid–CuI nanoparticles. Different conditions of CuI synthesis were examined at 500× magnification at 300 kV. In addition, SEM was used to investigate the morphology of the pure PVA film surface and PVA with CuI solution. The samples were coated with platinum using a sputtering device prior to SEM observation.

4.6. Functional Group Determination of CuI NPs and PVA Liquid Bandage

Fourier transform infrared spectroscopy (FTIR; Nicolet iS 5, Massachusetts, United States) was used to determine the functional group containing gallic acid–CuI NPs. Scans were conducted from 4000 to 600 cm^{-1} with the iD1 Transmission module, with the FTIR spectrum scanned 64 times. For characterization of PVA liquid bandage, the PVA liquid bandage with and without CuI solution was also analyzed with Fourier transform infrared spectroscopy (FTIR; Nicolet iS5).

X-ray photoelectron spectroscopy (XPS) analysis (Kratos, Axis ultra DLD, New York, United States) was employed to investigate the exact chemical state (oxidant state) of the elements composing the synthesized compounds.

4.7. Size Determination of CuI NPs

A zeta potential analyzer (MALVERN Zetasizer Nano ZSP, Worcestershire, England) was used to determine the stability and surface properties of CuI nanoparticles synthesized with different amounts of gallic acid in deionized water. In addition, a nanosizer was utilized to measure the size distribution of synthesized CuI NPs.

4.8. Peroxidase-like Activity of CuI NPs

The protocol for measuring peroxidase-like properties was taken from Li Wang et al. [18] with some modifications. A UV–Vis spectrophotometer was used to determine the peroxide-like reaction. Additionally, 3,3′,5,5′-tetramethyl-benzidine (TMB) was used as a substrate to examine the peroxidase-like property of CuI and GA-CuI NPs. An acetate buffer (pH = 5.5) was used as a control. There were four groups in the peroxidase test, including TMB + H_2O_2 + acetate buffer (control), TMB + CuI NPs + H_2O_2, TMB + 2.5 GA-CuI NPs + H_2O_2 and TMB + 5.0 GA-CuI NPs + H_2O_2. The peroxidase-like property of CuI NPs was investigated at a pH condition of 5.5 in acetate buffer, because these are the conditions during bacterial infection after a wound has occurred. In detail, TMB was dissolved in DMSO to prepare 1 mM of TMB solution; then, 150 μg of each sample (CuI, 2.5 GA-CuI and 5.0 GA-CuI NPs) was separately prepared in 1 mL of acetate buffer. Finally, 1 mM of H_2O_2 was prepared in deionized water. After the required solutions were completely prepared, 20 μL of TMB, 200 μL of each sample and 180 μL of H_2O_2 were mixed together in each group and kept in dark conditions for 10 min. The absorbance of TMB treated with CS-Cu-GA NCs at 650 nm was measured using a UV–Vis spectrometer.

4.9. Evaluation of Minimum Inhibitory Concentration (MIC) and Minimum Bactericidal Concentration (MBC) of CuI NPs

The antibacterial activity of gallic acid–CuI NPs against Gram-negative *E. coli* (ATCC 25922) and Gram-positive methicillin-resistant *Staphylococcus aureus* (MRSA, ATCC 33591) was tested by broth dilution method to evaluate minimum inhibitory concentration (MIC) and minimum bactericidal concentration (MBC). Gram-negative *E. coli* and Gram-positive MRSA were separately dispersed in saline solution; the optical density was measured to be 0.5 using a densitometer, resulting in 1.5×10^8 CFU/mL of bacteria. Test tubes containing 1.5×10^8 CFU/mL were diluted by pipetting 100 μL into 3 mL of saline solution; then, 5×10^6 CFU/mL of *E. coli* and of MRSA was obtained. After the bacterial suspension was prepared, serial dilution of the copper iodide solution was conducted. An amount of 512 mg of copper iodide was dissolved in 2 mL of DMSO, and 2 mL of MHB was added to this test tube. Then, serial dilution was conducted 10 times in 24-well plate. Bacterial suspension of 5×10^6 was transferred to a 24-well plate from column 1 to column 10 and was incubated in a shaking incubator with overnight cultures of bacteria.

4.10. Physical Properties of PVA Liquid Bandage Formulation

To evaluate the drying time of the PVA formulations on human skin, a test was applied on the inner side of the forearm. Different PVA formulations were spread on human forearm and let dry for a period of 1, 2 and 3 min. If there was no liquid visible on the skin, the PVA film was considered to be dry. The PVA liquid bandage with minimum drying time was chosen to create a liquid bandage.

Additionally, the water vapor transmission rate describes the amount of water passed through a unit area of film in a unit time. The water vapor transmission rate was evaluated to study the water permeability of the films, as this has an effect on skin temperature and the hydration of the wound. Following Zurdo Schroeder et al. [40], PVA liquid bandages loaded with different amounts of CuI suspension were cast on glass plates to evaporate the solvent, forming PVA films. These films were cut into circular samples with diameters of 0.03 m and covered on centrifuge tubes containing 30 mL of distilled water. The tubes were placed into an incubator with a temperature of 37 °C and relative humidity of 52%. Then, the tubes were weighed at each time point—0, 3, 6 and 24 h—to calculate the water vapor transmission rate. The weight loss of the glass tubes was determined as Δm (g), $m_{0\,h}$–$m_{24\,h}$.

The water vapor transmission rate is calculated by the amount of water transmitted through in PVA film related to surface area A (m^2) and time t (day), following the equation below:

$$\text{Water vapor transmission rate (WVTR)} = \frac{\Delta m\ (g)}{A\ (m^2)\ \times\ \text{time (day)}}$$

The swelling ratio is usually measured for polymer hydrogel or polymer films to determine water resistance. The study of swelling behavior was performed referring to J. Liu et al. [36]. Pure PVA, PVA/CuI (0.095%) and PVA/CuI (0.190%) were cast in glass Petri dishes to form films. Then, each film was cut into a circular film with a diameter of 12 mm, and the average thickness was measured. After that, PBS buffer solution was prepared with sodium chloride, potassium chloride, sodium phosphate dibasic and monopotassium phosphate to make a buffer solution with pH 7.2 ± 0.1. Each film was weighed before the swelling test; the weight was determined as m_0. Then, the samples were incubated at 37 °C. After the specified time, the samples were removed from PBS, then quickly cleaned with Kimwipes paper to remove water and weighed with an electronic balance. At time t, the weight was recorded as m_t. All samples were continuously incubated in PBS solution and weighed at each time point up to 48 h to ensure that equilibrium was reached. The swelling ratio was calculated by the equation below:

$$\text{Swelling ratio} = \frac{m_t - m_0}{m_0} \times 100$$

4.11. *In Vitro Indirect Cytotoxic Test of PVA Liquid Bandage Formulations*

The indirect cytotoxicity of PVA loaded with CuI suspension was tested following the ISO10993-5 (biological evaluation of medical devices) standard test using adult human dermal fibroblasts (GIBCO, Grand Island, NY, USA). Firstly, the culture medium was prepared and collected by exposing samples to UV radiation for 30 min for sterilization. Then, the samples were immersed in DMEM in a 96-well tissue culture polystyrene (TCPS) plate at 0.5, 5 and 10 mg/mL for 1 and 3 days to produce different concentrations of sample extractions. After, the samples were placed into a 24-well culture plate containing of Dulbecco's phosphate-buffered saline modified with Eagle's medium (Hyclone, Logan, UT, USA), with the addition of 10% fetal bovine serum (Gibco), 1% L-glutamine (Gibco), 100 µg/mL of streptomycin, 100 units/mL of penicillin (Gibco) and 5 µg/mL of amphotericin B (Gibco) for 24 h. Then, 10,000 cells of the human dermal fibroblast (HDFa) cell line were cultured in the prepared medium in a 96-well plate and incubated overnight in a humidity chamber with 5% CO_2 at 37 °C. The number of viable cells were counted using (4,5-dimethylthiazol-2-yl)-2,5-diphenyl-tetrazolium bromide (MTT) assay (USB Corporation, Cleveland, OH, USA). An absorbance microplate reader (BiotekELx800; Biotek Instruments Inc., Winooski, VT, USA) was used to determine the optical density (O.D.) at 570 nm of cells. The control group was 2D cell culture in normal conditions.

4.12. *Antibacterial Test by Time–Kill Assay*

The antibacterial activities of PVA liquid bandage with different concentration of CuI suspension were characterized through time–kill assay. It was necessary to determine the ability of the liquid bandage to kill bacteria in a fixed time. Briefly, the antibacterial activity of PVA liquid solutions was tested against *E. coli* (Gram-negative) and MRSA (Gram-positive) bacteria. The samples included pure PVA solution and PVA liquid bandages with different concentrations of CuI suspension. Firstly, both the *E. coli* and MRSA microorganisms were cultured in Mueller Hinton Broth (MHB) at 37 °C for 18 h. After, the bacterial solutions of *E. coli* and MRSA were adjusted to 0.5 McFarland standard solution using a densitometer, determined as 1.5×10^8 CFU/mL. Then, the bacterial solutions were diluted with MHB to 5×10^5 CFU/mL. Each sample contained 3 mL of 5×10^5 CFU/mL diluted inoculum with continual turbulence at 37 °C for 0, 1, 3, 6, 12 and 24 h. At each time, 20 µL of each inoculum was brought out and replaced with normal saline solution. Eventually, 5 drops of 10 μL of each serial dilution were dropped on prepared Mueller–Hinton agar (MHA) in a Petri dish and incubated overnight in an incubator with controlled relative humidity and temperature of 37 °C. The number of growing bacterial colonies at each time point was counted, averaged and compared with the number of the control culture by estimating the CFU/mL values. The percentage of bacterial reduction was calculated following this equation:

$$\text{Bacterial reduction } (\%) = \frac{N_{\text{control}} - N_{\text{specimen}}}{N_{\text{control}}} \times 100$$

where N_{control} is the number of bacterial colonies in the control (CFU/mL); N_{specimen} is the number of bacterial colonies in the specimen (CFU/mL).

Author Contributions: Conceptualization, supervision and funding acquisition, P.S. (Pitt Supaphol); data curation, formal analysis, investigation and methodology, P.P., C.C., S.C., P.S. (Pakakrong Sangsanoh), O.S. and P.C.; data curation, validation and review of the manuscript, O.S. and P.C.; visualization, P.P., C.C. and S.C.; writing—original draft, P.P. All authors have read and agreed to the published version of the manuscript.

Funding: This research was funded by the 90th Anniversary of Chulalongkorn University Scholarship Batch 50: GCUGR1125643048M No. 2–14 and Fundamental Fund 2565: FF65, Chulalongkorn University, Thailand.

Data Availability Statement: Not applicable.

Acknowledgments: The authors appreciate the full scholarship from the Petroleum and Petrochemical College (PPC), Chulalongkorn University. This work was performed by the support of Herbal Extract-Infused Advanced Wound Dressing Research Unit, Ratchadaphisekphot Endowment Fund, Chulalongkorn University, Thailand.

Conflicts of Interest: The authors declare no conflict of interest.

Declaration of Competing Interest: The authors declare that they have no known competing financial interests or personal relationships that could appear to influence the work reported in this paper.

References

1. Alven, S.; Aderibigbe, B.A. Fabrication of Hybrid Nanofibers from Biopolymers and Poly (Vinyl Alcohol)/Poly (ε-Caprolactone) for Wound Dressing Applications. *Polym. J.* **2021**, *13*, 2104. [CrossRef] [PubMed]
2. Li, S.; Dong, S.; Xu, W.; Tu, S.; Yan, L.; Zhao, C.; Ding, J.; Chen, X. Antibacterial Hydrogels. *Adv. Sci. (Weinh.)* **2018**, *5*, 1700527. [CrossRef] [PubMed]
3. Rezvani Ghomi, E.; Khalili, S.; Nouri Khorasani, S.; Esmaeely Neisiany, R.; Ramakrishna, S. Wound dressings: Current advances and future directions. *J. Appl. Polym. Sci.* **2019**, *136*, 47738. [CrossRef]
4. Kalantari, K.; Mostafavi, E.; Saleh, B.; Soltantabar, P.; Webster, T.J. Chitosan/PVA hydrogels incorporated with green synthesized cerium oxide nanoparticles for wound healing applications. *Eur. Polym. J.* **2020**, *134*, 109853. [CrossRef]
5. Chopra, H.; Bibi, S.; Kumar, S.; Khan, M.S.; Kumar, P.; Singh, I. Preparation and Evaluation of Chitosan/PVA Based Hydrogel Films Loaded with Honey for Wound Healing Application. *Gels* **2022**, *8*, 111. [CrossRef] [PubMed]
6. Yudaev, P.; Mezhuev, Y.; Chistyakov, E. Nanoparticle-Containing Wound Dressing: Antimicrobial and Healing Effects. *Gels* **2022**, *8*, 329. [CrossRef] [PubMed]
7. Mei, L.; Lu, Z.; Zhang, X.; Li, C.; Jia, Y. Polymer-Ag Nanocomposites with Enhanced Antimicrobial Activity against Bacterial Infection. *ACS Appl. Mater. Interfaces* **2014**, *6*, 15813–15821. [CrossRef]
8. Mei, L.; Xu, Z.; Shi, Y.; Lin, C.; Jiao, S.; Zhang, L.; Li, P. Multivalent and synergistic chitosan oligosaccharide-Ag nanocomposites for therapy of bacterial infection. *Sci. Rep.* **2020**, *10*, 10011. [CrossRef]
9. El-Sherif, H.; El-Masry, M.; Kansoh, A. Hydrogels as template nanoreactors for silver nanoparticles formation and their antimicrobial activities. *Macromol. Res.* **2011**, *19*, 1157. [CrossRef]
10. Lee, J.W.; Choi, S.-R.; Heo, J.H. Simultaneous Stabilization and Functionalization of Gold Nanoparticles via Biomolecule Conjugation: Progress and Perspectives. *ACS Appl. Mater. Interfaces* **2021**, *13*, 42311–42328. [CrossRef]
11. Sánchez-Sanhueza, G.; Fuentes-Rodríguez, D.; Bello-Toledo, H. Copper Nanoparticles as Potential Antimicrobial Agent in Disinfecting Root Canals: A Systematic Review. *Int. J. Odontostomatol.* **2016**, *10*, 547–554. [CrossRef]
12. Sunada, K.; Minoshima, M.; Hashimoto, K. Highly efficient antiviral and antibacterial activities of solid-state cuprous compounds. *J. Hazard. Mater.* **2012**, *235–236*, 265–270. [CrossRef] [PubMed]

13. Thurman, R.B.; Gerba, C.P.; Bitton, G. The molecular mechanisms of copper and silver ion disinfection of bacteria and viruses. *Crit. Rev. Environ. Sci. Technol.* **1988**, *18*, 295–315. [CrossRef]
14. Imlay, J.A. Pathways of oxidative damage. *Annu. Rev. Microbiol.* **2003**, *57*, 395–418. [CrossRef] [PubMed]
15. Pramanik, A.; Laha, D.; Bhattacharya, D.; Pramanik, P.; Karmakar, P. A novel study of antibacterial activity of copper iodide nanoparticle mediated by DNA and membrane damage. *Colloids Surf. B* **2012**, *96*, 50–55. [CrossRef]
16. Wang, L.; Hu, C.; Shao, L. The antimicrobial activity of nanoparticles: Present situation and prospects for the future. *Int. J. Nanomed.* **2017**, *12*, 1227–1249. [CrossRef]
17. Alven, S.; Peter, S.; Mbese, Z.; Aderibigbe, B.A. Polymer-Based Wound Dressing Materials Loaded with Bioactive Agents: Potential Materials for the Treatment of Diabetic Wounds. *Polym. J.* **2022**, *14*, 724. [CrossRef] [PubMed]
18. Pal, G.; Rai, P.; Pandey, A. Chapter 1—Green synthesis of nanoparticles: A greener approach for a cleaner future. In *Green Synthesis, Characterization and Applications of Nanoparticles*; Shukla, A.K., Iravani, S., Eds.; Elsevier: Amsterdam, The Netherlands, 2019; pp. 1–26. [CrossRef]
19. Badhani, B.; Sharma, N.; Kakkar, R. Gallic acid: A versatile antioxidant with promising therapeutic and industrial applications. *RSC Adv.* **2015**, *5*, 27540–27557. [CrossRef]
20. Khan, M.; Ahmad, F.; Koivisto, J.T.; Kellomäki, M. Green synthesis of controlled size gold and silver nanoparticles using antioxidant as capping and reducing agent. *Colloid Interface Sci. Commun.* **2020**, *39*, 100322. [CrossRef]
21. Singh, J.; Dutta, T.; Kim, K.-H.; Rawat, M.; Samddar, P.; Kumar, P. 'Green' synthesis of metals and their oxide nanoparticles: Applications for environmental remediation. *J. Nanobiotechnol.* **2018**, *16*, 84. [CrossRef]
22. Amini, S.M.; Akbari, A. Metal nanoparticles synthesis through natural phenolic acids. *IET Nanobiotechnol.* **2019**, *13*, 771–777. [CrossRef] [PubMed]
23. Archana, K.M.; Yogalakshmi, D.; Rajagopal, R. Application of green synthesized nanocrystalline CuI in the removal of aqueous Mn(VII) and Cr(VI) ions. *SN Appl. Sci.* **2019**, *1*, 522. [CrossRef]
24. Byranvand, M.M.; Kharat, A.N. Triangular-like Cuprous Iodide Nanostructures: Green and Rapid Synthesis Using Sugar Beet Juice. *Rom. J. Biochem.* **2014**, *51*, 101–107.
25. Haridas, M.; Srivastava, S.; Basu, J.K. Tunable variation of optical properties of polymer capped gold nanoparticles. *Eur. Phys. J. D* **2008**, *49*, 93–100. [CrossRef]
26. Chinnakutti, K.K.; Panneerselvam, V.; Govindarajan, D.; Soman, A.K.; Parasuraman, K.; Thankaraj Salammal, S. Optoelectronic and electrochemical behaviour of γ-CuI thin films prepared by solid iodination process. *Prog. Nat. Sci. Mater. Int.* **2019**, *29*, 533–540. [CrossRef]
27. Akhtar, F.Z.; Archana, K.M.; Krishnaswamy, V.G.; Rajagopal, R. Remediation of heavy metals (Cr, Zn) using physical, chemical and biological methods: A novel approach. *SN Appl. Sci.* **2020**, *2*, 267. [CrossRef]
28. Yao, K.; Chen, P.; Zhang, Z.; Li, J.; Ai, R.; Ma, H.; Zhao, B.; Sun, G.; Wu, R.; Tang, X.; et al. Synthesis of ultrathin two-dimensional nanosheets and van der Waals heterostructures from non-layered γ-CuI. *NPJ 2D Mater. Appl.* **2018**, *2*, 16. [CrossRef]
29. Das, P.E.; Abu-Yousef, I.A.; Majdalawieh, A.F.; Narasimhan, S.; Poltronieri, P. Green Synthesis of Encapsulated Copper Nanoparticles Using a Hydroalcoholic Extract of Moringa oleifera Leaves and Assessment of Their Antioxidant and Antimicrobial Activities. *Molecules* **2020**, *25*, 555. [CrossRef]
30. Selvamani, V. Chapter 15—Stability Studies on Nanomaterials Used in Drugs. In *Characterization and Biology of Nanomaterials for Drug Delivery*; Mohapatra, S.S., Ranjan, S., Dasgupta, N., Mishra, R.K., Thomas, S., Eds.; Elsevier Amsterdam, The Netherlands, 2019; pp. 425–444. [CrossRef]
31. Sun, X.; Dong, M.; Guo, Z.; Zhang, H.; Wang, J.; Jia, P.; Bu, T.; Liu, Y.; Li, L.; Wang, L. Multifunctional chitosan-copper-gallic acid based antibacterial nanocomposite wound dressing. *Int. J. Biol. Macromol.* **2021**, *167*, 10–22. [CrossRef]
32. Slavin, Y.N.; Asnis, J.; Häfeli, U.O.; Bach, H. Metal nanoparticles: Understanding the mechanisms behind antibacterial activity. *J. Nanobiotechnol.* **2017**, *15*, 65. [CrossRef]
33. Kharazmi, A.; Faraji, N.; Mat Hussin, R.; Saion, E.; Yunus, W.M.; Behzad, K. Structural, optical, opto-thermal and thermal properties of ZnS-PVA nanofluids synthesized through a radiolytic approach. *Beilstein J. Nanotechnol.* **2015**, *6*, 529–536. [CrossRef]
34. Zeimaran, E.; Pourshahrestani, S.; Pingguan-Murphy, B.; Kadri, N.A.; Rothan, H.A.; Yusof, R.; Towler, M.R.; Djordjevic, I. Fabrication and characterization of poly(octanediol citrate)/gallium-containing bioglass microcomposite scaffolds. *J. Mater. Sci.* **2015**, *50*, 2189–2201. [CrossRef]
35. Liu, J.; Zheng, X.J.; Tang, K. Study on the Gravimetric Measurement of the Swelling Behaviors of Polymer Films. *Rev. Adv. Mater. Sci.* **2013**, *33*, 452–458.
36. Shahi, N.; Joshi, G.; Min, B. Effect of Regenerated Cellulose Fibers Derived from Black Oat on Functional Properties of PVA-Based Biocomposite Film. *Processes* **2020**, *8*, 1149. [CrossRef]
37. Wen, Y.-H.; Tsou, C.-H.; de Guzman, M.R.; Huang, D.; Yu, Y.-Q.; Gao, C.; Zhang, X.-M.; Du, J.; Zheng, Y.-T.; Zhu, H.; et al. Antibacterial nanocomposite films of poly(vinyl alcohol) modified with zinc oxide-doped multiwalled carbon nanotubes as food packaging. *Polym. Bull.* **2022**, *79*, 3847–3866. [CrossRef]
38. Abdullah, Z.W.; Dong, Y.; Han, N.; Liu, S. Water and gas barrier properties of polyvinyl alcohol (PVA)/starch (ST)/ glycerol (GL)/halloysite nanotube (HNT) bionanocomposite films: Experimental characterisation and modelling approach. *Compos. B Eng.* **2019**, *174*, 107033. [CrossRef]

39. Nooeaid, P.; Chuysinuan, P.; Pengsuk, C.; Dechtrirat, D.; Lirdprapamongkol, K.; Techasakul, S.; Svasti, J. Polylactic acid microparticles embedded porous gelatin scaffolds with multifunctional properties for soft tissue engineering. *J. Sci. Adv. Mater. Dev.* **2020**, *5*, 337–345. [CrossRef]
40. Zurdo Schroeder, I.; Franke, P.; Schaefer, U.F.; Lehr, C.M. Development and characterization of film forming polymeric solutions for skin drug delivery. *Eur. J. Pharm. Biopharm.* **2007**, *65*, 111–121. [CrossRef] [PubMed]

gels

MDPI

Article

Robocasting and Laser Micromachining of Sol-Gel Derived 3D Silica/Gelatin/β-TCP Scaffolds for Bone Tissue Regeneration

María V. Reyes-Peces [1], Eduardo Félix [1,2], Francisco J. Martínez-Vázquez [3], Rafael Fernández-Montesinos [4,5], Óscar Bomati-Miguel [1,2], María del Mar Mesa-Díaz [2,6], Rodrigo Alcántara [2,7], José Ignacio Vilches-Pérez [5], Mercedes Salido [4,5], Nicolás De la Rosa-Fox [1,2] and Manuel Piñero [1,2,*]

1 Departamento de Física de la Materia Condensada, Facultad de Ciencias, Universidad de Cádiz, 11510 Puerto Real, Spain
2 Institute of Research on Electron Microscopy and Materials (IMEYMAT), Universidad de Cádiz, 11510 Puerto Real, Spain
3 Departamento de Ingeniería Mecánica, Energética y de los Materiales, Universidad de Extremadura, 06006 Badajoz, Spain
4 Departamento de Histología, Facultad de Medicina, Universidad de Cádiz, 11004 Cádiz, Spain
5 Instituto de Biomedicina de Cádiz, INIBICA, Universidad de Cádiz, 11009 Cádiz, Spain
6 Departamento de Ingeniería Química, Facultad de Ciencias, Universidad de Cádiz, 11510 Puerto Real, Spain
7 Departamento de Química-Física, Facultad de Ciencias, Universidad de Cádiz, 11510 Puerto Real, Spain
* Correspondence: manolo.piniero@gm.uca.es

Citation: Reyes-Peces, M.V.; Félix, E.; Martínez-Vázquez, F.J.; Fernández-Montesinos, R.; Bomati-Miguel, Ó.; Mesa-Díaz, M.d.M.; Alcántara, R.; Vilches-Pérez, J.I.; Salido, M.; De la Rosa-Fox, N.; et al. Robocasting and Laser Micromachining of Sol-Gel Derived 3D Silica/Gelatin/β-TCP Scaffolds for Bone Tissue Regeneration. *Gels* **2022**, *8*, 634. https://doi.org/10.3390/gels8100634

Academic Editors: Arish Dasan, Ashokraja Chandrasekar and Nupur Kohli

Received: 12 September 2022
Accepted: 3 October 2022
Published: 7 October 2022

Abstract: The design and synthesis of sol-gel silica-based hybrid materials and composites offer significant benefits to obtain innovative biomaterials with controlled porosity at the nanostructure level for applications in bone tissue engineering. In this work, the combination of robocasting with sol-gel ink of suitable viscosity prepared by mixing tetraethoxysilane (TEOS), gelatin and β-tricalcium phosphate (β-TCP) allowed for the manufacture of 3D scaffolds consisting of a 3D square mesh of interpenetrating rods, with macropore size of 354.0 ± 17.0 μm, without the use of chemical additives at room temperature. The silica/gelatin/β-TCP system underwent irreversible gelation, and the resulting gels were also used to fabricate different 3D structures by means of an alternative scaffolding method, involving high-resolution laser micromachining by laser ablation. By this way, 3D scaffolds made of 2 mm thick rectangular prisms presenting a parallel macropore system drilled through the whole thickness and consisting of laser micromachined holes of 350.8 ± 16.6-micrometer diameter, whose centers were spaced 1312.0 ± 23.0 μm, were created. Both sol-gel based 3D scaffold configurations combined compressive strength in the range of 2–3 MPa and the biocompatibility of the hybrid material. In addition, the observed Si, Ca and P biodegradation provided a suitable microenvironment with significant focal adhesion development, maturation and also enhanced in vitro cell growth. In conclusion, this work successfully confirmed the feasibility of both strategies for the fabrication of new sol-gel-based hybrid scaffolds with osteoconductive properties.

Keywords: 3D scaffold; hybrid; sol-gel ink; robocasting; laser micromachining; osteoblasts; regenerative medicine; bone tissue engineering; cytoskeleton; focal adhesion

1. Introduction

Three-dimensional (3D) porous scaffolds are widely used in various biomedical applications such as drug delivery [1], cell culture studies [2] and regenerative medicine [3–6]. Therefore, many different scaffold fabrication methods have been explored to produce biomaterials with the required 3D architecture including not only conventional subtractive approaches such as solvent-casting [7], particulate-leaching [8] and gas-foaming processes [9] but also advanced techniques, such as electrospinning [10] and additive manufacturing (AM) technologies [5,11,12]. Among the various AM procedures, special attention has been paid to the robocasting of porous scaffolds with customized design for bone tissue engineering [13–17]. In general, their architectures mimicking the extracellular matrix (ECM) of trabecular bone are expected to perform better as bone artificial implants because their composition can be adjusted by selecting appropriate ink precursors [18]. In this context, functional requirements such as biocompatibility and bioactivity, that are needed to accommodate cells and guide their growth can be facilitated in a variety of ways depending on the preferred precursors [19,20].

First approaches to the robocasting of such 3D structures were devoted to the processing of ceramics, mainly calcium phosphate compounds due to their osteoconductive and bioactive nature [21,22]. Hydroxyapatite (HA) and β-tricalcium phosphate (β-TCP) were used to prepare aqueous pastes for robocasting [13,14,21–26]. To improve the stabilization of the ceramic particles in these water-based inks, some dispersant additives such as polyacrylate (PAA), polyvinyl alcohol (PVA) and polyethyleneimine (PEI) were used [27,28]. Similarly, dispersant or plastifying agents, such as ethyl cellulose or carboxymethyl cellulose, have been incorporated to enhance the printability [29]. As a consequence, many of these inks turned out to be unstable, due to pH and chemistry environment variations. In addition, thermal calcination must be conducted after printing to eliminate undesired organics, followed by sintering to obtain the final ceramic products. These post-process heat treatments are usually accompanied by significant shrinkage of the scaffolds.

Gel-embedded suspensions based on physical hydrogels and organogels from natural biopolymers have been widely reported as suitable solvents in formulating robocasting inks due to their surfactant properties and easy handling [15]. Pluronic® F127 (SigmaAldrich, Saint Louis, MO, USA) commercial sol-gel agent and some cellulose-derived compounds, both exhibiting thermal reversible gelation, are among the most suitable hydrogel compounds utilized in the development of dense ceramic structures by robocasting [20,27,29,30]. Additionally, organogels, whose solidification mechanism depends on the sol-gel transition through solvent evaporation just after printing, have been widely studied in the formulation of ceramic pastes for DIW, including alginate [31,32], chitosan [33–35] and gelatin [31,36,37]. Similarly, hybrid organic/inorganic materials [19,38] and polymer/ceramic composite scaffolds, such as polylactide acid or polycaprolactone with high HA content [33] and chitosan/bioglass [39], were fabricated using robotic-assisted deposition at room temperature as bone substitutes, providing favorable substrate conditions for osteoconduction.

Additionally, while being less common, an alternative approach based on sol-gel inorganic polymerization, involving irreversible gelation to determine the printing window, was proposed as solvent media to develop new sol-gel inks for DIW. However, a stiff gel was rapidly formed avoiding an effective deposition of the filaments, according to CAD design [40]. Nevertheless, some different 3D oxide and hybrid organic/inorganic structures have been developed by robocasting using sol-gel inks based on organometallic and biopolymer precursors, with critical control of the irreversible gelation process involving chemical polymerization reactions [41,42]. By this way, silica/chitosan/HA hybrids [42] and TiO_2 3D scaffolds for in vitro cell growth have been obtained [43].

Although, in a bulk monolithic state, similar types of sol-gel hybrid materials (xerogels and aerogels) are of particular interest because they have previously demonstrated both appropriate mechanical and biological properties for tissue engineering applications without needing to perform further calcination or sintering heat treatments [44,45]. Hence, the hybridization of silica and biopolymer precursors in the sol state (such as chitosan, gelatin, etc.), can lead to robocasting inks undergoing irreversible gelation, whose viscosity and rheological properties can be tuned without intermediation of chemical binders, just by adding ceramic solid reinforcing phases (e.g., calcium phosphate compounds). By this way, 3D organic–inorganic hybrid/ceramic composite scaffolds, which cannot be achieved by either organic polymers or ceramics separately, can be obtained. To our knowledge, the consideration of whether sol-gel systems exhibiting irreversible gelation can be used to promote solidification in robocasting, thus facilitating the stabilization of printed structures, is a subject that has been scarcely investigated, and this work aims to contribute to this research topic by introducing the first results based on silica/gelatin/β-tricalcium phosphate (TCP) 3D scaffolds.

Otherwise, an alternative rapid prototyping production of 3D microstructures that has also been widely used for biomedical applications is surface micromachining by applying intense ultrashort pulsed laser ablation [46]. It is essentially a technique that has been used for selective material removal to generate precise biomimetic patterning on biomaterials [46,47]. Moreover, the creation of surface micro-channels or micro-holes can be achieved without excessive heating or the appearance of cracks in the annular melting zone [48]. Recently, functional surfaces from collagen gel [49], biopolymer/ceramic composite thin films [50] and metals [51] have been modified through femtosecond laser treatment, leading to 3D structures with potential interest in the improvement of cellular adhesion and tissue engineering applications [52]. Hydrogel 3D structures for biomedical applications have been successfully processed via femtosecond laser pulses [53]. However, although laser-assisted fabrication offers high spatial resolution on surface biomaterial patterning and cell guidance, it suffers from low processing speed and limited scaffold size (usually restricted to the millimeter or micrometer scales) [52,54].

In summary, the aim of this work is to introduce two different strategies to easily obtain 3D sol-gel-derived ceramic/biopolymer porous scaffolds. To this end, first, we present a direct ink writing (DIW) procedure to print 3D porous structures from a suspension of β-TCP microsized powder (2 μm) in a silica/gelatin hybrid sol. A crosslinking agent, 3-glycidoxypropyltrimethoxysilane (GPTMS), is incorporated into the hybrid for enhancing the chemical stability and mechanical properties of the resulting hybrid material and for controlling the degradation rate in aqueous media. As reported in previous results, a similar silica/gelatin hybrid system in combination with the GPTMS coupling agent presented the benefit of biocompatibility, as well as chemical and mechanical stability [55,56]. Next, the ink formulation is optimized to become a paste with adequate viscosity and shear-thinning behavior so that it can easily flow through the nozzle to form continuous rod-like filaments under room temperature conditions, while the resulting scaffolds should have the ability to retain their shapes. The solidification of the printed objects is achieved by irreversible gelation through acid-catalyzed condensation and polymerization of the hybrid matrix sol, followed by drying at ambient temperature. Immediately after, the scaffolds are available for different characterization experiments and cell culture test. Alternatively, the second method makes use of an ultrafast laser characterized by ultrahigh peak power pulses, to introduce a spatially patterned macropore system by laser drilling into a silica/gelatin/β-TCP sol-gel-derived monolith composite. According to the existing literature, although the technique has been previously utilized for the removal of organically modified silicate coatings [57], this is the first time that a complete perforation by laser micromachining of a sol-gel-processed hybrid polymer/ceramic monolith composite, is planned. Both the hole quality and maximum ablated depth are optimized by controlling the laser machining parameters for the selected solid target. Likewise, the mechanical behavior of the scaffolds is studied under uniaxial compression. In addition, this work investigates the textural properties of the sol-gel biomaterial utilized in scaffolding, as well as its biodegradation in saline solution and osteoblastic response in vitro in the presence of biomaterials.

2. Results and Discussion

2.1. Synthesis of the Sol-Gel Ink and Fabrication of Scaffolds

In our study, hybrid silica/gelatin matrix sols (SG) were prepared by the sol-gel method, using TEOS as the silica source under acidic conditions. Hybridization between the organic and inorganic phases was facilitated by adding GPTMS, an organosilane crosslinking agent. Gelatin, previously functionalized with GPTMS, is covalently linked to silica thanks to the condensation reaction of the carboxylic acid groups of gelatin with the epoxy ring of GPTMS and condensed into the silica network via Si-O-Si covalent bonding to form hybrid mechanically strong materials [55]. Hybrid sols (SGx) with three different gelatin contents (x = 20, 40 and 60 wt%) were thus prepared. Total gelation time is a critical variable for the solidification process in robocasting [53,58–60]. In the current study, it decreased with an increase in the gelatin content. Sols containing the highest gelatin content (SG60) gelled completely after ~5 min at room temperature, while the gelation of sols with 20 wt% gelatin (SG20) took over 60 min. Furthermore, a gel time of ~30 min was observed for the SG40 hybrid matrix, which best fits the robocasting process. Subsequently, to provide appropriate rheological behavior for extrusion to the ink precursor solution, different amounts of β-TCP microsized powder were added (40, 50 and 60 wt%), and after several trial experiments, a loading content of 60 wt% was found to be optimal for printing at room temperature.

The rheological properties of the three resulting homogenized pastes were investigated, and Figure 1 shows the apparent viscosity as a function of the shear rate immediately after its preparation. As observed, all of the pastes exhibit a shear-thinning behavior from 0.1 to 100 s^{-1}, mainly for higher solid contents. Similarly, the slight increase in the apparent viscosity of the 40 wt% sample at 0.3 s^{-1} shear rate could be attributed to some structural remodeling in the hybrid sol suspension requiring an extra applied stress before it drops due to shear-thinning [30]. This considered, the increase in the viscosity with increasing additions of β-TCP, which can be observed specifically within the 1–100 s^{-1} shear rate range, confirms that, as expected, the sol-gel ink with the highest β-TCP content (60 wt%) was the best option for direct writing.

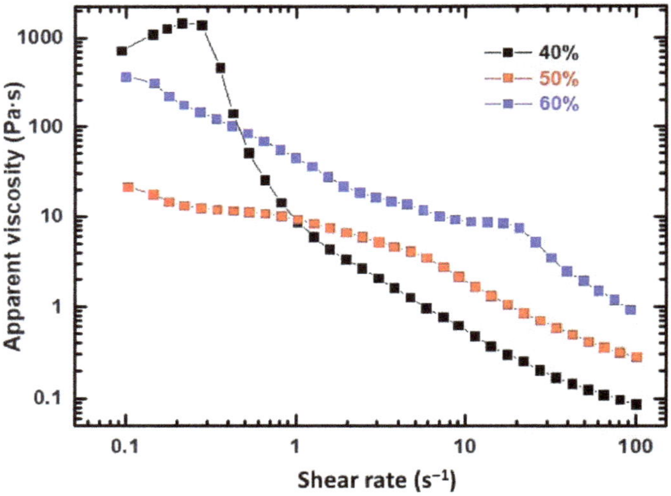

Figure 1. Apparent viscosity of SG40TCPy (y = 40, 50 and 60 wt% β-TCP) pastes vs shear rate.

By this way, 3D scaffolds with dimensions of $10 \times 10 \times 5$ mm^3 named as SG40TCP60-RC were obtained by the deposition of perpendicular layers of parallel rods with a 610.0 ± 18.0-micrometer diameter in a paraffin bath to reduce non-uniform drying throughout the printing process.

Alternatively, a bulk monolith was also produced by solvent casting and evaporative drying using the SG40TCP60 sol-gel paste. After that, it was further subjected to laser micromachining (LM) by ablation in air, in order to determine the minimum thickness to achieve a complete perforation of the sample, aiming to create a regular 350-micrometer diameter macropore system with a rectangular geometry and high accuracy. Some examples of the fabricated RC and LM scaffolds are shown in Figure 2. Both presented the benefit of moderate temperature treatment when using these types of biomaterials for biomedical applications, considering that sterilization treatments involve temperatures lower than 130 °C.

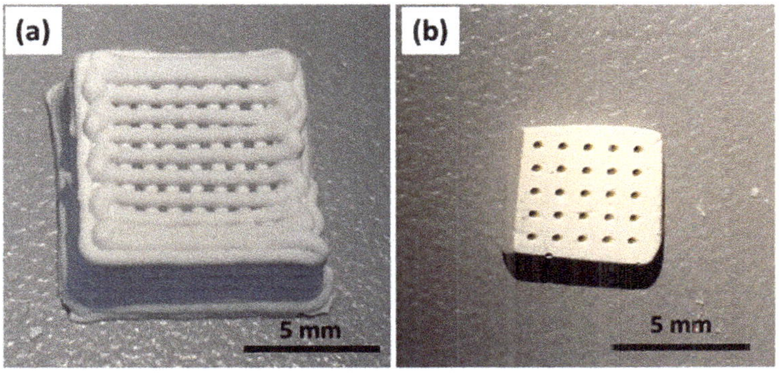

Figure 2. Optical images of 3D macropore structures. (**a**) SG40TCP60-RC scaffold, manufactured by robocasting and (**b**) SG40TCP60-LM, fabricated by laser micromachining (2-millimeter thickness).

The SEM micrographs in Figure 3 show the structure and organization of both RC and LM scaffolds at low and high resolution. As observed in Figure 3a,b, the cylindrical cross-sections of the rods remained almost unchanged after extrusion. Therefore, no layer collapsing was observed, and the RC scaffolds maintained their predesigned 3D structure, exhibiting an interconnected macroporosity, according to previous investigations [61,62].

Otherwise, the resulting macropore system introduced by laser machining through a massive SG40TCP60 monolith composite can be seen in Figure 3c,d. Figure 3c shows a cross-sectional view of the cylindrical holes with 350.8 ± 16.6-micrometer diameter oriented parallel to the cylinder axes, whose centers are regularly spaced by 1312.0 ± 23.0 µm, created by laser ablation on a 1 cm^2 flat surface of a 2 mm thick rectangular composite specimen. Moreover, Figure 3d shows the axial-section of the cylinder holes, passing through and maintaining its diameter along the entire thickness. This procedure is preliminary research of a deeper investigation in which the progression of the porosity through thicker specimens as well as its interconnection is under development.

Figure 3e,f provide visual clues about the surface texture of the solid phase. It appears to consist of agglomerates of ceramic microparticles, embedded by a thin film of silica–gelatin hybrid polymer wetting the contact surfaces between ceramic particles, presumably filling and sealing the interparticle pore walls and, therefore, provoking pore blocking effects in the characteristic micro-mesoporosity of this type of gel. This structure agrees with the physisorption results.

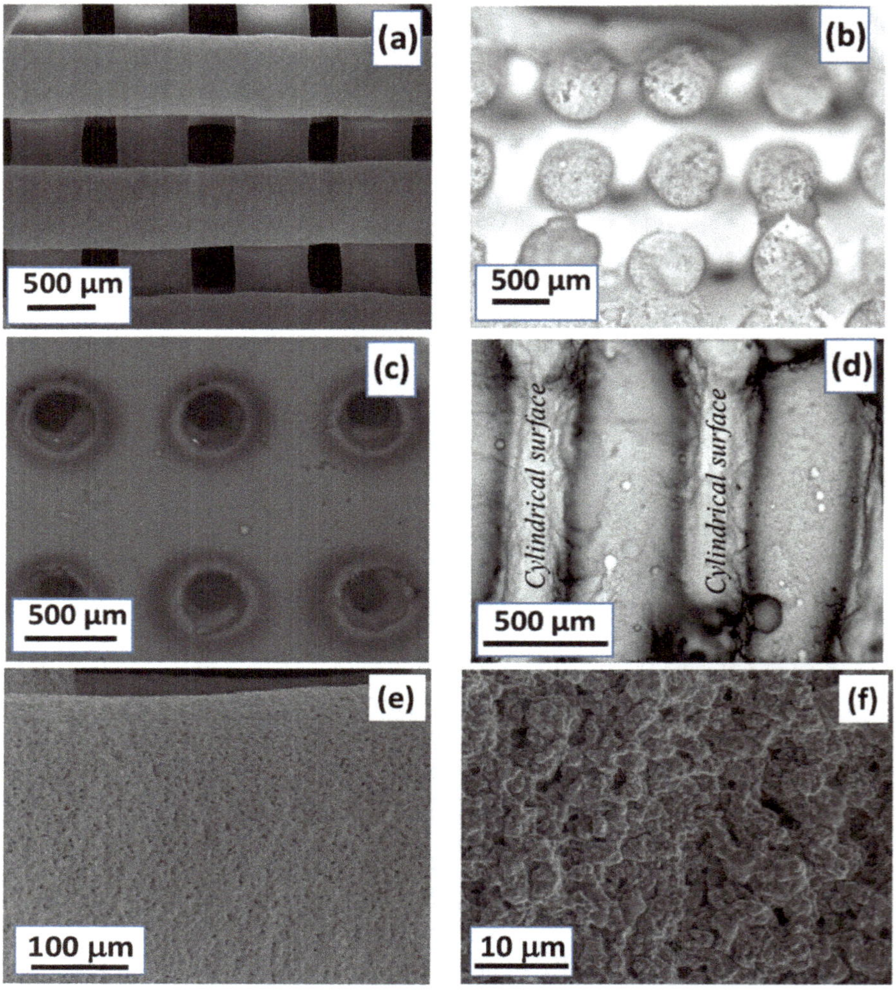

Figure 3. Micrographs show upper scaffold surfaces taken by SEM prepared from: (**a**) robocasting (RC) and (**c**) laser micromachining (LM). Images (**b**,**d**) were made using a confocal microscope and correspond, respectively, to cross-sectional areas of rods obtained by robocasting and axial-section area of the cylindrical pore cavity created by laser ablation (cylindrical surfaces); (**e**,**f**) SEM magnifications of the cylinder rod RC scaffold surface previously shown in Figure 3a.

2.2. Physical and Textural Characterization

RC and LM 3D scaffolds were fabricated displaying negligible volume shrinkage (see Figure 2) using the same hybrid sol-gel paste composition. Therefore, the macroporous interconnected structure of the RC scaffold fabricated by the deposition of perpendicular layers of parallel rods was preserved with minimal deviations from its respective CAD design. The same observation applies for the LM scaffold under laser processing, for which a system of parallel cylindrical pores by the ablation of material was produced. A highly uniform surface microstructure of both RC and LM scaffolds was observed by SEM (see Figure 3), confirming the homogeneity of the scaffold skeleton at the microstructural level.

The apparent densities for the RC and LM scaffolds were calculated from geometric measurements and similar results were observed: 1.00 ± 0.01 g cm^{-3} for RC scaffolds and 1.11 ± 0.01 g cm^{-3} for LM scaffolds. Otherwise, the skeletal density measured by He pycnometry for an SG40TCP60 cylinder monolith was 1.97 ± 0.31 g cm^{-3}, which underestimates by 20% the value determined by the rule of mixtures, bearing in mind the composition of the resulting material: 40 wt% [SiO_2 Gelatin (40 wt%)] and 60 wt% β-TCP. This result revealed the existence of a closed non-connected porosity within the material in the micro-mesopore range. As a consequence, the total open porosity of the samples calculated from Equation (1) was 50 ± 5% for the RC scaffolds and 44 ± 4% for the LM scaffolds.

N_2 physisorption experiments revealed some microstructural features of the SG40TCP60 material used for scaffolding. According to IUPAC classification, the sample exhibited reversible Type II isotherm, characteristic of macroporous or non-porous solids, as observed in Figure 4. The corresponding curve shape is attributed to unrestricted monolayer–multilayer adsorption up to high p/p_0, while the thickness of the adsorbed multilayer normally develops an apparent increase without limit when $p/p_0 = 1$ [63]. This type II N_2 isotherm usually displays a similar form over a wide range of multilayer coverage. However, differences arising in the curvature can be attributed to a significant overlap of monolayer coverage and multilayer adsorption, caused by the simultaneous presence of both high- and low-energy surface sites [64]. As a result, the N_2 adsorption process in SG40TCP60 can be described in terms of the BET parameter C data collected in Table 1. Consequently, higher values of C (~80–120) reveal that the adsorption process on energetic surfaces is predominant and the nitrogen monolayer is completed at low p/p_0 [63]. By contrast, low-energy adsorbents exhibit lower C values reflecting an appreciable monolayer–multilayer overlap [64]. These situations apply in the case of SG40TCP60, with a C value becoming even negative and, therefore, meaningless, thus invalidating the application of the BET method due to the difficulty of knowing the specific monolayer capacity. Special attention has been given by other authors to the preparation of new gels and composites in the silica–gelatin system, and the mutual interactions of organic and inorganic components on the formation of the hybrid network have almost been completely elucidated [65]. Research efforts have been primarily focused on the preparation of highly silica–gelatin porous materials, including crosslinkers such as GPTMS, mainly by controlling the sol-gel drying post-processing steps (supercritical drying) [66,67] or by thorough extraction of gelatin with water from xerogels [68] without using any crosslinker. In all these cases, specific surfaces in the range of 400–700 m^2g^{-1} were obtained for gelatin content up to 30 wt%. However, by varying the pH and processing conditions, xerogels with surface areas as low as 4–6 m^2g^{-1} were obtained, consisting of silica nanoparticles embedded in a gelatin network [68]. The textural properties measured for the SG40TCP60 material point in this direction, which means that both ceramic β-TCP microsized and silica nanosized particles are embedded in a gelatin percolating network.

Table 1. Bulk density and textural data from N_2 physisorption experiments for the SG40TCP60 material used in scaffold manufacture by robocasting and laser micromachining.

Sample	Bulk Density, ρ_{Bulk} (g^{-1} cm^{-3})	S_{BET} (m^2 g^{-1})	Pore Volume (cm^3 g^{-1})	Pore Size (nm)	C (BET)
SG40TCP60	1.11 ± 0.01	2.80 ± 1.30	0.03 ± 0.01	13.0 ± 0.1	<0

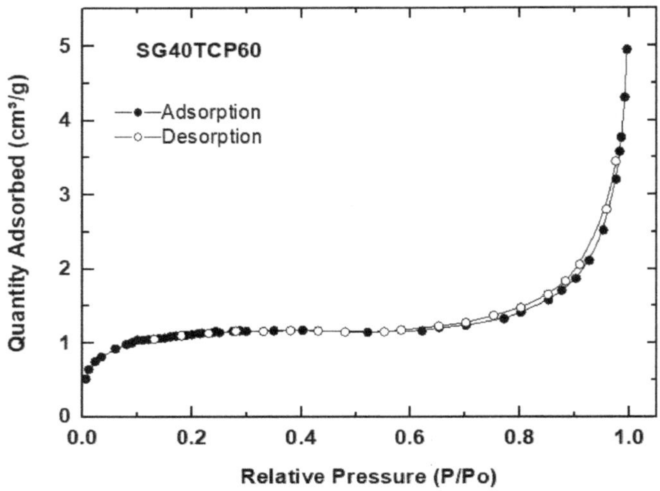

Figure 4. N_2 physisorption isotherm measured for the SG40TCP60 material.

Hence, as a general conclusion, it can be stated that regardless of its content, gelatin drastically alters the microstructure of the SG40TCP60 hybrid sample during drying in air by sealing the micro-mesopore distribution, thus provoking a severe reduction in the pore volume and causing the almost total loss of the specific surface area. The results and descriptions are consistent with the above-mentioned skeletal density measurements carried out by He pycnometry.

2.3. Mechanical Characterization

Compressive strength, compressive modulus of elasticity and other aspects of the stress–compression relationship were investigated by the uniaxial compressive stress on rectangular prisms obtained by robocasting (SG40TCP60-RC). To evaluate the directional mechanical strength and to determine the effect of structural compaction on its mechanical anisotropy, the samples were loaded in two different directions (perpendicular and parallel to the printing plane), as described in Figure 5a. Figure 5b shows the representative nominal stress–strain curve from the uniaxial compressive test performed on the perpendicular loading direction. It consisted of a short linear stage (0–22% strain), followed by load stabilization (2.0 ± 0.3 MPa) until 30% strain, corresponding to the collapse of the printed layers. Then, as the structure gets more compacted causing the characteristic pile-up of layers, the stress–strain curve increases progressively even after very large strains (~70%) until the whole structure is totally destroyed, and maximum load does not correspond to the compressive strength. This relationship between the stress and strain is related to the 3D structure of the scaffold, revealing differences in the mechanism of fracture and properties reported for similar samples [61], underlining the relevance of this new fabrication procedure. The inset in Figure 5b shows a representative stress–strain curve registered in the parallel direction, presenting heterogeneous transfer and distribution of the applied load, attributed to layer debonding due to a weak structural compaction of the sample. An abrupt brittle failure behavior at about 1.2 ± 0.3 MPa and 6.7% strain reports the existence of some mechanical anisotropy between the two loading directions.

In order to determine the mechanical behavior of the bulk samples being subjected to laser ablation (SG40TCP60-LM), additional compressive tests were also performed on the SG40TCP60 composite cylinder specimens. A representative stress–strain curve is shown in Figure 5c for test specimens of cylindrical geometry (10-millimeter diameter, 20-millimeter height), thus implementing the ASTM D7012 standard. Typical elasto-plastic behavior of sol-gel hybrid matrices reinforced with a high content of solid microparticles is displayed, where the relatively low compressive strength is provoked by a weak interfacial interaction between the matrix and the filler. A maximum stress of 3.30 ± 0.3 MPa was observed at 12% strain while Young's moduli obtained at the beginning of the curve showed marked stiffening in comparison with the printed sample. Thus, Young's moduli of 12.7 ± 5.2 MPa and 18.0 ± 2.7 MPa were estimated for perpendicular and parallel loading directions of the printed scaffold, respectively, while 56.3 ± 14.5 MPa was observed for the cylinder monolith sample. Young's modulus, as well as compressive strength and maximum compressive strength results of the different mechanical configurations, is reported in Table 2.

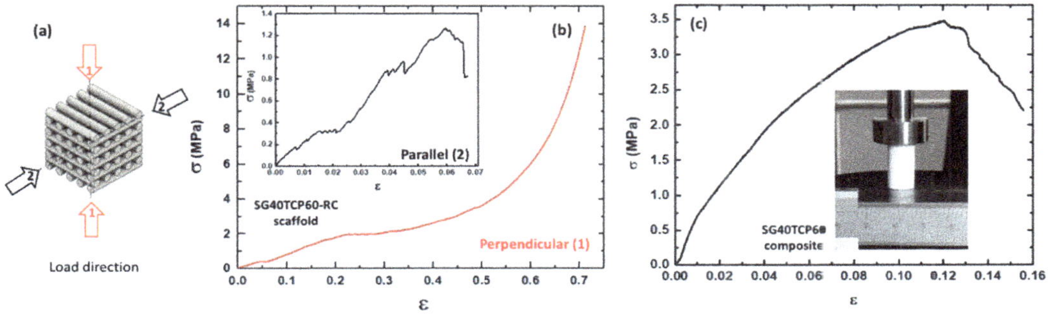

Figure 5. Stress–strain curves from the uniaxial compression test: (**a**) Load directions; (**b**) SG40TCP60-RC scaffold in perpendicular and parallel (inset) load directions to the printing plane; (**c**) SG40TCP60 composite cylinder specimen.

Table 2. Mechanical properties from the uniaxial compression test of the SG40TCP60-RC scaffold and SG40TCP60 composite cylinder; mean values ± standard deviation (n = 3 in all cases).

Sample	Young's Modulus, E (MPa)	Compressive Strength, σ (MPa)	Maximum Compressive Strain, ε (%)
SG40TCP60-RC (Perpendicular)	12.7 ± 5.2	2.00 ± 0.3 *	30 ± 2.2 *
SG40TCP60-RC (Parallel)	18.0 ± 2.7	1.00 ± 0.3	5.3 ± 0.1
SG40TCP60 composite cylinder	56.3 ± 14.5	3.3 ± 0.3	15.2 ± 2.6

* Compressive strength and maximum strain in the perpendicular direction of the RC sample were taken upon the collapse of the layered structures.

The scaffold obtained by laser micromachining (2-millimeter thickness) was not tested in compression because, according to the ASTM D7012 norm, the specimen must have a diameter of 1 mm to be tested. Hence, the related circular area does not allow for the creation of a hole size distribution system large enough to be representative of the laser-ablated scaffold design (350-micrometer diameter hole, whose centers must be spaced by 1015 μm). Nevertheless, further investigation of the laser processing to progress through thicker specimens would facilitate the measurement of the compressive strength.

2.4. Degradation Behavior in PBS

As bone regeneration is facilitated by the release of calcium and phosphorus ions that regulates the activation of osteoblasts and osteoclasts, the study of the biodegradation rates of these biomaterials is of crucial importance in future clinical applications. Figure 6 shows the dependence of the weight loss of the SG40TCP60 material on the incubation time for over 9 weeks. It occurs in two stages: a relatively fast first step during the first two weeks of soaking in PBS, where the scaffold exhibited an approximately 40 wt% weight loss and a slower second 10 wt% weight loss that took place during the next two weeks. Following this, the sample weight remained almost constant over the remaining test period until reaching the end of the test, in agreement with the results reported for similar hybrid gelatin-based materials [55,69]. This considered, the pH values were monitored over the same experimental time point as the weight losses, and they are also shown in Figure 6. As revealed, only small variations in the pH value were observed after the degradation process, varying from 7.54 at the starting point of the experiment to 7.18 at the end (week 9) of the test.

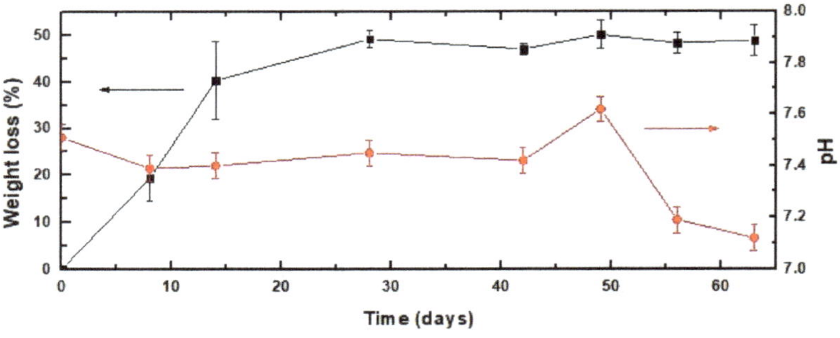

Figure 6. Weight loss (-■-) of the SG40TCP60 material in PBS and pH evolution (-■-) of the solution for 9 weeks.

The release profiles of Si, Ca and P from the 3D-printed scaffold soaked into the PBS buffer solution were evaluated using ICP_MS, and their corresponding results are presented in Figure 7a–c, respectively. A continuous increase in the Si concentration in time in the buffer medium was observed, from zero to approximately 23 mg L^{-1}, during 7 days of soaking (Figure 7a). This was attributed to a slow hydrolytic erosion process of the material in contact with the environment. These things considered, the quantified Si ion concentration release was three times lower than the Si concentration observed for other mesoporous silica-based hybrid xerogels made without gelatin using a similar experimental procedure [70]. The results obtained for both Ca and P release were obtained by subtracting the initial Ca and P concentrations of PBS [71] [3.23 $\pm$ 0.30 mg/L (Ca) and 273.00 $\pm$ 6.00 mg/L (P)] from the ICP_MS measurements in the studied mediums after different immersion periods. The amount of Ca released is presented in Figure 7b, exhibiting a slight increase from above 0.8 mg L^{-1}, after 24 h of exposure to PBS, up to 1.3 mg·L^{-1} at the end of the 7th week. Figure 7c shows the release profile of P, revealing a high dissolution from 0 to above 34 mg L^{-1} during the first 24 h, then increasing during the next 24 h up to 38 mg L^{-1} and remaining almost constant until reaching above 40 mg L^{-1} at the conclusion of the test procedure.

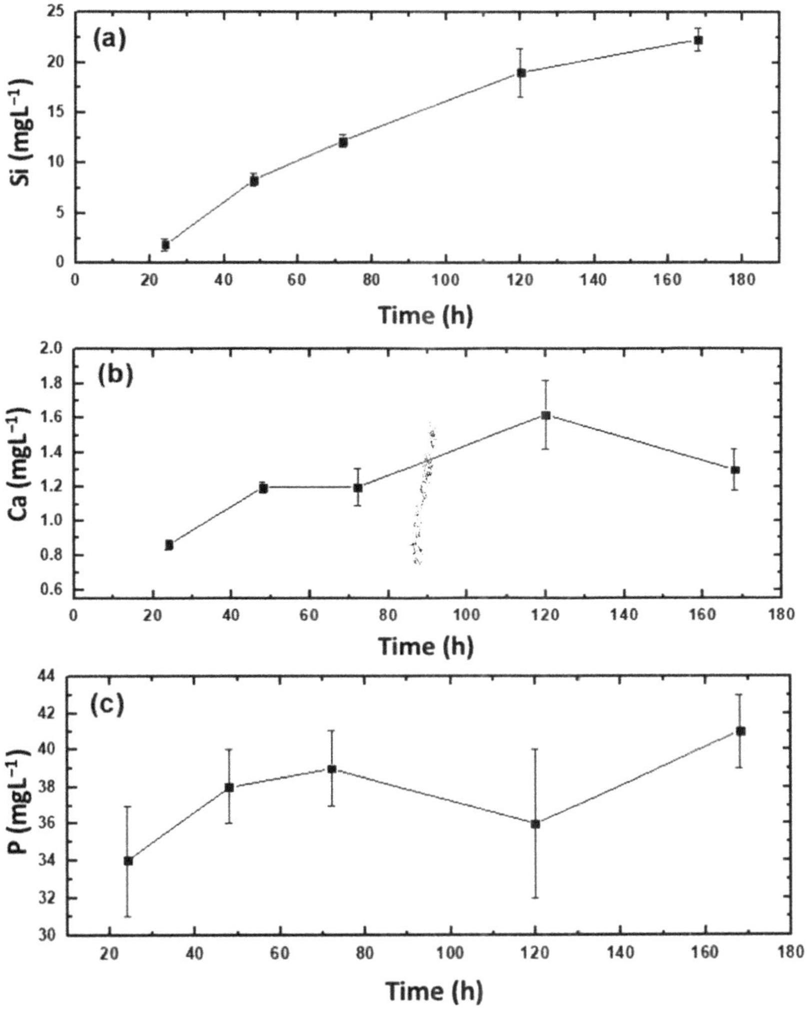

Figure 7. Profiles of cumulative ion release of (**a**) Si; (**b**) Ca; (**c**) P from the hydrolytic degradation of the SG40TCP60 material after immersion in PBS for 1 week. Both free Ca- and P-plotted concentrations were obtained by subtracting the initial Ca and P concentrations of PBS from the ICP_MS measurements after different immersion periods.

Basically, the Si-released ion concentration found in PBS was in good agreement with previous reports from silica–gelatin-similar hybrid materials [55], a result that potentially enhances osteogenesis [72]. Otherwise, Ca- and P-released concentrations were found in a typical range regulated by the small solubility of β-TCP (0.25 mg/L at 25 °C) [73], known well as a resorbable bone repair [6].

2.5. Cell Culture

Culture cells were tested on the SG40TCP60 composite cylinder biomaterial in order to assess the effect on cell viability, cell–biomaterial interactions and osteoblasts growth and maturation of the base compound used in scaffold manufacturing. Cell viability at seeding was up to 98%, and no significant apoptotic phenomena were detected either in the control or experimental groups. The scaffolds were designed to allow and improve cellular growth and ideally to favor positive interactions between the cells and materials, leading to the formation of functional tissue for medical purposes [6,74].

Cell live dead viability assays (live dead®) revealed higher percentages of viable (green) than dead (red) cells after 48 h (see Figure 8). In the osteoblast population, a majority of cells migrate towards and then adhere to biomaterial, crowding on top while a few osteoblasts remained adhered to the well bottom. No significant differences with positive control were observed either in terms of the cell viability or cell density (see Figure 9).

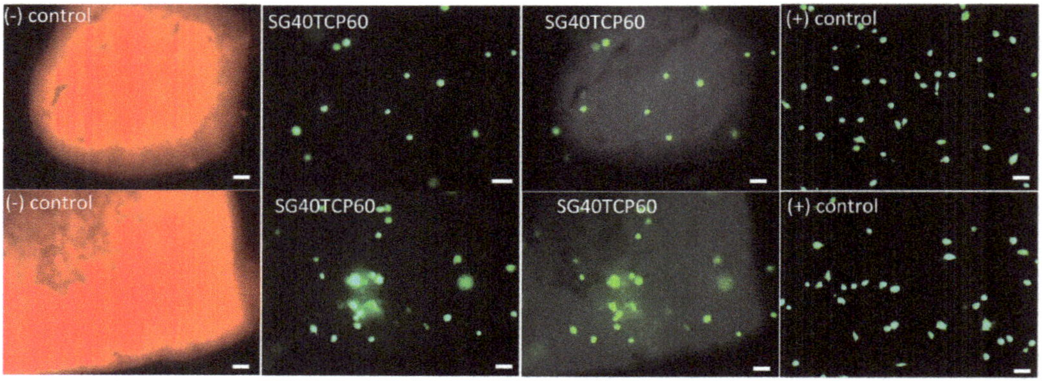

Figure 8. Live/Dead staining of HOB® cells after 48 h in culture. Positive controls for control groups shown in last column. Images acquired in the fluorescence and Nomarski modes of a Leica DMLI LED inverted microscope. Live cells (green); dead cells (red); biomaterials (gray). Scale bar equals 20 µm.

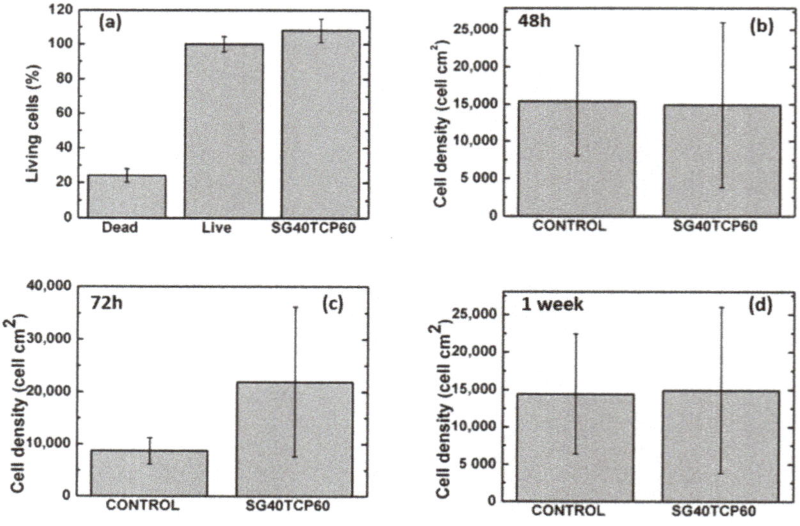

Figure 9. (**a**) Quantitative data for live dead assay after 48 h in culture and cell density for experimental times of (**b**) 48 h, (**c**) 72 h and (**d**) 1 week. One-way analysis of variance. Statistical significance $p < 0.05$.

2.6. Cell Morphology, Cytoskeletal Organization and Focal Adhesion Distribution

The ideal scaffold properties should allow for an adequate microenvironment for cell attachment, proliferation and differentiation, but also provide an adequate pathway for the nutrients to approach the cells. Furthermore, the scaffold properties should guarantee an optimal mechanical support for the cells, together with an adequate control of the degradation rate and absence of cytotoxicity. As described, live dead assays revealed a careful cell distribution on top and along the SG40TCP60 material pieces (see Figure 8). It was confirmed that osteoblasts seeded on the scaffolds presented an initial polarization from the first 24 h, with subsequent cell adhesion and changes in cell morphology compatible with osteoblast differentiation [44,74–80]. This was confirmed by changes in shape parameters, mainly those observed in area, roundness and circularity quantifications (Figure 10).

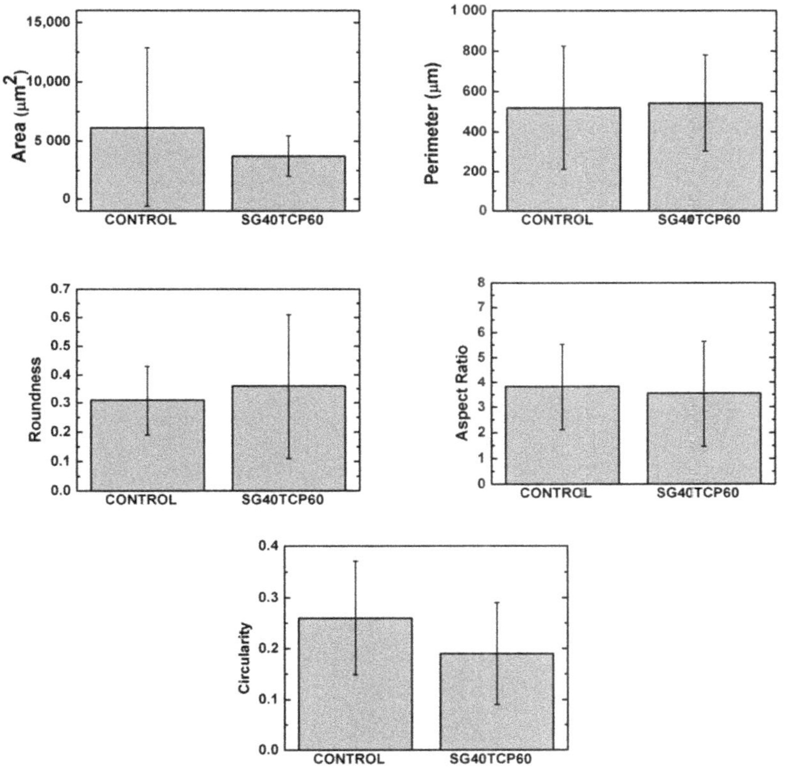

Figure 10. Quantitative data for shape parameters after 48 h. One-way analysis of variance. Statistical significance $p < 0.05$.

Immunolabeling for actin cytoskeleton after 48 h revealed that in comparison to control cells, osteoblasts elongated in the presence of SG40TCP60, while actin cytoskeleton developed and small focal adhesions became distributed along the cell (Figure 11). After 72 h in culture and in the presence of SG40TCP60, actin cytoskeleton mainly develops in the periphery with small focal adhesions and filopodial and lamellipodial emissions, while some cells elongate and some others grow wider. A remarkable number of stress fibers arise into a well-developed actin cytoskeleton mostly enhanced in the periphery with focal adhesions widely distributed, predominantly small-sized and located on the tips of stress fibers. A progressive approach of osteoblasts to the biomaterial was observed with time, and the material surface appeared to be covered, while cells contacted to neighbors and to the material surface. No stress fibers or polarization were found in the control cells. After 1 week in the culture, filopodial emissions were predominant and mature focal adhesions on the tips of widely distributed and well-developed stress fibers became evident. The phase contrast mode reveals that osteoblasts, independent of number or size, delicately distributed on the biomaterial at any experimental time.

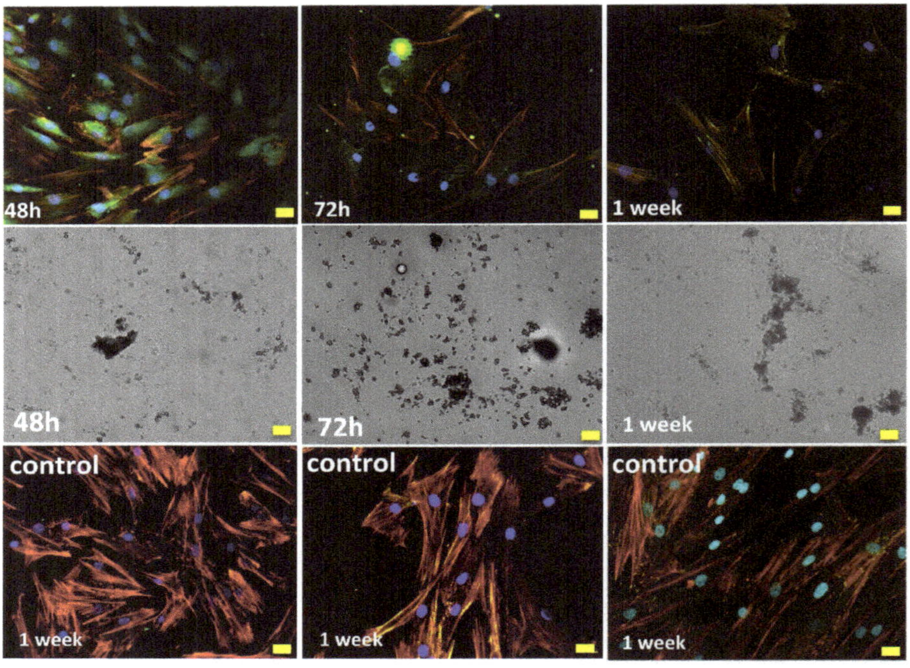

Figure 11. HOB® Osteoblasts growing in the presence of the SG40TCP60 biomaterial. Images acquired in the confocal microscope in the fluorescence (upper row) and Nomarski mode (mid row). In red: actin cytoskeleton immunolabeled with rhodamine phalloidin, in green: vinculin immunolabeling for focal adhesions and in blue: DAPI-labeled nuclei. Representative images of control groups growing on glass are shown in the lower row. Scale bar equals 20 μm.

3. Conclusions

A novel biocompatible 3D scaffold was successfully fabricated by robocasting using a sol-gel paste ink, consisting of 60 wt% β-TCP powder dispersed in a silica/40 wt% gelatin hybrid matrix (SG40TCP60), undergoing an irreversible gelling process. The printing procedure was performed without needing to incorporate chemical additives, which simplifies, to a great extent, the overall printing process. Moreover, the technique is adaptable enough to allow ink formulation from a broad range of materials, just by adjusting its gelation time and selecting an appropriate rheology of the resulting sol-gel paste because the processing is performed at room temperature. SG40TCP60 3D scaffolds with accurately controlled pore size (about 350 μm) with interconnected porous structures and negligible volume shrinkage were thus obtained. In addition, a novel 3D scaffold consisting of parallel regularly spaced cylindrical holes of about 350-micrometer diameter created on 2 mm thick SG40TCP60 bulk monoliths was successfully fabricated by laser micromachining. The method is being researched as an alternative manufacturing method for 3D structures from a variety of sol-gel materials. The compressive behavior of the laser-assisted scaffold was found to be in the range of the trabecular bone of similar skeletal density, with significant mechanical anisotropy in the case of the robocasted scaffold. Moreover, the SG40TCP60 biomaterial underwent in vitro biodegradation, and the release of Si, Ca and P ions to the free medium while its osteoblast response demonstrated that it is composed of a biocompatible material. In addition, and for clinical use, these scaffolds are easily sterilizable and their 3D architecture facilitates cell attachment, proliferation and initial markers of differentiation such as morphological changes and stress fiber and focal adhesion development. Further studies are needed to discern how the surface topography could influence osteoblast response in the culture for personalized scaffold design.

4. Materials and Methods

4.1. Materials

Tetraethylortosilicate (TEOS, 99%) and hydrochloric acid (HCl) (37%) were purchased from Alfa Aesar (Haverhill, MA, USA). Gelatin (from porcine skin, Type A, gel strength 300) and 3-glycidoxypropyltrimethoxysilane (GPTMS, >98%) were both purchased from Sigma Aldrich (St. Louis, MI, USA). Microsized β-tricalcium phosphate (Captal β-TCP, 2.0 μm average particle size) was purchased from Plasma Biotal Ltd. (Buxton, Derbyshire, UK). Ethanol absolute (99.5%) was purchased from Panreac (Spain). HOB® human osteoblasts, foetal calf serum and Osteoblast Growing Medium were purchased from Promocell (Heidelberg, Germany). PBS, methanol, Triton x-100, bovine serum albumin, paraformaldehyde, rhodamine phalloidin and monoclonal anti-vinculin FITC conjugate were all acquired from Sigma, (St Louis, MI, USA) and Hard Set Vectashield with DAPI ® from Vector (Burlingame, CA, USA).

4.2. Scaffold Fabrication

In Figure 12, a flow chart summarises the process used in the scaffolds fabrication.

4.2.1. Preparation and Characterization of the Silica/Gelatin/β-TCP Sol-Gel Ink for Robocasting

A gelatin solution in 80 °C distilled water (100 mg mL^{-1}) was prepared under vigorous agitation for 10 min. After cooling at room temperature, functionalization of gelatin was promoted by reaction with GPTMS to give a molar ratio of 750 with respect to gelatin. The silica sol was obtained separately by mixing TEOS with a stoichiometric quantity of acidified water (HCl, 0.1 N) to obtain a transparent silica sol. Next, both gelatin-GPTMS and TEOS solutions were mixed and homogenized to produce silica/gelatin hybrid sol matrices, labeled as SGx, with x = 20, 40 and 60 wt% gelatin content. All chemical reactions were carried out at room conditions with mechanical stirring. The gelation time of SGx gels decreased from 60 to ~5 min with increasing gelatin content from 20 to 60 wt%, respectively. Therefore, given that the setting time and rheological properties are critical for the direct writing process, the SG40 hybrid was considered to exhibit the best value of gelation time desirable for successful printing (~30 min), while SG20 and SG60 with, respectively, longer or shorter gelation times were not used in the ink formulation. Next, three different microsized β-TCP high powder contents (40, 50, 60 wt%) were added to the selected SG40 hybrid sol, until three different solid suspensions were obtained, named as SG40TCP40, SG40TCP50 and SG40TCP60, which were placed in a planetary centrifugal mixer (ARE-250, Thinky Corp., Tokyo, Japan) for 5 min to obtain stable homogeneous solid suspensions. Next, the rheological characterization of the suspensions was performed at room temperature, and the best composition for the ink formulation was selected. The apparent viscosities of the three hybrid pastes prior to aging and gelation were measured with the Discovery HR10 rheometer (TA instruments, USA), using a cone-plate geometry and a gap of 50 μm, for shear rates ranging from 1 to 100 s^{-1} at 25 °C. All measurements were conducted in triplicate.

4.2.2. Fabrication of Silica/Gelatin/β-TCP Hybrid Scaffolds by Robocasting

The optimized ink (SG40/β-TCP 60 wt%, as detailed in Section 2.1) was introduced into the printing syringe and then placed on the three-axis motion stage of an A3200 robotic deposition device (Aerotech/3D inks, Stillwater, OK, USA), which was controlled by a computer-aided direct-write program (Robocad 3.0, 3-D Inks, Stillwater, OK). Deposition of the ink was carried out by extrusion of filament rods layer-by-layer at room temperature with a speed of 20 mm s^{-1}, using a conical nozzle with a tip diameter of 600 μm. The scaffolds were created from a CAD model using a control software (RoboCAD 4.1, 3D inks LLC, OK, USA) and consisted of a 10 × 10 mm^2 squared base, where the distance between filaments was 400 μm. The model comprises 5 layers, always rotated 90° from the previous one, with a distance between layers of 480 μm. Deposition was performed in a paraffin oil bath to prevent non-uniform drying during assembly, while the hybrid paste underwent chemical polymerization to produce solid composite filament rods. This way, a 3D scaffold was obtained, as described in Figure 13a. The resulting sample, labeled as SG40TCP60-RC, was dried in air at room temperature for 24 h and then at 100 °C for 48 h to evaporate organics. Given that the viscosity of the ink increased in time due to the sol-gel condensation and that gelation is the mechanism associated with the shape retention and solidification of the 3D-deposited structures, a 30-minute printing window interval, as found for the SG40 hybrid matrix, was considered long enough to perform the robocasting procedure.

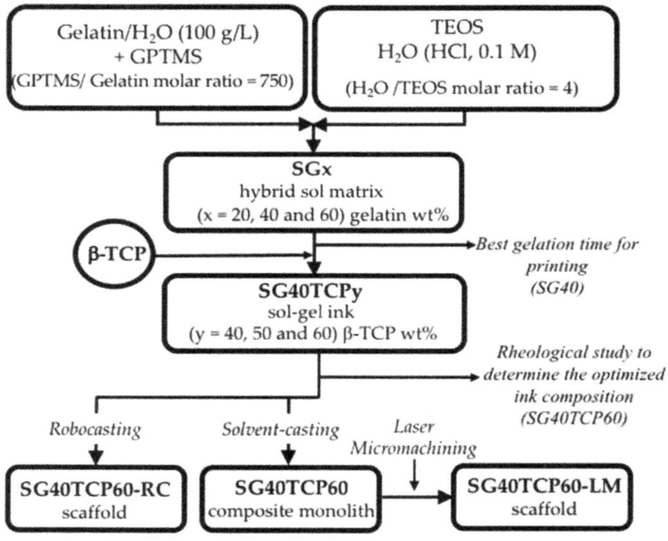

Figure 12. Schematic showing the different processing methods used to prepare SG40TCP60 3D sol-gel-derived scaffolds by robocasting (RC) and laser micromachining (LM). The last-mentioned scaffold was fabricated by laser irradiation of the corresponding monolith composites, which in turn were obtained by solvent casting of SG40TCP60 ink, followed by evaporative drying, (y = 60 wt% represents the β-TCP content for the optimized ink composition according to rheological studies.)

4.2.3. Fabrication of Silica/Gelatin/β-TCP Hybrid Scaffolds by Laser Micromachining

The sol-gel slurry was allowed to polymerize at room temperature in hermetically closed plastic containers, where gelation took place in about 30 min. The resultant gel was aged for 1 week and then dried by evaporation at 50 °C for another 48 h to obtain the related SG40TCP60 monolith composite to be used as the solid target for laser irradiation. Next, laser micromachining of the composite surface was conducted in the lasing processing laboratory of the Condensed Matter Physics Department of the University of Cadiz by using a commercial high repetition rate pulsed laser model Carbide CB3-40W of light conversion (nominal pulse energy beam (Ep) of 38 μJ, changeable pulse width (τp) from 190 fs to 10 ps, an adjustable repetition rate from one single pulse to 1 MHz) equipped with a Carbide harmonics module to generate ultrashort laser pulses with variable wavelength in the range between 1035 nm and 343 nm. This laser system was equipped with galvanometric mirrors, which allowed the movement of the laser radiation along the surface of the target; a mechanical Z-axis that allowed the position of the laser beam focus to be modified with respect to the surface of the sample. This laser device was controlled with a CAD-like LS-PRO software, which allows a precise control of the laser working parameters. In these laser micromachining, the laser beam was incident perpendicular to the longitudinal section of a 10.0 mm × 10.0 mm × 2.0 mm composite rectangular prism in order to generate a pattern of cylinders 350 μm in diameter, whose centers were separated by 1050 μm. To perform this, the laser working parameters were established at: 343 nm (UV) for the pulse laser wavelength, 38 μJ for the average laser pulse energy (0.76 W of beam power), 20 kHz for the pulse repetition rate, 220 fs for the pulse width and 10 mm·s^{-1} for the scanning speed. In order to drill through the whole thickness of the rectangular prism, the pattern was repeated 3 times in the same spot, and then, the Z-axis was lowered 150 μm. A geometrical model of the perforated monoliths, labeled as SG40TCP60-LM, is shown in Figure 13b.

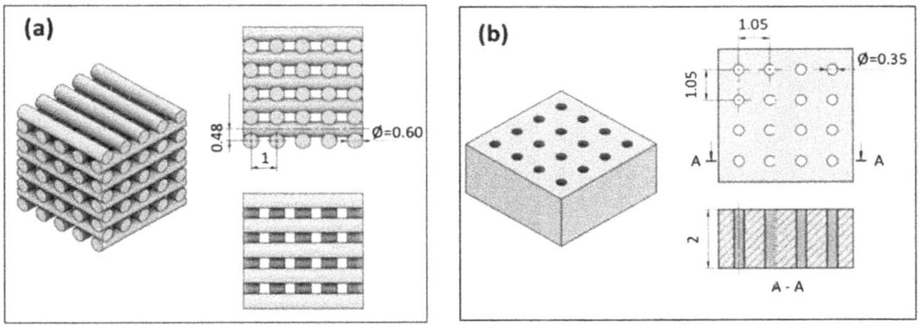

Figure 13. Design section views of the scaffold shapes and its corresponding pore structure: (**a**) layer-by-layer of the interconnected macroporous pattern planned by using robocasting; (**b**) arrangement of the cylindrical holes for a macroporous structure generated by applying laser ablation in a rectangular prism monolith composite. All dimensions are presented in mm.

4.3. Characterization Techniques

Physical, textural and mechanical properties as well as biodegradation and biological responses were investigated for the SG40TCP60 composition optimized for extrusion.

The bulk density (ρ_{Bulk}) was obtained by measuring the dimensions with a sliding caliper and the mass with a microbalance (precision of ±0.1 mg). In addition, the skeleton density (ρ_{Skel}) of the SG40TCP60 monolith composite was measured using helium pycnometry, according to method described by Weinberger et al. [81], so that the porosity of the scaffolds (ε) could be estimated from Equation (1). Values were obtained from the average of three replicates.

$$\varepsilon = \left(1 - \frac{\rho_{Bulk}}{\rho_{Skel}}\right) \times 100 \quad (1)$$

Specific surface area, pore volume and pore size were investigated using nitrogen physisorption experiments, considering BET and BJH standard models for the analysis. A Micromeritics ASAP2010 working at 77 K and equipped with a pressure transducer with resolution of 10^{-4} mm Hg was used for the analysis.

The mechanical behavior was characterized through uniaxial compression tests in samples obtained by robocasting (SG40TCP60-RC) using as-printed right-angled prisms with 10 × 10 × 5 mm^3 dimensions as test specimens, according to ASTM D7012 standard [82]. Compressive loads were applied in the direction perpendicular and parallel to the printing plane of the RC samples. The compressive strength and maximum strain were computed from the maximum stress and deformation before the fracture of the sample, and the Young's modulus was obtained from the slope at the beginning of the stress–strain curve.

For comparison, the compressive behavior of the SG40TCP60 cylinder-shaped monoliths (18.00-millimeter height and 10-millimeter diameter) was also examined. All the tests were conducted in air with a Shimadzu AG-I Autograph 5 kN using a load cell of 500 N within a ±1% of the indicated test force at 1/250 load cell rating and a constant crosshead speed of 0.1 mm min^{-1}. For each testing condition, a minimum of 3 samples were examined.

4.4. Biodegradation

The in vitro biodegradation study of the SG40TCP60 material was performed in phosphate buffer saline (PBS; pH 7.4), an effective physiological buffer with pH, osmolarity and ion concentrations similar to human plasma [83–85]. The weight loss of the samples was determined gravimetrically at room temperature, using a microbalance (precision of ±0.1 mg). The pH time-dependence of the PBS-incubated solution was also monitored using a pH meter probe (HACH sensIONTM + pH = 3, with pH resolution of 0.01). Basically, printed samples were cut with a sterile scalpel knife and then immersed in PBS (10 mg mL^{-1}) under physiological conditions at 37 °C. In the next stage, samples were removed from buffer media weekly and then weighed after wiping off any liquid on the surface. Likewise, the pH of the incubated solution was recorded. Last, samples were immersed again to continue with the degradation test in a totally refreshed PBS solution. This procedure was repeated over 9 weeks of incubation period, and the weight loss time-dependence was calculated according to Equation (2), as follows:

$$Weight\ loss\ (\%) = \frac{(W_i - W_t)}{W_i} \times 100 \tag{2}$$

where W_i and W_t are the weights of the samples before and after the degradation experiment, respectively, determined at regular intervals of 1 week.

In addition, to provide a more detailed picture of the biodegradation process, the concentrations of accumulated of Si, Ca and P ions in the incubated PBS solution were also measured. The corresponding ion release profile analysis was performed by inductively coupled plasma mass spectrometry with an ICP-MS mass spectrometer, Series X2, Thermo Elemental. The ion concentrations were quantified by collecting 4-milliliter aliquots of the scaffold/PBS-incubated solution (1 mg mL^{-1}), at intervals of 12 h, 24 h, 48 h, 72 h, 120 h and 168 h after starting the experiment and without refreshing solution. The aliquot parts were filtered with a 0.45-millimeter Millipore membrane filter, subsequently placed in plastic vials to prevent any type of contamination and stored at 4 °C until the ICP-MS analysis. Weight loss, pH and Si, Ca and P ion-released concentrations were calculated using the average of all triplicates of the samples.

4.5. Cell Culture

HOB® cells were seeded, under sterile conditions, on the preselected scaffolds. Osteoblasts were detached when the optimal confluence was reached, then counted to optimal cell density and studied for cell viability in an automated Luna® cell counter, Invitrogen. All experiments were performed with cells under ten population doublings. In order to achieve optimal sterilization of the biomaterials prior to cell seeding, a clinically standardized autoclave (under European standard DIN EN ISO 13060 recommendations for class B autoclaves) was employed. Sterilized samples were placed on Mattek® glass bottom wells in a laminar flow chamber. A drop of 50 μL of cell suspension containing 15,000 HOB® cells/cm^2 was then added to each sample and then kept for 30 min under incubation at 37 °C and 5%. Afterwards, wells were filled with supplemented OGM® (final concentration of 0.1 mL/mL of fetal calf serum). Experiments were performed at 37 °C and 5% CO_2. The growth medium was changed every two days. Experimental groups included: SG40TCP60 sample and HOB® cells grown on glass under the conditions before mentioned, used as the control.

4.6. Live/Dead Cell Assay

In order to evaluate the viability/cytotoxicity of HOB cells grown on SG40TCP60 sample and also in the controls, live /dead cell assay was performed as follows: after being incubated for 48 h, the cell/scaffold constructs were twice PBS-rinsed and then exposed to calcein-AM (0.5 µL/mL) in PBS to display the live cells and then to ethidium homodimer-1 (EthD-1) (2 µL/mL) in PBS to display dead cells, correspondingly. The samples were then observed in the fluorescence and Nomarski modes of a Leica DMLI LED inverted microscope.

4.7. Cell Morphology and Spreading

Cell morphology, alignment, distribution and spreading of osteoblasts were daily assessed after examination with the phase-contrast microscope and at the end of any experimental period. Furthermore, the Nomarski mode of both fluorescence and confocal microscopes was combined with the fluorescence mode to assess the simultaneous imaging of both the material and growing cells.

4.8. Actin Cytoskeletal Organization

Rhodamine-phalloidin and vinculin immunolabeling was performed after incubation for 48 h, 72 h and 7 days in order to assess cytoskeletal changes and focal adhesion development. Briefly, cells were washed with prewarmed PBS, keeping pH to 7.4 and fixing the constructs with paraformaldehyde (3.7%) at room temperature, and then permeabilized with 0.1% Triton x-100. After twice washing, preincubation with 1% bovine serum albumin in PBS for 20 min was performed, prior to cell immunolabeling with rhodamine phalloidin for 20 min. Then, samples were rinsed twice with prewarmed PBS prior to Vectashield ® mounting. At least five samples were seeded and analyzed per experiment and experimental group: SG40TCP60 sample and control. HOB® cells grown on glass under the conditions above-mentioned were used in the last case.

4.9. Confocal Examination

An Olympus confocal microscope was employed to assess the surface influence on the following parameters: cell density, cytoskeletal features and organization, focal adhesion distribution and changes in cell morphology. A total of 50 cells per sample were analyzed at least in each group, for a time interval not higher than 5 min to avoid photobleaching using a pinhole of 1 Airy unit.

4.10. Image Analysis

Sample images were collected as frames obtained at 40x magnification at a resolution of 1024 × 1024 Image J software (NIH, http://rsb.info.nih.gov/ij, accesed on 9 October 2020) was used for image processing. Shape variables analyzed were: area, roundness, circularity, perimeter and aspect ratio. A minimum of 40 regions of interest (ROIs) were measured under the following criteria: clear identification of nucleus, well-defined limits and absence of intersection with neighboring cells. All experiments were repeated in triplicates. Data were analyzed with SPSS and expressed as the mean ± standard deviation. After confirmation of normality and homoscedasticity, a one-way analysis of variance, Brown–Forsythe and Games-Howell tests were employed to assess the difference between the mean values. Statistical significance was defined as $p < 0.05$.

Author Contributions: Conceptualization, M.V.R.-P., F.J.M.-V., E.F., Ó.B.-M., M.S. and M.P.; methodology, M.V.R.-P., F.J.M.-V., R.F.-M., M.d.M.M.-D., E.F., Ó.B.-M., N.D.l.R.-F., M.S. and M.P.; formal analysis, M.V.R.-P., M.d.M.M.-D., N.D.l.R.-F., R.F.-M., M.S. and M.P.; investigation, M.V.R.-P., F.J.M.-V., E.F., R.F.-M., M.S., R.A., N.D.l.R.-F. and M.P.; resources, F.J.M.-V., Ó.B.-M., J.I.V.-P., R.A., M.S., N.D.l.R.-F. and M.P.; data curation, M.V.R.-P., E.F., R.F.-M., Ó.B.-M., N.D.l.R.-F., M.S. and M.P.; writing—original draft preparation, M.V.R.-P., M.S., N.D.l.R.-F. and M.P.; writing—review and editing, M.V.R.-P., M.S. and M.P.; supervision, Ó.B.-M., M.S., N.D.l.R.-F. and M.P.; project administration, M.S.; funding acquisition, M.S. All authors have read and agreed to the published version of the manuscript.

Funding: This research was 80% supported by Andalucía FEDER/ITI 2014-2020 Grant for PI 013/017 and Junta de Andalucía TEP115 and CTS 253 PAIDI research groups (Spain). The work has also been co-financed by the 2014–2020 ERDF Operational Program and by the Department of Economy, Knowledge, Business and University of the Junta de Andalucía. FEDER-UCA18_106598. OBM thanks the Project for Scientific Equipment Acquisition: "Sistema Láser de Generación de Nanomateriales (NANO-GLAS): Fabricación y procesado de materiales nanoestructurados y síntesis directa de dispersiones coloidales de nanopartículas funcionalizadas", from the Ministerio de Ciencia, Investigación y Universidad EQC2018-004979.

Acknowledgments: Authors acknowledge the use of instrumentation as well as the technical advice provided by the GEMA-Uex research group from Universidad de Extremadura (UNEX) with robocasting equipment, as well as SCCYT (UCA) for SEM, ICP and EA divisions as well as SCBM at the University of Cadiz. The authors would also like to thank. J. Vilches-Troya, retired Professor of Histology and Pathology of the University of Cadiz, for his expert advice and supervision, and Enrique Gallero-Rebollo for his assistance in figure design. All individuals included in this section have consented the acknowledgment.

Conflicts of Interest: The authors declare no conflict of interest.

References

1. Gross, B.C.; Erkal, J.L.; Lockwood, S.Y.; Chen, C.; Spence, D.M. Evaluation of 3D printing and its potential impact on biotechnology and the chemical sciences. *Anal. Chem.* **2014**, *86*, 3240–3253. [CrossRef] [PubMed]
2. Tibbitt, M.W.; Anseth, K.S. Hydrogels as extracellular matrix mimics for 3D cell culture. *Biotechnol. Bioeng.* **2009**, *103*, 655–663. [CrossRef]
3. Karageorgiou, V.K.D. Porosity of 3D biomaterial scaffolds and osteogenesis. *Biomaterials* **2005**, *26*, 5474–5491. [CrossRef]
4. Rezwan, K.; Chen, Q.Z.; Blaker, J.J.; Bocaccini, A.R. Biodegradable and bioactive porous polymer/inorganic composite scaffolds for bone tissue engineering. *Biomaterials* **2006**, *27*, 3413–3431. [CrossRef]
5. Roseti, L.; Parisi, V.; Petretta, M.; Cavallo, C.; Desando, G.; Bartolotti, I.; Grigolo, B. Scaffolds for Bone Tissue Engineering: State of the art and new perspectives. *Mater. Sci. Eng. C* **2017**, *78*, 1246–1262. [CrossRef]
6. Turnbull, G.; Clarke, J.; Picard, F.; Riches, P.; Jia, L.; Han, F.; Li, B.; Shu, W. 3D bioactive composite scaffolds for bone tissue engineering. *Bioact. Mater.* **2018**, *3*, 278–314. [CrossRef]
7. Taboas, J.M.; Maddox, R.D.; Krebsbach, P.H.; Hollister, S.J. Indirect solid free form fabrication of local and global porous, biomimetic and composite 3D polymer-ceramic scaffolds. *Biomaterials* **2003**, *24*, 181–194. [CrossRef]
8. Cao, H.; Kuboyama, N. A biodegradable porous composite scaffold of PGA/β-TCP for bone tissue engineering. *Bone* **2010**, *46*, 386–395. [CrossRef] [PubMed]
9. Janik, H.; Marzec, M. A review: Fabrication of porous polyurethane scaffolds. *Mater. Sci. Eng. C* **2015**, *48*, 586–591. [CrossRef]
10. Jang, J.; Castano, O.; Kim, H. Electrospun materials as potential platforms for bone tissue engineering. *Adv. Drug Deliv. Rev.* **2009**, *61*, 1065–1083. [CrossRef]
11. Bose, S.; Vahabzadeh, S.; Bandyopadhyay, A. Bone tissue engineering using 3D printing. *Mater. Today* **2013**, *16*, 496–504. [CrossRef]
12. Gmeiner, R.; Deisinger, U.; Schönherr, J.; Lechner, B.; Detsch, R.; Boccaccini, A.R.; Stampfl, J. Additive manufacturing of bioactive glasses and silicate bioceramics. *J. Ceram. Sci. Technol.* **2015**, *6*, 75–86. [CrossRef]
13. Miranda, P.; Pajares, A.; Saiz, E.; Tomsia, A.P.; Guiberteau, F. Mechanical properties of calcium phosphate scaffolds fabricated by robocasting. *J. Biomed. Mater. Res. Part A* **2007**, *85A*, 218–227. [CrossRef]
14. Franco, J.; Hunger, P.; Launey, M.E.; Tomsia, A.P.; Saiz, E. Direct write assembly of calcium phosphate scaffolds using a water-based hydrogel. *Acta Biomater.* **2010**, *6*, 218–228. [CrossRef]
15. Feilden, E.; García-Tuñón, E.; Giuliani, F.; Saiz, E.; Vandeperre, L. Robocasting of structural ceramic parts with hydrogel inks. *J. Eur. Ceram. Soc.* **2016**, *36*, 2525–2533. [CrossRef]
16. Saadi, M.A.S.R.; Maguire, A.; Pottackal, N.T.; Thakur, M.S.H.; Ikram, M.M.; Hart, A.J.; Ajayan, P.M.; Rahman, M.M. Direct Ink Writing: A 3D Printing Technology for Diverse Materials. *Adv. Mater.* **2022**, *34*, 2108855. [CrossRef]
17. Ben-Arfa, B.A.E.; Pullar, R.C. A comparison of bioactive glass scaffolds fabricated by robocasting from powders made by sol-gel and melt-quenching methods. *Processes* **2020**, *8*, 615. [CrossRef]
18. Lewis, B.J.A. Direct Ink Writing of 3D Functional Materials. *Adv. Funct. Mater.* **2006**, *16*, 2193–2204. [CrossRef]
19. Russias, J.; Saiz, E.; Deville, S.; Gryn, K.; Liu, G.; Nalla, R.K.; Tomsia, A.P. Fabrication and in vitro characterization of three-dimensional organic/inorganic scaffolds by robocasting. *J. Biomed. Mater. Res.-Part A* **2007**, *83*, 434–445. [CrossRef] [PubMed]
20. Ben-Arfa, B.A.E.; Neto, A.S.; Palamá, I.E.; Salvado, I.M.M.; Pullar, R.C.; Ferreira, J.M.F. Robocasting of ceramic glass scaffolds: Sol–gel glass, new horizons. *J. Eur. Ceram. Soc.* **2019**, *39*, 1625–1634. [CrossRef]
21. Cesarano, J.; Dellinger, J.G.; Saavedra, M.P.; Gill, D.D.; Jamison, R.D.; Grosser, B.A.; Sinn-Hanlon, J.M.; Goldwasser, M.S. Customization of Load-Bearing Hydroxyapatite Lattice Scaffolds. *Int. J. Appl. Ceram. Technol.* **2005**, *2*, 212–220. [CrossRef]
22. Miranda, P.; Saiz, E.; Gryn, K.; Tomsia, A.P. Sintering and robocasting of -tricalcium phosphate sca V olds for orthopaedic applications. *Acta Biomater.* **2006**, *2*, 457–466. [CrossRef] [PubMed]

23. Saiz, E.; Gremillard, L.; Menendez, G.; Miranda, P.; Gryn, K.; Tomsia, A.P. Preparation of porous hydroxyapatite scaffolds. *Mater. Sci. Eng. C* **2007**, *27*, 546–550. [CrossRef]
24. Miranda, P.; Pajares, A.; Saiz, E.; Tomsia, A.P.; Guiberteau, F. Fracture modes under uniaxial compression in hydroxyapatite scaffolds fabricated by robocasting. *J. Biomed. Mater. Res. Part A* **2007**, *83A*, 646–655. [CrossRef] [PubMed]
25. Houmard, M.; Fu, Q.; Genet, M.; Saiz, E.; Tomsia, A.P. On the structural, mechanical, and biodegradation properties of HA/β-TCP robocast scaffolds. *J. Biomed. Mater. Res. Part B Appl. Biomater.* **2013**, *101*, 1233–1242. [CrossRef]
26. Hajian, M.; Esmaeili, F.; Torbati, M.; Poursamar, S.A.; Lotfibakhshaiesh, N.; Ai, J.; Ebrahimi-Barough, S.; Azami, M. Preparation and characterization of 3D nanocomposite scaffold from bioactive glass/β-tricalcium phosphate via Robocasting method for bone tissue engineering. *J. Non-Cryst. Solids* **2022**, *593*, 121769. [CrossRef]
27. Houmard, M.; Fu, Q.; Saiz, E.; Tomsia, A.P. Sol–gel method to fabricate CaP scaffolds by robocasting for tissue engineering. *J. Mater. Sci. Mater. Med.* **2012**, *23*, 921–930. [CrossRef]
28. Peng, E.; Zhang, D.; Ding, J. Ceramic robocasting: Recent achievements, potential and future developments. *Adv. Mater.* **2018**, *30*, 1802404. [CrossRef] [PubMed]
29. Del-Mazo-Barbara, L.; Ginebra, M.P. Rheological characterisation of ceramic inks for 3D direct ink writing: A review. *J. Eur. Ceram. Soc.* **2021**, *41*, 18–33. [CrossRef]
30. Bento, R.; Gaddam, A.; Oskoei, P.; Oliveira, H.; Ferreira, J.M.F. 3D Printing of Macro Porous Sol-Gel Derived Bioactive Glass Scaffolds and Assessment of Biological Response. *Materials* **2021**, *14*, 5046–15962. [CrossRef]
31. Kumar, A.; Akkineni, A.R.; Basu, B.; Gelinsky, M. Three-dimensional plotted hydroxyapatite scaffolds with predefined architecture: Comparison of stabilization by alginate cross-linking versus sintering. *J. Biomater. Appl.* **2016**, *30*, 1168–1181. [CrossRef] [PubMed]
32. Abramov, A.; Tsygankov, P.; Lovskaya, D. Extrusion-Based 3D Printing for Highly Porous Alginate Materials Production. *Gels* **2021**, *7*, 7030092.
33. Dorj, B.; Park, J.; Kim, H. Robocasting chitosan/nanobioactive glass dual-pore structured scaffolds for bone engineering. *Mater. Lett.* **2012**, *73*, 119–122. [CrossRef]
34. Ramirez Caballero, S.S.; Saiz, E.; Montembault, A.; Tadier, S.; Maire, E.; David, L.; Delair, T.; Grémillard, L. 3-D printing of chitosan-calcium phosphate inks: Rheology, interactions and characterization. *J. Mater. Sci. Mater. Med.* **2019**, *30*, 5726–5732. [CrossRef]
35. Marques, C.F.; Olhero, S.M.; Torres, P.M.C.; Abrantes, J.C.C.; Fateixa, S.; Nogueira, H.I.S.; Ribeiro, I.A.C.; Bettencourt, A.; Sousa, A.; Granja, P.L.; et al. Novel sintering-free scaffolds obtained by additive manufacturing for concurrent bone regeneration and drug delivery: Proof of concept. *Mater. Sci. Eng. C* **2019**, *94*, 426–436. [CrossRef]
36. Motealleh, A.; Eqtesadi, S.; Pajares, A.; Miranda, P. Enhancing the mechanical and in vitro performance of robocast bioglass scaffolds by polymeric coatings: Effect of polymer composition. *J. Mech. Behav. Biomed. Mater.* **2018**, *84*, 35–45. [CrossRef]
37. Maazouz, Y.; Montufar, E.B.; Guillem-Marti, J.; Fleps, I.; Öhman, C.; Persson, C.; Ginebra, M.P. Robocasting of biomimetic hydroxyapatite scaffolds using self-setting inks. *J. Mater. Chem. B* **2014**, *2*, 5378–5386. [CrossRef]
38. Chung, J.J.; Yoo, J.; Sum, B.S.T.; Li, S.; Lee, S.; Kim, T.H.; Li, Z.; Stevens, M.M.; Georgiou, T.K.; Jung, Y.; et al. 3D Printed Porous Methacrylate/Silica Hybrid Scaffold for Bone Substitution. *Adv. Healthc. Mater.* **2021**, *10*, 2100117. [CrossRef]
39. Iglesias-Mejuto, A.; García-González, C.A. 3D-printed alginate-hydroxyapatite aerogel scaffolds for bone tissue engineering. *Mater. Sci. Eng. C* **2021**, *131*, 112525. [CrossRef]
40. Gao, C.; Rahaman, M.N.; Gao, Q.; Teramoto, A.; Abe, K. Robotic deposition and in vitro characterization of 3D gelatin-bioactive glass hybrid scaffolds for biomedical applications. *J. Biomed. Mater. Res. Part A* **2013**, *101A*, 2027–2037. [CrossRef]
41. Duoss, B.E.B.; Twardowski, M.; Lewis, J.A. Sol-Gel Inks for Direct-Write Assembly of Functional Oxides. *Adv. Mater.* **2007**, *19*, 3485–3489. [CrossRef]
42. Dong, Y.; Liang, J.; Cui, Y.; Xu, S.; Zhao, N. Fabrication of novel bioactive hydroxyapatite-chitosan-silica hybrid scaffolds: Combined the sol-gel method with 3D plotting technique. *Carbohydr. Polym.* **2018**, *197*, 183–193. [CrossRef]
43. Wang, R.; Zhu, P.; Yang, W.; Gao, S.; Li, B.; Li, Q. Direct-writing of 3D periodic TiO_2 bio-ceramic scaffolds with a sol-gel ink for in vitro cell growth. *Mater. Des.* **2018**, *144*, 304–309. [CrossRef]
44. Reyes-Peces, M.V.; Pérez-Moreno, A.; De-los-Santos, D.M.; Mesa-Díaz, M.d.M.; Pinaglia-Tobaruela, G.; Vilches-Pérez, J.I.; Fernández-Montesinos, R.; Salido, M.; de la Rosa-Fox, N.; Piñero, M. Chitosan-GPTMS-Silica Hybrid Mesoporous Aerogels for Bone Tissue Engineering. *Polymers* **2020**, *12*, 2723. [CrossRef]
45. Perez-Moreno, A.; Reyes-Peces, M.V.; de los Santos, D.M.; Pinaglia-Tobaruela, G.; de la Orden, E.; Vilches-Pérez, J.I.; Salido, M.; Piñero, M.; de la Rosa-Fox, N. Hydroxyl Groups Induce Bioactivity in Silica/Chitosan Aerogels Designed for Bone Tissue Engineering. In Vitro Model for the Assessment of Osteoblasts Behavior. *Polymers* **2020**, *12*, 2802 [CrossRef]
46. Krüger, J.; Kautek, W.; Newesely, H. Femtosecond-pulse laser ablation of dental hydroxyapatite and single-crystalline fluoroapatite. *Appl. Phys. A. Mater. Sci. Process.* **1999**, *69*, 403–408. [CrossRef]
47. Shirk, M.D.; Molian, P.A. A review of ultrashort pulsed laser ablation of materials. *J. Laser Appl.* **1998**, *10*, 18–28. [CrossRef]
48. Kautek, W.; Krüger, J. Femtosecond pulse laser ablation of metallic, semiconducting, ceramic, and biological materials. In *Laser Materials Processing: Industrial and Microelectronics Applications*; SPIE: Bellingham, DC, USA, 1994; Volume 2207, pp. 600–611. [CrossRef]
49. Liu, Y.; Sun, S.; Singha, S.; Cho, M.R.; Gordon, R.J. 3D femtosecond laser patterning of collagen for directed cell attachment. *Biomaterials* **2005**, *26*, 4597–4605. [CrossRef]

50. Daskalova, A.; Bliznakova, I.; Angelova, L.; Trifonov, A.; Declercq, H.; Buchvarov, I. Femtosecond laser fabrication of engineered functional surfaces based on biodegradable polymer and biopolymer/ceramic composite thin films. *Polymers* **2019**, *11*, 378–399. [CrossRef]
51. Negrea, R.; Busuioc, C.; Constantinoiu, I.; Miu, D.; Enache, C.; Iordache, F.; Jinga, S.I. Akermanite-based coatings grown by pulsed laser deposition for metallic implants employed in orthopaedics. *Surf. Coat. Technol.* **2019**, *357*, 1015–1026. [CrossRef]
52. Koo, S.; Santoni, S.M.; Gao, B.Z.; Grigoropoulos, C.P.; Ma, Z. Laser-assisted biofabrication in tissue engineering and regenerative medicine. *J. Mater. Res.* **2017**, *32*, 128–132. [CrossRef]
53. Jiang, H. Laser-Assisted Micromachining of Hydrogel Films for Biomedical Applications. Ph.D. Thesis, Purdue University Graduate School, West Lafayette, IN, USA, 2018.
54. Zhao, W.; Mei, X. Optimization of trepanning patterns for holes ablated using nanosecond pulse laser in Al_2O_3 ceramics substrate. *Materials* **2021**, *14*, 3834. [CrossRef]
55. Mahony, O.; Tsigkou, O.; Ionescu, C.; Minelli, C.; Ling, L.; Hanly, R.; Smith, M.E.; Stevens, M.M.; Jones, J.R. Silica-gelatin hybrids with tailorable degradation and mechanical properties for tissue regeneration. *Adv. Funct. Mater.* **2010**, *20*, 3835–3845. [CrossRef]
56. Gao, C.; Gao, Q.; Li, Y.; Rahaman, M.N.; Teramoto, A.; Abe, K. In vitro Evaluation of Electrospun Gelatin-Bioactive Glass Hybrid Scaffolds for Bone Regeneration. *J. Appl. Polym. Sci.* **2012**, *127*, 2588–2599. [CrossRef]
57. Metroke, T.L.; Stesikova, E.; Dou, K.; Knobbe, E.T. Application of laser ablation technique for removal of chemically inert organically modified silicate coatings. *Prog. Org. Coat.* **2003**, *46*, 250–258. [CrossRef]
58. Montheil, T.; Maumus, M.; Valot, L.; Martinez, J.; Amblard, M.; Mehdi, A.; Subra, G. Inorganic Sol–Gel Polymerization for Hydrogel Bioprinting. *ACS Omega* **2020**, *5*, 2640–2647. [CrossRef]
59. Shukrun, E.; Cooperstein, I.; Magdassi, S. 3D-Printed Organic–Ceramic Complex Hybrid Structures with High Silica Content. *Adv. Sci.* **2018**, *5*, 1800061. [CrossRef]
60. Echalier, C.; Levato, R.; Garric, X.; Pinese, C.; Engel, E.; Martinez, J.; Mehdi, A.; Subra, G. Modular bioink for 3D printing of biocompatible hydrogels hydrogels: Sol–gel polymerization of hybrid. *RSC Adv.* **2017**, *7*, 12231–12235. [CrossRef]
61. Martínez-Vázquez, F.J.; Cabañas, M.V.; Paris, J.L.; Lozano, D.; Vallet-Regí, M. Fabrication of novel Si-doped hydroxyapatite/gelatine scaffolds by rapid prototyping for drug delivery and bone regeneration. *Acta Biomater.* **2015**, *15*, 200–209. [CrossRef]
62. Paredes, C.; Martínez-Vázquez, F.J.; Pajares, A.; Miranda, P. Development by robocasting and mechanical characterization of hybrid HA/PCL coaxial scaffolds for biomedical applications. *J. Eur. Ceram. Soc.* **2019**, *39*, 4375–4383. [CrossRef]
63. Thommes, M.; Kaneko, K.; Neimark, A.V.; Olivier, J.P.; Rodriguez-Reinoso, F.; Rouquerol, J.; Sing, K.S.W. Physisorption of gases, with special reference to the evaluation of surface area and pore size distribution. *Pure Appl. Chem.* **2015**, *87*, 1052. [CrossRef]
64. Gregg, S.J.; Sing, K.S.W. *Adsorption, Surface Area and Porosity*; Academic Press: London, UK, 1982.
65. Coradin, T.; Bah, S.; Livage, J. Gelatine/silicate interactions: From nanoparticles to composite gels. *Colloids Surf. B Biointerfaces* **2004**, *35*, 53–58. [CrossRef] [PubMed]
66. Connell, L.S.; Gabrielli, L.; Mahony, O.; Russo, L.; Cipolla, L.; Jones, J.R. Functionalizing natural polymers with alkoxysilane coupling agents: Reacting 3-glycidoxypropyl trimethoxysilane with poly(γ-glutamic acid) and gelatin. *Polym. Chem.* **2017**, *8*, 1095–1103. [CrossRef]
67. Mahony, O.; Yue, S.; Hanna, J.V.; Smith, M.E.; Lee, P.D.; Jones, J.R. Silica–gelatin hybrids for tissue regeneration: Inter-relationships between the process variables. *J. Sol-Gel Sci. Technol.* **2014**, *69*, 288–298. [CrossRef]
68. Retuert, J.; Quijada, R.; Yazdani-Pedram, M.; Martı, Y. Highly porous silica networks derived from gelatin/siloxane hybrids prepared starting from sodium metasilicate. *J. Non-Cryst. Solids* **2004**, *347*, 273–278. [CrossRef]
69. Houaoui, A.; Szczodra, A.; Lallukka, M.; El-Guermah, L.; Agniel, R.; Pauthe, E.; Massera, J.; Boissiere, M. New Generation of Hybrid Materials Based on Gelatin and Bioactive Glass Particles for Bone Tissue Regeneration. *Biomolecules* **2021**, *11*, 444. [CrossRef]
70. Antonio, P.; Vilches, P.I.; Fern, R.; Pinaglia-Tobaruela, G.; Salido, M.; Piñero, M. Effect of Washing Treatment on the Textural Properties and Bioactivity of Silica/Chitosan/TCP Xerogels for Bone Regeneration. *Int. J. Mol. Sci.* **2021**, *22*, 8321–8448.
71. Bai, Y.; Gai, X.; Li, S.; Zhang, L.; Liu, Y.; Hao, Y.; Zhang, X. Improved corrosion behaviour of electron beam melted Ti-$_6$Al–$_4$V alloy in phosphate buffered saline. *Corros. Sci.* **2017**, *123*, 289–296. [CrossRef]
72. Gao, C.; Peng, S.; Feng, P.; Shuai, C. Bone biomaterials and interactions with stem cells. *Bone Res.* **2017**, *5*, 17059. [CrossRef]
73. Sheikh, Z.; Abdallah, M.; Hanafi, A.A.; Misbahuddin, S.; Rashid, H.; Glogauer, M. Mechanisms of in Vivo Degradation and Resorption of Calcium Phosphate Based Biomaterials. *Materials* **2015**, *8*, 7913–7925. [CrossRef]
74. Jonathan, M.; Biggs, P.; Richards, R.G.; Dalby, M.J. Nanotopographical modification: A regulator of cellular function through focal adhesions. *Nanomedicine* **2010**, *6*, 619–633. [CrossRef]
75. Natale, C.F.; Ventre, M.; Netti, P.A. Tuning the material-cytoskeleton crosstalk via nanocon fi nement of focal adhesions. *Biomaterials* **2014**, *35*, 2743–2751. [CrossRef]
76. Unbehau, R.; Gemming, T.; Kruppke, B.; Wiesmann, H.; Hanke, T. Calcite incorporated in silica/collagen xerogels mediates calcium release and enhances osteoblast. *Sci. Rep.* **2020**, *10*, 128–137. [CrossRef]
77. Lamers, E.; van Horssen, R.; te Riet, J.; van Delft, F.C.M.J.M.; Luttge, R.; Walboomers, X.F.; Jansen, J.A. The influence of nanoscale topographical cues on initial osteoblast morphology and migration. *Eur. Cells Mater.* **2010**, *20*, 329–343. [CrossRef]

78. Maggi, A.; Li, H.; Greer, J.R. Three-dimensional nano-architected scaffolds with tunable stiffness for efficient bone tissue growth. *Acta Biomater.* **2017**, *63*, 294–305. [CrossRef]
79. Gardel, M.L.; Schneider, I.C.; Aratyn-Schaus, Y.; Waterman, C.M. Mechanical Integration of Actin and Adhesion Dynamics in Cell Migration. *Annu. Rev. Cell Dev. Biol.* **2010**, *26*, 315–335. [CrossRef]
80. Sukul, M.; Sahariah, P.; Lauzon, L.; Mano, F.; Haugen, H.J.; Reseland, J.E. In vitro biological response of human osteoblasts in 3D chitosan sponges with controlled degree of deacetylation and molecular weight. *Carbohydr. Polym.* **2021**, *254*, 117437. [CrossRef] [PubMed]
81. Weinberger, C.; Vetter, S.; Tiemann, M.; Wagner, T. Microporous and Mesoporous Materials Assessment of the density of (meso) porous materials from standard volumetric physisorption data. *Microporous Mesoporous Mater.* **2016**, *223*, 53–57. [CrossRef]
82. *ASTM D7012-14e1*; Standard Test Methods for Compressive Strength and Elastic Moduli of Intact Rock Core Specimens under Varying States of Stress and Temperatures. ASTM International: West Conshohocken, PA, USA, 2014.
83. Carrodeguas, R.G.; Aza, S. De alpha-Tricalcium phosphate: Synthesis, properties and biomedical applications. *Acta Biomater.* **2011**, *7*, 3536–3546. [CrossRef]
84. Owens, G.J.; Singh, R.K.; Foroutan, F.; Alqaysi, M.; Han, C.; Mahapatra, C.; Kim, H. Progress in Materials Science Sol–gel based materials for biomedical applications. *Prog. Mater. Sci.* **2016**, *77*, 137–182. [CrossRef]
85. Vallet-Regí, M.; Izquierdo-Barba, I.; Colilla, M. Structure and functionalization of mesoporous bioceramics for bone tissue regeneration and local drug delivery. *Philos. Trans. R. Soc. A Math. Phys. Eng. Sci.* **2012**, *370*, 1400–1421. [CrossRef] [PubMed]

gels

Article

Processing of Calcium Magnesium Silicates by the Sol–Gel Route

Andrada-Elena Alecu, Claudiu-Constantin Costea, Vasile-Adrian Surdu, Georgeta Voicu, Sorin-Ion Jinga and Cristina Busuioc *

Department of Science and Engineering of Oxide Materials and Nanomaterials, Faculty of Chemical Engineering and Biotechnologies, University Politehnica of Bucharest, RO-060042 Bucharest, Romania

* Correspondence: cristina.busuioc@upb.ro

Abstract: In this work, calcium magnesium silicate ceramics were processed through the sol–gel method in order to study the crystalline and morphological properties of the resulting materials in correlation with the compositional and thermal parameters. Tetraethyl orthosilicate and calcium/magnesium nitrates were employed as sources of cations, in ratios specific to diopside, akermanite and merwinite; they were further subjected to gelation, calcination (600 °C) and thermal treatments at different temperatures (800, 1000 and 1300 °C). The properties of the intermediate and final materials were investigated by thermal analysis, scanning electron microscopy, energy dispersive X-ray spectroscopy, Fourier transform infrared spectroscopy, X-ray diffraction and Rietveld refinement. Such ceramics represent suitable candidates for tissue engineering applications that require porosity and bioactivity.

Keywords: silicates; diopside; akermanite; merwinite; sol–gel; tissue engineering

check for updates

Citation: Alecu, A.-E.; Costea, C.-C.; Surdu, V.-A.; Voicu, G.; Jinga, S.-I.; Busuioc, C. Processing of Calcium Magnesium Silicates by the Sol–Gel Route. *Gels* **2022**, *8*, 574. https://doi.org/10.3390/gels8090574

Received: 25 August 2022
Accepted: 7 September 2022
Published: 9 September 2022

1. Introduction

One of the most recent challenges for researchers in the scientific field is obtaining new biomaterials for tissue engineering. In this sense, bioceramics based on calcium magnesium silicates are increasingly studied, following their use in medicine due to their properties, such as high biocompatibility, bioactivity and biodegradability [1–3], superior mechanical properties [4–6] and appropriate degradability rate [7–9], being often compared with calcium silicates ($CaSiO_3$) and calcium phosphates ($Ca_3(PO_4)_2$) [10]. The class of calcium silicates also includes the ceramic components of the ternary system $CaO–SiO_2–MgO$ [11–13], such as diopside ($CaMgSi_2O_6$), akermanite ($Ca_2MgSi_2O_7$) and merwinite ($Ca_3MgSi_2O_8$). Their multifunctional properties recommend them as candidates for the development of materials suitable for the treatment of bone tissue injuries, as well as its regeneration [12–17]; this is due to Ca and Mg ions [1,18] that promote the process of mineralization through apatite deposition [3,19] and enhance cell proliferation and differentiation [1,20,21]. Some researchers prepared larnite and rankinite through the sol–gel combustion method [22], but also monticellite and diopside from eggshell waste via the combustion route [23], with good results in terms of mechanical strength, bioactivity, antibacterial activity, as well as cell adhesion, proliferation and differentiation. Sodium calcium silicate is another bioactive ceramic that was synthesized by the combustion technique and whose bioactivity was found to be rapid when compared with that of calcium silicates and calcium magnesium silicates [24].

There are several methods of preparing these silicates, such as sol–gel route [25,26], solid-state reaction [27–29], co-precipitation approach [10,30], and the spray pyrolysis technique [31,32], but the most used and the simplest one is the well-known wet-chemistry protocol that involves the transition from a sol to a gel, subsequently completed by a thermal treatment [32,33]. Sol–gel is a method that starts by mixing some precursors, either salts or alkoxides [34], continues with the pH adjustment and ends up with the achievement of a homogeneous and stable solution, which will undergo hydrolysis and gelation processes [7]. Usually, after the maturation stage of the transparent gel, a heating program is applied, determining the formation of predominantly white powders [35]. The main advantages of the sol–gel technique are the morphological control and good adhesion during deposition of thin films [30,33,36], as well as the uniform mixing of the precursors, which leads to the obtaining of homogeneous products at low temperatures [37] and, in addition, the powders when processed in such a way can be further sintered at high temperatures [38].

Recently, it was reported that diopside powder synthesized through the sol–gel method, followed by calcination at 950 °C, consists of particles with sizes ranging between 22 and 38 nm [39]. In the case of merwinite prepared by the sol–gel technique, at calcination temperatures of 850 or 1400 °C, the particle dimensions were between 25 nm and 3.5 μm [19,40]. By applying the same approach for akermanite, but with calcination at 1300 °C, values belonging to the 5–40 μm range were obtained [41].

Various papers showed that the field is also developing in the direction of doping these silicates with different types of oxides containing Sr, Zn, Cu, Ti, Zr, etc. metal ions to give them superior or new properties [8,28,42,43]. It was demonstrated that diopside, akermanite and merwinite powders can be integrated in composite scaffolds together with other materials, thus forming biomaterials such as: diopside/PLA [16], diopside/PCL [44], diopside/graphene [45], diopside/silk [46], akermanite/PLA [47], akermanite/PGLA [14], merwinite/PLGA [17], merwinite/PCL [48]. Likewise, many studies confirmed the ability of these three silicates to trigger the formation of a surface bone-like apatite and the high rate of biodegradability during exposure to biomimetic environments for several days. In the case of diopside and akermanite, it was observed that apatite crystals increase with soaking time (from 9 to 28 days) [5,10,49]; the significant increase in the content of Ca, Mg and Si ions in the simulated body fluid solution after immersion for 28 days indicated that merwinite can be hydrolyzed fast, providing a connection with the living bone and a high biodegradability [50].

Considering the great potential of calcium magnesium silicates in the field of medical applications, especially for the development of bone tissue substitutes or bone defect fillers, in this study, three different systems were designed starting from the compositions of diopside, akermanite and merwinite and subsequently produced in the form of thermally treated ceramic powders. These were characterized from multiple perspectives with the purpose of understanding the effects of different calcium content, which increases in the series: diopside < akermanite < merwinite. Both morphology and mineralogical composition were correlated with the processing parameters.

2. Results and Discussion

According to the definition of the sol–gel method, it implies the primary preparation of solutions or suspensions that are further processed in order to ensure the hydrolysis and polycondensation/polymerization of the constituent entities, followed by the occurrence of bridging oxygens and finally the formation of a three-dimensional structure with increasing viscosity as the gel ages; the latter is able to embed a large quantity of solvent, which opens new perspectives towards its subsequent processing. Thus, the gel can be converted into particles, fibers, film or scaffold. Moreover, the solvent content can be removed in a slow or fast manner, allowing completely different morphologies to be reached. Overall, the sol–gel approach ensures the obtaining of oxide powders with complex composition, high purity, increased homogeneity and last, but not least, small particle size.

2.1. Gels Characterization

In the first part, three gels with compositions specific to Dy, Ak and Mw were obtained. These were transparent and bulky immediately after gelation, but white, contracted and cracked after drying. Small amounts of ground gels were subjected to complex thermal analysis, the resulting curves being available in Figure 1. Analyzing comparatively the thermogravimetric behavior (WL), it can be stated that the highest weight loss corresponds to Mw, namely 66%, followed by Ak with 64% and Dy with 60%. This can be explained based on the quantity of calcium nitrate added to the precursor solution, which increased in the series: Dy < Ak < Mw. As well, the mass loss is mainly recorded below the temperature of 600 °C, since the gas-generating components are taken out by volatilization, burning or decomposition at such temperatures [51]. After this value, the solid-state reactions take place and the reduction was insignificant (3–5%), which led to the selection of 600 °C as the calcination temperature. Otherwise, the general features revealed by this investigation technique are quite similar for all three specimens, with small shifts or variations in intensity that do not change the general appearance.

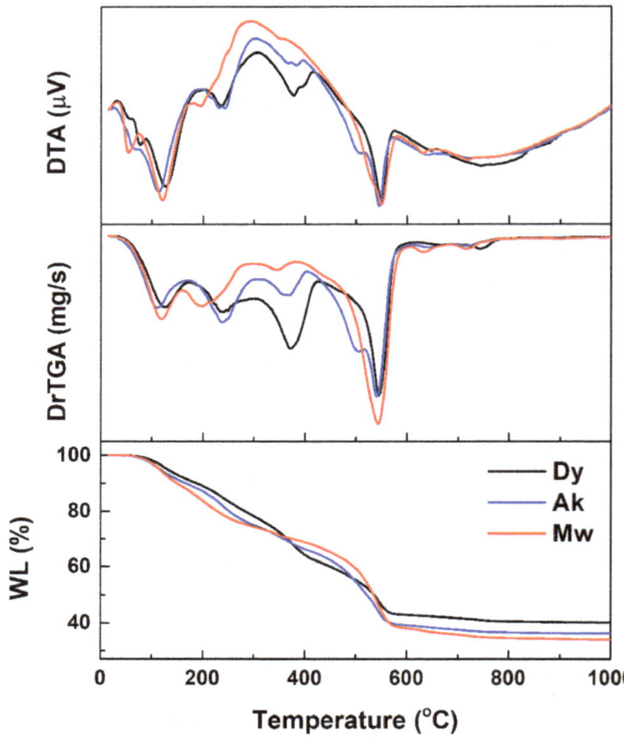

Figure 1. Complex thermal analyses of the dry gels corresponding to: diopside (Dy), akermanite (Ak) and merwinite (Mw) compositions. WL represents weight loss, DrTGA is the derivative thermogravimetric analysis and DTA stands for differential thermal analysis.

According to the DrTGA curves, there are four weight loss stages, highlighted by the function minima. Moreover, DTA curves indicate at least three endothermic processes and an exothermic one below 600 °C, if we do not take into account the shoulders or the peaks with low intensity. The first stage of mass loss occurs between 40 and 170 °C, which is associated with an endothermic process, probably attributed to the removal of residual solvents or adsorbed water. The second stage of mass loss takes place between 170 and 300 °C, which is linked with an endothermic process, mainly the elimination of chemically bound water (dehydration of recrystallized nitrates and their incipient decomposition). The third stage of mass loss occurs between 300 and 420 °C and is associated with an overlap of endothermic and exothermic processes, which could indicate the first phase of nitrate decomposition and the organic residue burning. The last stage of mass loss takes place between 420 and 600 °C and is linked with an endothermic process, certainly due to the completion of nitrate decomposition. At higher temperatures, between 600 and 800 °C, several weight variations occur, presumably generated by the accidental carbonation of some species in the sample with CO_2 from the atmosphere, resulting in compounds that are weakly crystalline and decarbonate in this temperature range [52,53].

Figure 2 shows the SEM images captured on the dry and ground gels with compositions specific to Dy, Ak and Mw. Irregularly branched formations with a specific morphology of polymeric material are visible; these consist of blocks with sizes between 5 and 50 μm. In places, granular entities embedded in a continuous matrix, uniform in size, can be detected. No important differences are discernible from one specimen to another because the morphology is induced at this stage by the 3D skeleton built on the oxygen bridges (Si–O–Si) arising after the hydrolysis and polycondensation/polymerization processes, with Ca and Mg ions trapped in this gel matrix.

Figure 2. SEM images of the dry gels corresponding to: (**a**) diopside, (**b**) akermanite and (**c**) merwinite compositions.

2.2. Ceramics Characterization

In the second part, the gels were calcined at 600 °C in order to remove the gas-generating constituents and possibly attain a preliminary crystallization of the silicate compounds. Afterward, the calcined powders were subjected to a second thermal treatment with the aim of studying the phase composition evolution at higher temperatures (800, 1000 and 1300 °C) and the implications of different calcium content. The morphology of the samples with compositions specific to Dy, Ak and Mw is displayed in Figures 3–5. The SEM images at different magnifications evidence the presence of nanoparticles, which appear in the form of agglomerates with sizes between 1 and 5 μm in the case of calcination at 600 °C; in the second image in each case, individual particles with quasi-spherical shape, an average diameter of 25 nm and narrow size distribution can be identified. The temperature increase to 800 or 1000 °C leads to a different microstructure, in which the particles built sponge structures, having thin walls and encapsulating large pores. The maximum temperature (1300 °C) causes partial sintering of the powder, resulting in consolidated blocks with a high percentage of open porosity, especially in the case of Mw composition. Otherwise, the growth of particle/grain dimension with temperature increase is obvious, but also the distinct aspect of Ak-1300 and Mw-1300 samples: the first one presents well-packed faceted robust grains, while the second one consists of perfectly joint round grains of different sizes. Thus, the superior temperature promotes material diffusion and enables the initiation of sintering with positive repercussions on the mechanical properties of a material that works under permanent loads.

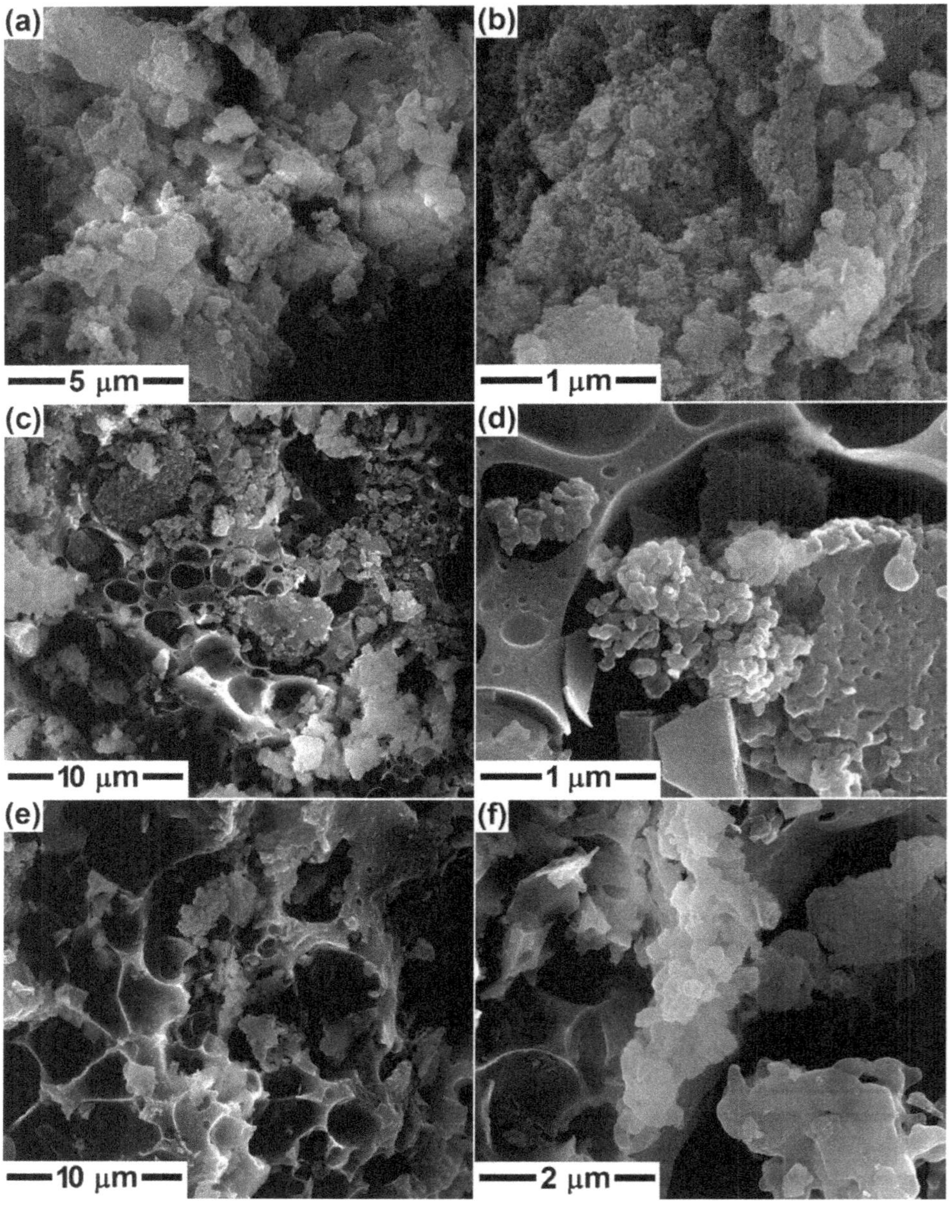

Figure 3. SEM images of the ceramics corresponding to diopside composition, thermally treated at: (**a**,**b**) 600 °C; (**c**,**d**) 800 °C; and (**e**,**f**) 1000 °C.

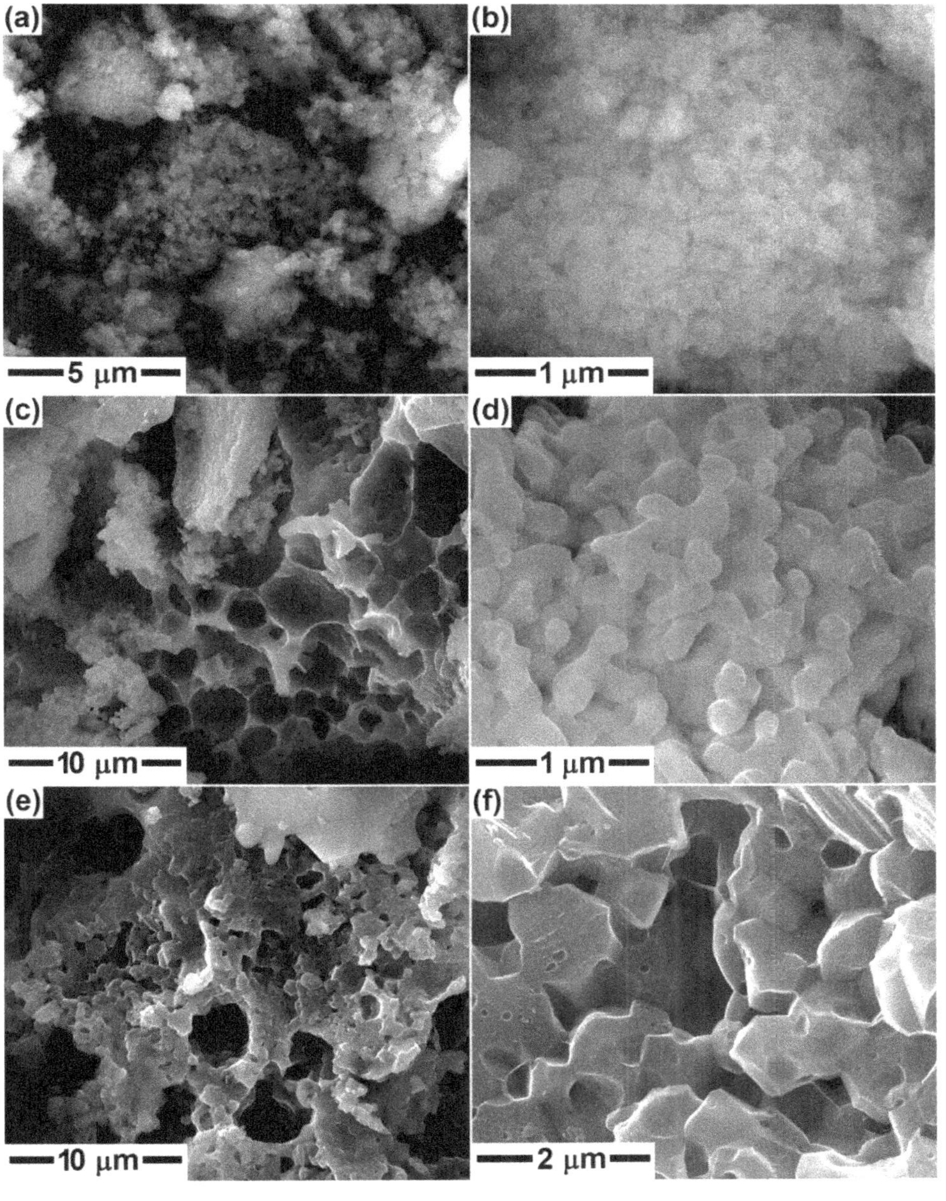

Figure 4. SEM images of the ceramics corresponding to akermanite composition, thermally treated at: (**a**,**b**) 600 °C; (**c**,**d**) 1000 °C; and (**e**,**f**) 1300 °C.

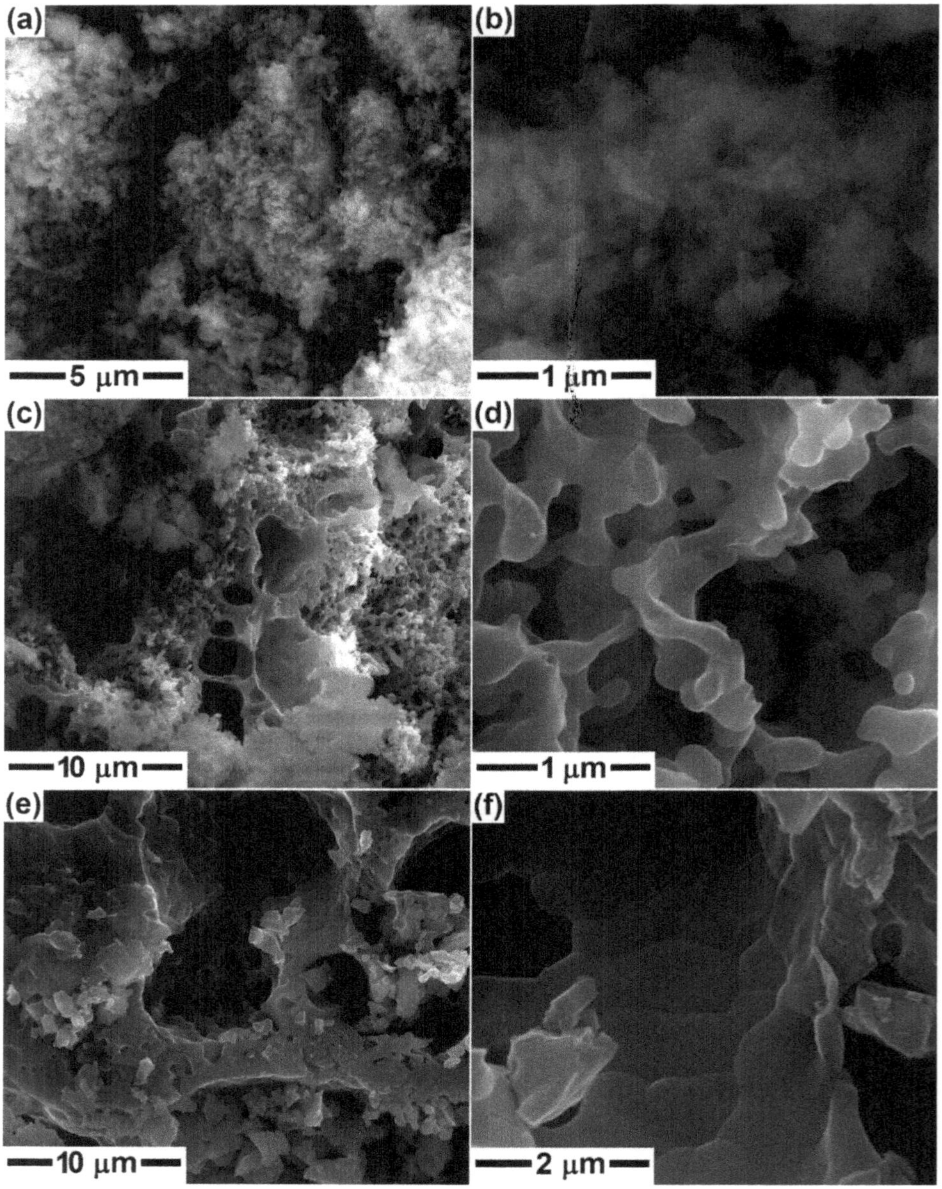

Figure 5. SEM images of the ceramics corresponding to merwinite composition, thermally treated at: (**a**,**b**) 600 °C; (**c**,**d**) 1000 °C; and (**e**,**f**) 1300 °C.

The EDX spectra registered on the gels and powders provide information about the chemical elements present in each sample. Figure 6 centralizes all the data corresponding to the compositions of Dy, Ak and Mw. Maxima specific to all elements of interest (Ca, Mg, Si and O) can be observed, but also differences in intensity, probably related to local inhomogeneities. The presence of Au is justified by the deposition of a conductive layer on the sample's surface before microscopy so that the quality of SEM images is high. The powders underwent a thermal treatment, which eliminated the gas-generating components, leaving clean EDX spectra; the presence of C and N is noticeable only for the gels due to the use of an alkoxide and two nitrate-type precursors.

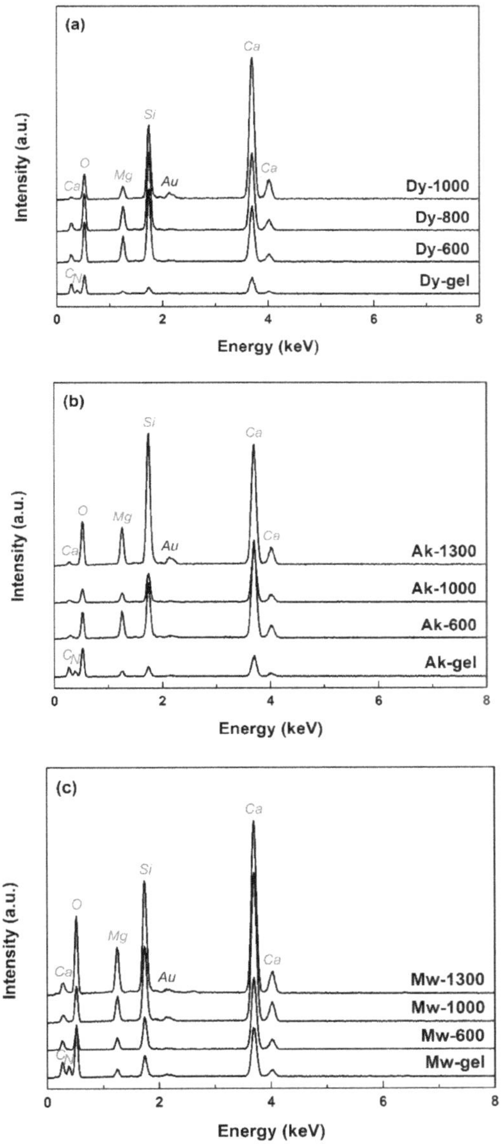

Figure 6. EDX spectra of the dry gels and ceramics corresponding to: (**a**) diopside; (**b**) akermanite; and (**c**) merwinite compositions.

Figure 7 shows the FTIR spectra obtained on the gels and powders with compositions of Dy, Ak and Mw. Within the gels, the broad band located at around 3370 cm^{-1} is attributed to the stretching vibrations of adsorbed and bound water and OH^- groups, while the narrow one placed at around 1630 cm^{-1} is assigned to the bending vibrations of water. According to the scientific literature, around the values of 1450, 1050, 820 and 740 cm^{-1}, the asymmetric and symmetric stretching, as well as the out-of-plane and in-plane bending vibrations of C–O bonds from CO_3^{2-} groups, can be observed [33]; these were generated through the contamination with atmospheric CO_2 during synthesis. The sharp band associated with the symmetric stretching vibrations of NO_3^- groups originating from the nitrate-type precursors introduced into the precursor solution can be found around 1300 cm^{-1} [33].

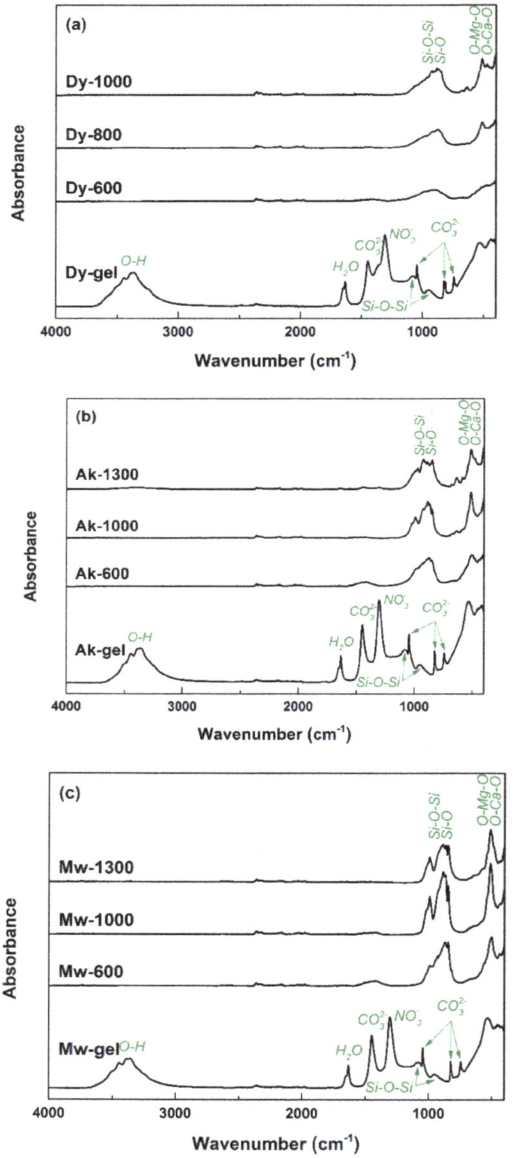

Figure 7. FTIR spectra of the dry gels and ceramics corresponding to: (**a**) diopside; (**b**) akermanite; and (**c**) merwinite compositions.

It is well-known that the signals of Si–O bonds can be identified in the range of 800–1100 cm^{-1}. If, in the FTIR spectra of the gel, there are two wide and low bands centred at around 1080 and 950 cm^{-1}, associated with the stretching vibrations of Si–O–Si bonding, the situation is different in the FTIR spectra of the thermally treated ceramics: both Si–O–Si and Si–O bonding are visible, integrated into a complex band with multiple peaks; more precisely, antisymmetric and symmetric stretching vibrations of Si–O–Si bridging oxygen bonds in SiO_4^{4-} tetrahedra, as well as Si–O–Ca and Si–O–Mg non-bridging oxygen bonds overlap and give rise to a broad and jagged band [53,54]. In the case of the signals that emerged below the value of 600 cm^{-1}, the fingerprints of O–Ca–O and O–Mg–O bonding can be identified as bending vibrations at 510 and 540 cm^{-1}, respectively; Mg–O contribution is sometimes integrated into Ca–O contribution or detected as a shoulder [55,56]. The bands seem to be better defined, with well-separated maxima when it goes from Dy to Ak and then to Mw, in strong correlation with the particularities of the crystalline structure and degree of ordering.

The mineralogical composition and type of crystalline structure were investigated with the help of XRD analysis. The corresponding XRD patterns are shown in Figure 8, for all three compositions (Dy, Ak and Mw). In the case of Dy, the evolution is obvious from the gel stage to the powder thermally treated at 1000 °C. Thus, the gel contains calcium nitrate ($Ca(NO_3)_2$, cubic crystal system, ICDD 00-007-0204) recrystallized from solution and different types of ordered silicon dioxide (SiO_2, tetragonal crystal system—ICDD 00-081-1666, hexagonal crystal system—ICDD 00-083-2471, etc.), which turn into a less crystalline mass after calcination at 600 °C; the last mainly consists of a type of diopside ($CaMgSi_2O_6$, monoclinic crystal system, ICDD 00-083-1821) and a mixture of dicalcium silicates (Ca_2SiO_4, monoclinic crystal system—ICDD 00-083-0464 and orthorhombic crystal system—ICDD 00-083-2457). The heating at 800 °C promotes the crystallization of all three calcium magnesium silicates: another type of diopside ($CaMgSi_2O_6$, monoclinic crystal system, ICDD 00-083-1817), akermanite ($Ca_2MgSi_2O_7$, tetragonal crystal system, ICDD 00-083-1815) and merwinite ($Ca_3MgSi_2O_8$, monoclinic crystal system, ICDD 00-074-0382), but also the quantitative growth of dicalcium silicate previously mentioned. The temperature of 1000 °C seems to be the optimal temperature for the binary compound (Ca_2SiO_4) removal, since the associated ceramic presents only diopside, akermanite and merwinite, with increased crystallinity, as the high intensity and low width of the peaks indicate.

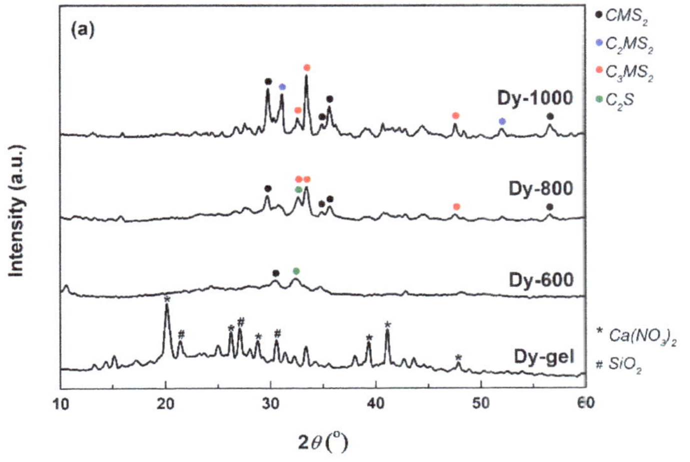

Figure 8. *Cont.*

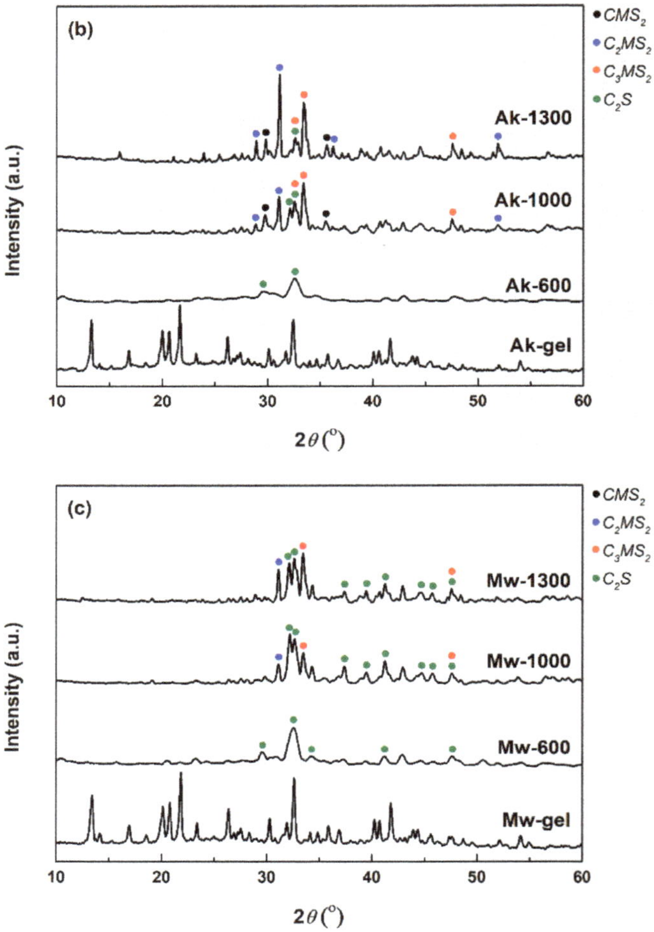

Figure 8. XRD patterns spectra of the dry gels and ceramics corresponding to: (**a**) diopside; (**b**) akermanite; and (**c**) merwinite compositions.

The XRD patterns of the samples with Ak and Mw compositions are quite similar in the stages of gel and 600 °C calcinated powders. In both cases, the crystalline phases from the gel could not be identified, probably because they are non-stoichiometric compounds. Then, the calcined samples exhibit a small crystallinity, conferred by dicalcium silicates formerly indicated. Going further, the two systems evolve slightly differently at higher temperatures (1000 and 1300 °C). In the case of Ak composition and the 1000 °C temperature, merwinite, followed by akermanite seem to be the major phases, accompanied by diopside and dicalcium silicate as minor phases; at 1300 °C, the same four crystalline compounds are maintained, but akermanite becomes predominant, a fact that was expected and desired since this composition was targeted. When it comes to Mw composition, the number of phases is reduced to three: akermanite, merwinite and dicalcium silicate; the temperature of 1000 °C favors the prevalent crystallization of dicalcium silicate, which changes at 1300 °C when merwinite and diopside increase quantitatively.

Concluding, the temperature increase ensures the rise of crystallinity degree, as well as the conversion of the primary or intermediate compounds to the desired ternary compounds (diopside, akermanite and merwinite) through a continuous change of phase ratio. The general reactions can be summarized as displayed below.

$$2 \times C + S \rightarrow C_2S \quad (1)$$

$$y \times C_2S + 2 \times M + (4 - y) \times S \rightarrow 2 \times C_yMS_2 \quad (2)$$

A more detailed insight into the phase composition of the samples thermally treated at the highest temperatures (1000 and 1300 °C) was possible due to a Rietveld refinement on the recorded XRD patterns. The values obtained after performing this type of processing are listed in Table 1. Thus, the numbers confirm the previous statements regarding the ratios between different crystalline compounds. Indeed, the ceramic corresponding to Dy composition has diopside as the main ordered phase (almost 60%), followed by merwinite (about 30%) and akermanite (around 10%). All three ternary compounds (diopside, akermanite and merwinite) are present in fairly equivalent amounts in the sample with Ak composition. The behavior is completely different for the material corresponding to Mw composition, namely the quantitative superiority of dicalcium silicate (almost 60%), compared to merwinite (about 30%) and akermanite (around 10%).

Table 1. Composition of the ceramics corresponding to diopside (Dy); akermanite (Ak); and merwinite (Mw) compositions, thermally treated at the highest temperature (1000 or 1300 °C).

Phases/Sample	CMS_2	C_2MS_2	C_3MS_2	C_2S
		(wt%)		
Dy-1000	59.2	11.4	29.4	-
Ak-1300	25.4	26.5	**31.6**	16.5
Mw-1300	-	9.6	32.0	**58.4**

It is obvious that all three oxide systems follow different pathways of crystallization even though the processing and thermal history are identical, calcium content being the determining parameter; this will also have important implications in the next stage when the biomineralization ability of these materials will be investigated.

3. Conclusions

Starting from diopside, akermanite and merwinite compositions and employing the sol–gel method, crystalline ceramics were obtained after applying thermal treatments at different temperatures. If the powders calcined at 600 °C were in an incipient state of crystallization, mainly given by the dicalcium silicate binary compound, the materials thermally treated at higher temperatures (800, 1000 and 1300 °C) displayed distinct phase compositions as a result of multiple and competitive solid-state reactions defined by specific thermodynamic conditions. Only the sample corresponding to Dy composition presented diopside as the leading crystalline compound, while Ak composition ended up as a balanced mixture of diopside, akermanite and merwinite and Mw composition exhibited a majority of dicalcium silicate.

The approached ceramics contain ternary compounds in the oxide system $CaO–MgO–SiO_2$ and present tremendous potential as biomaterials in the field of hard tissue engineering. They are well-known as biocompatible and bioactive bioceramics, with some differences in terms of mineralization kinetics due to the different calcium content. Future research will be dedicated to the development of calcium magnesium silicate-based scaffolds with controlled properties, as well as to an extended biological evaluation of all the developed materials. Such silicate powders can be further processed by 3D printing, but only after a severe selection of the organic counterpart, necessary for ensuring appropriate rheological properties, as well as working parameters, essential for acquiring high-quality 3D porous materials.

4. Materials and Methods

4.1. Materials

Calcium nitrate tetrahydrate ($Ca(NO_3)_2{\cdot}4H_2O$, 99%, Sigma-Aldrich, Burlington, MA, USA), magnesium nitrate hexahydrate ($Mg(NO_3)_2{\cdot}6H_2O$, 99%, Sigma-Aldrich) and tetraethyl orthosilicate ($Si(OC_2H_5)_4$, TEOS, 98%, Sigma-Aldrich) were used as cationic precursors, while nitric acid (HNO_3) had the role of pH regulator.

4.2. Sol–Gel Method

Basically, the necessary amounts of raw materials were determined starting from the composition of three calcium magnesium silicates, as follows: diopside ($CaO{\cdot}MgO{\cdot}2SiO_2$, CMS_2, Dy), akermanite ($2CaO{\cdot}MgO{\cdot}2SiO_2$, C_2MS_2, Ak) and merwinite ($3CaO{\cdot}MgO{\cdot}2SiO_2$, C_3MS_2, Mw). As a result, three oxide systems having variable content of calcium oxide (CaO) were established as targeted materials. Going to the experimental part, the measured volume of TEOS was solubilized in ethanol, the pH stabilized around 2 with HNO_3, and the solution homogenized on a magnetic stirrer for 1 h. The nitrates were dissolved in distilled water by ultrasonication for 30 min. The two solutions were mixed and homogenized on a magnetic stirrer for another 1 h. Afterward, the final solution was kept at 60 °C, for 12 h, so as to allow the gelation and aging processes to take place, followed by drying at 80 °C for another 48 h. The dry gel was mortared and calcined at 600 °C so as to ensure the removal of the unwanted components: solvent molecules, organic fraction and nitrate groups; the calcination temperature was chosen based on the complex thermal analysis of all prepared gels (Figure 1). Furthermore, each composition was subjected to secondary thermal treatments: 800 and 1000 °C for Dy, 1000 and 1300 °C for Ak and Mw, for 2 h, with 10 °C/min heating rate and equilibrium cooling, resulting in the powders of interest. It was not possible to apply the same temperature values for all three calcined powders since Dy has a melting point of about 1390 °C, while Ak and Mw melt around 1450 °C; this means that heating up to 1300 °C makes Dy prone to melting, being well-known that the sol–gel method triggers a significant reduction of threshold temperatures (up to 200 °C). As a consequence, 1000 °C was maintained as maximum treatment temperature for Dy.

4.3. Materials Characterization

The dry gels were investigated from thermal and morphological points of view. The complex thermal analysis was recorded from room temperature to 1000 °C, with a rate of 5 °C/min, in air, on Shimadzu DTG-60 equipment (Shimadzu Corporation, Kyoto, Japan). The morphology was visualized by scanning electron microscopy (SEM), with a Quanta Inspect F microscope (FEI Company, Hillsboro, OR, USA) equipped with an energy-dispersive X-ray spectroscopy (EDX) probe; the operating parameters were: 30 kV accelerating voltage, 10 mm working distance and gold coating by DC magnetron sputtering for 40 s.

The thermally treated ceramics were assessed in terms of composition, structure and morphology. The elemental composition was determined with the help of the EDX probe, while the chemical bonds and groups were studied by Fourier transform infrared (FTIR) spectroscopy, with a Thermo Scientific Nicolet iS50 spectrophotometer (Thermo Fisher Scientific, Waltham, MA, USA), in the attenuated total reflection (ATR) mode; the working conditions were: 400–4000 cm^{-1} wavenumber range, 4 cm^{-1} resolution and 64 scans/sample. The phase composition and crystal structure were revealed by X-ray diffraction (XRD), with a Shimadzu XRD 6000 diffractometer (Shimadzu Corporation, Kyoto, Japan), using Ni-filtered Cu Kα radiation (λ = 1.54 Å); the procedure involved: 10–60° 2θ range, 2°/min scan speed, 0.02° step size and 0.6 s preset time. To have a deeper view, a Rietveld refinement in HighScore Plus v3.0e software (Malvern Panalytical, Royston, UK) was applied for the powders thermally treated at the highest temperatures (1000 °C for Dy, 1300 °C for Ak and Mw); a polynomial function was employed for the background fit, with flat background coefficient, coefficient 1, coefficient 2, coefficient 3 and 1/x, a pseudo-Voigt profile function and Caglioti function for FWHM approximation. The samples crystallinity was evaluated through the intensity ratio of the diffraction peaks and of the sum of all measured intensity, as it is presented in Equation (3), where I_{net} is the net intensity (intensity of the crystalline peaks), I_{tot} is the total intensity and I_{bgr} is the background intensity (intensity which arises also for completely crystalline sample from imperfections of the sample).

$$\text{Crystallinity} = 100 \times \Sigma\, I_{net}/(\Sigma\, I_{tot} - \Sigma\, I_{bgr})\ (\%) \quad (3)$$

Author Contributions: Conceptualization, C.B.; methodology, C.B.; software, V.-A.S.; validation, C.B.; investigation, A.-E.A., C.-C.C., G.V. and C.B.; resources, S.-I.J. and C.B.; writing—original draft preparation, A.-E.A. and C.B.; writing—review and editing, C.B.; supervision, C.B. All authors have read and agreed to the published version of the manuscript.

Funding: This research received no external funding.

Informed Consent Statement: Not applicable.

Conflicts of Interest: The authors declare no conflict of interest.

References

1. Ba, Z.; Chen, Z.; Huang, Y.; Feng, D.; Zhao, Q.; Zhu, J.; Wu, D. Nanoporous diopside modulates biocompatibility, degradability and osteogenesis of bioactive scaffolds of gliadin-based composites for new bone formation. *Int. J. Nanomed.* **2018**, *13*, 3883–3896. [CrossRef] [PubMed]
2. Dasan, A.; Kraxner, J.; Grigolato, L.; Savio, G.; Elsayed, H.; Galusek, D.; Bernardo, E. 3D printing of hierarchically porous lattice structures based on akermanite glass microspheres and reactive silicone binder. *J. Funct. Biomater.* **2022**, *13*, 8. [CrossRef] [PubMed]
3. Hafezi-Ardakani, M.; Moztarzadeh, F.; Rabiee, M.; Talebi, A.R.; Abasi-shahni, M.; Fesahat, F.; Sadeghian, F. Sol-gel synthesis and apatite-formation ability of nanostructure merwinite ($Ca_3MgSi_2O_8$) as a novel bioceramic. *J. Ceram. Proc. Res.* **2010**, *11*, 765–768.
4. Han, Z.; Gao, C.; Feng, P.; Shen, Y.; Shuai, C.; Peng, S. Silicon carbide whiskers reinforced akermanite scaffolds for tissue engineering. *RSC Adv.* **2014**, *4*, 36868–36874. [CrossRef]
5. Liu, T.; Deng, Y.; Gao, C.; Feng, P.; Shuai, C.; Peng, S. Analysis of 3D printed diopside scaffolds properties for tissue engineering. *Mater. Sci.* **2015**, *21*, 590–594. [CrossRef]
6. Dasan, A.; Elsayed, H.; Kraxner, J.; Galusek, D.; Bernardo, E. Hierarchically porous 3D-printed akermanite scaffolds from silicones and engineered fillers. *J. Eur. Ceram. Soc.* **2019**, *39*, 4445–4449. [CrossRef]
7. Han, Z.; Feng, P.; Gao, C.; Shen, Y.; Shuai, C.; Peng, S. Microstructure, mechanical properties and in vitro bioactivity of akermanite scaffolds fabricated by laser sintering. *Biomed. Mater. Eng.* **2014**, *24*, 2073–2080. [CrossRef]
8. Shahrouzifar, M.R.; Salahinejad, E. Strontium doping into diopside tissue engineering scaffolds. *Ceram. Int.* **2019**, *45*, 10176–10181. [CrossRef]
9. Pang, S.; Wu, D.; Kamutzki, F.; Kurreck, J.; Gurlo, A.; Hanaor, D.A.H. High performing additively manufactured bone scaffolds based on copper substituted diopside. *Mater. Des.* **2022**, *215*, 110480. [CrossRef]
10. Collin, M.S.; Venkatraman, S.K.; Mohana, S.; Sumathi, S.; Drweesh, E.A.; Elnagar, M.M.; Mosa, E.S.; Sasikumar, S. Solution combustion synthesis of functional diopside, akermanite, and merwinite bioceramics: Excellent biomineralization, mechanical strength, and antibacterial ability. *Mater. Today Commun.* **2021**, *27*, 102365. [CrossRef]
11. Ansari, M.; Malmir, F.; Salati, A. Preparation and characterization of akermanite/merwinite scaffolds for bone tissue repair. *J. Biomim. Biomater. Biomed. Eng.* **2020**, *44*, 73–81. [CrossRef]
12. Razavi, M.; Fathi, M.; Savabi, O.; Vashaee, D.; Tayebi, L. In vivo biocompatibility of Mg implants surface modified by nanostructured merwinite/PEO. *J. Mater. Sci. Mater. Med.* **2015**, *26*, 184. [CrossRef] [PubMed]
13. Hosseini, Y.; Emadi, R.; Kharaziha, M. Surface modification of PCL-diopside fibrous membrane via gelatin immobilization for bone tissue engineering. *Mater. Chem. Phys.* **2017**, *194*, 356–366. [CrossRef]
14. Dong, X.; Li, H.; Lingling, E.; Cao, J.; Guo, B. Bioceramic akermanite enhanced vascularization and osteogenic differentiation of human induced pluripotent stem cells in 3D scaffolds: In vitro and vivo. *RSC Adv.* **2019**, *9*, 25462–25470. [CrossRef] [PubMed]
15. Arastouei, M.; Khodaei, M.; Atyabi, S.M.; Nodoushan, M.J. Poly lactic acid-akermanite composite scaffolds prepared by fused filament fabrication for bone tissue engineering. *J. Mater. Res. Technol.* **2020**, *9*, 14540–14548. [CrossRef]
16. Birhanu, G.; Doosti-Telgerd, M.; Zandi-Karimi, A.; Karimi, Z.; Daryasari, M.P.; Javar, H.A.; Seyedjafari, E. Enhanced proliferation and osteogenic differentiation of mesenchymal stem cells by diopside coated Poly-L-lactic Acid-Based nanofibrous scaffolds. *Int. J. Polym. Mater. Polym. Biomater.* **2022**, *71*, 707–716. [CrossRef]
17. Nadernezhad, A.; Torabinejad, B.; Hafezi, M.; Baghaban-Eslaminejad, M.; Bagheri, F.; Najafi, F. Poly (lactic-co-glycolic)/nanostructured merwinite porous composites for bone tissue engineering: Structural and in vitro characterization. *J. Adv. Mater. Proc.* **2014**, *2*, 13–24.
18. Bafandeh, M.R.; Mojarrabian, H.M.; Doostmohammadi, A. Poly (vinyl alcohol)/chitosan/akermanite nanofibrous scaffolds prepared by electrospinning. *J. Macromol. Sci. B Phys.* **2019**, *58*, 749–759. [CrossRef]

19. Ou, J.; Yin, G.F.; Zhou, D.L.; Chen, X.C.; Yao, Y.D.; Yang, W.Z.; Wu, B.L.; Xue, M.; Cui, J.; Zhu, W.F.; et al. Preparation of merwinite with apatite-forming ability by sol-gel process. *Key Eng. Mater.* **2007**, *330*, 67–70. [CrossRef]
20. Shuai, C.; Han, Z.; Feng, P.; Gao, C.; Xiao, T.; Peng, S. Akermanite scaffolds reinforced with boron nitride nanosheets in bone tissue engineering. *J. Mater. Sci. Mater. Med.* **2015**, *26*, 188. [CrossRef]
21. Nezafati, N.; Hafezi, M.; Zamanian, A.; Yasaei, M.; Mohammadi, M.B. Preparation and characterization of a novel nano-structured merwinite scaffold prepared by freeze casting method. In Proceedings of the 5th International Conference on Nanostructures, Kish Island, Iran, 11–13 August 2014; pp. 58–60.
22. Venkatraman, S.K.; Choudhary, R.; Krishnamurithy, G.; Raghavendran, H.R.B.; Murali, M.R.; Kamarul, T.; Suresh, A.; Abraham, J.; Swamiappan, S. Biomineralization, mechanical, antibacterial and biological investigation of larnite and rankinite bioceramics. *Mater. Sci. Eng. C* **2021**, *118*, 111466. [CrossRef] [PubMed]
23. Venkatraman, S.K.; Choudhary, R.; Krishnamurithy, G.; Raghavendran, H.R.B.; Murali, M.R.; Kamarul, T.; Suresh, A.; Abraham, J.; Praharaj, S.; Swamiappan, S. Comparative investigation on antibacterial, biological and mechanical behaviour of monticellite and diopside derived from biowaste for bone regeneration. *Mater. Chem. Phys.* **2022**, *286*, 126157. [CrossRef]
24. Reddy, P.M.; Lakshmi, R.; Dass, F.P.; Swamiappan, S. Synthesis, characterization and formulation of sodium calcium silicate bioceramic for drug delivery applications. *Sci. Eng. Compos. Mater.* **2014**, *23*, 375–380. [CrossRef]
25. Yamamoto, S.; Kawamura, N.; Nonami, T. Diopside synthesized by sol-gel method as phosphorus adsorption material: Evaluation of apatite deposition in pseudo body solution. *Trans. Mater. Res. Soc. Jpn.* **2019**, *44*, 17–23. [CrossRef]
26. Choudhary, R.; Koppala, S.; Swamiappan, S. Bioactivity studies of calcium magnesium silicate prepared from eggshell waste by sol-gel combustion synthesis. *J. Asian Ceram. Soc.* **2015**, *3*, 173–177. [CrossRef]
27. Schumacher, T.C.; Volkmann, E.; Yilmaz, R.; Wolf, A.; Treccani, L.; Rezwan, K. Mechanical evaluation of calcium-zirconium-silicate (baghdadite) obtained by a direct solid-state synthesis route. *J. Mech. Behav. Biomed. Mater.* **2014**, *34*, 294–301. [CrossRef]
28. Mohammadi, H.; Ismail, Y.M.B.; Shariff, K.A.; Noor, A.F.M. Effect of substitutional strontium on mechanical properties of akermanite ceramic prepared by solid-state sintering. *Mater. Today Proc.* **2019**, *17*, 929–936. [CrossRef]
29. Sharafabadi, A.K.; Abdellahi, M.; Kazemi, A.; Khandan, A.; Ozada, N. A novel and economical route for synthesizing akermanite ($Ca_2MgSi_2O_7$) nano-bioceramic. *Mater. Sci. Eng. C* **2017**, *71*, 1072–1078. [CrossRef]
30. Iwata, N.Y.; Tsunakawa, S.; Tanaka, M.; Utsu, T.; Matsumoto, K. Improvements of apatite-forming abilities on pure and sodium-containing diopside substrates using porous diopside thin films as nucleating agent. *MRS Online Proc. Libr.* **1999**, *599*, 169–174. [CrossRef]
31. Gheisari Dehsheikh, H.; Karamian, E. Characterization and synthesis of hardystonite (HT) as a novel nanobioceramic powder. *Nanomed. J.* **2016**, *3*, 143–146.
32. Gheisari, H.; Karamian, E.; Soheily, A. Survey and evaluation of merwinite (MW) as a new nanobioceramic powder. *J. Nanoanal.* **2020**, *7*, 225–229.
33. Negrea, R.; Busuioc, C.; Constantinoiu, I.; Miu, D.; Enache, C.; Iordache, F.; Jinga, S.I. Akermanite-based coatings grown by pulsed laser deposition for metallic implants employed in orthopaedics. *Surf. Coat. Technol.* **2019**, *357*, 1015–1026. [CrossRef]
34. Hafezi-Ardakani, M.; Moztarzadeh, F.; Rabiee, M.; Talebi, A.R. Synthesis and characterization of nanocrystalline merwinite ($Ca_3Mg(SiO_4)_2$) via sol-gel method. *Ceram. Int.* **2011**, *37*, 175–180. [CrossRef]
35. Iwata, N.Y.; Lee, G.H.; Tsunakawa, S.; Tokuoka, Y.; Kawashima, N. Preparation of diopside with apatite-forming ability by sol-gel process using metal alkoxide and metal salts. *Colloids Surf. B Biointerfaces* **2004**, *33*, 1–6. [CrossRef]
36. Lombardi, M.; Cacciotti, I.; Bianco, A.; Montanaro, L. RKKP bioactive glass-ceramic material through an aqueous sol-gel process. *Ceram. Int.* **2015**, *41*, 3371–3380. [CrossRef]
37. Voicu, G.; Ene, V.L.; Sava, D.F.; Surdu, V.A.; Busuioc, C. Sol-gel derived vitroceramic materials for biomedical applications. *J. Non-Cryst. Solids* **2016**, *449*, 75–82. [CrossRef]
38. Duman, S.; Bulut, B. Effect of akermanite powders on mechanical properties and bioactivity of chitosan-based scaffolds produced by 3D-bioprinting. *Ceram. Int.* **2021**, *47*, 13912–13921. [CrossRef]
39. Naga, S.M.; El-Maghraby, H.F.; Mahmoud, E.M.; Killinger, A.; Gadow, R. Hydroxyapatite/diopside porous scaffolds: Preparation and in vitro study. *Interceram* **2019**, *68*, 22–29. [CrossRef]
40. Bigham, A.; Hassanzadeh-Tabrizi, S.A.; Khamsehashari, A.; Chami, A. Surfactant-assisted sol–gel synthesis and characterization of hierarchical nanoporous merwinite with controllable drug release. *J. Sol-Gel Sci. Technol.* **2018**, *87*, 618–625. [CrossRef]
41. Wu, C.; Chang, J. Synthesis and apatite-formation ability of akermanite. *Mater. Lett.* **2004**, *58*, 2415–2417. [CrossRef]
42. No, Y.; Li, J.; Zreiqat, H. Doped calcium silicate ceramics: A new class of candidates for synthetic bone substitutes. *Materials* **2017**, *10*, 153. [CrossRef] [PubMed]
43. Nezafati, N.; Hafezi, M.; Zamanian, A.; Naserirad, M. Effect of adding nano-titanium dioxide on the microstructure, mechanical properties and in vitro bioactivity of a freeze cast merwinite scaffold. *Biotechnol. Prog.* **2015**, *31*, 550–556. [CrossRef] [PubMed]
44. Hosseini, Y.; Emadi, R.; Kharaziha, M.; Doostmohammadi, A. Reinforcement of electrospun poly(e-caprolactone) scaffold using diopside nanopowder to promote biological and physical properties. *J. Appl. Polym. Sci.* **2017**, *134*, 44433. [CrossRef]
45. Shuai, C.; Liu, T.; Gao, C.; Feng, P.; Xiao, T.; Yu, K.; Peng, S. Mechanical and structural characterization of diopside scaffolds reinforced with graphene. *J. Alloys Compd.* **2016**, *655*, 86–92. [CrossRef]
46. Teimouri, A.; Ghorbanian, L.; Dabirian, I. Preparation and characterization of silk/diopside composite nanofibers via electrospinning for tissue engineering application. *Int. J. Chem. Mol. Eng.* **2016**, *10*, 791–794.

47. Arastouei, M.; Khodaei, M.; Atyabi, S.M.; Nodoushan, M.J. Improving the properties of the porous polylactic acid scaffold by akermanite nanoparticles for bone tissue engineering. *J. Adv. Mater. Proc.* **2020**, *8*, 11–19.
48. Chen, C.; Watkins-Curry, P.; Smoak, M.; Hogan, K.; Deese, S.; McCandless, G.T.; Chan, J.Y.; Hayes, D.J. Targeting calcium magnesium silicates for polycaprolactone/ceramic composite scaffolds. *ACS Biomater. Sci. Eng.* **2015**, *1*, 94–102. [CrossRef]
49. Wu, C.; Chang, J.; Zhai, W.; Ni, S.; Wang, J. Porous akermanite scaffolds for bone tissue engineering: Preparation, characterization, and in vitro studies. *J. Biomed. Mater. Res. B Appl. Biomater.* **2006**, *78*, 47–55. [CrossRef]
50. Abdollahi, M.; Ghomi, H. Fabrication of highly porous merwinite scaffold using the space holder method. *Int. J. Mater. Res.* **2020**, *111*, 711–718. [CrossRef]
51. Koppala, S.; John, S.P.; Balan, R.; Lokesh, B.; Munusamy, S.; Karthikeyan, P.; Godiya, C.B.; Chandragiri, S.Y.; Aminabhavi, T.M.; Duan, K.; et al. Glowing combustion synthesis, characterization and biomedical properties of Sr-hardystonite ($Sr_2ZnSi_2O_7$) powders. *Ceram. Int.* **2022**, *48*, 23649–23656. [CrossRef]
52. Jinga, S.I.; Constantinoiu, I.; Surdu, V.A.; Iordache, F.; Busuioc, C. Sol-gel-derived mineral scaffolds within SiO_2–P_2O_5–CaO–MgO–ZnO–CaF_2 system. *J. Sol.-Gel. Sci. Technol.* **2019**, *90*, 411–421. [CrossRef]
53. Jinga, S.I.; Anghel, A.M.; Brincoveanu, S.F.; Bucur, R.M.; Florea, A.D.; Saftau, B.I.; Stroe, S.C.; Zamfirescu, A.I.; Busuioc, C. Ce/Sm/Sr-incorporating ceramic scaffolds obtained via sol-gel route. *Materials* **2021**, *14*, 1532. [CrossRef] [PubMed]
54. Prefac, G.A.; Milea, M.L.; Vadureanu, A.M.; Muraru, S.; Dobrin, D.I.; Isopencu, G.O.; Jinga, S.I.; Raileanu, M.; Bacalum, M.; Busuioc, C. CeO_2 containing thin films as bioactive coatings for orthopaedic implants. *Coatings* **2020**, *10*, 642. [CrossRef]
55. Draghici, D.A.; Mihai, A.A.; Aioanei, M.O.; Negru, N.E.; Nicoara, A.I.; Jinga, S.I.; Miu, D.; Bacalum, M.; Busuioc, C. Strontium-substituted bioactive glass-ceramic films for tissue engineering. *Bol. Soc. Esp. Ceram. Vidr.* **2022**, *61*, 184–190. [CrossRef]
56. Schitea, R.I.; Nitu, A.; Ciobota, A.A.; Munteanu, A.L.; David, I.M.; Miu, D.; Raileanu, M.; Bacalum, M.; Busuioc, C. Pulsed laser deposition derived bioactive glass-ceramic coatings for enhancing the biocompatibility of scaffolding materials. *Materials* **2020**, *13*, 2615. [CrossRef] [PubMed]

gels

MDPI

Article

Gelatin Nanoparticles for Targeted Dual Drug Release out of Alginate-di-Aldehyde-Gelatin Gels

Sophie Schrade [1,2], Lucas Ritschl [1], Regine Süss [2], Pia Schilling [1] and Michael Seidenstuecker [1,*]

[1] G.E.R.N. Center of Tissue Replacement, Regeneration & Neogenesis, Department of Orthopedics and Trauma Surgery, Medical Center—Albert-Ludwigs-University of Freiburg, Faculty of Medicine, Albert-Ludwigs-University of Freiburg, Hugstetter Straße 55, 79106 Freiburg, Germany; sophie.schrade@uniklinik-freiburg.de (S.S.); lucas.ritschl@uniklinik-freiburg.de (L.R.); pia.schilling@uniklinik-freiburg.de (P.S.)

[2] Institute of Pharmaceutical Sciences, Albert-Ludwigs-University of Freiburg, Sonnenstr. 5, 79104 Freiburg, Germany; regine.suess@pharmazie.uni-freiburg.de

* Correspondence: michael.seidenstuecker@uniklinik-freiburg.de

Abstract: The aim of the present work was to develop a dual staged drug release of an antibiotic (clindamycin) and a growth factor: bone morphogenetic protein-2 (BMP-2) from a biodegradable system consisting of hydrogel and gelatin nanoparticles (GNP). Two-step de-solvation allowed us to prepare GNPs (~100 nm) as drug carriers. Fluorescein isothiocyanate (FITC)-conjugated protein A was used as a model substance for BMP-2. A 28-day release experiment was performed to determine the release kinetics from GNP for both FITC-protein A and BMP-2, and for clindamycin (CLI) from the hydrogel. The size, structure, and overall morphology of GNP samples (empty, loaded with FITC-protein A and BMP-2) were examined using an environmental scanning electron microscope (ESEM). Cell culture assays (Live/dead; cell proliferation; cytotoxicity) were performed with MG-63 cells and BMP-2-loaded GNPs. Drug release experiments using clindamycin-loaded alginate-di-aldehyde (ADA) gelatin gels containing the drug-loaded GNPs were performed for 28 days. The resulting GNPs showed an empty size of 117 ± 29 nm, 176 ± 15 nm and 216 ± 36 nm when containing 2% FITC-protein A and 1% BMP-2, respectively. No negative effects of BMP-2-loaded GNPs on MG-63 cells were observed in live/dead staining. In the proliferation assay, an increase in cell proliferation was observed for both GNPs (GNP + BMP-2 and controls). The cytotoxicity assay continuously showed very low cytotoxicity for GNPs (empty; loaded). Clindamycin release showed a concentration of 25-fold higher than the minimum inhibitory concentration (MIC) against Staphylococcus aureus throughout the 28 day period. BMP-2 showed a reduced burst release and a steady release (~2 μg/mL) over a 28 day period.

Keywords: gelatin nanoparticle; controlled drug release; antibiotics; growth factor; ADA-gelatin

Citation: Schrade, S.; Ritschl, L.; Süss, R.; Schilling, P.; Seidenstuecker, M. Gelatin Nanoparticles for Targeted Dual Drug Release out of Alginate-di-Aldehyde-Gelatin Gels. *Gels* **2022**, *8*, 365. https://doi.org/10.3390/gels8060365

Academic Editor: Esmaiel Jabbari

Received: 10 May 2022
Accepted: 1 June 2022
Published: 8 June 2022

1. Introduction

Bone infections, such as osteomyelitis, often present a challenge in trauma surgery and orthopaedic settings. The therapy is often lengthy and requires a high degree of diagnostic and therapeutic experience on the part of the treating physician. Osteomyelitis is an infection of the bone and surrounding soft tissues. As a rule, the infection is bacterial in nature, but fungi and viruses can also cause osteomyelitis [1,2]. Osteitis can be divided into acute post-traumatic, chronic post-traumatic and acute haematogenic osteomyelitis, the latter being of endogenous origin and rare in adults. Acute haematogenic osteomyelitis is more common in children and adolescents, with one third of cases occurring in children under 2 years of age. The germs usually enter the bloodstream through infections in the nasopharynx and can thus attack the bone from the inner medullary cavity [3]. In an exogenous infection the germs penetrate via an injury to the exterior of the bone. Post-traumatic or post-operative infections occur mainly after open fractures and surgical interventions such as the insertion of endoprostheses into the body. Open fractures present an increased risk of developing a bone infection. The risk of post-traumatic infection can rise to 25% depending on the type of fracture, the amount of bone loss, bacterial contamination, the degree of soft tissue injury and other injuries, such as damage to local vasculature [4,5]. Periprosthetic infection is a common complication after implantation of an endoprosthesis [6]. Due to the increasing age of the population and the growing desire not to lose mobility even in old age, the number of implantations of such total endoprostheses (TEP) is rising sharply [7] and thus also the number of postoperative infections. 2019 alone (as this, COVID-19 related, is the most reliable data of the last years [8]), 243,477 TEPs were implanted in the hip joint and 193,759 in the knee joint in Germany [9]. In the USA, 498,000 TEPs were implanted in the hip joint and 1,065,000 total knee replacement surgeries were performed during the same period [10]. Further implantations of such TEPs are performed on the shoulder, elbow and ankle joint. It is expected that infection will manifest itself in 17% of the procedures of such TEPs. As a rule, the bacteria usually reach the surface of the implants during the surgical procedure, although less frequently they may also stem from endogenous sources [11]. The bacterium *Staphylococcus epidermidis*, which is part of the normal skin flora, often colonizes the surface of medical devices such as catheters and implants with a biofilm [12]. The surface of such implants offers the bacteria the opportunity to form a biofilm by changing from a planktonic to a sessile form. In the sessile form the bacteria do not grow at all or grow very slowly and are therefore not attacked by most antibiotics. The biofilm protects the bacteria from phagocytosis of the immune cells and represents a diffusion barrier for the antibiotics [13]. Periprosthetic osteomyelitis develops from a periprosthetic infection in 20% of cases. Depending on the risk factors present, about 10–30% of acute osteomyelitis develop into chronic forms [14]. Patients with chronic osteomyelitis often have a long history of suffering with frequent recurrences and sometimes a life-long illness [15]. Frequent hospitalisation and loss of employment severely restrict the quality of life of patients and the economic impact of such chronic infections is also severe. The most common pathogen causing osteomyelitis is *Staphylococcus aureus* (*S. aureus*) [16], a gram-positive, spherical bacterium which, like *Staphylococcus epidermidis*, is part of the skin and mucous membrane flora and can also form a biofilm. Besides *Escherichia coli* (*E. coli*), *S. aureus* causes the most common bacterial infections in humans [17]. In addition to the many resistances to common antibiotics, the bacterium brings along further "tools" that considerably increase its pathogenicity. Cell wall-associated proteins such as fibronectin binding proteins and collagen binding proteins help *S. aureus* adhere to cells, tissue and foreign bodies. A number of extracellular enzymes such as hyaluronidases, lipases and plasma coagulases degrade tissue and ensure its dissemination within the tissue. In addition to osteomyelitis, other enzymes and toxins can cause many other clinical pictures such as toxic shock syndromes, endocarditis and sepsis [18]. The current treatment strategy for chronic osteomyelitis includes surgical removal of the infected area (radical debridement) followed by systemic or local antibiotic therapy. In order to prevent resistance and to ensure a targeted therapy, it is important to detect the pathogen and to determine the sensitivity of the bacterium to the respective antibiotic by means of an antibiogram. The antibiotic used should be bone-compatible and well tolerated by the patient. However, systemic antibiotic therapy reaches its limits in case of a bone infection. On the one hand, the biofilm protects the bacteria from the penetration of antibiotics, on the other hand, the progressing course of the disease often makes antibiotic therapy ineffective. The progressive inflammation leads to necrotic tissue and thus to a reduced blood supply. So called sequesters are formed, which as avital bone tissue create ideal conditions for biofilm colonization [19]. All of these factors ensure that at the site of the infection no drug levels above the minimum inhibitory concentration (MIC) can be achieved by systemic antibiotic therapy.

Local antibiotic therapy, on the other hand, has the advantage that higher drug levels can be achieved at the site of infection, which if administered systemically would lead to toxic serum levels. Thus, with a suitable choice of active ingredient carrier, the active ingredient can be released over a longer period of time, for example over 4 weeks. Systemic side effects can thus be prevented and the concentration of the active substance at the site of infection is above the MIC for a sufficient amount of time to eliminate all bacteria, even in deeper areas of the bone. Local drug carriers can be divided into non-biodegradable and biodegradable systems. The current standard is the implantation of gentamycin-containing bone cement spheres. These polymethylmethacrylate (PMMA) spheres (e.g., Septopal®) are strung together in a chain and are not biodegradable. The PMMA chains are inserted into the bone cavities after debridement and usually remain there for 7–10 days [20]. After this period, the chain must be removed again and a second surgical procedure is performed, which in turn can lead to complications. A further disadvantage is the incomplete release of the gentamycin from the hydrophobic polymer matrix [21]. Biodegradable local drug carriers have the advantage that a second operation is not necessary. Biodegradable collagen fleeces loaded with gentamycin (Septocoll®) or collagen lyophilisates loaded with teicoplanin (Targobone®) can be used for local antibiotic therapy of bone infections [22]. However, a complete release of gentamycin-loaded collagens was determined after 4 days [23]. A constant release over several weeks however, is more desirable. Due to the fact that the current therapeutic options for the treatment of bone infections are still not free of weak points, further research on biodegradable, local drug carriers with adequate and controllable drug release will be required in the future.

The aim of the present work was to develop a dual, staggered drug release of both an antibiotic and a growth factor utilizing such a biodegradable system. The lincosamide clindamycin (CLI) was used as an antibiotic and should be released first to eliminate the pathogens causing the infection. Subsequently, the growth factor Bone Morphogenetic Protein 2 (BMP-2) should be released, which promotes the formation of new bone and cartilage. CLI was to be released uniformly over a period of 4 weeks, so it was placed in the hydrogel for delayed release. The ADA-GEL was then placed in a microporous ceramic (β-TCP, as described in earlier studies) that was intended to serve as a drug delivery device. The hydrogel used in the present study was alginate-dialdehyde (ADA) combined with gelatin through the formation of Schiff's bases. The combination of ADA with gelatin has established itself as particularly suitable, as it shows good stability and adequate degradation behavior. The prolonged release of clindamycin can be achieved by the molecules of the antibiotic first diffusing through the gel to the edge of the ceramic and then gradually being released. The BMP-2 was intended to be released with a delay and was therefore enclosed in gelatin nanoparticles (GNPs), which in turn were incorporated into the gel containing CLI. The ceramic was loaded with the CLI-containing ADA-gelatin gel, which contained gelatin nanoparticles with BMP-2 enclosed, via a vacuum-induced flow. Initially, a method was developed to produce GNPs in uniform shape and size (having a diameter of approximately 100 nm) reproducibly by two-step de-solvation [14,22]. In a further step, these nanoparticles were loaded with proteins. The morphology of GNPs was studied by means of electron microscopy. Due to the high cost of BMP-2, protein A coupled with fluorescein isothiocyanate (FITC) was initially used as a model substance. The release kinetics from the GNPs were determined for both FITC-protein A and BMP-2 in a 28 day release experiment. The GNPs loaded with FITC-protein A and then in the further course of the experiments with BMP-2 were placed in ADA gelatin gels, which additionally contained CLI. The release experiments were performed directly from beads cast from this ADA gelatin gel containing CLI and FITC-protein A or BMP-2. The quantitative determination of clindamycin was done by high pressure liquid chromatography (HPLC) measurements and FITC-protein A levels were determined by fluorimeter and BMP-2 using an ELISA immunoassay. Biocompatibility was investigated by means of cell culture experiments. A live dead assay was performed on the MG-63 cells cultured with the GNP beads.

2. Results

2.1. Manufacturing Process of GNPs

The manufacturing method, in which the second de-solvation was performed at a pH of 2.5, provided the best results in terms of size and shape (117 ± 29 nm) and aggregation behavior. GNPs prepared at pH 3 could be resuspended very well but were more inconsistent in size and shape. Furthermore, the ESEM images also showed structures that were not of nanoparticulate form (see red arrows in Figure 1). GNPs produced at pH 2 were either too small (less than 100 nm) or the morphology was not spherical (see Figure 1).

(a) (b) (c)

Figure 1. (**a**) ESEM Images of GNPs at a pH 2.0 which leads to GNPs smaller than 100 nm; (**b**) pH 2.5 with well-shaped GNPs having a size of 100 nm and (**c**) pH 3.0 with non-uniform size and shape of the GNPs, as well as structures not representing GNP (red arrow) Images taken with FEI Quanta 250 FEG at 12 kV acceleration voltage, all samples were sputter coated (JEOL, JFC-1200) with a 25 nm Gold layer.

The short precipitation time of 5 min after the first de-solvation was decisive for the success of the GNPs. The longer the wait after the first addition of acetone, the more likely more LMG precipitated into the HMG precipitate. This led to reticular and needle-like structures, but not to GNP production.

GNPs with FITC-protein A differed strongly in morphology and size. Hence, the GNPs of the mixture prepared with 1% FITC-protein A solution were very small (less than 50 nm), whereas those of the 2% mixture were on average about 176 ± 15 nm in size and were all uniformly spherical (see Figure 2).

(a) (b)

Figure 2. ESEM images of GNP with (**a**) 1% *w*/*v* FITC-protein A and (**b**) 2% *w*/*v* FITC-protein A. Images taken with FEI Quanta 250 FEG at 12 kV acceleration voltage, all samples were sputter coated (JEOL, JFC-1200) with a 25 nm Gold layer.

The GNPs loaded with BMP-2 were on average larger at 216 ± 36 nm than those with FITC-protein A (176 nm). The SEM images in Figure 3 show the formation of aggregates.

(a) (b)

Figure 3. ESEM images of BMP-2 loaded GNP; (**a**) 1% *w*/*v* BMP-2; (**b**) 2% *w*/*v* BMP-2; images taken with FEI Quanta 250 FEG at 12 kV acceleration voltage, all samples were sputter coated (JEOL, JFC-1200) with a 25 nm Gold layer.

2.2. *Inclusion Capacity*

The inclusion capacity (IC) was determined as described in 2.3.2. A 6-point calibration curve was drawn (R^2 = 0.9946) and the amount of FITC-protein A in the supernatant was determined. The inclusion capacity of the GNPs was 45.98%, with 250 μg FITC-protein A and 0.625 g GEL. The inclusion capacity of BMP-2 in the GNPs was 99.5% with 1 μg BMP-2 and 0.625 g GEL.

2.3. *Biocompatibility*

2.3.1. Live Dead Staining

The cell count of MG-63 cells was found to increase continuously over the entire 7 day maintenance period (3, 7 and 10 days). Only isolated dead cells could be observed, especially within the first 24 h. Thereafter, only a living cell layer was observed. In Figure 4 the live dead staining of day 3 and day 7 is shown as an exemplar. The live/dead assay showed an increase in cell number over time. But in addition, the cell number with GNPs containing BMP2 was significantly lower compared to the "empty" non-loaded GNPs or control at all three time points (3, 7, 10 days). Nevertheless, a steady increase in the number of cells can be observed, both for the samples and the controls, while the number of dead cells remains at about the same level. The comparison of the living cells shows clear differences between the "empty" GNP, GNP + BMP2 and the controls. The control and the empty GNPs behaved in the same way, whereas the BMP2 loaded ones showed only 79.6 + 3.2% viable cells after 3 days instead of 99.5% for the controls and 99.9% for the empty GNPs (see Figure 5). From day 7 all three samples behaved the same way. Furthermore, agglomerates of GNPs could be detected by their strong fluorescence and subsequent ESEM investigations (see white arrow in Figures 4b and S1).

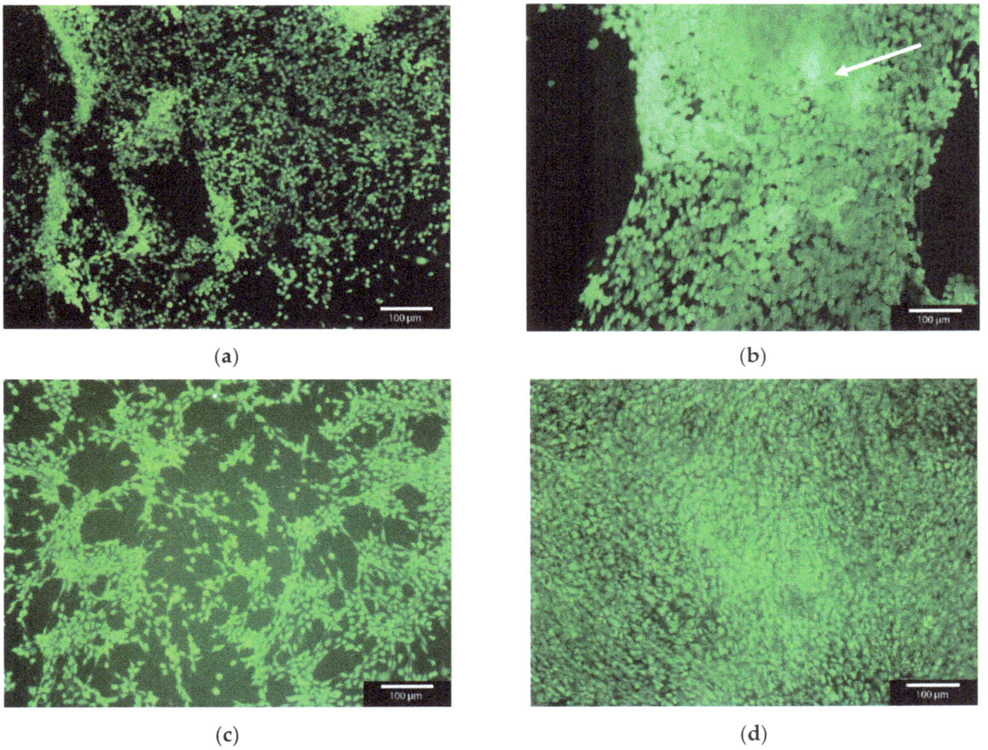

Figure 4. Live/Dead staining of MG-63 cells with GNPs versus MG-63 as control. The agglomerates of GNPs glowed slightly brighter in fluorescence microscopy (see white arrow); with GNPs after (**a**) 3 days; (**b**) 7 days; (**c**) control after 3 days; and (**d**) control after 7 days; green—living cells, red—dead cells. Images taken with Olympus BX-51 Fluorescence microscope, 5× magnification.

In the early stages of the cell culture experiments we found that the cells preferred the GNPs with BMP-2 and agglomerated directly onto the GNPs instead of onto the Thermanox™ cover slips. These were then washed away during the media changes over the duration of the experiment. This also resulted in a lower number of living cells on the Thermanox™ cover slip in Figure 5a.

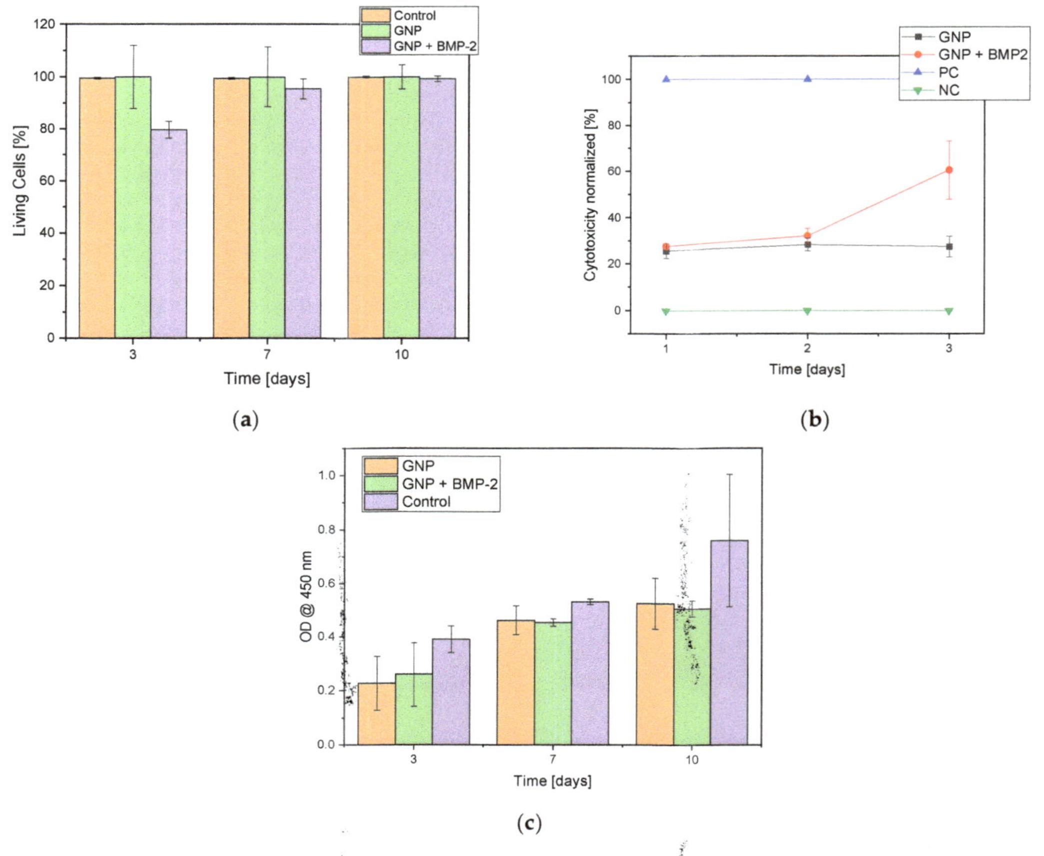

Figure 5. Overview of (**a**) live cells in the live/dead assay comparing "empty" GNPs, GNPs + BMP-2 with control; (**b**) cytotoxicity of negative control (NC = MG-63 cells), positive control (PC = Triton X), MG-63 cells with GNPs and MG-63 cells with GNP + BMP2; (**c**) Cell proliferation assay (WST-I). N = 3.

2.3.2. Cell Proliferation Assay (WST-I)

In the WST-1 assay an increase in cell proliferation was observed over the duration of the experiment. In addition, it was observed that MG-63 cells prefered GNPs with BMP-2 over Thermanox™ cover slips and formed agglomerates with GNP + BMP-2 (see Supplement Figure S1).

2.3.3. Lactate Dehydrogenase (LDH)

In the cytotoxicity assay, the constant cytotoxicity for the empty GNPs was shown to be 27.0 $\pm$ 0.1% compared to the cytotoxicity of GNPs with BMP-2, which increased after 2 days and reached a value of 60.5 $\pm$ 12.6% after 3 days (see Figure 5b).

2.4. Drug Release Experiments

For dual release, clindamycin was released from the ADA-GEL beads and the growth factor BMP-2 or its model substance FITC conjugated protein A from the GNPs within the beads.

2.4.1. Clindamycin Release out of ADA-GEL Beads

Initially only the release of clindamycin will be considered. Figure 6a shows the release of clindamycin-HCl over a period of 28 days. On the last day 3.1 ± 1.9 μg/mL of CLI has been released. This value was clearly above the MIC of 0.06 μg/mL for CLI against *Staph. aureus* [24]. In relation to the weighed amount, clindamycin was released with an initial burst release of 17% on average on the first day. After day 3, the amount released decreased sharply up to day 28 and was even below 1%. If the cumulative release curve was fitted according to Ritgers et al. [25] and the diffusion coefficient (n) was calculated, abnormal release rates were obtained for both the initial range (days 1–3) and the end of the release (days 6–28) (see supplement Figure S2a).

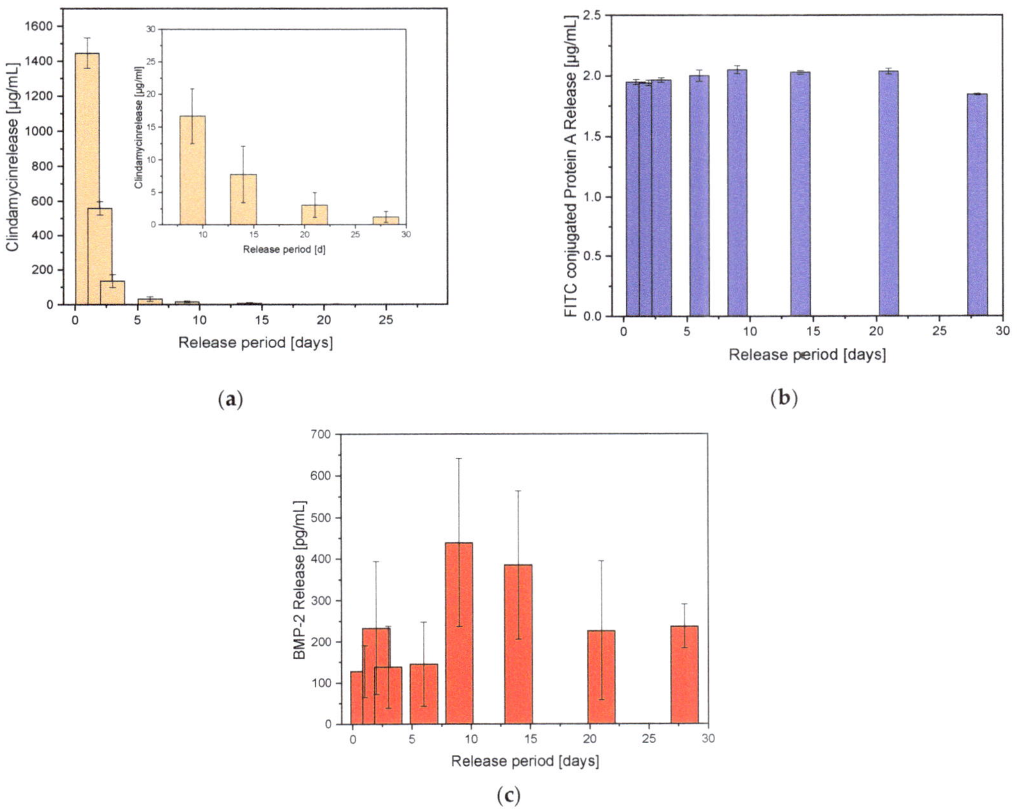

Figure 6. Dual release out of ADA-GEL beads containing GNPs.: (**a**) Clindamycin release over 28 days, the small figure represents an enlargement of the last release times. (**b**) FITC conjugated protein A released out of GNPs within the ADA-GEL beads. (**c**) BMP-2 release out of GNP within the ADA-GEL beads. N = 12.

2.4.2. FITC-Protein a Release out of GNPs within ADA-GEL Beads

In GNPs, the release of FITC conjugated protein-A was constant over the released period. A constant 1.98 ± 0.08 μg/mL was released at the measurement points. Looking at the cumulative release, a Fickian diffusion over the whole period (according to Ritger et al. [25]) with a diffusion coefficient of n = 0.5 was determined (see supplement Figure S2b). Figure 6b shows the FITC conjugated protein A release out of the GNPs within the ADA-GEL beads.

2.4.3. BMP-2 Release out of GNPs within ADA-GEL Beads

Due to the use of GNPs, the expected burst release failed to appear. Instead, the release took place over the entire 28 days. Figure 6c shows the BMP-2 release out of the GNPs within the ADA-GEL beads. The cumulative release showed an almost linear course with similar daily release rates. According to Ritger et al. [25] the diffusion coefficient was 0.96, which corresponded to an anomalous release (see supplement Figure S2c).

3. Discussion

3.1. Characterization of GNP

According to Hathout and Metwally [26], the double de-solvation method was the best method to date for the production of GNPs. According to Coester et al. [14], GNPs produced by this method have a reduced tendency to agglomerate. Nevertheless, the reproducible production of GNPs was the greatest challenge of this work. Since gelatin was available as a heterogeneous mixture with different compositions, it was important to develop the individual production steps in such a way that they can always be carried out consistently. However, even if two approaches were carried out identically, the individual results often varied. The step that was least reproducible was the first de-solvation of the gelatin by the rapid pouring of acetone. This often resulted in chewing gum-like agglutinations of the gelatin, which, due to entrapped LMG, severely impaired the formation of GNPs. Therefore, this step was worked out by letting the acetone run in at the edge of the Erlenmeyer flask, since this was where the best results were achieved and the precipitation was also visually always uniform. Hamarat Sanlier et al. [13] described similar GNPs, but with a size of 300 nm. In their work they showed a similar distribution and arrangement of GNPs. Azizian et al. [23] and Modaresifar et al. [27] also described in their work the production and characterization of nanoparticles and their loading with growth factors such as BSA-bFGF, but on a chitosan basis. Their particles were 266 ± 5 nm in size. Similar to this work, they observed an increase in particle size due to loading. The size of the chitosan nanoparticles increased by 150 nm to 415 ± 9 nm. Feng et al. [28] and Zhai [22] investigated the influence of the pH value of the gelatin solution on the particle size of the GNPs. Feng et al. [28] achieved GNPs with a size of 400–450 nm with pH values of 8–11.

3.2. Inclusion Capacity

Azizian et al. [23] reported an inclusion capacity of the BSA or BSA-bFGF used in their work of 20 ± 1% and 21.3%. The inclusion capacity of the FITC conjugated protein A used in the present study was about twice as high at 46%. This was due to the fact that the FITC conjugated protein A of Invitrogen was already dissolved and stabilized with 10 mg/mL BSA. The BMP-2 was obtained as a lyophilized product from SinoBiological and the solvent was produced by our group without additional protein. The calculated inclusion capacity of the BMP-2 was 99.5%. Similar results were discussed by Poth et al. [29] in their work. They spoke of a complete inclusion of BMP-2 in their chitosan nanoparticles, but unfortunately did not specify an inclusion capacity.

3.3. Biocompatibility

Lee et al. [30] described GNPs in a similar order of magnitude as the ones we used (~150 nm). In the MTT test, they could not detect any significant toxicity based on the GNPs alone. However, they adressed a different application, i.e., the transport of siRNA in which their GNPs were taken up by the cells used. This differed from our study, in which our GNPs were detectable as agglomerates between the cells. Narayanan et al. [31] also described GNPs on a similar scale (70–220 nm) to those we used. They investigated the lymphocyte activation that occurred with any foreign body reaction. With the highest concentration of GNPs (1 mg/mL), no lymphocyte proliferation and therefore no foreign body reaction could be detected. The influence of BMP-2 on cells was sufficiently described in the work of Ribeiro et al. [32]. When looking at the results of cytotoxicity assay, a similar release pattern between the samples "empty" GNP and GNP + BMP2 was observed. Only on the last day of the study GNP + BMP-2 showed increased values compared to the empty GNPs. In a similar study [33], no negative effects have been observed. However, there only alginate and gelatin gels were used for the encapsulation of BMP-2. The values for cytotoxicity were quite high for GNPs with 27% and GNPs with BMP-2 at day 3 with a value of 60%. However, since there was a steady increase in cell numbers in the proliferation experiments and a significant increase in cell numbers in the live/dead, we assume that this is an effect related to the BMP-2 concentration released. It has been sufficiently described that high concentrations of BMP-2 tend to have a negative effect on cell growth [34]. We assume exactly such an effect within the first 72 h, but then BMP-2 levels subsequently returned to normal and led to an increase in cell numbers in the live/dead assay after 7 and 10 days. However, an interaction of GNPs with the cytotoxicity kit used cannot be completely excluded due to the high baseline values which were even observed for the unloaded GNPs. Some authors such as Li et al. [35] use only cell proliferation and live/dead analysis for the detection of cytotoxicity and avoid cytotoxicity investigations using specific kits. Abdelrady et al. [36] described the same method fabricating GNPs as in our work. They used GNPs in the same size range as our GNPs, 160 ± 9.78 nm, and also loaded GNPs with a drug during the 2nd desolvation stage and proved the safety and biocompatibility of empty GNPs.

Minardi et al. [37] also investigated the controlled release of BMP-2. But they used microspheres with a size of 23 ± 3 µm. As a proliferation test, they used a MTT assay where no significant difference between the control and the BMP-2 loaded microspheres could be detected. Similarly, Kim et al. [38] did not find any negative effects of the BMP-2-loaded PLGA nanoparticles on the surface of an HA scaffold when using human MSCs. Tseng et al. [39] used comparable sized GNPs between 180.6 ± 45.7 nm and 230.7 ± 84.6 nm in their work. The manufacturing process was very similar to ours, including the use of glutaraldehyde as a crosslinking agent. The Quick WST-1 test showed no significant difference between the different sizes of GNPs on cell proliferation of human corneal cells. But the proliferation was measured only at one time point, instead of three different time points compared to our work. Kuo et al. [40] also used GNPs in a similar size range (289.7 ± 6.8 nm to 360.0 ± 6.0 nm). They also used a WST-1 assay to determine cell proliferation. However, this test was only performed at two time points after 1 and 2 days only. In contrast to the MG-63 cells used by us, A-549 and H292 cell lines were used in their work.

3.4. Drug Release Experiments

The release experiments from the ADA-GEL beads showed similar results in terms of the release of CLI to those we have published in the past [41,42]. However, the use of the ADA-GEL gel resulted in an increase in burst release compared to our previously published work. By using alginate, the burst release could be reduced to 35% of the clindamycin amount weighed in. The fact that the ADA-GEL gel structure was different from that of the pure alginate changed the burst release and the total release. Nevertheless, the application of the drug release system, in another part of the project, showed very good antimicrobial efficacy [42]. Sarker et al. [43] also described in their work differences in the mechanical properties of the ADA-GEL gels compared to pure alginate gels. The release from the GNP beads showed a continuous release with a reduced burst release, as we intended. In comparison to other authors like Modaresifar et al. [27], where 75% of the loaded protein was released within the first 2 h, or Azizian et al. [23], where 80% of the release was completed after 4 h, we could show a continuous release over 4 weeks from the ADA-GEL beads with GNPs. However, we used bidest. water for our release experiments and Modaresifar et al. [27] and Azizian et al. [23] used PBS. In addition, the way we sampled (complete exchange of liquid) was different from sampling only a limited volume. Thus, we achieved higher values due to a higher diffusion pressure compared to authors who used only a part of the volume for the analysis.

However, the concentrations released were also very low in our work. In the case of BMP-2, very low concentrations are desirable due to the high potential for side effects [44]. Kim et al. [38] used BMP-2 loaded PLGA nanoparticles at the surface of a HA scaffold for their experiments. At 544 ± 39 nm the nanoparticles were significantly larger than those used by us. They were also able to demonstrate a release over 30 days in which 66% of the BMP-2 was released.

4. Conclusions

In the present work, GNPs were shown to be well suited for the release of BMP-2 or other large molecules such as FITC-conjugated protein A. A constant release (~2 µg/mL) corresponding to Fick's diffusion was shown. In addition, dual release of clindamycin and BMP-2 from ADA-GEL beads was shown to be possible over an extended period of time (up to 4 weeks) with antimicrobial effective concentrations (CLI 25-fold above the MIC). In addition, the tissue compatibility of GNPs was demonstrated using various biocompatibility tests (L/D, WST-I, LDH) in cell culture with MG-63 cells.

5. Materials and Methods

5.1. Reagents and Materials

Gelatin type A (GEL), 175 Bloom, Alginic Acid for microbiological applications (Art.No. 71238), Ethylen glycol (Art.No. 324558) and Clindamycin-HCl (Art.No. PHR1159) (CLI) were obtained from Sigma-Aldrich (St. Louis, MO, USA). Acetone, HCl 0.1 M, Acetonitrile (HPLC grade) and Calcium chloride were obtained from Carl Roth (Karlsruhe, Germany). FITC conjugated protein A (molar ratio FITC/ protein A = 1.8; 2.5 mg/mL Protein A) was obtained by Invitrogen (Thermo Fisher Scientific, Waltham, MA, USA) and BMP-2 and the Human BMP-2 ELISA kit (KIT10426) from Sino Biological (Sino Biological, Inc., Beijing, China). Glutaraldehyde (GTA) was obtained from EMS (Electron Microscopy Sciences, Hatfield, PA, USA), gelatin type A 300 Bloom was kindly provided by GELITA (GELITA AG, Eberbach, Germany). Gelatin and Alginic acid were sterilized by means of a plasma sterilization process.

5.2. Manufacturing of the Gelatin Nanoparticles (GNPs)

The production of the GNPs was based on the protocol of Coester [14]. For the precipitation of the GEL, acetone was used according to Zhai [22] to produce GNPs with a diameter of 100 nm. For the production of GNPs, gelatin type A, 175 g Bloom was used. GEL was weighed into a 100 mL Erlenmeyer flask (Kern precision balance PCB250-2, Balingen, Germany) and a 5% aqueous GEL solution was mixed with aqua bidest. (0.625 g gelatin to 12.5 mL bidest. water). The GEL solution was homogenized on the heated magnetic stirrer (RCT basic, IKA®-Werke GmbH & CO. KG, Staufen, Germany) for 30 min, at 50 °C and 300 rpm. Parafilm was used to seal the Erlenmeyer flask. The first de-solvation was performed to separate the low molecular gelatin (LMG) from the high molecular gelatin (HMG). For this purpose, the temperature was switched off, the thermometer and magnetic stirring rod were removed and acetone (12.5 mL) was added to the gelatin solution evenly and rapidly at the slowest possible rate using a 25 mL disposable pipette, with a pipetboy (Integra Pipetboy 2, INTEGRA Biosciences AG, Zizers, Switzerland). It is important that the tip of the pipette touches the rim of the Erlenmeyer flask and that acetone flows into the solution via the inside of the glass flask. If acetone hits the GEL solution from above or if turbulence occurs in the vessel, flocculation and clumping of the GEL may occur. However, a uniform hydrogel-like, transparent precipitation of the HMG was desired, as the precipitated GEL lumps most probably also contained LMG, which could not form GNP. The Erlenmeyer flask was left standing for exactly 5 min, covered with parafilm. The supernatant was decanted and carefully rinsed with 1–2 mL bidest. water to remove any remaining LMG. Thermometer and magnetic stirring fish were replaced in the flask. HMG was solved in H_2O. (12.5 mL) at 40 °C, 45 °C and 50 °C, 500 rpm for 30 min on the heated magnetic stirrer. With 0.1 M HCl the pH was adjusted to 2, 2.5 and 3 using a pH meter (Mettler Toledo Education line EL20, Schwerzenbach, Switzerland). The acetone was added drop by drop (40 mL) using a 100 mL dropping funnel. The connection between the dropping funnel and the Erlenmeyer flask was sealed with parafilm to prevent acetone from escaping. The opening of the funnel was also sealed and only shortly before the dripping process small openings were cut with a scalpel. If acetone escaped, no GNPs were produced or a large amount of acetone was needed for the particles to form. During the addition of acetone, the stirring speed was increased to 650 rpm. The formation of the nanoparticles can be recognized by the fact that the suspension has a bluish opalescent (Tyndall effect) and slightly turbid appearance. The nanoparticle suspension was stirred for another 10 min at 400 rpm, with the heater switched off, after the addition of acetone. To stabilize the resulting particles, 8% GTA (400 µL) was added. It is important here that the GTA is added drop by drop and over a period of at least 5 min, otherwise irreversible aggregation and cross-linking between the GNPs may occur. The nanoparticle suspension was stirred for 12 h, at room temperature and 350 rpm. The purification was performed by centrifugation (Avanti JXN-26, Beckman Coulter, Krefeld, Germany) with a fixed angle rotor (JLA-16.250, Beckman Coulter, Krefeld, Germany). Centrifuged at 15,000× *g*, 20 °C for 15 min, the supernatant was decanted and the pellet was resuspended in H_2O by vortexing (Mini Vortex, Heathrow Scientific, Vernon Hills, IL, USA). The washing step was repeated twice, this time centrifuging at 16,000× *g*, 20 °C for 10 min. The purified GNP was dried with a lyophilisator (FreeZone 2.5 plus, Labconco, Kansas City, MO, USA), white powder was obtained when all of the water was sublimated.

Manufacturing of FITC-Protein A-Loaded GNPs and BMP-2 Loaded GNPs

The FITC-protein A GNPs were prepared as described, the addition of the FITC conjugated protein A was performed shortly before the second de-solvation step. It is important to note that FITC is light sensitive, therefore, the addition of the FITC-protein A was carried out in the absence of light. Concentrations of 1 and 2% *v*/*v* FITC-protein A in the GEL solution were prepared. For the final release experiments, the BMP-2-GNP (1 and 2% *v*/*v*) were prepared as described above, just like the FITC-protein A, the BMP-2 was added directly before the second de-solvation step. Therefore 0.4 mL the BMP-2 stock solution with 2.5 µg/mL was diluted with 0.6 mL bidest. water to reach a final concentration of 1 µg/mL. This solution was directly added before the second de-solvation step to the 125 mL solution (containing 0.625 g GEL).

5.3. *Characterization of the GNP*

5.3.1. Characterization of GNP by ESEM

Size, structure and overall morphology of the GNP samples (empty, FITC-protein A and BMP-2 loaded), were examined using ESEM (FEI quanta 250 FEG, FEI, Hilsboro, OR, USA). The impact of variations in the production method (temperature, pH, stirring speeds and times of addition of the crosslinker) on the formation of GNPs, size, shape and aggregation were investigated. For this purpose, double-sided adhesive polycarbonate-based conductive tabs were glued to pin sample plates made of aluminium. The freeze-dried samples in powder form could thus be fixed on the adhesive surface and were sputtered with gold before ESEM measurement (JFC-1200 fine coater, Jeol, Freising, Germany) to increase the conductivity and stability of the samples in high vacuum.

5.3.2. Determination of the Inclusion Capacity

To assess how much protein was actually present in the nanoparticles, the supernatant obtained after centrifugation was measured with the fluorimeter (Ensight Multimode Plate Reader, PerkinElmer, Rodgau, Germany). The Kaleido software was used to obtain the data for quantitative analysis. A calibration curve was generated based on levels within the supernatant, which could then be used to approximate protein content within the solvent. Acetone and water in a ratio of 4:1 (*v*/*v*), plus a quarter of the amount of GTA were used. The greatest uncertainty was found for the amount of GTA, since the amount of unreacted GTA after the crosslinking process was unknown. The difference between the amount of FITC-protein A used and the amount in the supernatant was defined as the amount included in the GNPs. In order to quantify how much BMP-2 was present in the produced GNPs, the supernatant obtained after the first centrifugation was analyzed using an ELISA. The evaluation of the color reaction was performed with the UV/Vis spectrometer (Spectrostar Nano, BMG Labtech, Ortenberg, Germany). A calibration curve was generated using the BMP-2 standard of the kit. The difference between the amount of BMP-2 used and the amount in the supernatant was defined as included BMP-2 in the GNPs.

$$IC = \frac{m(FITC_{total}) - m(FITC_{supernatant})}{m(FITC_{total})} \quad (1)$$

IC: Inclusion capacity;
$m(FITC_{supernatant})$: Mass of FITC-protein A at supernatant;
$m(FITC_{total})$: Total mass of FITC-protein A.

5.4. *Biocompatibility*

In order to show that the GNPs were biocompatible with living cells and that after purification no cytotoxic residues such as acetone or GTA from the production steps were present, a cell culture experiment was carried out. MG-63 cells (ATCC CRL 1427) were used. For each sample 50,000 cells were placed in a 12-well plate on glass plates, which were previously coated with polylysine. The cells were cultivated overnight with Dulbecco's Modified Eagle Medium (DMEM F12 containing F12 nutrient and the additions of 1% penicillin/streptomycin (P/S, Sigma Aldrich (now Merck), Darmstadt, Germany) and 10% fetal bovine serum (FBS, Merck, Darmstadt, Germany) in a New Brunswick Galaxy 170 R incubator (Eppendorf, Hamburg, Germany) at 37 °C and a CO_2 saturation of 5%. The next day, a production batch (from originally 1.25 g gelatin) GNP were washed in 1 mL ethanol (60%) to prevent cell culture contamination. The GNP were dissolved in the culture medium and pipetted to the cells. Nunc™ Thermanox™ Coverslip (Thermo Fisher Scientific, Waltham, MA, USA) membranes were used as a control. For each time point at least three samples were used and all experiments were repeated at least three times.

5.4.1. Live/Dead Assay

A total of 3 samples for each day and one blank sample (without GNP) for day 3, 7 and 10 were used for the live/dead assay. For the live/dead assay (L/D), the cells were stained with Calcein-AM and EthD-III (out of Live/Dead Cell Staining Kit II (PromoCell, Heidelberg, Germany)) on the respective days and observed under the fluorescence microscope (Olympus BX 51, Olympus, Hamburg, Germany), the evaluation was performed with the Stream Motion Software Version 1.7.1 from Olympus. Living cells exhibited a green fluorescence under blue light, and dead cells a red fluorescence. In a further experiment the influence of the released BMP-2 on the MG-63 cells was investigated. The L/D assay was repeated with GNPs loaded with BMP-2. This was followed by incubation in the incubator for the various experiments.

5.4.2. Cell Proliferation Assay

Thermanox™ Coverslip membranes were used as a control. All samples and controls were equally covered with 50,000 cells in 200 µL. The cells were incubated for 4 h at 37 °C with a CO_2 saturation of 5% in the incubator for 3, 7 and 10 days. At the end of this period, 400 µL of the DMEM-F12 complete medium were added to each sample and incubated. After 24 h at day 0, 100 µL resuspended GNP + BMP2 were added. A medium change with the DMEM-F12 with the 10% FBS and 1% P/S additives was performed for days 7 and 10. For each WST evaluation the medium was aspirated, and the wells were washed three times with PBS. The samples and the Thermanox™ Coverslips were then transferred to a new well, and then 400 µL of the DMEM-F12 phenol red free (Art. No. 11039-021, Gibco, Grand Island, NE, USA) with the 1% P/S and 1% fetal bovine serum (FBS, Merck,

Darmstadt, Germany) additives were added to the wells with the sample. 400 µL of the medium were added to the previously used empty sample wells, Positive Control (C + R), Empty Control Well (C+), and Blank. The blank contained only the DMEM medium without phenol red and was measured to account for background absorption. 10% WST reagent (Art. No. 05015944001, Roche, Basel, Switzerland) was added to the corresponding volume of medium. Thus, 250 µL WST were added to the sample wells, and 40 µL were added to the old wells, the blank wells, and the positive control (C + R). This was incubated in an incubator at 37 °C for 2 h. After this time, the liquids were transferred into a 96 well plate. 100 µL of each solution were added to three separate wells. The absorption was then measured at 450 nm using a Spectrostar Nano microplate reader (BMG Labtech, Ortenberg, Germany). In a further experiment the influence of the released BMP-2 on the MG-63 cells was investigated. The WST-I was repeated with GNPs loaded with BMP-2.

5.4.3. Lactate Dehydrogenase (LDH) Assay

Each experiment assessed three Thermanox™ cover slips, a negative control (cells only), a positive control (Triton X), and a blank to account for background absorbance in the ELISA reader. The experiments were repeated at least three times. A 200 µL cell solution containing 50,000 cells was seeded onto each scaffold, and a 100 µL cell solution containing 50,000 cells was seeded onto the Thermanox™ cover slips. One well was left empty for use as a blank. The well plate was placed in an incubator at 37 °C with 5% CO_2 for 4 h. Following incubation, 400 µL of DMEM-F12 phenol red free with the 1% P/S and 1% FBS additives were added into the sample-wells and control-wells. Since FBS itself contains LDH, a concentration of 10% in the medium might have triggered background absorption. Therefore, only a concentration of 1% FBS was added to the medium. For the positive controls, 1% Triton × 100 (Art. No. X100, Sigma Aldrich, Saint Louis, MO, USA) was added to the DMEM-F12 medium with 1% P/S and 1% FBS to kill the cells. The LDH experiments were carried out at 24, 48 and 72 h following seeding and the same procedure was repeated at each interval: Three 100 µL samples were taken from each well into a 96 well plate. An LDH reagent (100 µL) was added to each well in use, and the plate was incubated in darkness at room temperature for 30 min. Following incubation, the plate was placed in a Spectrostar Nano microplate reader (BMG Labtech, Ortenberg, Germany), and absorbance was measured at a λ of 490 nm with a reference λ of 600 nm. In a further experiment the influence of the released BMP-2 on the MG-63 cells was investigated. The LDH was repeated with GNPs loaded with BMP-2.

5.5. ADA-GEL Hydrogel

5.5.1. Preparation of Alginate-di-Aldehyde (ADA)

ADA was produced according to the method developed by Sarker et al. [45]. For this purpose, sodium alginate was dissolved in ethanol (99.8%) and sodium periodate dissolved in aqua bidest. was added. For the oxidation reaction, stirring was carried out for 6 h under exclusion of light (beaker was wrapped with aluminum foil). The reaction was stopped with ethylene glycol and stirring was continued for 30 min. The ADA was dialysed for 7 days against aqua bidest. to remove any remaining sodium periodate by using the dialysis system Spectra/Por (Repligen, Boston, MA, USA) with standard RC dialysis membranes (6–8 kD MWCO). The bidest. water was changed twice a day. After dialysis, the ADA was dried in a lyophilisator (FreeZone 2.5, Labconco, Kansas City, MO, USA) for another seven days.

5.5.2. Crosslinking ADA and GEL

The cross-linking of the aldehyde groups produced by oxidation with the amino groups of the gelatin always took place shortly before the gel was used and is described in the following experiments.

5.6. ADA-GEL Beads with Drugs

5.6.1. ADA-GEL Beads with Clindamycin and FITC-Protein A-Containing GNPs

A 5% *w*/*v* ADA solution was prepared by homogenizing freeze-dried ADA with aqua bidest. water for several hours on a magnetic stirrer in a beaker. Meanwhile, a 5% GEL solution was prepared from GEL A, 300 g Bloom (GELITA AG, Eberbach, Germany) with aqua bidest. water at 37 °C on the magnetic stirrer. Clindamycin was added to the gelatin after homogenization to achieve a final concentration in the beads of 50 mg/mL. Since clindamycin is light sensitive, the beaker was wrapped in aluminum foil. The amount of FITC-protein A GNPs from 2 production batches of 1.25 g each of initial GEL mass was added. GNP prepared with 2% FITC-protein A was used. The GNPs were added to the homogenized ADA. The clindamycin-containing gelatin and the ADA containing GNP were mixed together (1:1) and carefully dripped with the beaker into a 30 mM calcium chloride solution [19]. In addition to the active substance containing beads, hydrogel beads were prepared as blank samples without drugs.

5.6.2. ADA-GEL Beads with CLI and BMP-2 Containing GNP

The beads were prepared with CLI and BMP-2 containing GNPs as described in 2.6.1. Before adding the gelatin, GNPs with BMP-2 from a batch of 1.25 g of gelatin (twice as much as described in 2.2) was added to the ADA. After the addition of the gelatin containing CLI, the mixture was homogenized under light for 30 s. The beads were then prepared by dropping the ADA-GEL into 30 mM $CaCl_2$ solution.

5.7. Drug Release Experiments

All drug release experiments were performed with a minimum of 10 samples and repeated at least three times.

5.7.1. Drug Release from ADA-GEL Beads

1 g of the CLI loaded beads were stored, after being weighed with a precision balance (Secura 125-1CEU, Sartorius, Göttingen, Germany), with 3 mL bidest. water in 5 mL Eppendorf Tubes, at 37 °C for 28 days in an oven (Memmert IN 75, Schwabach, Germany). After 1, 2, 3, 6, 9, 14, 21 and 28 days, samples were taken and the liquid was completely removed and replaced by aqua bidest. The obtained liquid was frozen at −20 °C until analysis by HPLC.

5.7.2. Dual Drug Release from ADA-GEL Beads

The dual release (CLI out of the ADA-GEL beads, BMP-2 out of GNPs within the ADA-GEL beads) was carried out here according to the same principle as described above. 1 g of loaded ADA-GEL beads was placed in a 5 mL Eppendorf tube with 3 mL bidest. water. The release was tested at 37 °C for 28 days. After 1, 2, 3, 6, 9, 14, 21 and 28 days, samples were taken, the complete liquid was removed once again and replaced by bidest. water. The samples were frozen at −20 °C in the same way as the release tests from the beads until they were analyzed by HPLC or fluorimeter (FITC-prot A) and ELISA kit (BMP-2).

5.7.3. Quantitative Analysis

The released clindamycin from the ADA gelatin beads was quantitatively determined by HPLC. The HPLC System (Shimadzu, Kyoto, Japan) consisting of 2 Nexera XR LC-20AD pumps and a SIC-30AC autosampler, CTO 20 AC column oven, DGU-20A5R Degasser, SPD-M20A PDA detector, RF 20A fluorescence detector and a CBM-20A controller. A reversed-phase column of butyl-modified silica gel was used as a separation column (NUCLEOSIL 300-5 C4, 5 μm, 250 × 4.6 mm, Macherey-Nagel, REF 761989.30, Düren, Germany). An acetonitrile/sodium hydrogen phosphate solvent with pH of 3.5 at a the ratio 29:71 was used according to Batzias et al. [46]. Flow rate was 0.66 mL/min at 25 °C. Clindamycin was eluted after 3.5 min and registered by a PDA detector at 193 nm. The released FITC-protein A concentration was measured with a fluorimeter (Perkin Elmer EnSight, Waltham, MA, USA) at 490 nm excitation and 525 nm emission. The determination of the BMP-2 concentration was performed with an ELISA kit from Sino Biological according to their protocol. The 96 well plate already coated with a BMP-2 antibody by the manufacturer was first washed 3 times with 200 µL wash buffer. 100 µL of the samples were then pipetted into the wells and incubated at room temperature for two hours. The washing step was repeated 3 times before 100 µL of the detection antibody against BMP-2 were added conjugated with horseradish peroxidase. This was incubated for one hour at room temperature. Washing was repeated 3 times before the dye solution was added. After 20 min 50 µL of the Stop Solution were added to stop the color reaction. Using a UV/Vis spectrometer (SpectroStar Nano, BMG Labtech, Ortenberg, Germany), the color reaction was quantitatively evaluated using a calibration curve (2500 ... 31.5 pg/mL) at λ = 450 nm.

5.7.4. Kinetics Model

The release kinetics of CLI, FITC, and BMP-2 were fitted to a cumulative diagram according to the following Ritger [47] exponential relationship:

$$\frac{M_t}{M_\infty} = kt^n \quad (2)$$

where M_t/M_∞ = fractional solute release; t = release time; k = a constant; and n = diffusional exponent characteristic of the release mechanism. In the case of pure Fickian release, the exponent n has limited values 0.50, 0.45 and 0.43 for release from slabs, cylinders and spheres.

5.8. Statistical Analysis

Data was expressed as mean values ± standard deviation of the mean and analyzed by one-way analysis of variance (ANOVA). The level of statistical significance was set at $p < 0.05$. For statistical calculations, Origin 2020 Professional SR1 (OriginLab, Northampton, MA, USA) was used.

Supplementary Materials: The following supporting information can be downloaded at: https://www.mdpi.com/article/10.3390/gels8060365/s1. Figure S1: Image (live/dead staining) of agglomerated GNP-BMP2 with MG-63 cells; Figure S2: Cumulative releases, fit according to Ritger et al. [25]; (a) Clindamycin; (b) Conjugated FITC-protein A, with a diffusion coefficient n = 0.5 for Fickian' diffusion; (c) BMP-2 in relation to the BMP-2 amount used. According to previous published work [41], the fitting for the CLI release was split into a beginning part (n_B) and final part (n_F), both with anomalous diffusion (n_B = 0.35 and n_F = 0.01).

Author Contributions: M.S. and R.S. conceived and designed the experiments; M.S., L.R., P.S. and S.S., performed the experiments; L.R. sterilized and characterized the ADA-GEL Gels, M.S., L.R., P.S. and S.S. analyzed the data; M.S. and R.S. contributed reagents/materials/analysis tools; M.S. and S.S. wrote the paper. All authors have read and agreed to the published version of the manuscript.

Funding: This research was funded by German Research Foundation (DFG) grant number 388988890. The article processing charge was funded by the Baden-Wuerttemberg Ministry of Science, Research and Art and the University of Freiburg in the funding programme Open Access Publishing.

Institutional Review Board Statement: Not applicable.

Informed Consent Statement: Not applicable.

Data Availability Statement: The data presented in this study are available on request from the corresponding author.

Acknowledgments: The authors would like to thank GELITA for providing the gelatin 300 bloom. We would also like to thank Logan Poehlman, Isabelle Caseley and Melanie Lynn Hart for proofreading the manuscript.

Conflicts of Interest: The authors declare no conflict of interest.

References

1. Lew, D.P.; Waldvogel, F.A. Osteomyelitis. *Lancet* **2004**, *364*, 369–379. [CrossRef]
2. Tiemann, A.; Hugo, H. Osteomyelitis. *Orthopädie Unf. Up2date* **2020**, *15*, 67–82. [CrossRef]
3. Schmelz, A.; Kinzl, L.; Einsiedel, T. Osteitis. Infections of the locomotive system. *Unfallchirurg* **2007**, *110*, 1039–1058. [CrossRef] [PubMed]
4. Suedkamp, N.P.; Barbey, N.; Veuskens, A.; Tempka, A.; Haas, N.P.; Hoffmann, R.; Tscherne, H. The incidence of osteitis in open fractures: An analysis of 948 open fractures (a review of the hannover experience). *J. Orthop. Trauma* **1993**, *7*, 473–482. [CrossRef]
5. Zimmerli, W.; Ryu, S.Y.; Patel, R.; Landersdorfer, C.B.; Bulitta, J.B.; Soergel, F.; Richards, R.; Moriarty, F. *Bone and Joint Infections from Microbiology to Diagnostics and Treatment*; John Wiley & Sons Inc.: Oxford, UK, 2015; ISBN 978-1-118-58177-3.
6. Zimmerli, W.; Sendi, P. Orthopaedic biofilm infections. *APMIS* **2017**, *125*, 353–364. [CrossRef]
7. Wengler, A.; Nimptsch, U.; Mansky, T. Hip and knee replacement in germany and the USA. *Dtsch. Arztebl. Int.* **2014**, *111*, 407–416. [CrossRef]
8. Farid, Y.; Schettino, M.; Kapila, A.K.; Hamdi, M.; Cuylits, N.; Wauthy, P.; Ortiz, S. Decrease in surgical activity in the COVID-19 pandemic: An economic crisis. *Br. J. Surg.* **2020**, *107*, e300. [CrossRef]
9. Destatis. *Gesundheit—Fallpauschalenbezogene Krankenhausstatistik (Drg-Statistik) Operationen und Prozeduren der Vollstationären Patientinnen und Patienten in Krankenhäusern (4-Steller)*; Statistisches Bundesamt (Destatis): Wiesbaden, Germany, 2020.
10. Singh, J.A.; Yu, S.; Chen, L.; Cleveland, J.D. Rates of total joint replacement in the united states: Future projections to 2020–2040 using the national inpatient sample. *J. Rheumatol.* **2019**, *46*, 1134–1140. [CrossRef]
11. Steckelberg, J.M.; Osmon, D.R. Prosthetic joint infections. In *Infections Associated with Indwelling Medical Devices*; ASM Press: Washington, DC, USA, 2000; pp. 173–209. [CrossRef]
12. Oliveira, W.F.; Silva, P.M.S.; Silva, R.C.S.; Silva, G.M.M.; Machado, G.; Coelho, L.C.B.B.; Correia, M.T.S. Staphylococcus aureus and staphylococcus epidermidis infections on implants. *J. Hosp. Infect.* **2018**, *98*, 111–117. [CrossRef]
13. Hamarat Sanlıer, S.; Yasa, M.; Cihnioglu, A.O.; Abdulhayoglu, M.; Yılmaz, H.; Ak, G. Development of gemcitabine-adsorbed magnetic gelatin nanoparticles for targeted drug delivery in lung cancer. *Artif. Cells Nanomed. Biotechnol.* **2016**, *44*, 943–949. [CrossRef]
14. Coester, C.J.; Langer, K.; Von Briesen, H.; Kreuter, J. Gelatin nanoparticles by two step desolvation a new preparation method, surface modifications and cell uptake. *J. Microencaps.* **2000**, *17*, 187–193.
15. Renner, L.; Perka, C.; Trampuz, A.; Renz, N. Treatment of periprosthetic infections. *Chirurg* **2016**, *87*, 831–838. [CrossRef] [PubMed]
16. Tong, S.Y.C.; Davis, J.S.; Eichenberger, E.; Holland, T.L.; Fowler, V.G. Staphylococcus aureus infections: Epidemiology, pathophysiology, clinical manifestations, and management. *Clin. Microbiol. Rev.* **2015**, *28*, 603–661. [CrossRef] [PubMed]
17. Pollitt, E.J.G.; Szkuta, P.T.; Burns, N.; Foster, S.J. Staphylococcus aureus infection dynamics. *PLoS Path.* **2018**, *14*, e1007112. [CrossRef]
18. Kayser, F.H.; Böttger, E.C. *Medizinische Mikrobiologie*; Georg Thieme Verlag: Stuttgart, Germany; New York, NY, USA, 2005; ISBN 978-3-13-1514431.
19. Distler, T.; McDonald, K.; Heid, S.; Karakaya, E.; Detsch, R.; Boccaccini, A.R. Ionically and enzymatically dual cross-linked oxidized alginate gelatin hydrogels with tunable stiffness and degradation behavior for tissue engineering. *ACS Biomater. Sci. Eng.* **2020**, *6*, 3899–3914. [CrossRef]
20. Klemm. The use of antibiotic-containing bead chains in the treatment of chronic bone infections. *Clin. Microbiol. Infect.* **2000**, *7*, 28–31. [CrossRef]
21. Diez-Pena, E.; Frutos, G.; Frutos, P.; Barrales-Rienda, J.M. Gentamicin sulphate release from a modified commercial acrylic surgical radiopaque bone cement. I. Influence of the gentamicin concentration on the release process mechanism. *Chem. Pharm. Bull.* **2002**, *50*, 1201–1208. [CrossRef]
22. Zhai, X. Gelatin Nanoparticles & Nanocrystals for Dermal Delivery. Ph.D. Thesis, Freie Universität Berlin, Berlin, Germany, 2014.
23. Azizian, S.; Hadjizadeh, A.; Niknejad, H. Chitosan-gelatin porous scaffold incorporated with chitosan nanoparticles for growth factor delivery in tissue engineering. *Carbohydr. Polym.* **2018**, *202*, 315–322. [CrossRef]
24. Reeves, D.S.; Holt, H.A.; Phillips, I.; King, A.; Miles, R.S.; Paton, R.; Wise, R.; Andrews, J.M. Activity of clindamycin against staphylococcus aureus and staphylococcus epidermidis from four uk centres. *J. Antimicrob. Chemother.* **1991**, *27*, 469–474. [CrossRef]
25. Ritger, P.L.; Peppas, N.A. A simple equation for description of solute release ii. Fickian and anomalous release from swellable devices. *J. Control. Release* **1987**, *5*, 37–42. [CrossRef]
26. Hathout, R.M.; Metwally, A.A. Gelatin nanoparticles. In *Pharmaceutical Nanotechnology: Basic Protocols*; Weissig, V., Elbayoumi, T., Eds.; Springer: New York, NY, USA, 2019; pp. 71–78.
27. Modaresifar, K.; Hadjizadeh, A.; Niknejad, H. Design and fabrication of gelma/chitosan nanoparticles composite hydrogel for angiogenic growth factor delivery. *Artif. Cells Nanomed. Biotechnol.* **2018**, *46*, 1799–1808. [CrossRef] [PubMed]
28. Feng, X.; Dai, H.; Ma, L.; Yu, Y.; Tang, M.; Li, Y.; Hu, W.; Liu, T.; Zhang, Y. Food-grade gelatin nanoparticles: Preparation, characterization, and preliminary application for stabilizing pickering emulsions. *Foods* **2019**, *8*, 479. [CrossRef] [PubMed]
29. Poth, N.; Seiffart, V.; Gross, G.; Menzel, H.; Dempwolf, W. Biodegradable chitosan nanoparticle coatings on titanium for the delivery of bmp-2. *Biomolecules* **2015**, *5*, 3–19. [CrossRef] [PubMed]

30. Lee, S.J.; Yhee, J.Y.; Kim, S.H.; Kwon, I.C.; Kim, K. Biocompatible gelatin nanoparticles for tumor-targeted delivery of polymerized sirna in tumor-bearing mice. *J. Control. Release* **2013**, *172*, 358–366. [CrossRef] [PubMed]
31. Narayanan, D.; Geena, M.G.; Lakshmi, H.; Koyakutty, M.; Nair, S.; Menon, D. Poly-(ethylene glycol) modified gelatin nanoparticles for sustained delivery of the anti-inflammatory drug ibuprofen-sodium: An in vitro and in vivo analysis. *Nanomed. Nanotechnol. Biol. Med.* **2013**, *9*, 818–828. [CrossRef] [PubMed]
32. Ribeiro, F.O.; Gómez-Benito, M.J.; Folgado, J.; Fernandes, P.R.; García-Aznar, J.M. In silico mechano-chemical model of bone healing for the regeneration of critical defects: The effect of bmp-2. *PLoS ONE* **2015**, *10*, e0127722. [CrossRef]
33. Kissling, S.; Seidenstuecker, M.; Pilz, I.H.; Suedkamp, N.P.; Mayr, H.O.; Bernstein, A. Sustained release of rhbmp-2 from microporous tricalciumphosphate using hydrogels as a carrier. *BMC Biotechnol.* **2016**, *16*, 44. [CrossRef]
34. Zara, J.N.; Siu, R.K.; Zhang, X.; Shen, J.; Ngo, R.; Lee, M.; Li, W.; Chiang, M.; Chung, J.; Kwak, J.; et al. High doses of bone morphogenetic protein 2 induce structurally abnormal bone and inflammation in vivo. *Tissue Eng. Part A* **2011**, *17*, 1389–1399. [CrossRef]
35. Li, D.; Chen, K.; Duan, L.; Fu, T.; Li, J.; Mu, Z.; Wang, S.; Zou, Q.; Chen, L.; Feng, Y.; et al. Strontium ranelate incorporated enzyme-cross-linked gelatin nanoparticle/silk fibroin aerogel for osteogenesis in ovx-induced osteoporosis. *ACS Biomater. Sci. Eng.* **2019**, *5*, 1440–1451. [CrossRef]
36. Abdelrady, H.; Hathout, R.M.; Osman, R.; Saleem, I.; Mortada, N.D. Exploiting gelatin nanocarriers in the pulmonary delivery of methotrexate for lung cancer therapy. *Eur. J. Pharm. Sci.* **2019**, *133*, 115–126. [CrossRef]
37. Minardi, S.; Fernandez-Moure, J.S.; Fan, D.; Murphy, M.B.; Yazdi, I.K.; Liu, X.; Weiner, B.K.; Tasciotti, E. Biocompatible plga-mesoporous silicon microspheres for the controlled release of bmp-2 for bone augmentation. *Pharmaceutics* **2020**, *12*, 118. [CrossRef] [PubMed]
38. Kim, B.-S.; Yang, S.-S.; Kim, C.S. Incorporation of bmp-2 nanoparticles on the surface of a 3d-printed hydroxyapatite scaffold using an ε-polycaprolactone polymer emulsion coating method for bone tissue engineering. *Colloids Surf. B Biointerfaces* **2018**, *170*, 421–429. [CrossRef] [PubMed]
39. Tseng, C.-L.; Chen, K.-H.; Su, W.-Y.; Lee, Y.-H.; Wu, C.-C.; Lin, F.-H. Cationic gelatin nanoparticles for drug delivery to the ocular surface: In vitro and in vivo evaluation. *J. Nanomater.* **2013**, *2013*, 238351. [CrossRef]
40. Kuo, W.-T.; Huang, J.-Y.; Chen, M.-H.; Chen, C.-Y.; Shyong, Y.-J.; Yen, K.-C.; Sun, Y.-J.; Ke, C.-J.; Cheng, Y.-H.; Lin, F.-H. Development of gelatin nanoparticles conjugated with phytohemagglutinin erythroagglutinating loaded with gemcitabine for inducing apoptosis in non-small cell lung cancer cells. *J. Mater. Chem. B* **2016**, *4*, 2444–2454. [CrossRef] [PubMed]
41. Seidenstuecker, M.; Ruehe, J.; Suedkamp, N.P.; Serr, A.; Wittmer, A.; Bohner, M.; Bernstein, A.; Mayr, H.O. Composite material consisting of microporous β-tcp ceramic and alginate for delayed release of antibiotics. *Acta Biomater.* **2017**, *51*, 433–446. [CrossRef]
42. Kuehling, T.; Schilling, P.; Bernstein, A.; Mayr, H.O.; Serr, A.; Wittmer, A.; Bohner, M.; Seidenstuecker, M. A human bone infection organ model for biomaterial research. *Acta Biomater.* **2022**, *144*, 230–241. [CrossRef]
43. Sarker, B.; Rompf, J.; Silva, R.; Lang, N.; Detsch, R.; Kaschta, J.; Fabry, B.; Boccaccini, A.R. Alginate-based hydrogels with improved adhesive properties for cell encapsulation. *Int. J. Biol. Macromol.* **2015**, *78*, 72–78. [CrossRef]
44. Neovius, E.; Lemberger, M.; Docherty Skogh, A.C.; Hilborn, J.; Engstrand, T. Alveolar bone healing accompanied by severe swelling in cleft children treated with bone morphogenetic protein-2 delivered by hydrogel. *J. Plast. Reconstr. Aesthetic Surg.* **2013**, *66*, 37–42. [CrossRef]
45. Sarker, B.; Singh, R.; Silva, R.; Roether, J.; Kaschta, J.; Detsch, R.; Schubert, D.; Cicha, I.; Boccaccini, A. Evaluation of Fibroblasts Adhesion and Proliferation on Alginate-Gelatin Crosslinked Hydrogel. *PLoS ONE* **2014**, *9*, e107952. [CrossRef]
46. Batzias, G.C.; Delis, G.A.; Koutsoviti-Papadopoulou, M. A new hplc/uv method for the determination of clindamycin in dog blood serum. *J. Pharm. Biomed. Anal.* **2004**, *35*, 545–554. [CrossRef]
47. Ritger, P.L.; Peppas, N.A. A simple equation for description of solute release i. Fickian and non-fickian release from non-swellable devices in the form of slabs, spheres, cylinders or discs. *J. Control. Release* **1987**, *5*, 23–36. [CrossRef]

gels

MDPI

Article

Influence of Ceria Addition on Crystallization Behavior and Properties of Mesoporous Bioactive Glasses in the SiO_2–CaO–P_2O_5–CeO_2 System

Elena Maria Anghel , Simona Petrescu *, Oana Catalina Mocioiu , Jeanina Pandele Cusu and Irina Atkinson *

Institute of Physical Chemistry 'IlieMurgulescu' of Romanian Academy, Splaiul Independentei 202, 060021 Bucharest, Romania; manghel@icf.ro (E.M.A.); omocioiu@icf.ro (O.C.M.); jeaninamirea@yahoo.com (J.P.C.)
* Correspondence: simon_pet@icf.ro (S.P.); irinaatkinson@yahoo.com (I.A.)

Abstract: Knowledge of the crystallization stability of bioactive glasses (BGs) is a key factor in developing porous scaffolds for hard tissue engineering. Thus, the crystallization behavior of three mesoporous bioactive glasses (MBGs) in the $70SiO_2$-(26-x)CaO-$4P_2O_5$-$xCeO_2$ system (x stands for 0, 1 and 5 mol. %, namely MBG(0/1/5)Ce), prepared using the sol–gel method coupled with the evaporation-induced self-assembly method (EISA), was studied. A thermal analysis of the multiple-component crystallization exotherms from the DSC scans was performed using the Kissinger method. The main crystalline phases of $Ca_5(PO_4)2.823(CO_3)_{0.22}O$, $CaSiO_3$ and CeO_2 were confirmed to be generated by the devitrification of the MBG with 5% CeO_2, MBG5Ce. Increasing the ceria content triggered a reduction in the first crystallization temperature while ceria segregation took place. The amount of segregated ceria of the annealed MBG5Ce decreased as the annealing temperature increased. The optimum processing temperature range to avoid the crystallization of the MBG(0/1/5)Ce powders was established.

Keywords: sol–gel processes; spectroscopy; X-ray methods; thermal properties; bioactive glass; silicate; biomedical applications

Citation: Anghel, E.M.; Petrescu, S.; Mocioiu, O.C.; Cusu, J.P.; Atkinson, I. Influence of Ceria Addition on Crystallization Behavior and Properties of Mesoporous Bioactive Glasses in the SiO_2–CaO–P_2O_5–CeO_2System. *Gels* **2022**, *8*, 344. https://doi.org/10.3390/gels8060344

Received: 29 April 2022
Accepted: 25 May 2022
Published: 31 May 2022

1. Introduction

Since their discovery in the late 1960s, bioactive glasses have been intensively studied due to their excellent bioactive response in hard tissue engineering [1]. However, the main limitation of the use of bioactive glass (BG) in obtaining porous scaffolds that mimic the structure of human bones [2] consists of improper mechanical characteristics, especially brittleness. Additionally, crystallized glass-ceramics show a lower surface reactivity in physiological solutions as a consequence of the reduction in the surface Si-OH linkages in comparison with the glassy counterparts [3]. Both brittleness and bioactivity are influenced by the crystallization (devitrification) behavior of BGs [1]. To overcome the limitation of BG crystallization during scaffold preparation [2], few solutions are used, e.g., tailoring BG composition, sol–gel preparation, understanding crystallization behavior and obtaining polymer-bioactive glass composites. The silica content, type of glass modifier (Na, K, Ca, Mg and Ba [4]) and doping oxides (ZnO, Ce_2O_3, Ga_2O_3, Bi_2O_3, Nb_2O_5 [5–8], etc.) are critical factors determining the ability of BGs to crystallize [9,10]. Although BGs with a silica content of up to 80% are still bioactive [11], they are denser than the Hench's 45S5 Bioglass® with 45% SiO_2 [1]. A better workability was reported for alkali-free BGs developing wollastonite ($CaSiO_3$) during crystallization in comparison with sodium–calcium–silicate phases, such as combeite ($Na_2Ca_2Si_3O_9$) [12,13]. To inhibit the crystallization of sodium–calcium–silicate phases in 45S5 while preserving their sintering and fiber-drawing abilities, magnesium and zinc were partially substituted for calcium [14]. The latter ions enabled the processing range to be widened, namely, the temperature range within the glass transition

(T_g) and crystallization onset, T_x. Except for BG formulation, the synthesis method highly influences the surface area and pore architecture, which are essential for an adequate surface reactivity in physiological fluids required by scaffolds in bone regeneration [2,15,16]. In contrast with melt and sol–gel-derived BGs, an enhanced bioactivity of the MBGs obtained by the sol–gel method coupled with the surfactant method is influenced by their pore architecture [15–19]. The ability to form ordered mesopores (2–50 nm sized pores) is mainly affected by the SiO_2 content and selected surfactant [15]. The most well known method for tuning the pore architecture of a BG obtained using sol-gel processing is the evaporation-induced self-assembly method (EISA) [16,17]. No correlation between the ordered mesoporosity and devitrification tendency has been previously reported in the literature.

Differential thermal analysis (DTA) and differential scanning calorimetry (DSC) are useful techniques for studying the crystallization of BGs [1,9], while the kinetic analysis of the thermal data provides information on the reactivity and stability of BGs. Although richer silica BGs do not easily crystallize [11,20], the few kinetics reports on crystallization in BGs are mostly conducted on the 45S5 Bioglass® [1,12,21]. In comparison to richer-silica BGs 1-98($53SiO_2$-$22CaO$-$6Na_2O$-$11K_2O$-$5MgO$-$2P_2O_5$-$1B_2O_3$, wt.%) and 13-93($53SiO_2$-$20CaO$-$6Na_2O$-$12K_2O$-$5MgO$-$4P_2O_5$, wt. %) with surface nucleation [11], the crystallization of the 45S5 Bioglass® suddenly proceeds from the surface to bulk phase [1]. The kinetics of the thermally simulated devitrification of BGs and their corresponding energy barrier has been very often studied by using the Kissinger method for the nth order reactions [22,23].

Since the thermal behavior of MBGs obtained using sol-gel process helps when choosing parameters for fiber and bioactive scaffold preparation, this work presents the non-isothermal crystallization kinetics of Ce-containing MBGs in the $70SiO_2$-(26-x)CaO-$4P_2O_5$-$xCeO_2$ system (x stands for 0, 1 and 5 mole %) by using DSC data. The identification of the crystallization mechanism and crystalline phases was assessed.

2. Results and Discussion

2.1. Phase Evaluation in the G(0/1/5)Ce Gels

The FT-IR spectra of the dried G(0/1/5)Ce gels, which are the un-stabilized MBGs, as illustrated in Figure 1a and exhibited characteristics of a silicate structure at 473, 825 and 1049 cm^{-1} due to the rocking, bending and stretching modes of the Si-O-Si bonds that formed into coalesced silica particles during condensation processes. The band located at ~570 cm^{-1} [24] might be assigned to four-fold rings, i.e., $Si(OSi)_3(OR)$ (R stands for C_2H_5 or H) and/or four-fold silanol rings, as well as to the stretching and bending modes of the P-O bonds. Nitrate ions, depictable by the intense band at 1380 cm^{-1} and smaller bands at 822 and 740 cm^{-1} [25], indicated the non-incorporation of calcium ions into the silica network [26]. Calcium nitrate, covering silica nanoparticles, was reported to exist in dried gels of 70S30C ($70SiO_2$-$30CaO$, mol. %) up to 350 °C [26,27]. Hence, the IR spectra of the dried 70S30C gels resembled those of calcium nitrate, but the 1047 cm^{-1} band was thinner for the latter compound, and the shoulder at approximately 1080 cm^{-1} indicates the formation of Si-O-Si [27]. The coexistence of the 1049 and 1079 cm^{-1} spectral features is easier to observe for the G5Ce sample in Figure 1a. The O-H presence in H_2O and alcohols [24,28] was indicated by the 1635 and 3423 cm^{-1} bands. The small band at approximately 950 cm^{-1} of the G5Ce spectrum was due to Si-OH linkages [28]. The organic residue was identified by the C-H stretching modes of the CH_2 and CH_3 groups at 2938 and 2878 cm^{-1} [28].

UVRaman spectroscopy, enabling fluorescence avoidance, as well as the selectively enhanced detection of nitrates [29], was used for the first time, in the present study, to investigate the surface of the cerium-doped gels in the CaO-SiO_2-P_2O_5 system. The $\nu_1(NO_3^-)$ band [30] in Figure 1b (inset) was up-shifted for the G(1/5)Ce in comparison with that for cerium-free gel, which is very likely due to hydrated water and metal cation (calcium and cerium cations) effects [31]. This behavior confirmed a lack of calcium nitrate incorporation, as already depicted by IR, as well as the lack of cerium nitrate, which covers the already formed $Si(OSi)_3(OR)$ network. Additionally, hydroxyl from molecular H_2O (1600–1650 cm^{-1} and 3350–3500 cm^{-1} ranges), organic residue (symmetric and asymmetric C-H stretching modes of the methylene, CH_2, and methyl, CH_3, within 2600–3100 cm^{-1}) and the silicophosphate network of four-fold rings at approximately 490 cm^{-1} [24,32] are depicted in Figure 1b. Bands at approximately 1415 and 1460 cm^{-1} are attributable to bending vibrations of the CH_2 and CH_3 groups [32]. The tinny band at 1140 cm^{-1} indicated the formation of the Si-O-P bonds [32]. Symmetric vibrations of the P=O [32] were observed at 1279 cm^{-1}. In order to stabilize these dried gels, namely, to remove the organic and nitrate residue, their thermal behavior should be assessed.

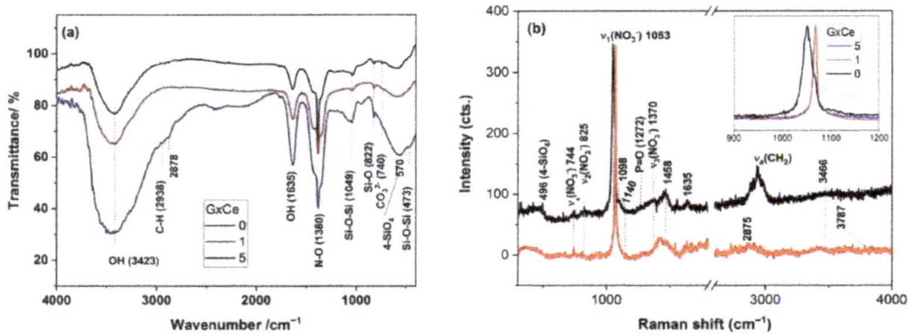

Figure 1. (**a**) IR and (**b**) UV-Raman spectra of the GxCe gels (x stands for 0, 1 and 5% ceria).

2.2. DTA/DTG/TG Analysis of the G(0/1/5)Ce Gels

The DTA/DTG/TG curves of the G(0/1/5)Ce materials are illustrated in Figure 2a,b. The TG curves in the inset of Figure 2 indicate the lowest thermal stability of the G1Ce sample up to 160 °C, intermediate behavior over the 160–250 °C range, while above 250 °C, its stability improved more than the other two gels. Hence, the rate of the mass loss was not solely dependent on the gel composition for the whole temperature range investigated. The overall mass loss ranged from 58.39% (G1Ce) to 60.80% (G5Ce). The first stage, at approximately 70 °C, with a mass loss <8wt. %, for all the gels investigated, corresponded to the physically adsorbed water [33–35] and ethanol [32,36]. The next two stages (Table 1) record a mass loss of a maximum of 37%, up to 300 °C, corresponded to the vaporization o water and decomposition of organic residue. Chemisorbed water resulted from precursor condensation was removed at approximately 230 °C [13]. The third stage was due to the alkoxy group decomposition [37,38].

Above 100 °C, the weight loss stages were accompanied by two exothermic events on the corresponding DTA curves (Figure 2b). The second DTA exotherm had left and/or right sided shoulders due to the complex decomposition process. Ethyl groups of the TEOS and TEP along with un-hydrolyzed precursor vaporization [32] were responsible for the first exothermic event at approximately 200 °C.

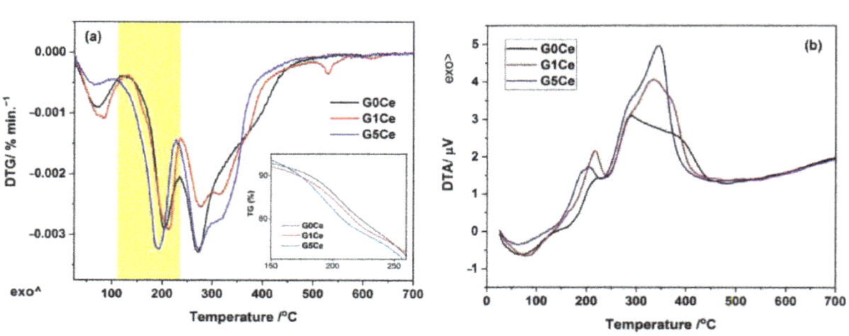

Figure 2. (**a**) DTG/TG curves of the G(0/1/5)Ce gels and (**b**) DTA curves of the G(0/1/5)Ce gels.

Table 1. Thermogravimetric decomposition data of the G(0/1/5)Ce gels.

Gel Code	Stage No	ΔT/°C	Mass/%	Assignment	Total Loss/%
0Ce	1	25–132	6.30	Physically absorbed water [34,35]	59.91
	2	132–234	16.16	Chemisorbed water and organic oxidation [13]	
	3	234–301	18.05		
	4	301–497	18.93	Pluronic and nitrate decomposition [26,35]	
1Ce	1	25–134	7.14	Physically absorbed water	58.39
	2	134–236	16.63	Chemisorbed water and organic oxidation	
	3	236–298	12.94		
	4	298–390	13.85	Pluronic and nitrate decomposition	
	5	390–640	6.96		
5Ce	1	25–106	3.39	Physically absorbed water	60.85
	2	106–227	20.17	Chemisorbed water and organic oxidation	
	3	227–293	16.11		
	4	293–620	20.04	Pluronic and nitrate decomposition	

The deegradation of Pluronic P-123 and nitrate by-products [26] accounted for the last (fourth and fifth) mass loss stages. The removal of the nitrate byproducts confirmed incorporation by the diffusion of the calcium [26] and cerium ions into the silica network. The calcium diffusion process in the 70S30C gels takes place above 400 °C. Thus, the thermal events at 531 and 608 °C, of the DTG curve for the G1Ce in Figure 2a, correspond, are very likely due to loss of the remnant surfactant and nitrates, respectively [26,35]. A tinny endotherm event at 600 °C was present on the DTG curve of the G5Ce (Figure 2a).

According to the thermal features above, the obtained gels were two-step thermally treated to obtain MBGs. The first step at 300 °C (1 h) was required for water and organic residue removal, while the second one was carried out at 700 °C (3 h) when no mass loss occurred and calcium and cerium incorporation in the phosphosilicate network took place. To evaluate the stabilization efficiency of the Ce-containing MBGs under discussion, FT-IR and XRD measurements were carried out. The XRD patterns of the MBG(0/1/5)Ce samples calcined at 700 °C [33] showed the presence of halos specific to the glassy phase.

2.3. Phase Identification in the Devitrified Ce-Containing MBGs

The DSC curves of the stabilized BGs powders, MBG(0/1/5)Ce, collected by heating with a 10 °C/min. rate (Figure 3a), had multiple crystallization exotherms. Thus, two crystallization effects were recorded for MBG1Ce, while three components were required for MBG(0/5)Ce (Figure 3b). The peculiar behavior of MBG1Ce regarding higher pore interconnectivity was determined [33] from the wider hysteresis loop of the nitrogen adsorption/desorption isotherms of the MBG(0/1/5) samples. Hence, except for MBG formulation, the textural data can influence thermal behavior. Three exotherms of crystallization were also reported for the 70S26C4P (70SiO_2-26CaO-4P_2O_5, mol. %) BG obtained using the sol–gel method in the absence of surfactants and stabilized in a single step at 700 °C [36]. Decreasing temperature for the first crystallization exotherm in Figure 3a was noticeable as ceria content increased. Jones et al. [39] reported a crystallization peak at 873 °C for β-wollastonite in the foamed and unfoamed monoliths of 70S30C, while Sigueira and Zanotto [36] found apatite in primary phase crystallizing at 900 °C. In the case of the MBG0Ce powders, the first exotherm of crystallization (onset temperature, T_x, of 863 °C and peak temperature, T_{c1},of 878 °C) was assignable to an apatite phase (XRD data in Figure 4a and Table 2), whereas the other two exotherms belonged to wollastonite (β-phase) and higher temperature wollastonite, pseudowollastonite or α-wollastonite [25]. The crystallization of wollastonite phases was accompanied by the enhancement of mechanical properties [40].

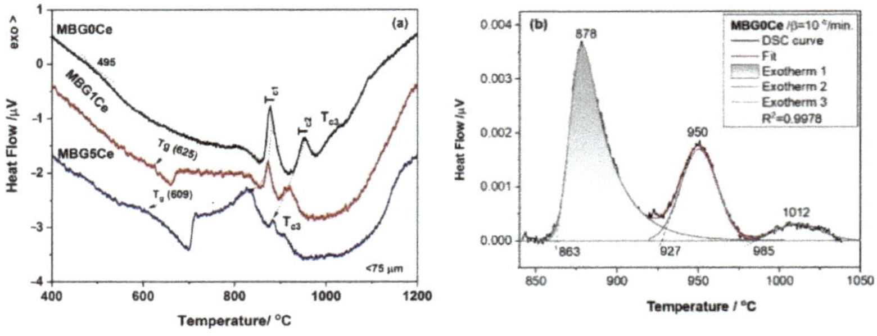

Figure 3. (**a**) DSC runs at a heating rate of 10 °C/min. for the stabilized MBG(0/1/5)Ce and (**b**) three-component fitted DSC curve of the MBG0Ce crystallization.

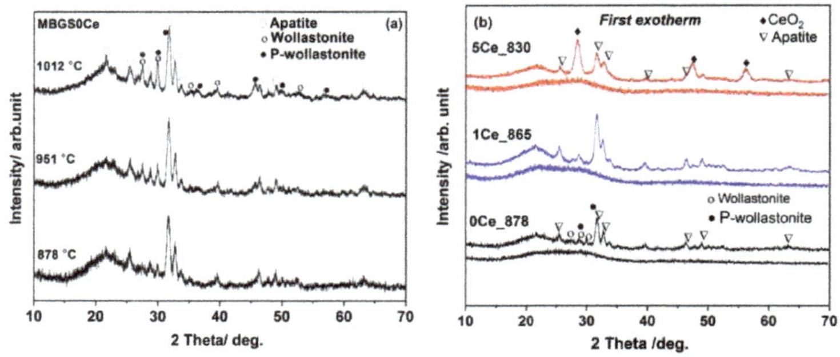

Figure 4. XRD patterns of the annealed: (**a**) MBG0Ce_(878/951/1012) and (**b**) MBG(0/1/5)_T_{c1} samples at first crystallization exotherm.

Table 2. List of thermal treatment holds for 24 h at DSC exotherms of crystallization and corresponding XRD data of MBGs (particle size 75 μm).

MBG Code	T_{c1} (°C/Phases)	T_{c2} (°C/Phases)	T_{c3} (°C/Phases)
0Ce	878/ A (68.01%), W (20.72%) PW (11.27%)	951/ A (68.34%), W (10.04%), PW (21.62%)	1012/ A (70.98), W (17.69%), PW (11.33%)
1Ce	865/ A (90.10%), W (9.90%)	900/ A (85.88%), W (15.12%)	-
5Ce	830/ A (38.21%), C (14.10%), PW (47.68)	876/ A (54.15%, C (13.41%) PW (32.44%)	906/ A (59.05%), C (7.80%), PW (33.14%)

A = apatite (JCPDS 01-073-1731), W = wollastonite (JCPDS 00-900-8151), PW = pseudowollastonite (JCPDS 01-074-0874), C = ceria (JCPDS 00-043.

The formation of the $Ca_5(PO_4)_{2.823}(CO_3)_{.22}O$ (64.45% crystallinity) was noticeable in the X-ray pattern of the MBG1Ce isothermally treated at 865 °C (MBG1Ce_865 in Table 2), which is similar to the MBG0Ce_878 sample (Figure 4b). Hence, the annealed samples at the first crystallization effect, MBG(0/1)_T_{c1}, showed the presence of an apatite phase, although only amorphous phases were shown prior to thermal treatments (Figure 4b). Ceria prevailed in the annealed MBG5Ce_T_{c1}(Figure 4b and Table 2). The annealing of the MBG(0/1/5)Ce BGs at approximately 870 °C (Table 2) induced the formation of apatite phase.

Crystalline phosphates were observed in the IR spectra of the two annealed BGs with three crystallization effects, MBGS(0/5)_(T_{c1}/T_{c2}/T_{c3}), as shown in Figure 5, due to its bands at ~570 and 609 cm^{-1} as a result of the P-O bending vibrations of the orthophosphate

PO_4^{3-} groups [39,40]. These bands were used to monitor glass bioactivity [41] since the other P-O vibrations overlapped with the Si-O vibrations. Only the 565 cm^{-1} band was observed for the MBGS1Ce_(Tc1/TC2) in Figure S1. A small band at ~428 cm^{-1} might have belonged to Ce-O [42] in MBG5Ce_(830/876). The asymmetric Si-O-Si stretching mode at ~1092 cm^{-1} [43], and symmetric stretching and bending bands at 794 cm^{-1} and 465 cm^{-1} were also present in the IR spectra of all the thermally treated MBG0Ce as shown Figure 5a. Moreover, IR bands located at 1018, 937, 903, 720, 684, and 642 cm^{-1} of the samples treated at a higher temperature, MBG0Ce_(951/1012), belonged to β-wollastonite [44]. IR data reported for the sol–gel 70S26C4P bioglass annealed at 1000 °C [44] revealed the formation of quartz (798/780 doublet and 697 cm^{-1}). Quartz absence in the XRD pattern of MBG0Ce_1012 might be due to its low quantity under the XRD detection limit. Barely perceptible shoulder at 984 cm^{-1} for the MBG0Ce_1012 sample belonged to pseudowollastonite (PW). This finding, supported by the XRD pattern in Figure 5a, was also reported for the sintered BG of 58S (58 wt. % SiO_2, 33 wt. % CaO and 9 wt. % P_2O_5) at 1100 °C [44].

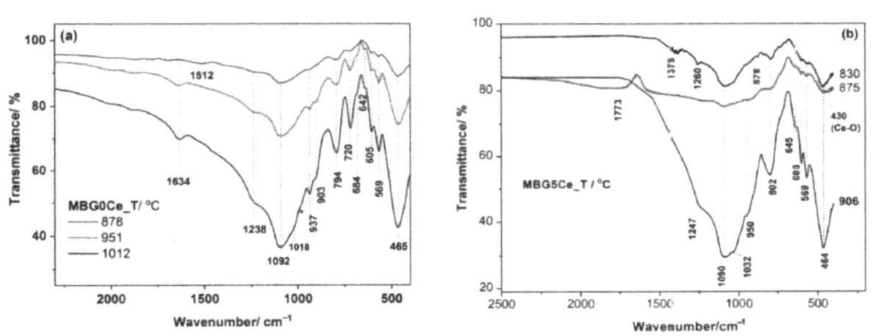

Figure 5. FTIR spectra of the annealed (**a**) MBG0Ce_T and (**b**) MBG5Ce_T samples at all crystallization exotherms (* stands for the 984 cm^{-1} shoulder).

Raman spectroscopy is a powerful technique for studying nucleation and growth in annealed glasses [45]. Moreover, UV-Raman spectroscopy accesses the surface information of BGs. The annealed MBGS5Ce_T samples showed distinct Raman features below 750 cm^{-1} (Figure 6 and Table S1). Thus, the band at 438 cm^{-1} of the MBG5Ce_830 is attributable to rocking modes of the bridging oxygen atoms located perpendicular to the P–O–P plane and of the F_{2g} modes of CeO_2 [46]. More intense bands of defects, D_1 and D_2, of the MBG5Ce_(830/876) samples in comparison with commercial ceria were due to oxygen vacancies associated with Ce^{3+} and Ce^{4+} sites [46]. The defect bands were either thermally and/or dopant activated [46]. The overtones (2LC) of ceria at approximately 1167 cm^{-1} were also shown for the first exotherm-annealed MBG5Ce_(830/876). The symmetric stretch of the Q^0(P) units (595 cm^{-1}) [8] indicated the formation of a crystalline phosphate phase in the MBG5Ce_(RT/830/876). The band at 464 cm^{-1} [47] revealed the presence of quartz in the MBG5Ce_906 sample. The wollastonite band (Si-O bending [48]) at approximately 635 cm^{-1} (Si-O-Si bonds [8] in Q^2(Si)) supported by XRD and IR data for the second exotherm-annealed sample MBG5Ce_876. A band at 620 cm^{-1} was reported [49] for cerium phosphate. UV-Raman spectra of the annealed MBGS5Ce_T sample showed the $\nu_1(PO_4^{3-})$ band at approximately 950 cm^{-1}, as observed in Figure 6 [24,45]. The same band could also belong to the Q^2(Si) units (SiO_4 tetrahedra with two non-bridging oxygen atoms, NBOs) of the metasilicate chains ($Si_2O_6^{2-}$) [24]. The abundance of the Q^2(Si) units influenced the glass bioactivity. The band at approximately 846 cm^{-1}, which was seen in all the samples (Figure 6), corresponds to the Q^0(Si) units that provided information about the devitrified glass structure.

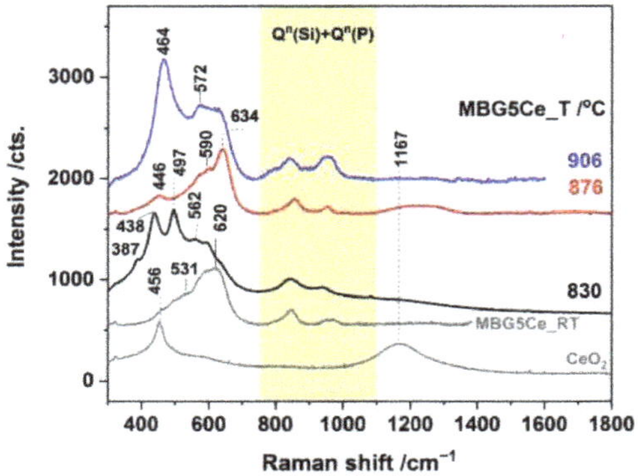

Figure 6. UV-Raman spectra of annealed MBG5Ce_T at crystallization exotherms and commercial CeO_2.

2.4. Crystallization Behavior of the Stabilized Ce-Containing MBGs

To obtain workability information (processing range, T_x-Tg, reactivity, and stability) of the MBGs kinetic approach of the DSC runs were collected at various rates (Figure 7). Although glasses can show multiple crystallization exotherms only the first one was used in the calculation of glass stability defined as the crystallization resistance of glass upon heating [9]. The most simplified estimation of the glass stability is to obtain the T_x-T_g quantity from DSC measurements [1], i.e., temperature range where crystallization is avoided. A value of higher than 100 °C of this quantity implies good thermal stability, which was the case for all MBGs presented here (Figure 7 and Table S2). The lowest T_x-T_g value was recorded for the MBG5Ce (149 °C at 10 K/min.). Other parameters (Hruby, Weinberg, etc.) for measuring glass stability imply the use of glass melting temperature [9]. However, the measurement of the melting temperature of the broad endothermal DSC effect is rather inaccurate.

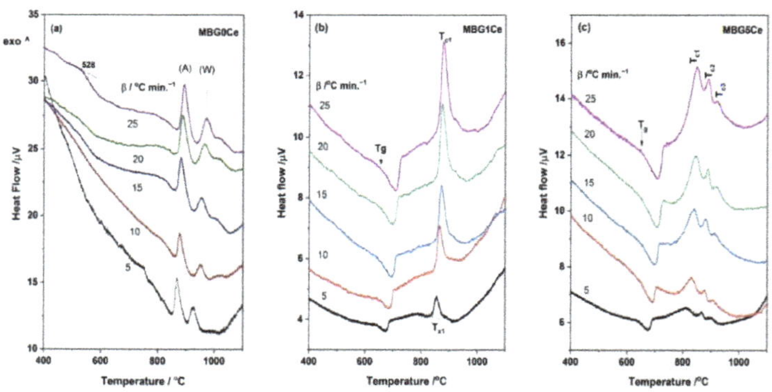

Figure 7. DSC runs with various heating rates of: (a) MBG0Ce, (b) MBG1Ce and (c) MBG5Ce (A and W stand for apatite and wollastonite).

Therefore, considering crystallization exotherm dependence on the DSC heating rate, activation energy (E_a) was derived using the Kissinger equation (Equation (1)) for the first exothermof crystallization. Plots of $[-\ln(\beta/T_{C1})]$ as functions of $[1000/T_{C1}]$ for the first crystallization exotherm of the MBG(0/1/5)Ce samples are presented in Figure 8. The straight lines used to fit these plots represent E_a/R activation energy for crystallization and gas constant. The high activation energy of apatite crystallization in MBG(0/1)Ce samples (Table 3) indicates the good thermal stability of the two samples. Larger E_{a1}values of the MBG(0/1)Ce than those reported for coarse and fine apatite particles, 514 and 482 KJ/mol

K, crystallized in lime- and magnesia- containing silicate glasses [40], indicate a lower tendency for the crystallization of these MBGs. The Avrami exponent, n, calculated using Equation (2) indicates the surface crystallization of the ceria (MBG5Ce) and apatite phase (MBG(0/1)Ce). As well as size, the shapes of the glass particles influence crystallization kinetics [9]. Thus, the wide FWHM of the first crystallization exotherm (~40 °C in Table S2) for the MBG5Ce might originate from spherical and/or cuboid glass particles, whereas much narrower MBG(0/1)Ce exotherms (~20 °C) can be due to prolate and needle-like apatite particles [9]. Smaller E_a and T_x-T_g values were obtained for the MBG5Ce compared with MBG(0/1)Ce samples. Surface area values, S_{BET}, as reported elsewhere [33], decreased in the same succession as E_a for the first crystallization, namely, MBG0Ce (307 m^2/g) > MBG1Ce(230 m^2/g) > MBG5Ce (223 m^2/g), and inversely proportional to the increase in ceria content.

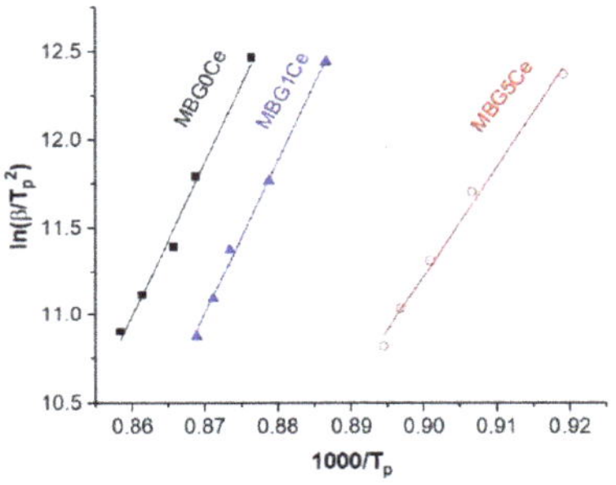

Figure 8. Linear fit of the Kissinger plots for the first exotherm of crystallization (T_{c1}) of the MBG(0/1/5)Ce.

Table 3. Linear fit of the Kissinger plots for the first crystallization exotherms of stabilized MBG(0/1/5)Ce.

Sample	Intercept	Slope	E_a	n	R^2
MBG0Ce	64.5969	87.8985 ± 5.22	730.82 ± 43.40	1.5979	0.9860
MBG1Ce	64.9096	87.2713 ± 3.23	725.61 ± 26.85	1.8879	0.9959
MBG5Ce	−44.7813	62.2325 ± 3.17	517.43 ± 26.35	1.1871	0.9922

Given the distinct thermal behavior and segregation of CeO_2 in the MBGS5Ce the bioactivity study focused on this composition.

2.5. Bioactivity of the MBG5Ce_T

Multiple steps were identified in developing a hydroxy-carbonate apatite (HCA) layer on the surface of BGs when treated with simulated biological fluid, SBF [1]. This complex process of obtaining an HCA layer that is well-matched with natural bones and teeth relies on three stages: leaching, dissolution, and precipitation. The leaching of Ca^{2+} from BGs and exchanging with H^+ and H_3O^+ from SBF was followed by the dissolution process when the breaking the Si-O-Si bonds caused the formation of Si-OH at the surface and as well as the release of $Si(OH)_4$ into SBF. Superficial Si-OH underwent polycondensation into a silica gel layer. Glass-released and SBF-originating calcium and phosphate ions migrated to the silica gel layer and precipitated as an amorphous Ca-P-rich layer which subsequently crystallized into HCA due to carbonate incorporation. The dissolution process is highly dependent on glass and/or glass-ceramic composition, structure, and morphology. For instance, Ce^{3+} ions released by the glass surface can compete with Ca^{2+} for phosphate ions in SBF to form insoluble $CePO_4$ and hence delay HCA formation [50,51].

The surface reactions of the MBG particles can be also monitored using UV-Raman spectroscopy with a low penetration depth. To assess mineralization ability, SEM, UV-

Raman, and IR investigations were carried out after 14-day immersion of the annealed MBG5Ce_(RT/830/876) samples in SBF. The FTIR spectrum shown in Figure 9a of the 14-day soaked MBG5Ce_830 in SBF was dominated by the 1060 and 466 cm^{-1} bands of the Si-O-Si stretching, and bending modes. The lack of bands at approximately 602 cm^{-1} for the MBG5Ce_(830/875)_14 samples showed either that the crystalline phosphate phase existing in these samples prior to immersion was soluble in SBF or an amorphous phosphate phase was formed, despite the segregation of CeO_2 (see XRD data in Figure 4b and Table 2) in the MBG5Ce sample, which is known to hinder bioactivity [33,52]. IR and Raman spectra of the sample without thermal treatment indicated the formation of HCA (intense peak of PO_4^{3-} and CO_3^{2-} at 954 and 1100 cm^{-1} [41,51,52] in Figure 9b and 602 cm^{-1} in Figure 9a) after 14 days of immersion in SBF (MBG5Ce_RT_14). Instead, the wide band peaking at approximately 1200 cm^{-1} (Figure 9b) in the MBG5Ce_(830/876)_14 samples provided evidence of ceria. Except fluorescence removal, UV-Raman spectroscopy is a very sensitive technique to identify the defective structure of ceria at a lower penetration depth.

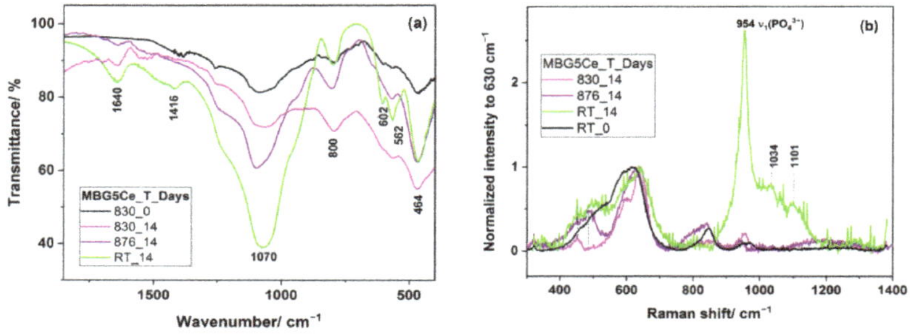

Figure 9. (**a**) FT-IR and (**b**) Raman of the MBG5Ce_(RT/830/875) soaked in SBF for 14.

The silica gel layer obtained in the early stages of the bioglass immersed in SBF [38] was subsequently penetrated by calcium and phosphate ions and an amorphous calcium phosphate layer covered the sample surface (Stage 4) [53]. Hydroxyl, phosphate, carbonate and calcium ions formed an outer hydroxyapatite layer.

Delayed bioactivity was confirmed by IR, UV-Raman and SEM data (Figure 10) collected on the annealed MBG5Ce at the second crystallization event corresponding to wollastonite crystallization. Shoulders at approximately 642 and 953 cm^{-1} of the 14-day soaked MBG5Ce_875 spectrum indicated β-wollastonite phase.

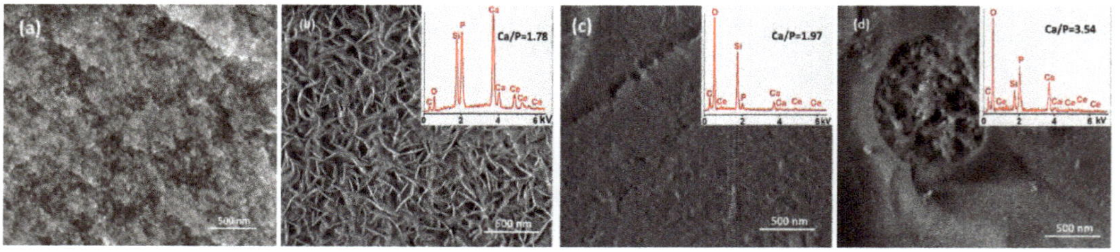

Figure 10. SEM images of the MBG5Ce sample: (**a**,**b**) untreated, and 14-day SBF soaked; (**c**) MBG5Ce _830_14 and (**d**) MBG5Ce _876_14.

The SEM micrograph of the sample without annealing treatment immersed for 14 days in SBF, MBG5Ce_14, showed the formation of hydroxyapatite. Moreover, its Ca/P ratio of 1.78 (EDS spectrum illustrated in the inset of Figure 10b) was closer to that of the natural hydroxyapatite (1.67). Cerium and phosphorous depicted in the inset of Figure 10 confirmed the presence of segregated ceria and apatite phase crystallized at the first crystallization peak of the MBG5Ce_830 prior to immersion in SBF. Tinny needle crystals on the surface of theMBG5Ce_830_14 sample might point out incipient apatite phase formation. Conversely, the MBG5Ce_876_14 micrograph (Figure 10d) is evidence of the prevention of hydroxyapatite formation in samples containing crystallized wollastonite at the surface. The highest Ca/P ratio of 3.54 is further proof of this.

Delayed bioactivity was confirmed by IR, UV-Raman and SEM data (Figure 10) collected for the annealed MBG5Ce at the second crystallization event corresponding to wollastonite crystallization. Shoulders at approximately 642 and 953 cm^{-1} of the 14-day immersed MBG5Ce_875 spectrum indicated β-wollastonite phase.

3. Conclusions

Information on the crystallization avoidance, i.e., thermal stability and temperature processing window (Tx-Tg range) of some cerium-containing MBGs. MBG(0/1/5), was obtained in this study. Double and triple crystallization events of the DSC curves were recorded for the MBG(0/1/5)Ce powders. The XRD findings of the annealed samples at the crystallization peak temperatures revealed the crystallization of apatite, wollastonite, and ceria phases. The large activation energy (>500 KJ/mol) corresponding to the first crystallization event, derived by using the Kissinger method, indicated a low crystallization tendency which is appropriate for the thermal processing of the MBGs. The addition of 5% ceria to the MBGs, MBG5Ce, caused the lowering of the first crystallization exotherm from 876 °C to 830 °C and the narrowing of the processing window to ~100 °C while ceria segregation took place. Conversely slower bioactivity was seen for the annealed MBG5Ce at its first crystallization exotherm. According to these results, the MBG(0/1/5)Ce powders presented here are suitable for the production of porous biomedical scaffolds. Further cellular tests are nedeed to investigate the biological response of the annealed MBGs.

4. Materials and Methods

4.1. Materials

Tetraethylorthosilicate (TEOS of >98% from Merck, Darmstadt, Germany), pluronic® P123 (Sigma-Aldrich, Darmstadt, Germany)), triethylphosphate (TEP of 99% from Aldrich), calcium nitrate tetrahydrate (99% p.a., Carl Roth, Karlsruhe, Germany) and cerium nitrate hexahydrate (99%, Aldrich) were the starting materials. Cerium oxide (99.9% from Loba-Chemie, Mumbai, India) was used for comparison.

4.2. Sol–Gel Preparation Methods

The sol–gel synthesis coupled with the evaporation-induced self-assembly method (EISA) using Pluronic® P123 as a structure-directing agent was employed to obtain cerium-doped MBGs in the $70SiO_2$-(26-x)CaO-$4P_2O_5$-$xCeO_2$ system (x stands for 0, 1 and 5 mole %) as described elsewhere [33]. The obtained dried gels, denominated G(0/1/5)Ce gels according to the molar percent of CeO_2, were two-step thermally treated according to the thermal effects indicated by the corresponding TG/DTG/DTA curves in Figure 2.

Further, the glass powders were isothermally crystallized separately in a Pt crucible in air at the temperature (T) corresponding the DSC exotherms of crystallization, MBG(0/1/5)_T for 3h and 24 h, respectively.

4.3. Bioactivity of the Devitrified Glasses

The in vitro bioactivity of the devitrified MBG(0/1/5)_T powders was checked in simulated body fluid (SBF) for 14 days, as described elsewhere [5].

4.4. Characterization

4.4.1. Thermal Characterization

The thermal behavior of the G(0/1/5)Ce gels was determined through differential thermal analysis and thermo-gravimetric analysis using Mettler Toledo TGA/SDTA 851e equipment, in Al_2O_3 crucibles and in a flowing air atmosphere. The maximum temperature was set at 1000 °C and the heating rate was 10 °C/min.

Thermal stability analysis of the calcined MBGs, MBG(0/1/5)Ce, was performed by means of a LabSysEvo in a platinum crucible in an argon atmosphere. Approximately 26 mg of grounded glass with a particle size below 75 µm was used for DSC measurement within 25–1200 °C. Crystallization kinetics analysis was conducted with heating rates, β, of 5, 10, 15, 20 and 25 °C/min. after appropriate equipment calibration for each rate with In, Sn, Zn, Al and Ag. The glass transition temperature (T_g) established based on the inflection point and peak temperature of crystallization (T_p) were determined by using Calisto 1.051 software from the DSC curves of the respective glasses.

The activation energy (E_a) was calculated using the Kissinger equation [22,23] according to the DSC data:

$$\ln\left(\frac{\beta}{T_C^2}\right) = \text{constant} - \frac{E_a}{RT_C} \quad (1)$$

where: T_c is crystallization temperature measured at various heating rates, β, and R is the gas constant. The activation energy, E_a, for crystallization is derived from slope ($-E_a/R$) of the straight line of the $\ln(\beta/T_C{}^2)$ versus ($1000/T_C$) representation.

The Avrami exponent (n), an indicator of crystal growth dimensionality, was determined using Equation (2) proposed by Augis and Bennett [54,55]:

$$n = \frac{2.5}{\Delta T_{FWHM}} \frac{T_C^2}{\frac{E_a}{R}} \quad (2)$$

where: ΔT_{FWHM} is the full width at the half maximum of the crystallization exotherm from the DSC curve.

4.4.2. Structural Characterization

The structures of the G(0/1/5) gels were investigated using IR and UV-Raman spectroscopies. Additionally, crystalline phases obtained through isothermal devitrification in accordance with the DSC exotherms of crystallization were identified by means of X-ray Diffraction (XRD), IR and UV-Raman spectroscopies. Thus, the XRD patterns of glassy and devitrified counterparts were recorded using a RigakuUltima IV diffractometer (Rigaku Corporation, Tokyo, Japan) equipped with CuKα radiation, with 2°/min and a step size of 0.02°. Fourier transform infrared (FTIR) spectra of the were recorded without additional slice preparations, in the 400–4000 cm^{-1} domain with a sensitivity of 4 cm^{-1} by using a Thermo Nicolet 6700 spectrometer (Thermo Fisher Scientific Inc., Waltham, MA, USA). UV-Raman spectra of the gels, stabilized and thermally devitrified MBGs were collected by means of a Labram HR 800 spectrometer (HORIBA FRANCE SAS, Palaiseau, France) equipped with a UV laser line (325 nm from Kimmon Koha Co., Ltd., Tokyo, Japan), grating of 2400 lines and a 40x/0.47 NUV objective, assuring a laser spot on the sample smaller than 1 µm.

The morphology and EDS characterization of the MBGs samples were carried out by using a FEI Quanta3DFEG (FEI, Brno, Czech Republic) microscope equipped with an Octane Elect EDS system. Secondary electron images were recorded at an accelerating voltage of 5 and 10 kV in high-vacuum mode. Samples were recorded without sputter coating with conductive material.

Supplementary Materials: The following supporting information can be downloaded at: https://www.mdpi.com/article/10.3390/gels8060344/s1, Figure S1: FT-IR spectra of the MBG1Ce annealed at 865 and 900 °C; Table S1: UV-Raman band positions of the annealed MBG5Ce_T spectra and assignments within 330–1200 cm^{-1} spectral range; Table S2: Results ($T_x$and T_{c1}) of fitted DSC curves of the MBG(0/1/5Ce).

Author Contributions: Conceptualization, E.M.A., S.P. and I.A.; investigation, O.C.M., J.P.C., E.M.A., S.P. and I.A.; writing—original draft preparation, E.M.A. and I.A.; writing—review and editing, I.A., S.P. and E.M.A.; supervision, I.A. and E.M.A.; project administration, I.A.; funding acquisition I.A. All authors have read and agreed to the published version of the manuscript.

Funding: The research and APC were funded by the Executive Agency for Higher Education, Research, Development and Innovation Funding (UEFISCDI), grant number PN-III-P2-2.1-PED-2019-0598, no. 258 PED/2020.

Institutional Review Board Statement: Not applicable.

Informed Consent Statement: Not applicable.

Data Availability Statement: The data presented in this study are contained within the article.

Conflicts of Interest: The authors declare no conflict of interest.

References

1. Clupper, D.C.; Hench, L.L. Crystallization kinetics of tape cast bioactive glass 45S5. *J. Non-Cryst. Solids* **2003**, *318*, 43–48. [CrossRef]
2. Jones, J.R. Reprint of: Review of bioactive glass: From Hench to hybrids. *Acta Biomater.* **2013**, *9*, 4457–4486. [CrossRef] [PubMed]
3. Filho, O.P.; LaTorre, G.P.; Hench, G.P. Effect of crystallization on apatite-layer formation of bioactive glass 45S5. *J. Biomed. Mater. Res.* **1996**, *30*, 509–514. [CrossRef]
4. Crovacea, M.C.; Soares, V.O.; Rodrigues, A.C.M.; Peitl, O.; Raucci, L.M.S.C.; Oliveira, P.T.; Zanotto, E.D. Understanding the mixed alkali effect on the sinterability and in vitro performance of bioactive glasses. *J. Eur. Ceram. Soc.* **2021**, *41*, 4391–4405. [CrossRef]
5. Atkinson, I.; Anghel, E.M.; Predoana, L.; Mocioiu, O.C.; Jecu, L.; Raut, I.; Munteanu, C.; Culita, D.; Zaharescu, M. Influence of ZnO addition on the structural, in vitro behavior and antimicrobial activity of sol–gel derived $CaO–P_2O_5–SiO_2$ bioactive glasses. *Ceram. Int.* **2016**, *42*, 3033–3045. [CrossRef]
6. Deliormanli, A.M. Synthesis and characterization of cerium- and gallium-containing borate bioactive glass scaffolds for bone tissue engineering. *J. Mater. Sci. Mater. Med.* **2015**, *26*, 67. [CrossRef]
7. Pazarçeviren, A.E.; Tahmasebifar, A.; Tezcaner, A.; Keskin, D.; Evis, Z. Investigation of bismuth doped bioglass/graphene oxide nanocomposites for bone tissue engineering. *Ceram. Int.* **2018**, *44*, 3791–3799. [CrossRef]
8. Siqueira, L.; Campos, T.M.B.; Camargo, S.E.A.; Thim, G.P.; Triches, E.S. Structural, crystallization and cytocompatibility evaluation of the 45S5 bioglass-derived glass-ceramic containing niobium. *J. Non-Cryst. Solids* **2021**, *555*, 120629. [CrossRef]
9. Zheng, Q.; Zhang, Y.; Montazerian, M.; Gulbiten, O.; Mauro, J.C.; Zanotto, E.D.; Yue, Y. Understanding Glass through Differential Scanning Calorimetry. *Chem. Rev.* **2019**, *119*, 7848–7939. [CrossRef]
10. Bellucci, D.; Cannillo, V.; Sola, A. An Overview of the Effects of Thermal Processing on Bioactive Glasses. *Sci. Sinter.* **2010**, *42*, 307–320. [CrossRef]
11. Fagerlund, S.; Masserax, J.; Hupa, L.; Hupa, M. T–T–T behaviour of bioactive glasses 1–98 and 13–93. *J. Eur. Ceram. Soc.* **2012**, *32*, 2731–2738. [CrossRef]
12. Arstila, H.; Vedel, E.; Hupa, L.; Hupa, M. Factors affecting crystallization of bioactive glasses. *J. Eur. Ceram. Soc.* **2007**, *27*, 1543–1546. [CrossRef]
13. Nawaz, Q.; Pablos-Martín, A.; Silva, J.M.S.; Hurle, K.; Contreras-Jaimes, A.T.; Brauer, D.S.; Boccaccini, A.R. New insights into the crystallization process of sol-gel–derived 45S5 bioactive glass. *J. Am. Ceram. Soc.* **2020**, *103*, 4234–4247. [CrossRef]
14. Wetzel, R.; Blochberger, M.; Scheffler, F.; Hupa, L.; Brauer, D.S. Mg or Zn for Ca substitution improves the sintering of bioglass 45S5. *Sci. Rep.* **2020**, *10*, 15964. [CrossRef]
15. Schumacher, M.; Habibovic, P.; Rijt, S. Mesoporous bioactive glass composition effects on degradation and bioactivity. *Bioact. Mater.* **2021**, *6*, 1921–1931. [CrossRef]
16. Brinker, C.J.; Lu, Y.; Sellinger, A.; Fan, H. Evaporation-induced self-assembly: Nanostructures madeeasy. *Adv. Mater.* **1999**, *11*, 579–585. [CrossRef]
17. Lalzawmliana, V.; Anand, B.; Roy, M.; Kundu, B.; Nandi, S.K. Mesoporous bioactive glasses for bone healing and biomolecules delivery. *Mater. Sci. Eng. C* **2022**, *106*, 110180. [CrossRef]
18. Baino, F.; Fiume, E. 3D Printing of Hierarchical Scaffolds Based on Mesoporous Bioactive Glasses (MBGs)—Fundamentals and Applications. *Materials* **2020**, *13*, 1688. [CrossRef]
19. Yan, X.; Yu, C.; Zhou, X.; Tang, J.; Zhao, D. Highly Ordered Mesoporous Bioactive Glasses with Superior In Vitro Bone-Forming Bioactivities. *Angew. Chem. Int. Ed.* **2004**, *43*, 5980–5984. [CrossRef]
20. Wen, C.; Bai, N.; Luo, L.; Ye, J.; Zhan, X.; Zhang, Y.; Sa, B. Structural behaviorand in vitro bioactivity evaluation of hydroxyapatite-like bioactive glass based on the SiO_2-CaO-P_2O_5system. *Ceram. Int.* **2021**, *47*, 18094–18104. [CrossRef]
21. Baranowska, A.; Lesniak, M.; Kochanowicz, M.; Zmojda, J.; Miluski, P.; Dorosz, D. Crystallization Kinetics and Structural Properties of the 45S5 Bioactive Glass and Glass-Ceramic Fiber Doped with Eu^{3+}. *Materials* **2020**. *13*, 1281. [CrossRef] [PubMed]
22. Kissinger, H.E. Reaction kinetics in differential thermal analysis. *Anal. Chem.* **1957**, *29*, 1702–1706. [CrossRef]
23. Vyazovkin, S. Kissinger Method in Kinetics of Materials: Things to Beware and Be Aware of. *Molecules* **2020**, *25*, 2813. [CrossRef] [PubMed]
24. Aguiar, H.; Serra, J.; González, P.; León, B. Structural study of sol–gel silicate glasses by IR and Raman spectroscopies. *J. Non-Cryst. Solid* **2009**, *355*, 475–480. [CrossRef]
25. Macon, A.L.B.; Lee, S.; Poologasundarampillai, G.; Kasuga, T.; Jones, J.R. Synthesis and dissolution behaviour of CaO/SrO-containing sol–gel-derived 58S glasses. *J. Mater. Sci.* **2017**, *52*, 8858–8870. [CrossRef]

26. Martin, R.A.; Yue, S.; Hanna, J.V.; Lee, P.D.; Newport, R.J.; Smith, M.E.; Jones, J.R. Characterizing the hierarchical structures of bioactive sol–gel silicate glass and hybrid scaffolds for bone regeneration. *Philos. Trans. R. Soc. A* **2012**, *370*, 1422–1443. [CrossRef] [PubMed]
27. Catauro, M.; Dell'Era, A.; Ciprioti, S.V. Synthesis, structural, spectroscopic and thermoanalytical study of sol–gel derived SiO_2–CaO–P_2O_5 gel and ceramic materials. *Thermochim. Acta* **2016**, *625*, 20–27. [CrossRef]
28. Innocenzi, P. Infrared spectroscopy of sol–gel derived silica-based films: A spectra-microstructure overview. *J. Non-Cryst. Solids* **2003**, *316*, 309–319. [CrossRef]
29. Ianoul, A.; Coleman, T.; Asher, S.A. UV Resonance Raman Spectroscopic Detection ofNitrate and Nitrite in Wastewater Treatment Processes. *Anal. Chem.* **2002**, *74*, 1458–1461. [CrossRef]
30. Brooker, M.H. Raman and i.r. spectra of zinc, cadmium and calcium nitrate: A study of the low temperature phase transitions in calcium nitrate. *Spectrochim. Acta* **1975**, *32*, 369–377. [CrossRef]
31. Zhu, F.; Zhou, H.; Zhou, Y.; Ge, H.; Liu, H.; Fang, C.; Fang, Y. Ab Initio Investigation of the Micro-species and Raman Spectra in $Ca(NO_3)_2$Solution. *J. Clust. Sci.* **2017**, *28*, 2293–2307. [CrossRef]
32. Todan, T.; Anghel, E.M.; Osiceanu, P.; Turcu, R.V.F.; Atkinson, I.; Simon, S.; Zaharescu, M. Structural characterization of some sol–gel derived phosphosilicate glasses. *J. Mol. Struct.* **2015**, *1086*, 161–171. [CrossRef]
33. Atkinson, I.; Anghel, E.M.; Petrescu, S.; Seciu, A.M.; Stefan, L.M.; Mocioiu, O.C.; Predoana, L.; Voicescu, M.; Somacescu, S.; Culita, D.; et al. Cerium-containing mesoporous bioactive glasses: Material characterization, in vitro bioactivity, biocompatibility and cytotoxicity evaluation. *Micropor. Mesopor. Mat.* **2019**, *276*, 76–88. [CrossRef]
34. Buriti, J.S.; Barreto, M.E.V.; Barbosa, F.C.; Buriti, B.M.A.B.; Souza, J.W.L.; Pina, H.V.; Rodrigues, P.L.; Fook, M.V.L. Synthesis and characterization of Ag-doped 45S5 bioglass and chitosan/45S5-Ag biocomposites for biomedical applications. *J. Therm. Anal. Calorim.* **2021**, *145*, 39–50. [CrossRef]
35. Martınez, A.; Izquierdo-Barba, I.; Vallet-Regı, M. Bioactivity of a CaO-SiO_2 Binary Glasses System. *Chem. Mater.* **2000**, *12*, 3080–3088. [CrossRef]
36. Siqueira, R.L.; Zanotto, E.D. The influence of phosphorus precursors on the synthesis and bioactivity of SiO_2–CaO–P_2O_5 sol–gel glasses and glass–ceramics. *J. Mater. Sci: Mater. Med.* **2013**, *24*, 365–379. [CrossRef] [PubMed]
37. Huang, K.; Cai, S.; Xu, G.; Ren, M.; Wang, X.; Zhang, R.; Niu, S.; Zhao, H. Sol–gel derived mesoporous 58S bioactive glass coatings on AZ31 magnesium alloy and in vitro degradation behavior. *Surf. Coat. Technol.* **2014**, *240*, 137–144. [CrossRef]
38. Zaharescu, M.; Predoana, L.; Pandele, J. Relevance of thermal analysis for sol–gel-derived nanomaterials. *J. Sol-Gel Sci. Technol.* **2018**, *86*, 7–23. [CrossRef]
39. Jones, J.R.; Ehrenfried, L.M.; Hench, L.L. Optimising bioactive glass scaffolds for bone tissue engineering. *Biomaterials* **2006**, *27*, 964–973. [CrossRef]
40. Likitvanichkul, S.; Lacourse, W.C. Apatite–wollastonite glass-ceramics. Part I Crystallization kinetics by differential thermal analysis. *J. Mater. Sci.* **1998**, *33*, 509–5904. [CrossRef]
41. Notingher, I.; Jones, J.R.; Verrier, S.; Bisson, I.; Embanga, P.; Edwards, P.; Polak, J.M.; Hench, L.L. Application of FTIR and Raman spectroscopy to characterisation of bioactive materials and living cells. *Spectroscopy* **2003**, *17*, 275–288. [CrossRef]
42. Senthilkumar, R.P.; Bhuvaneshwari, V.; Ranjithkumar, R.; Sathiyavimal, S.; Malayaman, V.; Chandarshekar, B. Synthesis, characterization and antibacterial activity of hybrid chitosan-cerium oxide nanoparticles: As a bionanomaterials. *Int. J. Biol. Macromol.* **2017**, *104*, 1746–1752. [CrossRef] [PubMed]
43. Padilla, S.; Roman, J.; Carenas, A.; Vallet-Regi, M. The influence of the phosphorus content on the bioactivity of sol–gel glass ceramics. *Biomaterials* **2005**, *26*, 475–483. [CrossRef] [PubMed]
44. Pereira, M.M.; Clark, A.E.; Hench, L.L. Calcium phosphate formation on sol-gel-derived bioFactive glasses in vitro. *J. Biomed. Mater. Res.* **1994**, *28*, 693–698. [CrossRef]
45. De Ligny, D.; Neuville, D.R. 12-Raman spectroscopy: A valuable tool to improve our understanding of nucleation and growth mechanism. In *From Glass to Crystal. Nucleation, Growth and Phase Separation: From Research to Applications*; Neuville, D.R., Cormier, L., Caurant, D., Montagne, L., Eds.; EDP Sciences: Les Ulis, France, 2017; pp. 319–344.
46. Taniguchi, T.; Watanabe, T.; Sugiyama, N.; Subramani, A.K.; Wagata, H.; Matsushita, N.; Yoshimura, M. Identifying defects in ceria-based nanocrystals by UV resonance Raman spectroscopy. *J. Phys. Chem. C* **2009**, *113*, 19789–19793. [CrossRef]
47. Olivares, M.; Zuluaga, M.C.; Ortega, L.A.; Murelaga, X.; Alonso-Olazabal, A.; Urteaga, M.; Amundaray, L.; Alonso-Martina, I.; Etxebarria, N. Characterisation of fine wall and eggshell Roman pottery by Raman spectroscopy. *J. Ram. Spectrosc.* **2010**, *41*, 1543–1549. [CrossRef]
48. Huang, E.; Chen, C.H.; T. Huang, T.; Lin, E.H.; Xu, J.-A. Raman spectroscopic characteristics of Mg-Fe-Ca pyroxenes. *Am. Mineral.* **2000**, *85*, 473–479. [CrossRef]
49. Uy, D.; O'Neill, A.E.; Xu, L.; Weber, W.H.; McCabe, R.W. Observation of cerium phosphate in aged automotive catalysts using Raman spectroscopy. *Appl. Catal. B* **2003**, *41*, 269–278. [CrossRef]
50. Awonusi, A.; Michael, A.E.; Morris, D.; Tecklenbur, M.M.J. Carbonate Assignment and Calibration in the Raman Spectrum of Apatite. *Calcif. Tissue Int.* **2007**, *81*, 46–52. [CrossRef]
51. Zambon, A.; Malavasi, G.; Pallini, A.; Fraulini, F.; Lusvardi, G. Cerium Containing Bioactive Glasses: A Review. *ACS Biomater. Sci. Eng.* **2021**, *7*, 4388–4401. [CrossRef]

52. Panah, N.G.; Atkin, R.; Sercombe, T.B. Effect of low temperature crystallization on 58S bioactive glass sintering and compressive strength. *Ceram. Int.* **2021**, *47*, 30349–30357. [CrossRef]
53. Martin, M.T.; Maliqi, L.; Keevend, K.; Guimond, S.; Ng, J.; Armagan, E.; Rottmar, M.; Herrmann, I.K. One-Step Synthesis of Versatile Antimicrobial Nano-Architected Implant Coatings for Hard and Soft Tissue Healing. *ACS Appl. Mater. Interfaces* **2021**, *13*, 33300–33310. [CrossRef]
54. Augis, J.A.; Bennett, J.E. Calculation of the Avrami Parameters for Heterogeneous Solid State Reactions Using a Modification of the Kissinger Method. *J. Thermal. Anal. Calorim.* **1978**, *13*, 283–292. [CrossRef]
55. Amorósa, J.E.; Blascoa, E.; Morenoa, A.; Marínc, N.; Feliu, C. Sinter-crystallisation kinetics of a SiO_2–Al_2O_3–CaO–MgO–SrO glass-ceramic glaze. *J. Non-Cryst. Solids* **2020**, *532*, 119900. [CrossRef]

gels

MDPI

Article

Photodynamic Therapy with Natural Photosensitizers in the Management of Periodontal Disease Induced in Rats

Laura Monica Dascalu (Rusu) [1], Marioara Moldovan [2], Codruta Sarosi [2,*], Sorina Sava [1,*], Alexandra Dreanca [3,†], Calin Repciuc [3], Robert Purdoiu [3], Andras Nagy [3], Mîndra Eugenia Badea [4], Ariadna Georgiana Paun [5], Iulia Clara Badea [4,†] and Radu Chifor [4]

1 Department of Prosthodontics and Dental Materials, Iuliu Hatieganu University of Medicine and Pharmacy, 31 Avram Iancu Str., 400083 Cluj-Napoca, Romania; dascalu.monica@umfcluj.ro

2 Raluca Ripan Institute of Chemistry, Babes-Bolyai University, 30 Fantanele Str., 400294 Cluj-Napoca, Romania; marioara.moldovan@ubbcluj.ro

3 Pathophysiology/Toxicology Department, Faculty of Veterinary Medicine, University of Agricultural Science and Veterinary Medicine, 3-5 Calea Manastur, 400372 Cluj-Napoca, Romania; alexandradreanca@gmail.com (A.D.); calin_c_repciuc@yahoo.com (C.R.); robert.purdoiu@usamvcluj.ro (R.P.); nagyandras26@gmail.com (A.N.)

4 Department of Preventive Dental Medicine, Iuliu Hatieganu University of Medicine and Pharmacy, 31 Avram Iancu Str., 400083 Cluj-Napoca, Romania; mebadea@umfcluj.ro (M.E.B.); iulia.badea@umfcluj.ro (I.C.B.); chifor.radu@umfcluj.ro (R.C.)

5 Department Community Medicine, Iuliu Hațieganu University of Medicine and Pharmacy, 31 Avram Iancu Str., 400083 Cluj-Napoca, Romania; badea.ariadna@umfcluj.ro

* Correspondence: codruta.sarosi@gmail.com (C.S.); savasorina@yahoo.com (S.S.); Tel.: +40-742010325 (C.S.)

† These authors contributed equally to this work.

Citation: Dascalu, L.M.; Moldovan, M.; Sarosi, C.; Sava, S.; Dreanca, A.; Repciuc, C.; Purdoiu, R.; Nagy, A.; Badea, M.E.; Paun, A.G.; et al. Photodynamic Therapy with Natural Photosensitizers in the Management of Periodontal Disease Induced in Rats. *Gels* **2022**, *8*, 134. https://doi.org/10.3390/gels8020134

Academic Editors: Arish Dasan, Ashokraja Chandrasekar and Nupur Kohli

Received: 19 January 2022
Accepted: 17 February 2022
Published: 20 February 2022

Abstract: This study aims to investigate the effect of new natural photosensitizers (PS) (based on oregano essential oil, curcuma extract, and arnica oil) through in vitro cytotoxicity and biological tests in rat-induced periodontal disease, treated with photodynamic therapy (aPDT). The cytotoxicity of PS was performed on human dental pulp mesenchymal stem cells (dMSCs) and human keratinocyte (HaCaT) cell lines. Periodontal disease was induced by ligation of the first mandibular molar of 25 rats, which were divided into 5 groups: control group, periodontitis group, Curcuma and aPDT-treated group, oregano and aPDT-treated group, and aPDT group. The animals were euthanized after 4 weeks of study. Computed tomography imaging has been used to evaluate alveolar bone loss. Hematological and histological evaluation showed a greater magnitude of the inflammatory response and severe destruction of the periodontal ligaments in the untreated group.. For the group with the induced periodontitis and treated with natural photosensitizers, the aPDT improved the results; this therapy could be an important adjuvant treatment. The obtained results of these preliminary studies encourage us to continue the research of periodontitis treated with natural photosensitizers activated by photodynamic therapy.

Keywords: gels; photosensitizers; photodynamic therapy; cytotoxicity; computed tomography

1. Introduction

There are a number of oral diseases that pose a major threat to public health worldwide, among which periodontal diseases and dental caries are the most invasive and difficult to treat. The need to develop an alternative prevention treatment with antibacterial agents is due to the side effects of conventionally used treatments (antibacterial, anti-inflammatory, or antibiotic drugs) and the increasing bacterial resistance to them. Thus, natural products and photodynamic therapy has become a topical and important research subject.

Periodontitis is a very common chronic inflammatory disease nowadays. It is caused by infection with the presence of various types of oral bacteria, the so-called "red complex", which includes, among others, Porphyromonas gingivalis, Tannerella forsythia,

Prevotella intermedia as well as other Gram-negative anaerobes wich are organized as dental biofilm [1–3].

For periodontal disease to occur, bacteria should be able to colonize the subgingival space and promote the appearance of virulent factors that could eventually affect the host tissue. There are several studies in the literature that have concluded that microbial etiology of periodontal disease has been attributed to a varied number of bacteria organized in biofilm and not just a single microorganism [4–6].

Conventional mechanical debridement (scaling and root planing) can achieve a temporary decrease in the subgingival levels of pathogens. However, organisms cannot be removed from the majority of periodontal pockets by mechanical therapy alone. In addition, various systemic and local chemical antimicrobial agents (chlorhexidine, enamel matrix derivative, and hyaluronic acid) have been introduced for the treatment of periodontitis, which suppress periodontal pathogens with greater efficacy than mechanical techniques and improve the results of conventional mechanical therapeutic techniques. Some disadvantages of antimicrobial agents' usage (such as antibiotics) include antibiotic resistance, immune suppression, and other unfavorable reactions. Considering the complications above, it is necessary to expand research in an attempt to find alternative antimicrobial techniques, such as natural agents for antimicrobial therapy. One of them is the use of lasers and photodynamic therapy, which might be effective in eliminating microbes in local and superficial infections in the presence of natural photosensitizers. The literature presents many studies regarding antimicrobial photodynamic therapy (aPDT). Antimicrobial chemotherapy may reduce periodontal pathogens and enhance the results of conventional mechanical treatment [1,7–10].

Antimicrobial photodynamic therapy is a new and promising alternative that aims to remove or reduce pathogenic microorganisms, both Gram-positive and Gram-negative bacteria, as well as viruses, parasites and fungi. Current research is looking for new approaches that can be bactericidal but also have advantages over traditional antibiotic therapy. PDT is a non-thermal photochemical reaction that requires the co-participation of three factors, namely visible light at an appropriate wavelength, oxygen and a photosensitizer. The properties of PSs have been studied in the literature, including their affinity for binding to the bacterial wall and thus efficiently generating reactive oxygen species (ROS) upon photostimulation. Various studies show that there are different types of ROS generated during aPDT, among which singlet oxygen (1O2) is considered the most potential, which means that it is mainly responsible for photo damage and cytotoxic reactions [7,9,11,12].

Turmeric (isolated from *Curcuma longa* L.), is generally known and used as a spice, but has also been shown to have therapeutic effects. Among them is worth mentioning the therapeutic effects in case of liver diseases, wounds and inflamed joints and it's also known for its blood purification and its antimicrobial effect. In terms of cytotoxicity, turmeric did not show toxic effects on cell cultures and animal studies. It has a wide absorption range of 300-500 nm (with maximum absorption at 430 nm) and produces strong phototoxic effects, thus being a suitable compound for use as PS. Being a fat-soluble material, this PS has certain restrictions, namely, it requires an oil or other synthetic material to make its water solubility possible, such as for example arnica oil [13].

Oregano essential oil is a volatile compound in which carvacrol, p-cymene, γ-terpinene, and α-humulene have been identified. Like other essential oils, it has antimicrobial and antioxidant properties, and is also able to inhibit the growth of *Escherichia coli* and *Staphylococcus aureus*. The compounds mainly responsible for its antioxidant and antibacterial properties are carvacrol and thymol. Its maximum absorption is at 270 nm [14].

aPDT is a new approach that involves combining a non-toxic PS and a low intensity visible light source. This method of treatment has been shown to have an important antimicrobial effect, so it's now an alternative for treating biofilm-related diseases. aPDT has been researched as an alternative and promising method for the eradication of oral pathogenic bacteria that over time can lead to endodontic disease, periodontitis, periimplantitis and caries [15–19].

The novelty of this study is the formulation and characterization of new photosensitizers used in photodynamic therapy, based on natural extracts and investigated through biological and cytotoxicity tests. The purpose of the study was to obtain new photosensitizers, using oxygen-enriched water and essential oils, to enhance the biological properties of photodynamic antimicrobial therapy used in the treatment of experimentally induced periodontal disease.

2. Results and Discussion

2.1. Cytotoxicity

Data are shown as the percentage of the average proliferation rate compared to the un-treated control cells. All tests were made three times in exactly the same way and data are presented in Figure 1. The statistic results presents no significant viability differences between the control group and that with experimental gels. As can be seen in Figure 1, the compounds studied had a reduced cytotoxic effect on cell cultures. The most significant decrease in viability was recorded for turmeric gel in human dental pulp mesenchymal stem cells (dMSCs). The toxicity results on the human keratinocyte cell lines (HaCaT) of investigated oregano gel present lower values than curcuma and control gels.

The ANOVA test had a significant interaction between the treated groups ($p = 0.40655$). Therefore, viability shows that the experimental gels were well tolerated by human dental pulp mesenchymal stem cells and human keratinocyte cell lines, with no signs of toxicity to any of the materials tested.

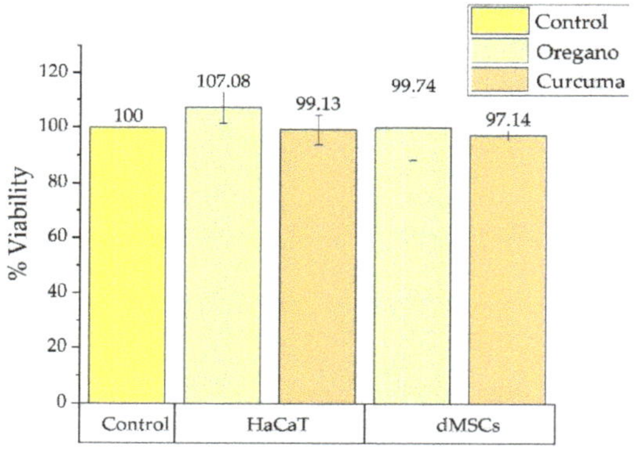

Figure 1. Cell viability of photosensitizers on human dental pulp mesenchymal stem cells (dMSCs) and human keratinocyte cell lines (HaCaT) compared with untreated control. Each bar represents mean and standard deviation ($n = 3$).

2.2. Clinical Evaluation

Body weight measurements were assessed in order to correlate the clinical status of the animal with the pathology induced. Statistical differences were interpreted (before and after treatment) by using Student t-tests. Even though all groups experienced gain in body weight, groups treated with photosensitizers and aPDT showed a more significant degree of weight gain (Table 1).

Table 1. Mean weight change (standard deviation) and weight variability (g) before and after treatment.

Crt. No.	Groups	Body Weight Initially (g)	Final (g)	Weight Gain (%)
1.	Control group (M)	452 ± 32.71	551 ± 24.14	21.9
2.	Periodontitis group (P)	491.4 ± 25.47	530.2 ± 29.07	7.85
3.	Curcuma group (GC)	475.6 ± 10.8	570 ± 14.85 **,^^	19.84
4.	Oregano group (GO)	516 ± 17.92	576.4 ± 26.73 *,^^	11.70
5.	Laser group (L)	438 ± 10.74	513.2 ± 27.39	17.16

(Mean ± SD) (*t*-tests, $n = 5$; * $p < 0.05$, ** $p < 0.005$). Compared with group M, (^^ $p < 0.005$).

2.3. Computed Tomography (CT) Analysis

CT images of experimental groups of Wistar rats on the 1st and 21st day of treatment after the periodontal disease was induced are presented in Figures 2 and 3.

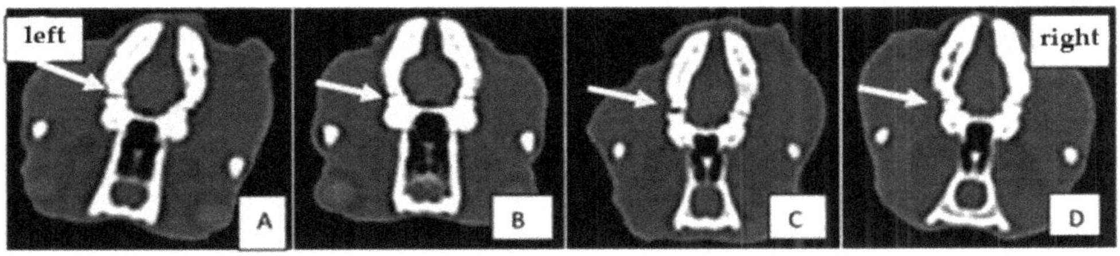

Figure 2. Comparative analysis of CT images of experimental groups on the first day of treatment: (**A**) positive control group; (**B**) aPDT + curcumin group; (**C**) aPDT + oregano group; (**D**) laser therapy group. The arrow indicates the first molar on the left hemimandible where periodontal disease was induced.

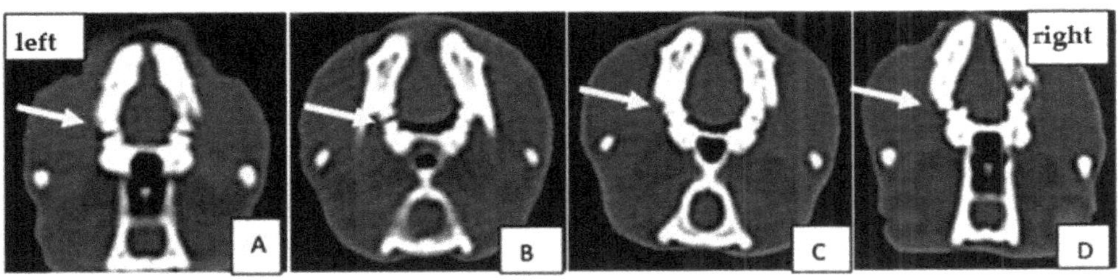

Figure 3. Comparative analysis of CT images of experimental groups on 21st day of treatment: (**A**) positive control group; (**B**) aPDT + curcumin group; (**C**) aPDT + oregano group; (**D**) laser therapy group. The arrow indicates the first molar on the left hemimandibula where periodontal disease was induced.

CT scans showed bone loss at the beginning of the experiment when periodontal disease was induced. At the end of the treatment, in the group treated with experimental photosensitizers, an improvement in bone loss and bone density was observed. The analyzed CT images show that after performing the laser treatment in the presence of natural photosensitizers (curcuma or oregano), both values of bone density and periodontal space improved. The groups treated with PS and light showed better results than the group treated with laser only. This demonstrates the treatment-enhancing effect when a PS is combined with laser therapy.

2.4. Histological Analysis

Following the histological analysis, the tooth and the periodontal ligament, respectively, and the dental alveolar bone (the dental support device) showed a normal appearance in the M group (Figure 4).

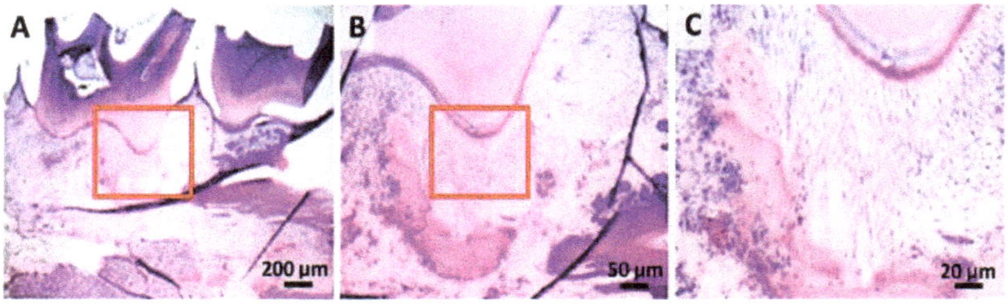

Figure 4. Control group: (**A**) histopathological images from the dental crown and partially from the dental root area (hematoxylin-eosin); (**B**,**C**) higher resolution, 50 µm and 20 µm, of the marked zone with red.

In the cervical and interdental space, moderate gingival retractions associated with chronic and superficial focal gingivitis were observed (group P). The superficial area of the inflammatory outbreak is covered by an abundant serum–leukocyte crust mixed with tissue and fodder debris. In addition, a moderate-segmental osteoclastic resorption of the alveolar bone was observed combined with suppurative (moderate) periodontitis extending from the previously described gingival defect. At the level of the sub-gingival area, abundant granulation tissue that partially delimits the septic focal point and replaces the dental ligament focal point was observed (Figure 5).

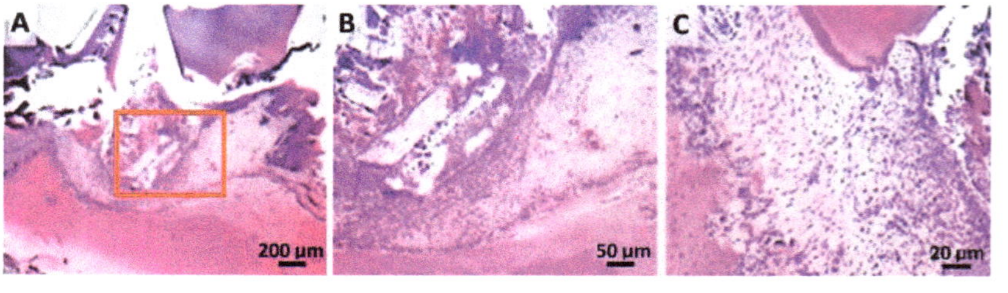

Figure 5. Periodontisis group: (**A**,**B**) the superficial area of the inflammatory outbreak covered by tissue and fodder debris; (**C**) detail of the dental support device, with moderate-segmental osteoclastic resorption of the alveolar bone and suppurated periodontitis; (hematoxylin–eosin).

In the L group (Figure 6), an important hyperplasia and hyperkeratosis (orthokeratotic) of the gingival epithelium with the formation of irregular, anatomic epithelial papillae, separated by a fibro-vascular inflammatory stroma (primarily neutrophils), moderate-segmental osteoclastic resorption of the alveolar bone, and suppurated periodontitis was observed. The inflammatory process was represented by bands and degenerated neutrophils in mixture with rarely mononuclear cells and reactive fibroblasts.

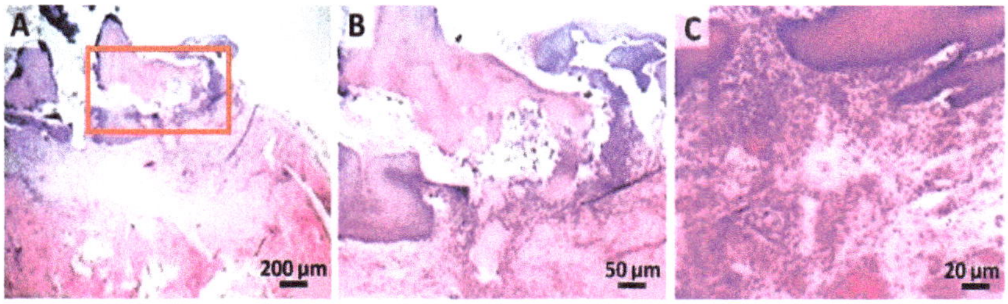

Figure 6. Laser group: (**A**) hyperplasia and marked hyperkeratosis of the gingival epithelium (hematoxylin–eosin); (**B**) high resolution, 50 µm, of the marked zone with red from image A; (**C**) numerous neutrophils with rare mononuclear and reactive fibroblasts.

In the GC group (Figure 7), an important hyperplasia and hyperkeratosis (orthokeratosis) in the gingival epithelium was noticed. Therefore, the formation of irregular, anastomosing epithelial papillae, separated by an abundant (inflammatory granulation tissue) fibro-vascular inflammatory stroma (primarily neutrophils and macrophages) was observed. Furthermore, the superficial gingival area presented a focal ulcer (minimal) covered by a serum cellular crust mixed with cellular debris and fodder. In Figure 7C, a detailed aspect of inflammatory granulation tissue with abundant neutrophils (viable and degenerate) mixed with rarely mononuclear cells and reactive fibroblasts were highlighted.

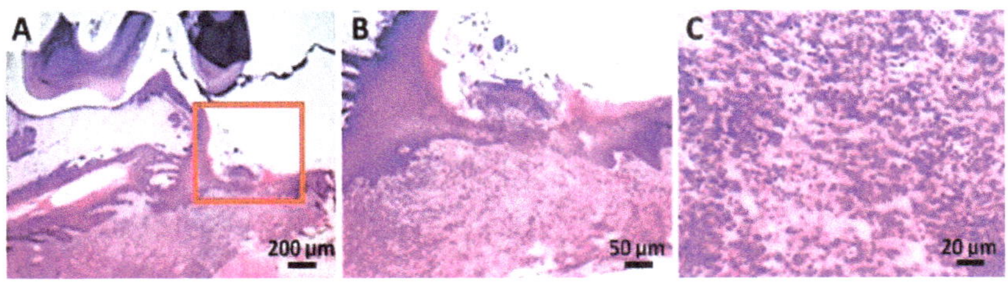

Figure 7. GC group: (**A**) hyperplasia and marked hyperkeratosis of the gingival epithelium. The superficial gingival area presents a focal ulcer (minimal), covered by a serocellular crust mixed with cellular debris and forage (the area demarcated by the rectangle); (**B**) higher resolution, 50 µm, of the marked zone with red; (**C**) detail of inflammatory granulation tissue, with the abundance of neutrophils (viable and degenerate) in combination with rare mononuclear and reactive fibroblasts.

In the GO group, marked gingival epithelial hyperplasia and hyperkeratosis (orthokeratosis) and partial replacement of the dental ligament with partially oriented fibro-vascular connective tissue was observed (Figure 8).

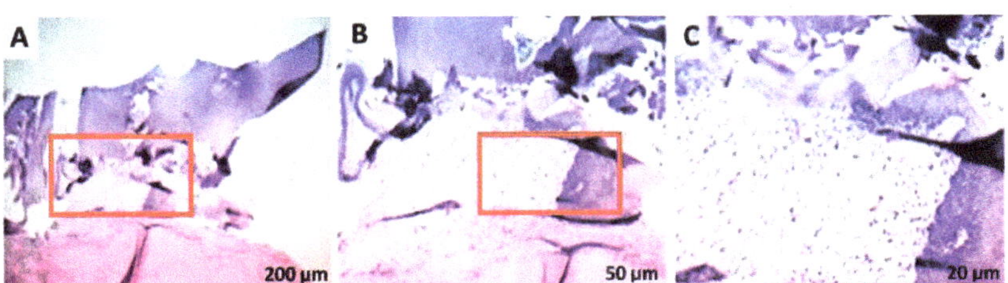

Figure 8. GO group: (**A**,**B**) hyperplasia and marked hyperkeratosis of the gingival epithelium, and partial replacement of the dental ligament with partially oriented fibro-vascularized connective tissue; (**C**) high resolution, 20 µm, of the marked zone with red from image 3.

2.5. Complete Blood Count

A significant increase in the total WBC count in the P group (11.16 ± 2.06) was observed compared to the M group (8.46 ± 1.13) (Figure 9) ($p < 0.05$); however, both recorded values were within the physiological limits of the species (4–12 × 10^9/L). At the same time, a statistically significant decrease ($p < 0.05$) of the cells mean number (monocytes) in both P (0.13 ± 0.11) and L groups (0.12 ± 0.10) compared to the M group (0.50 ± 0.28) was observed (Figure 9). Nevertheless, the values evidenced by monocytes are within the species' physical limits (0–0.98 × 10^9/L).

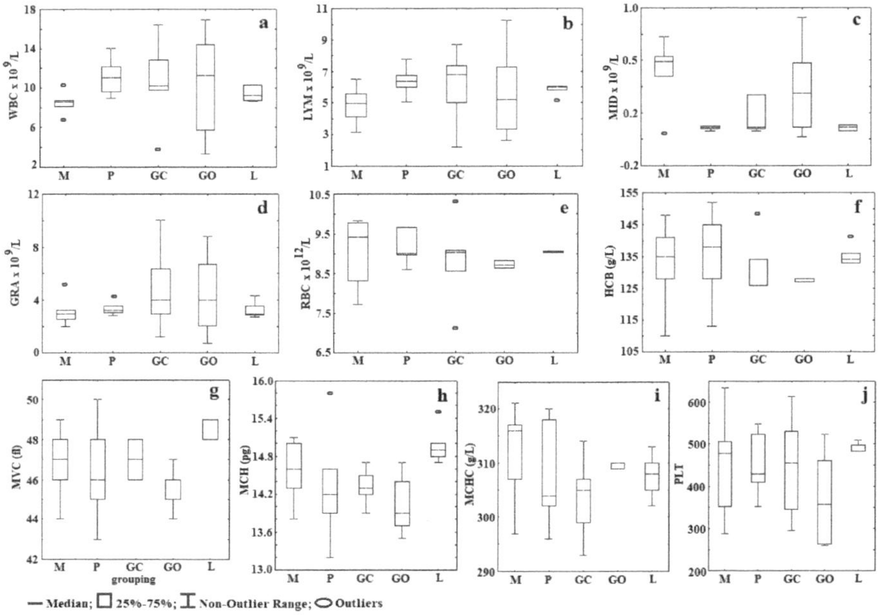

Figure 9. Means and standard deviations for all groups and variables without outliers: (**a**) WBC; (**b**) LYM; (**c**) MID; (**d**) GRA; (**e**) RBC; (**f**) HGB; (**g**) MVC; (**h**) MCH; (**i**) MCHC; (**j**) PLT. (Physiological values: WBC: 4–12 × 10^9/L, LYM: 2–14.1 × 10^9/L, MID: 0–0.98 × 10^9/L, GRA: 0.1–5.4 × 10^9/L, RBC: 9–15 × 10^{12}/L, HGB 90–150 mg/dL, HCT: 24–45%, PLT: 250–750 × 10^9/L).

Encouraging results was obtained in the treatment of some clinical pathologies using photodynamic therapy that involve a photochemical reaction between photosensitizer and laser light [20,21].

The antimicrobial effect of the photodynamic therapy is based on an oxidative explosion due to the light therapy and is based on the deterioration of the cellular structures and of the biomolecules, making, thus, a nonspecific mechanism. aPDT requires the presence of three components: (I) a non-toxic dye, the so-called photosensitizer; (II) visible light of a corresponding wavelength; (III) molecular oxygen. The absorption of light by the PS leads to a transition to its triple state, through which there are two reaction mechanisms to allow the PS to regain its baseline state. In the type I mechanism, the charge is transferred to a substrate or to molecular oxygen generating reactive oxygen species such as hydrogen peroxide, and oxygen radicals such as superoxide ions or free hydroxyl radicals. In the type II mechanism, only energy—not charged—is transferred directly to molecular oxygen, from which simple reactive oxygen ($^1O_2$) comes [22–24].

Therefore, combining photosensitizers with light and molecular oxygen species derived from enhanced water could generate a mechanism of cellular death by provoking cytotoxicity to bacterial etiological factors, which are implicated in periodontal disease.

Major active ingredients of turmeric include three curcuminoids: curcuma (diferuloylmethane, the primary constituent who gives the bright yellow color), desmethoxycurcumin, and bisdemethoxycurcumin, also a volatile oils (turmerone, atlantone and zingiberone), sugars, proteins, and resins [25].

The studies demonstrated that curcuma have important antibacterial, antiinflamatory, antioxidant and anticarcinogenic effects. The anti-infective effects makes curcuma appropriate in the wound-healing. Recent studies on the wound-healing properties of curcuma evidence its ability to increase the granulation tissue formation, collagen deposition, tissue changing structure, and wound reducing, favoring, in this way, the healing process [26–30].

Two of the major bioactive constituents, turmerone and atlantone, are presented in turmeric. Curcuma is a hydrophobic photosensitizer that is soluble in *dimethyl sulfoxide* (DMSO), acetone, ethanol, and oils; therefore, it was decided to add arnica oil to the curcuma-based gel.

Combined with the natural compounds isolated from Arnica Montana oil, a medicinal plant widely used as an herbal remedy containing terpenoids, sesquiterpene lactones, flavonoids, and tannins with anti-inflammatory, antifungal, antimicrobial, and antibiotic properties [31], this could provide a potential mechanism of action highlighting the benefits of our implemented therapy [32].

Essential oils of oregano (Origanum vulgare) are widely recognized for their antimicrobial activity, as well as their antiviral and antifungal properties. It is one of the most used aromatic plants, whose essential oils are particularly rich in mono- and sesquiterpenes [33]. In vitro studies of experimental gels with essential oils show anticariogenic and antibiofilm activities [34].

Nevertheless, recent investigations have demonstrated that these compounds are also potent antioxidant and anti-inflammatory agents. Carvacrol (CV), the main compound found in essential oils of oregano, is a phenolic monoterpenoid, which possesses a wide range of bioactive properties [35] and has been isolated within our experimental compound. These properties of oregano essential oils are a potential interest to the food, cosmetic, and pharmaceutical industries [36].

According to the literature, curcumin used as photosensitizer activated by blue light in the aPDT procedure has the potential for reducing the bacterial count in periodontal infection. Etemadi et al. states that several researchers have shown that curcuma as a photosensitizer activated by a blue wavelength is effective in the elimination of the various bacterial species involved in periodontal disease [33].

Belinello-Souza et al. found that to the treated animals with aPDT compared with scaling and root planning (SRP) group, the bone gain was approximately 30%, following 7 days after periodontal intervention. [37].

Pre-clinical study on animal models offer important information regarding the treatment methods in investigating the pathogenesis of periodontal disease. In addition to the immunological and microbiological characteristics of rats, the histological features of periodontal collagen fibrils, alveolar bone, cellular cement, connective tissue, junctional epithelium, oral gingival epithelium, and sulcular epithelium are similar to human periodontal tissues [14,38]. This model cannot completely represent all aspects of periodontitis in humans, but is considered an effective method for the exploration of its mechanisms [39], which is why this experimental protocol was chosen.

According to the IACUC standards, body weight is considered an indicator of the health and well-being of laboratory animals. This demonstrates that dental therapy with our materials tested in combination with photodynamic therapy does not negatively affect food intake, apprehension, and mastication. The animals in the treated groups (GO, GC, L) gained approximately 80 g of weight during the 4-week period, whereas those untreated gained only 40 g. Thus, this indicates a variable period of caloric restriction most likely induced by the pain caused by the periodontal disease, as well as dental and gingival mobility and bleeding [40]. According to Toth et al., weight gain is a positive indicator for animal welfare [41].

A complete blood count may reveal general pathological conditions of the body as evidenced by anemia, systemic infections, or blood neoplasm. This is also performed for monitoring a disease or a medical treatment [42].

At the end of the experiment, all values recorded in hematological analysis were within the physiological range of species [43]. This indicates the lack of an inflammatory or infectious chronic systemic process. After 4 weeks, an inflammatory reaction due to periodontitis or local treatment cannot be observed. The lack of adverse systemic hematological reactions is correlated with the clinical symptoms of the rats, having a good maintenance status. Different studies have investigated the effect of aPDT treatment on reducing the inflammation in the gingival tissue of Wistar rats with periodontal disease induced using ligature. Carvalho et al. observed the benefits of aPDT in periodontal disease, due to their immune modulator response in reducing the inflammatory effect and obtaining a resorption of bone tissue [44].

Histopathological analysis in this study demonstrated a greater magnitude of inflammatory response and severe destruction of periodontal ligaments in rats who did not receive treatment after ligation removal. It became obvious that ligation was effective in developing experimental periodontal disease. Our experimental protocol is reminiscent of that of Graves et al., which observed that the ligation favors bacterial plaque accumulation, epithelial ulceration, and periodontal tissue invasion by bacteria [45,46].

In the periodontitis group, bone loss due to the osteoclastic activity conferred by the local infections and inflammatory processes was observed. The same bone resorption, but on a smaller scale, was observed in the laser-treated group.

Remarkably, no bone resorption and minimal inflammatory local reactions were observed in the two experimental photosensitizers groups. However, it is worth mentioning that oregano-based treatment (protocol) has the most beneficial impact based on its capacity to provoke regeneration and the total absence of inflammation.

Knowledge about macro- and micro-structural characteristics of bone tissue may improve the ability to estimate in vivo its quality and quantity. For this purpose, computed tomography (CT) imaging techniques are appropriate and enable assessing the micro-architecture of bone with 2D and 3D quantitative evaluation [47]. CT investigation allows us to quantify different bone parameters, such as geometry, mass, and mineral density, simultaneously. The limitations of the study are: a lack of tests for the group with induced periodontal disease treated only with PS, without light therapy; MTT assay without phototoxicity on the cell lines, and the use of the same laser wavelength for curcuma and oregano, even if oregano absorbs on 370 nm and curcuma on 440 nm.

3. Conclusions

This study demonstrates that aPDT was an effective adjuvant treatment of induced periodontal disease in groups treated with natural photosensitizers based on oregano essential oil and curcuma extract. Histopathological analysis suggested the anti-inflammatory effect of aPDT in the presence of natural PS tested in this study.

Periodontal defects in rat mandibles were displayed using CT and compared with histological specimens. The results of histological and computed tomography analyses sustain that the bone loss at the alveolar level obtained through the induced periodontitis, was improved at the end of the treatment.

The cytotoxic effect of new photosensitizers on dental pulp mesenchymal stem cells (dMSCs) and human keratinocyte (HaCaT) in vitro was observed.

In this study, we evidenced the adjuvant effect of natural photosensitizers, based on curcuma extract and oregano essential oil, on induced periodontal disease. The obtained results of these preliminary studies encourage us to continue the research of periodontitis treated with natural photosensitizers activated by photodynamic therapy.

4. Materials and Methods

Newly developed gels based on natural compounds were used as photosensitizers in the aPDT experimental protocol. Natural revealers contain nanocapsules, which include an organic phase based on the essential oil of oregano (Young Living Europe B.V., Groningen, The Netherlands) and curcuma extract with arnica oil, having the active ingredient wrapped in a thin film of polycaprolactone. This technique ensures the controlled release of the active substance through the diffusion phenomenon.

The gels were prepared according to the following proportions: glycerol (Sigma–Aldrich Inc., St. Louis, MO, USA) in a weight ratio of 1:1, 60 mL Kaqun® water (KAQUN Distribution Kft., Nagytarcsa, Hungary), and 0.015% salicylic acid solution. The gels formed were divided into equal parts, in which oregano essential oil (20–50 μg/g according to GSMS analysis) and curcuma extract (70 μg/g according to HPLC Chromatography) were added.

The **cytotoxicity** assay of oregano essential oil and curcuma extract were performed using human dental pulp mesenchymal stem cells (dMSCs) and human keratinocyte HaCaT cell lines. The cells were cultured according to standard conditions. The potential cytotoxicity of the essential oils was assessed with (4,5-dimethylthiazol-2-yl)-2,5-diphenyltetrazolium bromide (MTT) assay. In order to obtain cell suspensions, the cells were treated with 0.25% trypsin-EDTA, and after centrifugation (1500 rpm for 5 min), 1×10^4 cells/well were seeded on 96-well plates in 200 μL complete culture medium. After 24 h, 0.1 g from each product was tested using a hanging cell culture insert with a pore size of 0.4 μm. Control samples were represented by untreated cells. Each experimental condition was performed in triplicate. Cell proliferation analysis was performed after 24 h. After 24 h, the medium was removed and 100 μL of 1 mg/mL MTT solution (Sigma–Aldrich, St. Louis, MO, USA) was added. After 3 h of incubation at 37 °C in dark, the MTT solution was removed from each well and 150 μL of DMSO (dimethyl sulfoxide) solution (Fluka, Buchs, Switzerland) was added. Spectrophotometric readings at 450 nm were performed with a BioTek Synergy 2 microplate reader (Winooski, VT, USA).

The cell viability of each medium was calculated using the following formula:

$$\%\ \text{Viability} = \left(\frac{\text{Absorbance of samples}}{\text{Absorbance of control}}\right) \times 100 \quad (1)$$

4.1. Statistical Analysis

The statistical analysis was completed with one-way ANOVA and Tukey Test, using GraphPad Prism version 4.00 for Windows (GraphPad Software, San Diego, CA, USA). The data obtained were presented as the mean of OD540 triplicate measurements ± standard deviation (SD), and a p value less than 0.05 was considered statistically significant.

4.2. Pre-Clinical Study

This study was performed on 25 adult male Wistar rats (14 months of age) that weighed 420–530 g. Animals were housed in the Establishment for Breeding and Use of Laboratory Animals of USAMV (Cluj-Napoca, Romania) in standard conditions, temperature 22–23 °C, humidity 55%, and 12-h light/dark cycle. The rats were kept in plastic cages with free access to standard rodent granular food (Cantacuzino Institute, Bucharest, Romania) and water ad libitum. The rats were allowed to acclimate to the laboratory environment for a period of 3 weeks. All procedures that involved the use of laboratory animals followed the European guidelines and rules 337, as established by the EU Directive 2010/63/EU and the Romanian law 43/2014 and were preformed by an experienced practitioner. The study protocol was approved by the Research Ethics Committee of the University of Agricultural Sciences and Veterinary Medicine Cluj-Napoca, Romania, and they were authorized by the State Veterinary Authority (aut. No. 52/30.03.2017).

4.3. Experimental Design

In order to investigate the natural photosensitizers, an experimental periodontal disease was induced [48,49]. For all procedures, the rats were anaesthetized with ketamine (60 mg/kg) and xylazine (6 mg/kg), which were administered via intramuscular injection, according to Flecknell et al. [50]. Access to the oral cavity was achieved with a retractor that provided constant opening of the mouth and that held away the cheeks and the tongue. The left-mandibular first molar from each rat, in all surgical groups, was selected to receive a cotton 4.0 ligature in a submarginal position to induce experimental periodontitis. The ligatures were removed after 7 days (Figure 10). Postoperatively, all animals received subcutaneous injections with tramadol (10 mg/bw) in order to achieve the analgesic effect.

Animals were randomly assigned to 5 groups. The groups (n = 5) were assigned according to the following treatments applied locally: group 1 was left without surgical intervention representing the control group (M, n = 5); group 2 (P, n = 5) received surgical intervention and was left untreated; group 3 (GC, n = 5) was treated with a curcuma photosensitizer and aPDT; group 4 (GO, n = 5) was treated with oregano photosensitizer and aPDT, and group 5 (L, n = 5) was treated with laser only.

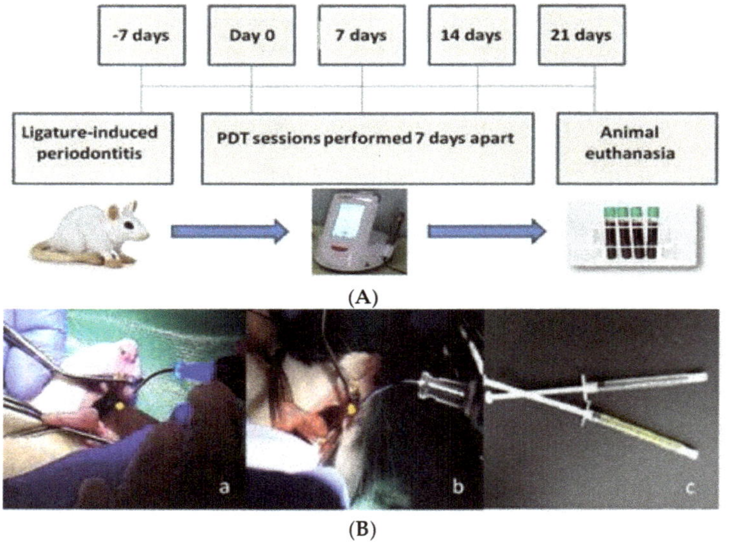

Figure 10. (**A**). The treatment regimen followed in the periodontal disease induced in rats; (**B**) (**a**,**b**) different stages during aPDT procedure; (**c**) oregano and curcuma-based gels used in the study.

Treatment was carried out with the use of low-intensity laser SiroLaser Blue (Sirona, 64625 Bensheim, Deutschland) at a wavelength of 445 nm and 200 mW, continuous wave (CW), contact mode, power density 400 mW/cm^2, using a periodontal tip with 320 μm diameter, attached to the hand piece. In the GC and GO groups, the PS was instilled into the subgingival area with a blunt needle in apical-coronal direction. After 60 s, the PS was rinsed with 1 mL of Kaqun® water (Harghita, Romania) using a graded syringe. Then in GC, GO and L groups, the laser beams were directed into the pockets for 40 s. Irradiation was maintained for 10 s in 4 equidistant sites (2 vestibular and 2 buccal sites). The procedures were repeated one week apart for 4 weeks (Figure 10A).

All treatments were applied locally following the procedure derived from the standard human treatment and were carried out by a specialist (Figure 10B).

The animals were closely monitored for the entire length of the study, focusing on infection prevention and analgesic therapy. At the end of the experimental study, blood samples were collected so that hematological and biochemical parameters could be determined. Additionally, body weight was closely monitored. The animals were euthanatized after 4 weeks of study by prolonged narcosis followed by cervical dislocation. Then, the left mandible was harvested from each rat for histological analysis.

4.4. Computed Tomography (CT) Analysis

In order to assess the induced periodontal disease and the effect of the applied treatment, bone density was followed on CT images as well as the size of the periodontal space at the treated hemiarchy compared to the opposite hemiarchy. Siemens Somatom Scope (Siemens Medical Solutions USA, Inc., Malvern, PA, USA) at 130 kV and 50 mAs was applied for CT analysis. Three-dimensional (3D) volume viewing and analysis software (RadiAnt Dicom Viewer, Poznań, Poland) were used to visualize and quantify the data obtained. Also a standardized gray-scale value was used to visualize and analyze the mineralized tissues. The mandibles were scanned after the first and last treatment session, on day 0 and day 21, respectively. Bone density was followed on the CT images, as well as the size of the periodontal space at the treated hemiarchy compared to the opposite hemiarchy.

4.5. Complete Blood Count and Biochemistry

Complete blood count was determined using the automatic Abacus Junior Vet hematology counter. From the blood tests were determined: white blood cell (WBC), lymphocyte (LYM), minimum inhibitory dilution (MID), granulocytes (GRA), red blood cell (RBC), hemoglobin (HGB), mean corpuscular volume (MCV), mean corpuscular hemoglobin (MCH), mean corpuscular hemoglobin concentration (MCHC), and platelet (PLT).

4.6. Histological Analysis

For the histological examination, bone samples from the mandibular lesion site were harvested. For decalcification, after fixation, the mandibula of the animals were kept in a mix of 8% formic and 8% clorhidric acid for 24 h and embedded in paraffin. The samples were fixed in 10% buffered neutral formalin, embedded in paraffin. The sections were made with a high-precision microtome Leica RM 2125 RT, at 5-µm thick, and stained by the hematoxylin–eosin method (HE). The slides were examined under a BX51 Olympus microscope, and images were taken with an Olympus UC 30 digital camera and processed using Olympus Basic Stream software. Sections were examined by an independent observer blinded to the experimental protocol.

4.7. Statistical Analysis of the Data

The Student *t*-test, Origine Pro 8 SRO (Origine Lab Corporation 2007, Northampton MA 01060, USA) was performed for the statistical analysis of the findings in this study. An analysis of variance was performed to determine whether there were significant differences between the different test conditions. The significance was evaluated at the level of $p \leq 0.05$.

Author Contributions: Conceptualization, M.M. and M.E.B.; methodology, R.C. and A.D.; validation, C.S., A.N., and S.S.; investigation, A.N., A.D., C.R., R.P., A.G.P., L.M.D. and I.C.B.; writing—original draft preparation, L.M.D.; writing—review and editing, C.S. and I.C.B.; visualization, S.S.; supervision, M.M. All authors have read and agreed to the published version of the manuscript.

Funding: This research received no external funding.

Institutional Review Board Statement: All procedures that involved the use of laboratory animals followed the European guidelines and rules 337 as established by the EU Directive 2010/63/EU and the Romanian law 43/2014, and were performed by an experienced practitioner. The study protocol was approved by the Research Ethics Committee of the University of Agricultural Sciences and Veterinary Medicine Cluj-Napoca, Romania, and they were authorized by the State Veterinary Authority (aut. No. 52/30.03.2017). This article does not contain any studies with human participants performed by any of the authors.

Informed Consent Statement: Not applicable.

Data Availability Statement: Not applicable.

Conflicts of Interest: The authors declare no conflict of interest.

References

1. Kömerik, N.; Nakanishi, H.; MacRobert, A.J.; Henderson, B.; Speight, P.; Wilson, M. In Vivo Killing of Porphyromonas gingivalis by Toluidine Blue-Mediated Photosensitization in an Animal Model. *Antimicrob. Agents Chemother.* **2003**, *47*, 932–940. [CrossRef] [PubMed]
2. Rajapakse, P.S.; O'Brien-Simpson, N.M.; Slakeski, N.; Hoffmann, B.; Reynolds, E.C. Immunization with the RgpA-Kgp Proteinase-Adhesin Complexes of Porphyromonas gingivalis Protects against Periodontal Bone Loss in the Rat Periodontitis Model. *Infect. Immun.* **2002**, *70*, 2480–2486. [CrossRef] [PubMed]
3. Nessa, N.; Kobara, M.; Toba, H.; Adachi, T.; Yamamoto, T.; Kanamura, N.; Pezzotti, G.; Nakata, T. Febuxostat Attenuates the Progression of Periodontitis in Rats. *Pharmacology* **2021**, *106*, 294–304. [CrossRef] [PubMed]
4. Prates, R.A.; Yamada, A.M.; Suzuki, L.C.; França, C.M.; Cai, S.; Mayer, M.P.A.; Ribeiro, A.C.; Ribeiro, M.S. Histomorphometric and Microbiological Assessment of Photodynamic Therapy as an Adjuvant Treatment for Periodontitis: A Short-Term Evaluation of Inflammatory Periodontal Conditions and Bacterial Reduction in a Rat Model. *Photomed. Laser Surg.* **2011**, *29*, 835–844. [CrossRef]
5. Graves, D.T.; Fine, D.; Teng, Y.T.A.; Van Dyke, T.E.; Hajishengallis, G. The use of rodent models to investigate host-bacteria interactions related to periodontal diseases. *J. Clin. Periodontol.* **2008**, *35*, 89–105. [CrossRef]
6. Wang, H.H.; Lee, H.M.; Raja, V.; Hou, W.; Iacono, V.J.; Scaduto, J.; Johnson, F.; Golub, L.M.; Gu, Y. Enhanced Efficacy of Chemically Modified Curcumin in Experimental Periodontitis: Systemic Implications. *J. Exp. Pharmacol.* **2019**, *11*, 1–14. [CrossRef]
7. Birang, E.; Talebi Ardekani, M.R.; Rajabzadeh, M.; Sarmadi, G.; Birang, R.; Gutknecht, N. Comparison of Er:YAG Laser andUltrasonic Scaler in the Treatment of Moderate Chronic Periodontitis: A Randomized Clinical Trial. *J. Lasers Med. Sci.* **2017**, *8*, 136–142. [CrossRef]
8. Hosseini, N.; Yazdanpanah, S.; Saki, M.; Rezazadeh, F.; Ghapanchi, J.; Zomorodian, K. Susceptibility of Candida albicans and Candida dubliniensis to photodynamic therapy using four dyes as the photosensitizer. *J. Dent.* **2016**, *17*, 354–360.
9. Sperandio, F.; Huang, Y.Y.; Hamblin, M. Antimicrobial photodynamic therapy to kill Gram-negative bacteria. *Recent Pat. Anti-Infect. Drug Discov.* **2013**, *8*, 108–120. [CrossRef]
10. Lara Alves, L.V.G.; Curylofo-Zotti, F.A.; Borsatto, M.C.; de Souza Salvador, S.L.; Valério, R.A.; Souza-Gabriel, A.E. Influence of antimicrobial photodynamic therapy in carious lesion. Randomized split-mouth clinical trial in primary molar. *Photodiagnosis Photodyn. Ther.* **2019**, *26*, 124–130. [CrossRef]
11. Ishiyama, K.; Nakamura, K.; Kano, T.; Niwano, Y. Bactericidal Action of Photodynamic Antimicrobial Chemotherapy (PACT) with Photosensitizers Used as Plaque-Disclosing Agents against Experimental Biofilm. *Biocontrol Sci.* **2016**, *21*, 187–191. [CrossRef]
12. Misba, L.; Zaidi, S.; Khan, A.U. A comparison of antibacterial and antibiofilm efficacy of phenothiazinium dyes between Gram positive and Gram negative bacterial biofilm. *Photodiagnosis Photodyn. Ther.* **2017**, *18*, 24–33. [CrossRef]
13. Dascalu (Rusu), M.L.; Sarosi, C.; Moldovan, M.; Badea, M.E. A Study on Revealing Agents in the Context of Photodynamic Therapy in Dental Medicine—A Literature Review. *Defect Diffus. Forum* **2017**, *376*, 54–65. [CrossRef]
14. Dascalu (Rusu), L.M.; Moldovan, M.; Prodan, D.; Ciotlaus, I.; Popescu, V.; Baldea, I.; Carpa, R.; Sava, S.; Chifor, R.; Badea, M.E. Assessment and Characterization of Some New Photosensitizers for Antimicrobial Photodynamic Therapy (aPDT). *Materials* **2020**, *13*, 3012. [CrossRef]
15. Mang, T.S.; Tayal, D.P.; Baier, R. Photodynamic therapy as an alternative treatment for disinfection of bacteria in oral biofilms. *Lasers Surg. Med.* **2012**, *44*, 588–596. [CrossRef] [PubMed]
16. Garcez, A.S.; Ribeiro, M.S.; Tegos, G.P.; Núñez, S.C.; Jorge, A.O.C.; Hamblin, M.R. Antimicrobial photodynamic therapy combined with conventional endodontic treatment to eliminate root canal biofilm infection. *Lasers Surg. Med.* **2007**, *39*, 59–66. [CrossRef]
17. De Almeida, J.M.; Theodoro, L.H.; Bosco, A.F.; Nagata, M.J.H.; Oshiiwa, M.; Garcia, V.G. In vivo effect of photodynamic therapy on periodontal bone loss in dental furcations. *J. Periodontol.* **2008**, *79*, 1081–1088. [CrossRef] [PubMed]
18. Wood, S.; Nattress, B.; Kirkham, J.; Shore, R.; Brookes, S.; Griffiths, J. An in vitro study of the use of photodynamic therapy for the treatment of natural oral plaque biofilms formed in vivo. *J. Photochem. Photobiol. B Biol.* **1999**, *50*, 1–7. [CrossRef]
19. Ghasemi, M.; Etemadi, A.; Nedaei, M.; Chiniforush, N.; Pourhajibagher, M. Antimicrobial efficacy of photodynamic therapy using two different light sources on the titanium-adherent biofilms of *Aggregatibacter actinomycetemcomitans*: An in vitro study. *Photodiagnosis Photodyn. Ther.* **2019**, *26*, 85–89. [CrossRef]
20. Tokubo, L.M.; Rosalen, P.L.; de Cássia Orlandi Sardi, J.; Freires, I.A.; Fujimaki, M.; Umeda, J.E. Antimicrobial effect of photodynamic therapy using erythrosine/methylene blue combination on Streptococcus mutans biofilm. *Photodiagnosis Photodyn. Ther.* **2018**, *23*, 94–98. [CrossRef] [PubMed]
21. Fang-Yen, C.; Gabel, C.V.; Samuel, A.D.T.; Bargmann, C.I.; Avery, L. Laser microsurgery in Caenorhabditis elegans. *Methods Cell Biol.* **2012**, *107*, 177–206. [PubMed]
22. Cieplik, F.; Tabenski, L.; Buchalla, W.; Maisch, T. Antimicrobial photodynamic therapy for inactivation of biofilms formed by oral key pathogens. *Front. Microbiol.* **2014**, *5*, 405. [CrossRef] [PubMed]

23. Yan, S.; Huang, Q.; Song, X.; Chen, Z.; Huang, M.; Zhang, J. A series of photosensitizers with incremental positive electric charges for photodynamic antitumor therapy. *RSC Adv.* **2019**, *9*, 24560–24567. [CrossRef]
24. Yoo, J.O.; Ha, K.S. New insights into the mechanisms for photodynamic therapy-induced cancer cell death. *Int. Rev. Cell. Mol. Biol.* **2012**, *295*, 139–174.
25. Jurenka, J.S. Anti-inflammatory properties of curcumin, a major constituent of Curcuma longa: A review of preclinical and clinical research. *Altern. Med. Rev.* **2009**, *14*, 141–153.
26. Akbik, D.; Ghadiri, M.; Chrzanowski, W.; Rohanizadeh, R. Curcumin as a wound healing agent. *Life Sci.* **2014**, *116*, 1–7. [CrossRef]
27. Mohanty, C.; Sahoo, S.K. Curcumin and its topical formulations for wound healing applications. *Drug Discov. Today* **2017**, *22*, 1582–1592. [CrossRef]
28. Zhang, Y.; McClain, S.A.; Lee, H.M. A Novel Chemically Modified Curcumin "Normalizes" Wound-Healing in Rats with Experimentally Induced Type I Diabetes: Initial Studies. *J. Diabetes Res.* **2016**, *2016*, 5782904. [CrossRef]
29. Nasri, H.; Sahinfard, N.; Rafieian, M.; Rafieian, S.; Shirzad, M.; Rafieian-Kopaei, M. Turmeric: A spice with multifunctional medicinal properties. *J. HerbMed Pharmacol.* **2014**, *3*, 5–8.
30. Xiao, C.J.; Yu, J.; Xie, J.L.; Liu, S.; Li, S. Protective effect and related mechanisms of curcumin in rat experimental periodontitis. *Head Face Med.* **2018**, *14*, 12. [CrossRef]
31. Judžentien, A.; Būdienė, J. Analysis of the chemical composition of flower essential oils from Arnica montana of Lithuanian origin. *Chemija* **2009**, *20*, 190–194.
32. Rostro-Alanis, M.; Báez-González, J.; Torres-Alvarez, C.; Parra-Saldívar, R.; Rodriguez-Rodriguez, J.; Castillo, S. Chemical Composition and Biological Activities of Oregano Essential Oil and Its Fractions Obtained by Vacuum Distillation. *Molecules* **2019**, *24*, 1904. [CrossRef] [PubMed]
33. Etemadi, A.; Hamidain, M.; Parker, S.; Chiniforush, N. Blue Light Photodynamic Therapy with Curcumin and Riboflavin in the Management of Periodontitis: A Systematic Review. *J. Lasers Med. Sci.* **2021**, *12*, e15. [CrossRef] [PubMed]
34. De Oliveira Carvalho, I.; Aparecida Purgato, G.; Soares Píccolo, M.; Ramos Pizziolo, V.; Ribeiro Coelho, R.; Diaz-Muñoz, G.; Alves Nogueira Diaz, M. In vitro anticariogenic and antibiofilm activities of toothpastes formulated with essential oils. *Arch. Oral Biol.* **2020**, *117*, 104834. [CrossRef] [PubMed]
35. Sharifi-Rad, M.; Varoni, E.M.; Iriti, M.; Martore, M.; Setzer, W.N.; Contreras, M.; Salehi, B.; Soltani-Nejad, A.; Rajabi, S.; Tajbakhsh, M.; et al. Carvacrol and human health: A comprehensive review. *Phytother. Res.* **2018**, *32*, 1675–1687.
36. Leyva-López, N.; Gutiérrez-Grijalva, E.P.; Vazquez-Olivo, G.; Heredia, J.B. Essential Oils of Oregano: Biological Activity beyond Their Antimicrobial Properties. *Molecules* **2017**, *22*, 989. [CrossRef]
37. Belinello-Souza, E.L.; Alvarenga, L.H.; Lima-Leal, C.; Almeida, P.; Leite, C.G.; Lima, T.R. Antimicrobial photodynamic therapy combined to periodontal treatment: Experimental model. *Photodiagnosis Photodyn. Ther.* **2017**, *18*, 275–278. [CrossRef]
38. Uslu, M.Ö.; Eltas, A.; Marakoğlu, I.; Dündar, S.; Şahin, K.; Özercan, I.H. Effects of diode laser application on inflammation and mpo in periodontal tissues in a rat model. *J. Appl. Oral Sci.* **2018**, *26*, e20170266. [CrossRef]
39. Lin, P.; Niimi, H.; Ohsugi, Y.; Tsuchiya, Y.; Shimohira, T.; Komatsu, K.; Liu, A.; Shiba, T.; Aoki, A.; Iwata, T.; et al. Application of Ligature-Induced Periodontitis in Mice to Explore the Molecular Mechanism of Periodontal Disease. *Int. J. Mol. Sci.* **2021**, *22*, 8900. [CrossRef]
40. Fernandes, L.A.; Theodoro, L.H.; Martins, T.M.; de Almeida, J.M.; Garcia, V.G. *J* Effects of diode laser application on inflammation and mpo in periodontal tissues in a rat model. *Appl. Oral Sci.* **2010**, *18*, 237–243. [CrossRef]
41. Robinson, M.; Hart, D.; Pigott, G.H. The effects of diet on the incidence of periodontitis in rats. *Lab. Anim.* **1991**, *25*, 247–253. [CrossRef]
42. Toth, L.A.; Gardiner, T.W. Food and water restriction protocols: Physiological and behavioral considerations. *Contemp. Top. Lab. Anim. Sci.* **2000**, *39*, 9–17. [PubMed]
43. Smith, C.; Jarecki, A. *Atlas of Comparative Diagnostic and Experimental Hematology*, 2nd ed.; Wiley-Blackwell: Hoboken, NJ, USA, 2013.
44. Giknis, M.L.A.; Charles, B.; Clifford, D.V.M. *Clinical Laboratory Parameters for Crl:CD(SD) Rats*; Charles River Lab: Middlesex County, MA, USA, 2008.
45. Carvalho, A.S.; Napimoga, M.H.; Coelho-Campos, J.; Silva-Filho, V.J.; Thedei, G. Photodynamic therapy reduces bone resorption and decreases inflammatory response in an experimental rat periodontal model. *Photomed. Laser Surg.* **2011**, *29*, 735–740. [CrossRef] [PubMed]
46. Graves, D.T.; Kang, J.; Andriankaja, O.; Wada, K.; Rossa, C. Animal models to study host-bacteria interactions involved in periodontitis. *Front. Oral Biol.* **2012**, *15*, 117–132.
47. Faot, F.; de Camargos G, C.; Duyck, J.; Vandamme, K. Micro-CT analysis of the rodent jaw bone micro-architecture: A systematic review. *Bone Rep.* **2015**, *2*, 14–24. [CrossRef] [PubMed]
48. Garcia, V.G.; Knoll, L.R.; Longo, M.; Novaes, V.C.N.; Assem, N.Z.; Ervolino, E. Effect of the probiotic Saccharomyces cerevisiae on ligature-induced periodontitis in rats. *J. Periodontal Res.* **2016**, *51*, 26–37. [CrossRef] [PubMed]
49. Garcia, V.G.; Erivan, C.G.V.; Fernandes, L.A.; Bosco, A.F.; Nagata, H.M.; Casatti, A.C.; Ervolino, E.; Theodoro, L.H. Adjunctive antimicrobial photodynamic treatment of experimentally induced periodontitis in rats with ovariectomy. *J. Periodontol.* **2013**, *84*, 556–565. [CrossRef]
50. Flecknell, P. *Laboratory Animal Anaesthesia*, 3rd ed.; Academic Press: Cambridge, MA, USA, 2009.

gels

MDPI

Article

Regenerative Activities of ROS-Modulating Trace Metals in Subcutaneously Implanted Biodegradable Cryogel

Abdulla A. Yergeshov [1,†], Mohamed Zoughaib [1,†], Rezeda A. Ishkaeva [1], Irina N. Savina [2] and Timur I. Abdullin [1,*]

1 Institute of Fundamental Medicine and Biology, Kazan (Volga Region) Federal University, 18 Kremlyovskaya St., 420008 Kazan, Russia; abdulla.ergeshov@mail.ru (A.A.Y.); zmokhamed@kpfu.ru (M.Z.); rezaahmadishina@kpfu.ru (R.A.I.)

2 School of Applied Sciences, University of Brighton, Huxley Building, Lewes Road, Brighton BN2 4GJ, UK; i.n.savina@brighton.ac.uk

* Correspondence: tabdulli@gmail.com or timur.abdullin@kpfu.ru

† These authors contributed equally to this work.

Abstract: Divalent trace metals (TM), especially copper (Cu), cobalt (Co) and zinc (Zn), are recognized as essential microelements for tissue homeostasis and regeneration. To achieve a balance between therapeutic activity and safety of administered TMs, effective gel formulations of TMs with elucidated regenerative mechanisms are required. We studied in vitro and in vivo effects of biodegradable macroporous cryogels doped with Cu, Co or Zn in a controllable manner. The extracellular ROS generation by metal dopants was assessed and compared with the intracellular effect of soluble TMs. The stimulating ability of TMs in the cryogels for cell proliferation, differentiation and cytokine/growth factor biosynthesis was characterized using HSF and HUVEC primary human cells. Multiple responses of host tissues to the TM-doped cryogels upon subcutaneous implantation were characterized taking into account the rate of biodegradation, production of HIF-1α/matrix metalloproteinases and the appearance of immune cells. Cu and Zn dopants did not disturb the intact skin organization while inducing specific stimulating effects on different skin structures, including vasculature, whereas Co dopant caused a significant reorganization of skin layers, the appearance of multinucleated giant cells, along with intense angiogenesis in the dermis. The results specify and compare the prooxidant and regenerative potential of Cu, Co and Zn-doped biodegradable cryogels and are of particular interest for the development of advanced bioinductive hydrogel materials for controlling angiogenesis and soft tissue growth.

Keywords: trace metals; cryogels; reactive oxygen species; tissue regeneration; skin; angiogenesis; immune cells

Citation: Yergeshov, A.A.; Zoughaib, M.; Ishkaeva, R.A.; Savina, I.N.; Abdullin, T.I. Regenerative Activities of ROS-Modulating Trace Metals in Subcutaneously Implanted Biodegradable Cryogel. *Gels* **2022**, *8*, 118. https://doi.org/10.3390/gels8020118

Academic Editors: Arish Dasan, Ashokraja Chandrasekar and Nupur Kohli

Received: 16 January 2022
Accepted: 10 February 2022
Published: 14 February 2022

1. Introduction

Treatment of severe organ/tissue injuries generally requires the replacement of a post-traumatic defect with a scaffold such as an autologous/decellularized graft or preferably biomimetic biodegradable material capable of supporting cell growth and functional activity while overcoming the limitations of donor grafts mostly related to their scarcity and host immunogenic responses [1]. Although a number of biomimetic scaffolds composed of synthetic or/and naturally occurring biopolymers have been proposed, these materials by themselves are not able to provide sufficient regenerative responses without special bioactivation [2].

Transplanted cells such as mesenchymal stem cells from different sources [3], skin fibroblasts [4], neural cells [5], as well as related products (e.g., platelets [6] and extracellular vesicles [7]), were proved to enhance tissue regeneration activity of biomaterials. However, these biological products unavoidably feature typical limitations of donor tissues, primarily, low availability, variability of characteristics, and health risks. Given that the regenerative potential of the transplanted cells is mainly attributed to secreted signaling molecules, recombinant growth factors can be used instead to improve tissue-replacing scaffolds [8], although this is complicated by the increased manufacturing cost of pure growth factors and their deactivation upon immobilization and storage. Therefore, more stable, reproducible and available active components of biomaterials are still demanded in tissue engineering and regeneration applications.

Trace metals (TM) are essential bioactive microelements involved in the maintenance and regulation of cell metabolism, the functioning of the immune system, turnover and the regeneration of soft and hard tissues. Since TM deficiency accompanies many degenerative and traumatic diseases, their local administration in combination with biomaterials represents a promising therapeutic strategy [9,10]. The regenerative activities of such TMs (primarily, divalent ions of Co, Cu, Zn, Mn, and Fe metals) were established mainly for solid osteoinductive materials based on inorganic scaffolds doped with metal ions or nanoparticles. For instance, collagen scaffolds functionalized with Cu-eluting bioactive glass particles possessed profound in vitro angiogenic activity toward rMSCs and HUVECs as well as antibacterial and osteogenic effects in vivo [11]. Implanted Co-containing borosilicate glass-based scaffolds remarkably enhanced bone regeneration and the vascularized network of the calvarial defective site in rats [12]. Likewise, Zn incorporation into Ca-silicate-based cements increased the osteostimulative activity of the composite material in a maxillofacial bone defect model in rabbits [13].

Hydrogels have been considered among the most promising materials for tissue repair, showing successful results in pre-clinical trials owing to their appropriate physicochemical and hydration properties similar to those of soft body tissues [14,15]. Earlier, gelatin methacrylate hydrogels embedded with Cu nanoparticles supported the attachment and proliferation of 3T3 fibroblasts and inhibited bacterial growth in vitro, in addition to promoting effective wound closure in mice without inflammatory response [16]. Zn cross-linked alginate-polyacrylamide hydrogel supported increased vascular growth, collagen deposition, granulation tissue formation and wound healing along with reduced inflammation [17]. The co-encapsulation of Co and Ca ions within gauze-alginate composite hydrogel resulted in enhanced local VEGF and TGF-β1 protein expression and accelerated wound healing in a mouse bacteria-infected wound model [18]. We have shown recently that macroporous hydrogels prepared by the cryogelation technique, namely, cryogels composed both of bio- and synthetic polymers, represent a promising type of scaffold for bulk bioactivation with TMs in a controllable and stable manner [19,20]. The advanced porous structure of cryogels ensures enhanced mammalian cell infiltration and activity of the incorporated metal dopant within the scaffold [19–21]. The increased healing of an excisional skin defect treated with the Zn-doped gelatin cryogel [19] as well as enhanced in vitro angiogenic responses of poly(2-hydroxyethyl methacrylate) cryogels modified with Cu^{2+} (via complexation with GHK peptide) [20] were demonstrated.

Further clarification of mechanisms of local regenerative and adverse effects of TM-containing hydrogels is demanded. These effects are often controversial, depending on biomaterial formulations, which may, for instance, show both prooxidant [22–24] and antioxidant [25,26] activities for the same TMs. Our previous studies prove cryogels as a relevant platform both for the examination of therapeutic effects of the TM compounds and for the development of advanced bioinductive materials [19,20].

In this work, we conducted a comparative investigation of regenerative activities of biodegradable gelatin cryogel doped with Zn, Cu or Co divalent metals as one of the most therapeutically relevant TMs [10]. Considering that TMs can participate in non-enzymatic redox-reactions, such as the Fenton-type generation of reactive oxygen species (ROS) involved in cell signaling [27–29], the TM-doped cryogels were assessed in relation to redox-modulating and cytokine-regulating in vitro activities of the metal dopants. To charactrize in vivo regenerative activities of the TM-doped cryogels, a subcutaneous implantation model with a comprehensive histological evaluation was optimized, considering the relevance of this model for understanding fundamental effects of biomaterials on host tissue responses related to cellularization, angiogenesis, and inflammation [30–32]. Thus, specific localized effects of the cryogel-formulated TMs on different skin structures and underlying tissues were studied and compared.

2. Materials and Methods

2.1. Materials

Bovine skin gelatin, 3-(4,5-dimethylthiazol-2-yl)-2,5-diphenyl tetrazolium bromide (MTT reagent), 4′,6-diamidino-2-phenylindole (DAPI), phenazine methosulfate (PMS), and Triton X-100 and 2′,7′-dichlorofluorescin diacetate (DCFDA) were purchased from Sigma-Aldrich. Monochlorobimane (MCB) was purchased from ThermoFisher Scientific. $CuSO_4{\cdot}5H_2O$, $ZnCl_2$, $CoCl_2{\cdot}6H_2O$, glutaric dialdehyde (GDA), and cresyl violet acetate were obtained from Acros Organics. Citrus pectin (classic CM 201) was obtained from Herbstreith&Fox.

3-(4,5-Dimethylthiazol-2-yl)-5-(3-carboxymethoxyphenyl)-2-(4-sulfophenyl)-2H-tetrazolium (MTS reagent) was purchased from Promega. Phalloidin CruzFluor™ 647 conjugate, anti-VEGF (C-1) mouse monoclonal, anti-ICAM-2 (S-16) goat polyclonal, anti-MMP-3 goat monoclonal and anti MMP-3 goat monoclonal antibodies were purchased from Santa Cruz Biotechnology. Anti-HIF-1a mouse monoclonal antibody, donkey anti-mouse IgG (H + L) highly cross-adsorbed secondary antibody, Alexa Fluor 647, and donkey anti-goat IgG (H + L) cross-adsorbed secondary antibody, Alexa Fluor 555, were obtained from ThermoFisher Scientific. Anti-CD31 (PECAM-1) rabbit monoclonal antibody was obtained from Abcam. Hematoxylin and Eosin, and Giemsa staining were purchased from BioVitrum (Russia). Cell culture media and reagents were purchased from Paneco (Russia).

2.2. Preparation and Characterization of Cryogels

Cryogels were prepared from bovine skin gelatin using cryotropic gelation method as previously described [19] with some modifications. Briefly, the reaction mixture contained gelatin (2.5 wt%), pectin (ca. 0.1 wt%) and TM (0.04–1 mM) in aqueous solution. The gelation was initiated by adding 0.25 wt% GDA to the solution upon stirring followed by its pouring into a glass Petri dish and cooling at a temperature of −12 °C for 4 h in a thermostat and then at −18 °C for additional 24 h in a freezer. The resultant ~3 mm thick cryogel sheet was thawed at room temperature, washed and stored in 25% ethanol solution in the fridge.

Rheological properties of the cryogels were analyzed using MCR 302 rotational rheometer (Anton Paar) at 25 °C. The strain sweep and frequency sweep tests were performed by applying 0.01–100% strain amplitude (ω = 10 rad s^{-1}) and 0.01–100 rad s^{-1} angular frequencies (γ = 1%), respectively. The storage (G') and loss (G'') modulus of the materials were presented as a function of strain and frequency. Frequency dependences of G' and G'' were detected within linear viscoelastic region (LVR). Porous structure of the cryogels was analyzed using laser scanning confocal microscopy (LSCM) using LSM 780 microscope (Carl Zeiss) equipped with argon laser excitation (488 nm). Zeiss ZEN black software was used for acquisition. Pore size of the cryogels was evaluated using ImageJ software (NIH, USA).

2.3. Cell Maintenance and Seeding

NIH 3T3 mouse embryonic fibroblasts (ATCC) and primary human skin fibroblasts (HSFs) isolated as described earlier [33] were grown in α-MEM supplemented with 10% FBS, penicillin (100 U/mL)/streptomycin (100 μg/mL) and L-glutamine (2 mM). Freshly isolated human umbilical vein endothelial cells (HUVECs) were kindly provided by Dr. Ilnur Salafutdinov (Kazan Federal University). HUVECs were grown in RPMI 1640 supplemented with 20% FBS, penicillin (100 U/mL)/streptomycin (100 μg/mL), L-glutamine (2 mM), sodium pyruvate (2 mM), heparin (100 μg/mL), and 30 μg/mL endothelial cell growth supplements (ECGS). The cells were cultured in a temperature- and humidity-controlled incubator at 37 °C. The culture medium was refreshed every 2 days. Primary cells (HSFs and HUVECs) were studied between passages 3 and 6.

Prior to cell seeding, round cryogel sheets (14 mm in diameter) were incubated in penicillin (2.5 kU/mL)/streptomycin (2.5 mg/mL) antibiotic mixture for 1 h, rinsed with HBSS and equilibrated in the culture medium. Cells were seeded onto the cryogel surface using top seeding method in 24-well plate at a density of 4.88×10^4 cells/cm^2 of cryogel area and incubated for 1.5 h under standard culture conditions to allow for cell attachment.

2.4. Cell Detection in Cryogels

The cryogels with cultured cells were collected at day 3, transferred into new wells containing 0.5 mL MTS/PMS reagents in fresh culture medium to assess cell metabolic activity [34]. After incubation for 1.5 h under standard culture conditions (37 °C, 5% CO_2), the absorbance was determined at 490 nm on an Infinite M200 PRO microplate analyzer (Tecan).

For bright-field microscopy analysis, the cryogels with cultured HUVECs were fixed with 4% *p*-formaldehyde for 2.5 h and gently washed with PBS. The fixed cells were subsequently stained with cresyl violet (0.1% *w*/*v* in ultrapure water) for 5 min and visualized using AxioObserver Z1 microscope (Carl Zeiss).

2.5. Immunocytochemistry

The fixed cryogel matrices were incubated in 0.1% Triton X-100 in PBS for 15 min for cell membrane permeabilization and washed three times with PBS. Non-specific binding sites in the materials were blocked using 1.5% bovine serum albumin (BSA) for 30 min at room temperature. The matrices were subsequently incubated with primary antibodies (1:500 in 1.5% BSA/PBS) against VEGF or ICAM-2 overnight at 4 °C followed by incubation with Alexa Fluor 647-conjugated donkey anti-mouse or Alexa Fluor 555-conjugated donkey anti-goat secondary antibodies (1:350 in 1.5% BSA/PBS) for 45 min at room temperature. Following washing with PBS, the cell nuclei were stained with 4′,6-diamidino-2-phenylindole (DAPI). For cytoskeleton visualization, F-actin was labeled using phalloidin CruzFluor™ 647 conjugate in 1% BSA for 30 min. LSCM images were acquired on an LSM 780 microscope.

2.6. Detection of ROS and Glutathione

To assess extracellular ROS-generating ability of the TM-doped cryogels, the materials in a 24-well plate were incubated with H_2O_2 (21.5 mM) in PBS for 60 min in the presence of 5 μM DCFDA. The probe fluorescence ($\lambda_{ex}/\lambda_{em}$ = 490/526) in the solution was monitored during the reaction using an Infinite M200 PRO microplate analyzer.

Intracellular effects of dissolved TMs on both ROS and reduced glutathione levels were additionally studied. 3T3 cells were seeded in 96-well plate at a density of 2×10^4 cells per well and grown overnight. The cells were exposed to dissolved $CuSO_4$, $ZnCl_2$ or $CoCl_2$ at a concentration of 1 or 10 μM in HBSS for 60 min in CO_2 incubator. Subsequently, the treated cells were stained with 20 μM DCFDA fluorescent probe for 40 min or with 5 μM monochlorobimane probe for 60 min to assess intracellular ROS and reduced glutathione, respectively. The cellular fluorescence of DCFDA ($\lambda_{ex}/\lambda_{em}$ = 490/526) and MCB ($\lambda_{ex}/\lambda_{em}$ = 380/480) was detected. The data are presented as mean ± SD.

2.7. Multiplexed Fluorescent Bead-Based Immunoassay

Top seeded HSFs were grown within non-doped and metal-doped cryogels as mentioned above in 2.3. At 24 h post-seeding, the conditioned culture medium containing cell secretion was collected and immediately frozen at −80 °C. The analysis of secreted levels of cytokines was performed using xMAP Luminex technology on a Bio-Plex MAGPIX analyzer (BioRad, USA) according to the manufacturer's recommendations. A commercially available MILLIPLEX MAP Human Cytokine/Chemokine Magnetic Bead Panel (HCYTMAG-60K-PX41) was used to quantitatively measure cytokine/chemokine levels.

Standard reference curve was used to determine the concentration of analytes in each sample according to their fluorescence intensities. Background levels of analytes in cell-free culture medium were subtracted. Bio-Plex Manager 4.1 software (Bio-Rad Laboratories) was used to analyze the data.

2.8. In Vivo Study

2.8.1. Animals

Wistar male rats (340 ± 38 g) were purchased from Vivarium of Academy of Medical and Technical Sciences (Russia). Animal care was performed according to European regulations on the protection of experimental animals (Directive 2010/63/UE) and Russian regulations (No. 742 from 13.11.1984, Ministry of Education and Science). The rats were divided into four groups and were kept in plastic cages under controlled conditions (at a temperature of 20 ± 3 °C and a humidity of 65 ± 10%) with running water and complete feed. The in vivo study was approved by the Institutional Ethical Review Board of the Kazan Federal University.

2.8.2. Subcutaneous Implantation Model

The animals were anesthetized using tiletamine-zolazepam-xylazine (30/20/10 mg/kg, respectively) administered via IP injection. The upper back skin was shaved and disinfected with 70% ethanol solution. Two symmetrical full-thickness skin incisions were made horizontally with a width of 1.5 cm using fine scalpel. A subcutaneous pocket with about 1.5 cm long from the lower incision border was created on each side by detaching the skin from the underlaying tissues employing anatomical forceps (Figure 7). The incision was disinfected with 0.05% chlorhexidine solution and washed with sterile isotonic solution. The cryogel sheets were cut into 1 × 1 cm square pieces, additionally decontaminated with penicillin (5 kU/mL)/streptomycin (5 mg/mL) solution and equilibrated with excess of sterile isotonic solution. The studied materials were aseptically implanted into subcutaneous pockets so that the control (non-doped) and metal-doped cryogels (TM concentration = 0.2 mM) were alternately placed at the right and left sides of an animal. The procedure allowed us to decrease the number of animals per group to $n = 6$. No manifestations of pain, infection or any worsening of animal behavior were observed during the experiment.

2.9. Histological Evaluation

On days 5 and 10 post-implantation, the animals were sacrificed using tiletamine-zolazepam-xylazine anesthesia and by applying incremental concentration of CO_2. The treated skin with implanted material was surgically excised, then subsequently fixed in 4% neutral buffered formalin solution in PBS at room temperature for 48 h, washed with distilled water, dehydrated in a graded series of ethanol solutions (50, 70, 90, 96, 99.8%) and cleared in xylene. The explants were embedded in paraffin blocks and cut on a microtome HM 355S (Thermo Fisher Scientific) into 10–14 μm sections. The tissue sections were stained with Giemsa, hematoxylin and eosin (H&E) or Picrosirius red and analyzed by bright-field and polarized light microscopy on an Axio Observer Z1 microscope (Carl Zeiss).

For immunohistochemical analysis, glass slide-adhered tissue sections were permeabilized using 0.1% Triton X-100 in PBS for 30 min, thoroughly washed with PBS, and blocked with 1.5% BSA. The sections were subsequently incubated with primary antibodies (diluted 1:300 in 1.5% BSA/PBS) against CD-31 (PECAM-1), HIF-1α, MMP-2 or MMP-3 overnight at 4 °C followed by incubation with proper Alexa Fluor 488-conjugated donkey anti-goat and Alexa Fluor 647-conjugated donkey anti-mouse secondary antibodies (1:350 in 1.5% BSA/PBS), for 45 min at room temperature. DAPI was used to stain cell nuclei. LSCM images were acquired on an LSM 780 microscope.

2.10. Statistical Analysis

Data were presented as mean ± SD. Statistical significance was determined by one-way analysis of variance (ANOVA) followed by Tukey's Multiple Comparison post-test (* $p < 0.05$, ** $p < 0.01$, *** $p < 0.001$).

3. Results

3.1. Characterization of Cryogels

Cryogels were made by the cryotropic gelation of bovine gelatin cross-linked with glutaraldehyde. Co^{2+}, Cu^{2+} and Zn^{2+} (further designated as Co, Cu and Zn) in the form of water-soluble salts were incorporated into the cryogel material during gelation. The addition of pectin to the gelatin cryogel has been used to facilitate the capture of metal ions by introducing additional anionic groups, such as galacturonic acid, into the polymer network. The metal content in the cryogels linearly depended on the concentration of TMs in the gel-forming solution from 0.04 to 5 mM; the TM-doped cryogels were designated by these concentrations. For the upper 1 mM concentration used in this study, the TM content in cryogels was previously shown to amount to ca. 3×10^3 ppm (Zn, Cu) and 1×10^3 ppm (Co) [19].

According to LSCM analysis, the TM-doped cryogels preserved an interconnected porous structure (Figure 1A), typical of cryogel scaffolds, with somewhat larger pores at the upper surface, commonly used for cell culture [34]. Non-doped cryogels possessed macropores with a calculated mean pore size of 80 ± 13 μm, which moderately decreased by 1.2–1.4 times in the TM-doped materials. According to rheological data, all cryogels displayed a linear viscoelastic region for a shear strain amplitude of $\gamma \leq 6\%$ (Figure 1B). Both storage (G′) and loss (G″) modulus relatively weakly depended on angular frequency, whereas G′ greatly prevailed over G″ (Figure 1B), indicating a well-structured hydrogel network with dominant elastic behavior. The incorporated TMs increased the G′/G″ ratio by 1.3–1.9-fold, with a relatively lower effect of Co, thus demonstrating that the metal component contributes to the elasticity and mechanical strength of the materials. The modulation of pore size and viscoelastic behavior of cryogels by the introduced TMs is explained by additional cross-linking of macromers (gelatin and pectin molecules) by the metal ions presumably involving not only ionic but also coordination bonding with biopolymer ligand groups.

3.2. Behavior of Fibroblasts in Metal-Doped Cryogels

3.2.1. Effect of Cryogel Composition on Cell Proliferation

Mouse embryonic fibroblasts (3T3 cells) were seeded on the top surface of cryogels and allowed to grow for 72 h followed by cell detection using the MTS proliferation assay [34]. At a concentration of 0.04 and 0.2 mM, the introduced TMs did not inhibit cell proliferation, whereas 1 mM Co and Zn, unlike 1 mM Cu, caused a moderate inhibitory effect of up to 24% (Figure 2A). When supplemented into the culture medium, soluble TMs had IC_{50} values of 170 ± 9 μM (Zn), 250 ± 17 μM (Co), and 407 ± 20 μM (Cu) (72 h). Assuming most of the amount of TMs in gelling solution to be attached to the cryogel [19], the above data together suggest that the entrapped metals are not readily released into the medium remaining less available and less cytotoxic toward the cells than dissolved metals. In total, 0.2 mM of Cu and 0.2 mM of Zn were found to noticeably stimulate cell proliferation, respectively, by 30 and 10% (Figure 2A); this intermediate concentration was, therefore, selected for further study and comparison of regenerative effects of the TM-doped cryogels.

Similar to 3T3 cells, human skin fibroblasts (HSFs) proliferated more rapidly in the cryogels with 0.2 mM Zn (by 29%) or Cu (by 40%) ($p < 0.05$), unlike Co (Figure 2B). Furthermore, binary TM compositions exhibited quite different effects on HSF behavior. In particular, Zn/Cu did provide additive stimulation of cell growth compared to the individual metals, whereas the stimulating effect was partially preserved for Zn/Co and disappeared for Cu/Co system (Figure 2B). Considering that such a mitogenic activity could be associated with Fenton-like reactions of TMs [29], the ability of cryogels to generate ROS in the reaction with hydrogen peroxide (H_2O_2) was analyzed with the aid of a DCFDA probe (Figure 3). The results show that the cryogel-formulated metal dopants are capable of reacting with H_2O_2, where Co and Cu, unlike Zn, effectively generate ROS (Co > Cu) in agreement with earlier observation for these TMs' behavior in solution [35]. When introduced together, Co and Cu showed additive prooxidant activity, whereas the individual activity of Co or Cu was profoundly inhibited in the presence of Zn co-dopant, reflecting its antioxidant/anticorrosive effect toward the metals with variable valency [28].

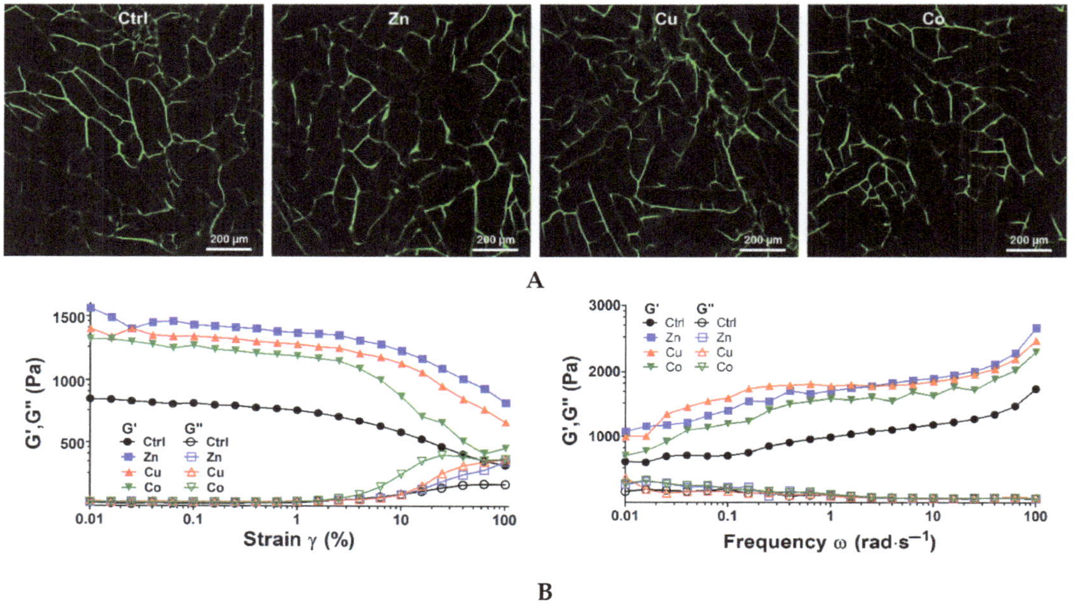

Figure 1. (**A**) LSCM images of non-doped (Ctrl) and TM-doped cryogel sheets (top surface) visualized by autofluorescence upon argon laser excitation (488 nm). (**B**) Strain amplitude sweep test (angular frequency ω = 10 rad s^{-1}) and frequency sweep test (strain deformation δ = 1%) data for the cryogels. TM-doped cryogels (1 mM) were analyzed.

3.2.2. Cytokine and Growth Factor Profile

The effect of metal dopants on the production of cytokines and growth factors by HSFs in the cryogels was assessed (Figure 4A). The secretion of FGF-2, VEGF, IL-6, and IL-8 tightly involved in paracrine stimulation of angiogenesis [26] was profoundly stimulated (by 1.2–5.6 times relative to the non-doped cryogel) generally as follows: Zn < Co < Cu (Figure 4B). Monocyte chemoattractant proteins (MCP-1, MIP-1b) were also overproduced in the presence of TMs. In comparison with Cu and Co, Zn weakly affected the level of pro-inflammatory cytokines IL-6/IL-8, whereas it greatly increased the level of MCPs comparably with Cu and Co. The TMs induced the overproduction of EGF and PDGF-AA, similar to other pleiotropic growth factors, namely, FGF-2 and VEGF (by 1.9–3.2 times). Exceptions were that Co and Zn exhibited a lack of effect, respectively, in the case of EGF and VEGF (Figure 4B). The results demonstrate a strong ability of the cryogel-formulated TMs to increase key signaling molecules involved in regeneration-related processes.

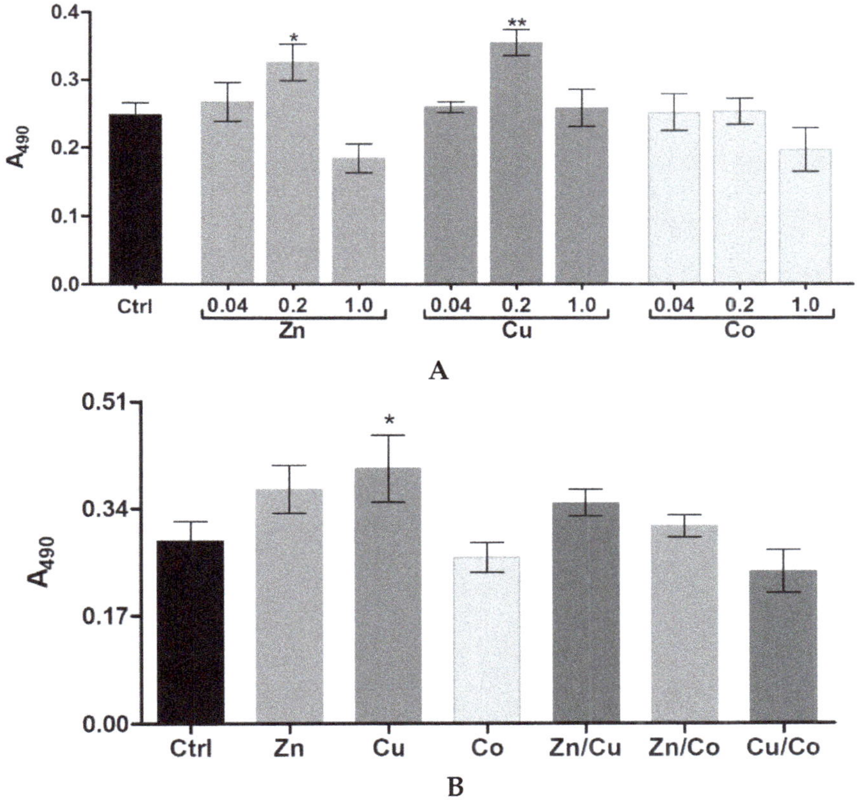

Figure 2. (**A**) Proliferation of 3T3 fibroblasts in TM-doped cryogels at different metal concentrations (MTS assay, 72 h). (**B**) Effect of metal dopants (0.2 mM) on HSF proliferation in the cryogels (MTS assay, 72 h). The data are presented as mean ± SD ($n = 3$, * $p < 0.05$, ** $p < 0.01$).

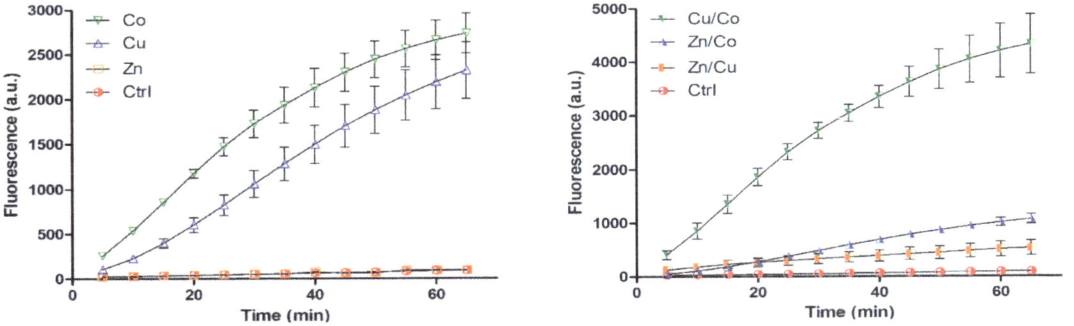

Figure 3. Generation of ROS in reaction of TM-doped cryogels with H_2O_2 according to DCFDA fluorescence ($\lambda_{ex} = 490$ nm, $\lambda_{em} = 526$ nm).

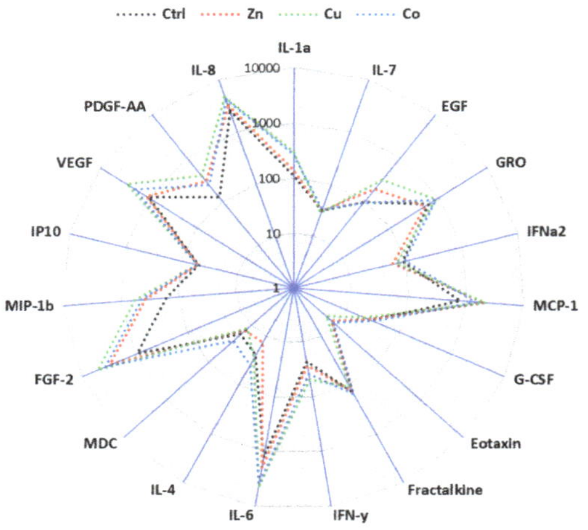

A

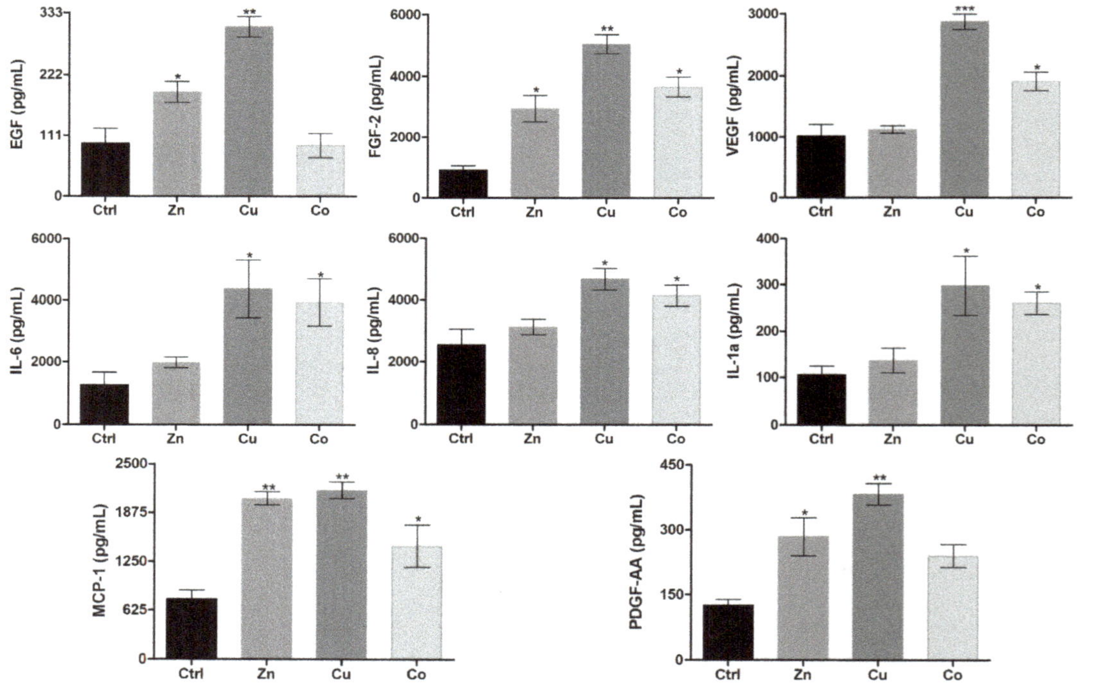

B

Figure 4. (**A**) Radar plot representation of secreted cytokine/growth factor profile for HSFs cultured in TM-doped cryogels (0.2 mM) for 24 h (pg/mL, log scale). (**B**) The corresponding levels of selected cytokines/growth factors. The data are presented as mean ± SD (n = 3, * $p < 0.05$, ** $p < 0.01$, *** $p < 0.001$).

3.3. Behavior of HUVECs in Metal-Doped Cryogels

3.3.1. Proliferation and Spreading

Human umbilical vein endothelial cells (HUVECs) were used as reference endothelial cells (EC) to compare angiogenic properties of the TM-doped cryogels. The metal dopants (0.2 mM) did not cause any inhibition of HUVEC proliferation, similar to that observed for HSFs, confirming the cytocompatibility of the materials for primary human cells. Furthermore, HUVECs proliferated much faster in the presence of TMs by a factor of ca. 1.3 (Zn), 1.5 (Co) and 1.9 (Cu) compared to the control cryogel (Figure 5A). In comparison with HSFs, HUVECs were characterized by higher sensitivity to Cu and Co, whereas Zn almost abolished the stimulating effect of Cu and Co in binary compositions (Figure 5A), presumably in relation to the ROS-modulating activity of the TMs (Figure 3).

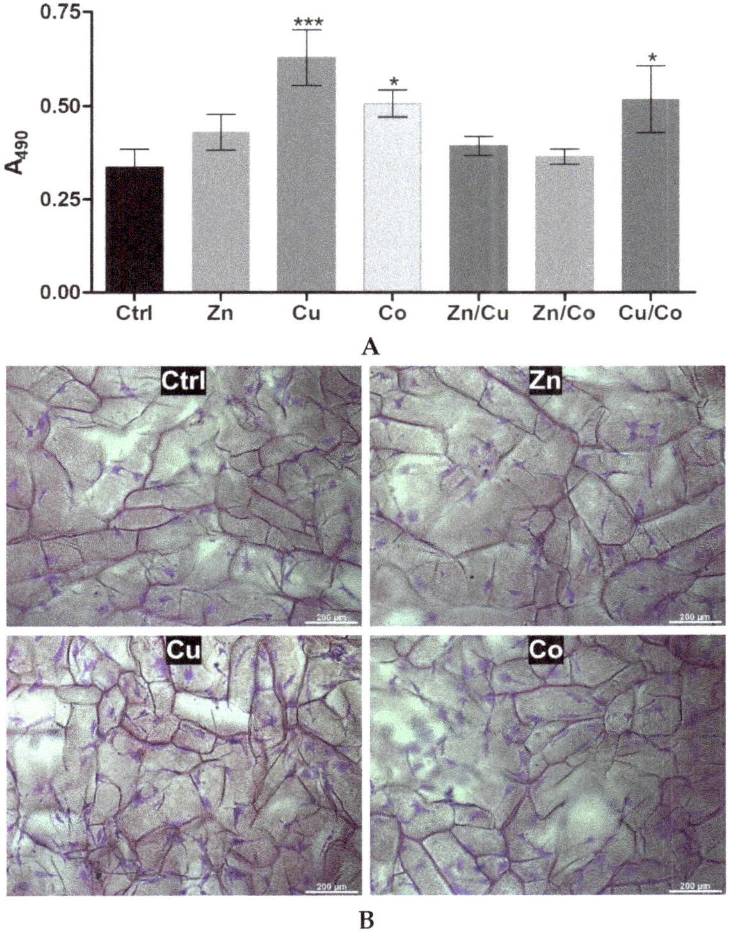

Figure 5. (**A**) Effect of metal dopants (0.2 mM) on proliferation of HUVECs cultured in TM-doped cryogels (MTS assay, 72 h). The data are presented as mean ± SD (n = 3, * $p < 0.05$, *** $p < 0.001$). (**B**) Corresponding bright-field microscopy images of HUVECs stained with cresyl violet in metal-free (Ctrl) and TM-doped cryogels.

In addition, HUVECs were visualized within the semi-transparent cryogels after cell staining with cresyl violet (Figure 5B). The calculated cell number per 1 mm^2 of the analyzed surface was as follows: 47 ± 8 (Ctrl), 53 ± 6 (Zn), 102 ± 16 (Cu) and 73 ± 11 (Co), supporting the fact that the corresponding MTS signals (Figure 4A) reflected the cell density in the matrix rather than the change in metabolic activity. Furthermore, HUVECs were well adhered and spread on the surface of all cryogels and their morphology was altered to spindle-shaped cells with developed extensions. Specifically, in the Cu-doped cryogel, up to 41% of the cells adopted more elongated morphology with at least three sprouts; the adjacent cells showed a tendency to migrate and connect to each other, which is attributed to the induction of tubulogenesis.

3.3.2. Angiogenic Differentiation

The markers of HUVEC differentiation in the TM-doped cryogels were detected by LSCM. In the presence of TMs, phalloidin CruzFluor™ 647-stained cells showed profound morphological changes with reorganization of actin cytoskeleton (Figure 6A).

In the Cu-doped cryogel, the cells were characterized by increased spreading with ca. 2.2-fold bigger area than control cells (mean cell area detected was 5807 ± 506 and 2595 ± 468 μm^2, respectively). According to immunofluorescence analysis, the expression of VEGF and ICAM-2 factors, which regulate EC viability, migration and microvasculature formation [36], was increased by the metal dopants as follows: Zn < Co < Cu, approximately 1.6–2.9-fold for VEGF and 2.6–5-fold for ICAM-2 (Figure 6B,C). This further suggests profound angiogenic activity of the TMs, especially Cu, under experimental conditions.

3.4. *Effects of Metal-Doped Cryogels upon Subcutaneous Implantation*

3.4.1. In Vivo Model Overview

Based on previous surgical procedures [30–32], subcutaneous implantation model in Wistar rats was optimized to assess well-defined effects of the cryogel-formulated TMs. Two subcutaneous square pockets (1.5 × 1.5 cm) were formed at the upper dorsal surface by means of incision at the lower side of the outlined square zone followed by the skin detachment from underlying tissues (Figure 7). Square cryogel sheets (1 × 1 cm) were aseptically inserted into the formed pockets followed by skin suturing. In comparison with skin excision, the model allows for informative analysis of localized effects of the materials on intact host tissues upon biodegradation and release of TMs. Furthermore, early-stage host tissue responses to the TM-doped cryogels at days 5 and 10 after implantation were selected to better compare regeneration and inflammation-related processes. The main skin layers and appendages from the subcutaneous muscle to the epidermis were histologically examined (Figure 7). The analysis of non-doped and TM-doped materials in the same animal enabled improved assessment of specific activities of TMs, as immune and regenerative responses in rats are intrinsically variable [37], allowing one to reduce the number of animals in each group.

3.4.2. Biodegradation of Cryogels

Lateral sections of the skin in contact with the cryogel were stained with Giemsa or Hematoxylin-eosin; the former stain was selected for histological differentiation between the main skin structures designated as numbers from **1** to **6** (Figure 8). Biodegradation of the implanted cryogels was analyzed by quantifying their residual area (Figure 8, **1**). The detected amount of the non-doped cryogel was reduced by ca. 38% from day 5 to day 10 post-implantation, suggesting initial resorption kinetics of the material. The components Zn and, to a lesser extent, Cu increased the rate of degradation of the cryogel, whereas Co had a relatively weak effect on the process. When the Zn dopant was used, ca. 80% of the cryogel implant was degraded at day 5, whereas comparable degradation was achieved for Cu at day 10 (Figure 8B, **1**). The acceleration of the biodegradation of Zn- and Cu-doped materials could be attributed to the TM-mediated increase in the catalytic activity of matrix metalloproteinases (MMP) [38].

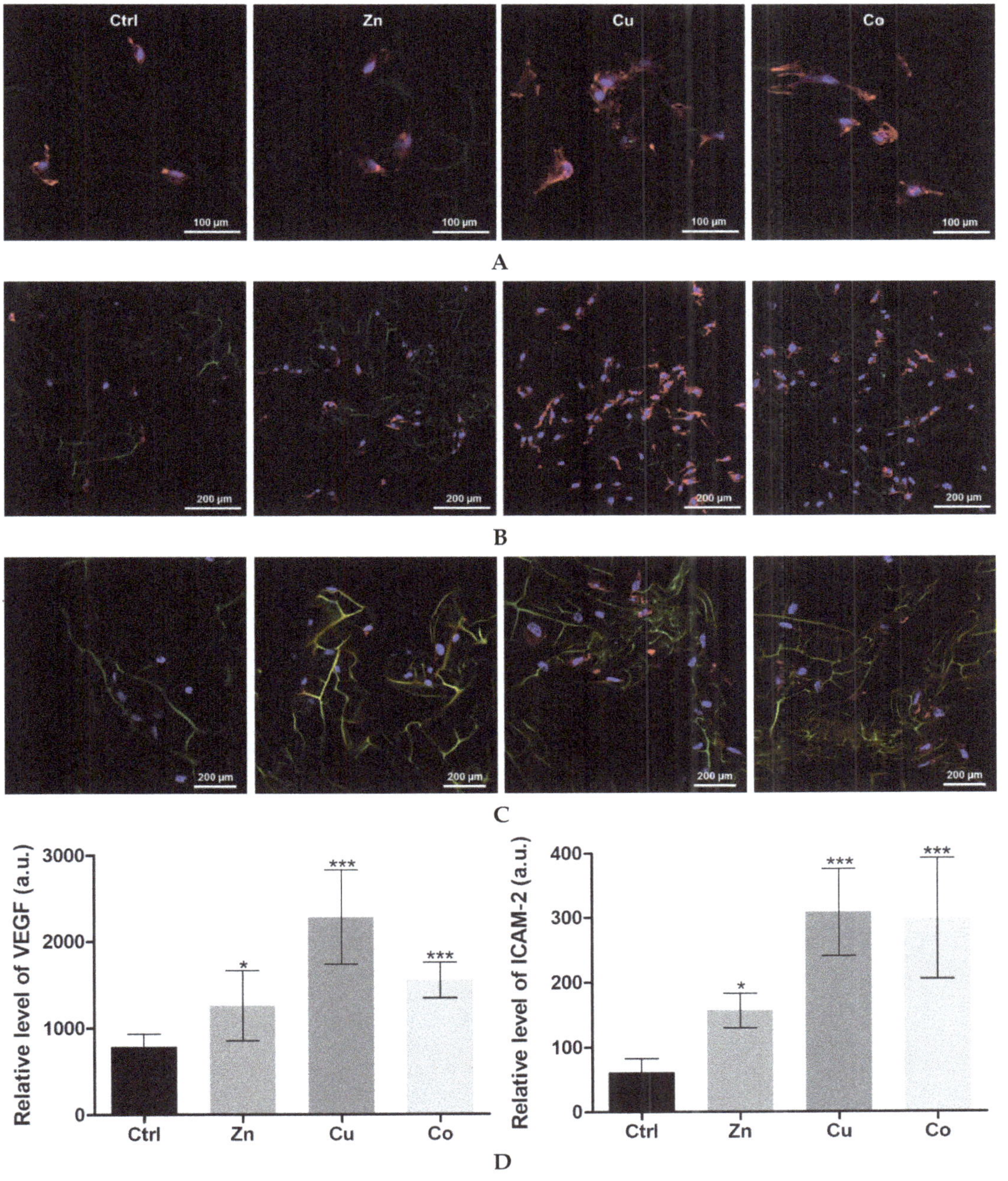

Figure 6. LSCM images of HUVECs grown in TM-doped cryogels (0.2 mM) at day 3 post-seeding. (**A**) Cells stained with phalloidin CruzFluor™ 647 conjugate for F-actin (red). (**B**) Immunofluorescence detection of VEGF. (**C**) Immunofluorescence detection of ICAM-2. Cell nuclei were stained with DAPI (blue). (**D**) Relative density of VEGF and ICAM-2 expression per field of view (mean ± SD, * $p < 0.05$, *** $p < 0.001$).

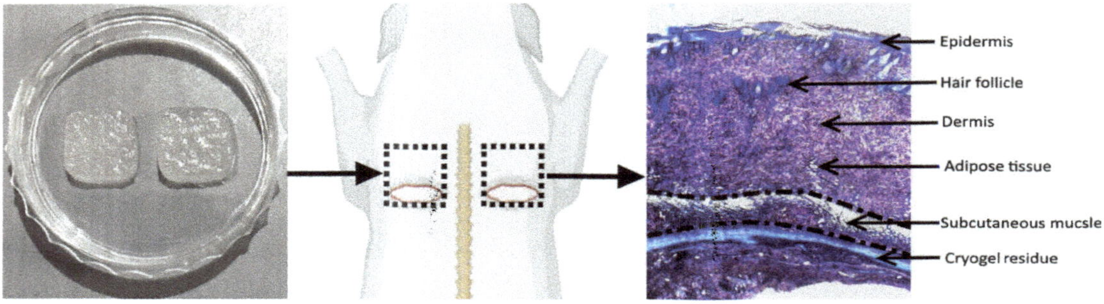

Figure 7. Schematic representation of implantation of cryogel sheets (1 × 1 cm) into subcutaneous pockets (1.5 × 1.5 cm, dotted line) in rats and main skin structures subjected to histological analysis. See Section 2.8.2 for details.

3.4.3. Subcutaneous Muscle and Adipose Tissue

The subcutaneous muscle (SM) and adipose are well-vascularized tissues that host different specialized and multipotent cells and play important roles in cell recruiting, nutrition, biosynthesis of extracellular matrix (ECM) and angiogenesis [39,40]. At day 5, the Zn dopant induced a profound 1.7-fold increase in the thickness of SM (Figure 8, **2**), which is in immediate contact with the cryogel, whereas this parameter was weakly affected by Cu and showed a tendency to decrease in the presence of Co. At day 10, the stimulating effect of Zn on SM decreased, probably in relation to rapid resorption of the Zn-doped cryogel, whereas Cu exhibited comparable stimulation to Zn at day 5. Furthermore, in the presence of Zn and Cu dopants, SM was reorganized so that round rolled muscle structures became unrolled; the latter SM configuration favors cell migration and proliferation upon skin regeneration [40]. Furthermore, SM was greatly disrupted and mostly replaced by an amorphous connective tissue when the Co-doped cryogel was applied (day 10), complicating corresponding analysis (Figure 8B, **2**).

Adipose tissue (Figure 8, **3**) detected as hollow reticular areas (corresponding to washed-out lipid contents of adipocytes) [41] was also enlarged and deeply penetrated into the dermis in the presence of Zn and Cu components, respectively, by ca. 1.5 times (day 5) and over 1.7 times (day 10), compared to the control (non-doped cryogel). This effect of TMs was observed along with morphological change of some adipocytes from round- to ellipsoid-shaped cells. In great contrast to Cu and Zn, Co generally disrupted the adipose structures by day 10 (Figure 8B, **3**).

3.4.4. Vascular System

In comparison with the control material, the Zn- and Cu-doped cryogels did not significantly change the distribution pattern of vascular structures (Figure 8A), which were predominantly localized in adipose and SM tissues, though causing noticeable expansion of these structures (Figure 8B, **4**). At day 5, the skin vascularization assessed by the relative area of the vasculature was increased by the Zn and Cu dopants (Cu > Zn), whereas at day 10 this effect was shown to be dissipated for Zn and maintained for Cu, presumably reflecting resorption properties of the corresponding materials (Zn > Cu). Co, though disintegrating intact SM and adipose layers, at day 10 induced a profound formation of many relatively small vessels in different skin layers, including upper dermis tissues (Figure 8A), which is not typical for intact rat skin. The results suggest all the metal dopants as angiogenic factors as follows: Co $\geq$ Cu > Zn. This assumes excessive activity of the Co component (also noting decreased degradation of the Co-doped cryogel).

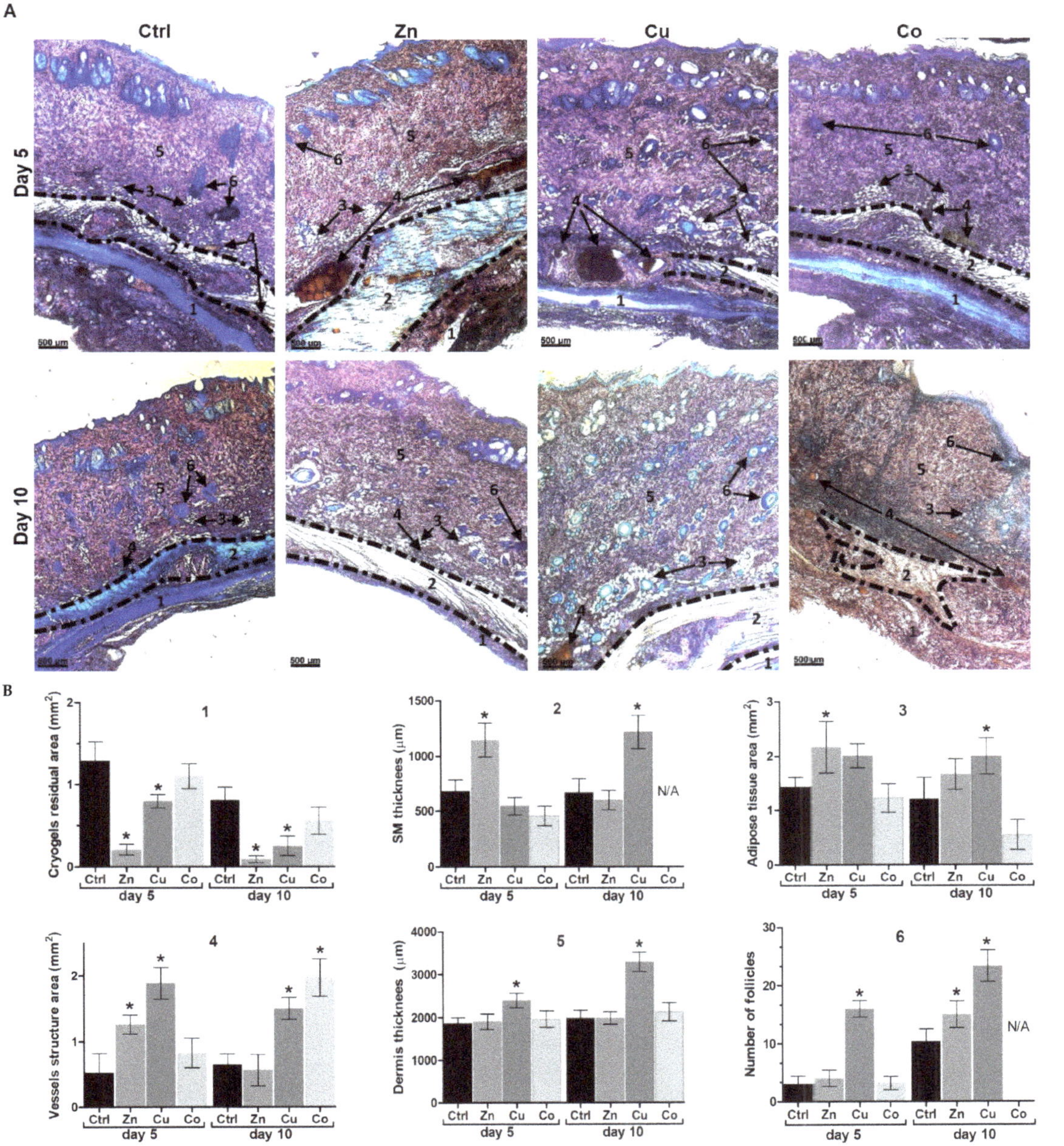

Figure 8. (**A**) Bright-field microscopy images of Giemsa-stained cross-sections of skin explants contacted with subcutaneously implanted TM-doped cryogels. (**B**) Mean morphometric parameters of the treated skin (designated as numbers from **1** to **6**) per cross-section (mean ± SD, * $p < 0.05$).

3.4.5. Dermis

The Cu dopant caused a noticeable thickening of the dermal layer (Figure 8B, **5**) by 1.3 and 1.7 times for days 5 and 10, respectively, attributed to extracellular matrix (ECM) overgrowth as a result of Cu-mediated tissue-vascularization. The Zn and Co dopants insignificantly impacted this parameter (Figure 8). Furthermore, the relative area of mature collagen in the dermis was quantified using polarization microscopy with Picrosirius red (Figure S4), which stains mature (type I) and immature (type III) collagens in yellow-red and green colors, respectively [42]. The parameter increased as follows: Co $\leq$ Ctrl < Zn < Cu, as if the Zn and Cu dopants enhanced collagen maturation by ca. 1.3 and 1.6 times, respectively. These data show the increased growth-promoting effect of Cu over Zn toward the dermis. Co, weakly affecting the thickness and collagenization of the dermis, prompted certain disorganization of its intact fibrillar structure, though to a lesser extent than that observed for SM and adipose tissues (Figure 8A), suggesting a specific distance-dependent disturbing effects of the Co dopant on surrounding tissues.

3.4.6. Hair Follicles and Epidermis

At day 5, the Cu-doped cryogel induced a noticeable increase in the number of hair follicles (**6**) in the dermis, which was 5-fold higher than in the control group, whereas the Zn-doped cryogel approached comparable effect at day 10 (Figure 8A). Most follicles in the Cu and Zn groups contained dermal papillary cells and had defined intensively stained outer covering, which is characteristic of the anagen phase [43,44]. The results show that the Cu and Zn dopants have folliculogenic activity (Cu > Zn), and the effect of Zn seems to be delayed, taking into account the rapid resorption of the corresponding matrix. The Co dopant demonstrated a lack of significant stimulation of folliculogenesis.

The skin treated with both the Cu- and Zn-doped cryogels was characterized by a well-structured epidermis, similar to that of the control group but with a somewhat more developed stratum spinosum layer and increased keratinization (Figure S3). In addition, the Cu dopant noticeably promoted epithelial invagination (Figure S3), which is involved in re-epithelization and folliculogenesis [45]. Under the same conditions, the Co-doped cryogel here and there disturbed the epidermis structure (Figure S3) presumably in relation to the disorganization of underlying skin layers.

3.4.7. Immune Cells Appearance

The TM-doped cryogels differently affected the appearance of immune cells with distinct morphological features [46] throughout the whole skin. In particular, large intensively stained cells were attributed to mononuclear macrophages having smoothly shaped elongated or roundish morphology with their interior predominantly occupied by a large nucleus (Figure 9, MM) and dendritic cells, which are irregularly shaped due to noticeable cytoplasmic extensions (Figure 9, DC). Furthermore, multinucleated giant cells usually formed as a result of macrophages fusion appeared as huge elongated cells with multiple nuclei [47] (Figure 9, GC).

Upon treatment with the non-doped and Zn-doped cryogels, mononuclear macrophages prevailed over dendritic cells, though the latter material significantly increased overall immune cell number at day 5 (Figure 9). Unlike the above materials, the Cu-doped cryogel (days 5 and 10) and the Co-doped cryogel (day 5) induced a predominant appearance of dendritic cells over macrophages. Furthermore, the Cu group was characterized by a somewhat decreased number of immune cells compared to the Zn and Co groups. The Co dopant in a specific manner resulted in the appearance of numerous giant cells in the dermis at day 10. The provided histological data (Figures 8 and 9) were additionally supported by the corresponding analysis of H&E-stained skin sections (Figure S2). In particular, the latter analysis of the Co group revealed that giant cells are closely located to small capillary structures filled with erythrocytes attributed to newly forming vessels (Figure 9, H&E).

3.4.8. Immunohistochemical Analysis

Additional immunohistochemical analysis of the treated skin (Figure 10 and Figure S5) showed that all TM-doped cryogels increased the number of CD31-positive cells attributed to ECs by a maximal factor of 3.8 (Cu) and 1.9 (Zn) observed at day 5 and 2.9 (Co, day 10) (Figure 10B). These data are consistent with the angiogenic activity of the metal dopants in vivo according to the histological analysis (Figure 8 and Figures S2–S4). Furthermore, the TMs differently affected the production of HIF-1α as follows: Ctrl $\leq$ Zn < Cu < Co. These data show the ability of prooxidant Co and Cu dopants to induce a hypoxia-like state in the surrounding tissues, which apparently underlies their enhanced angiogenic effects, whereas the Zn component somewhat increased the appearance of ECs and vascular structures without a significant overproduction of HIF-1α.

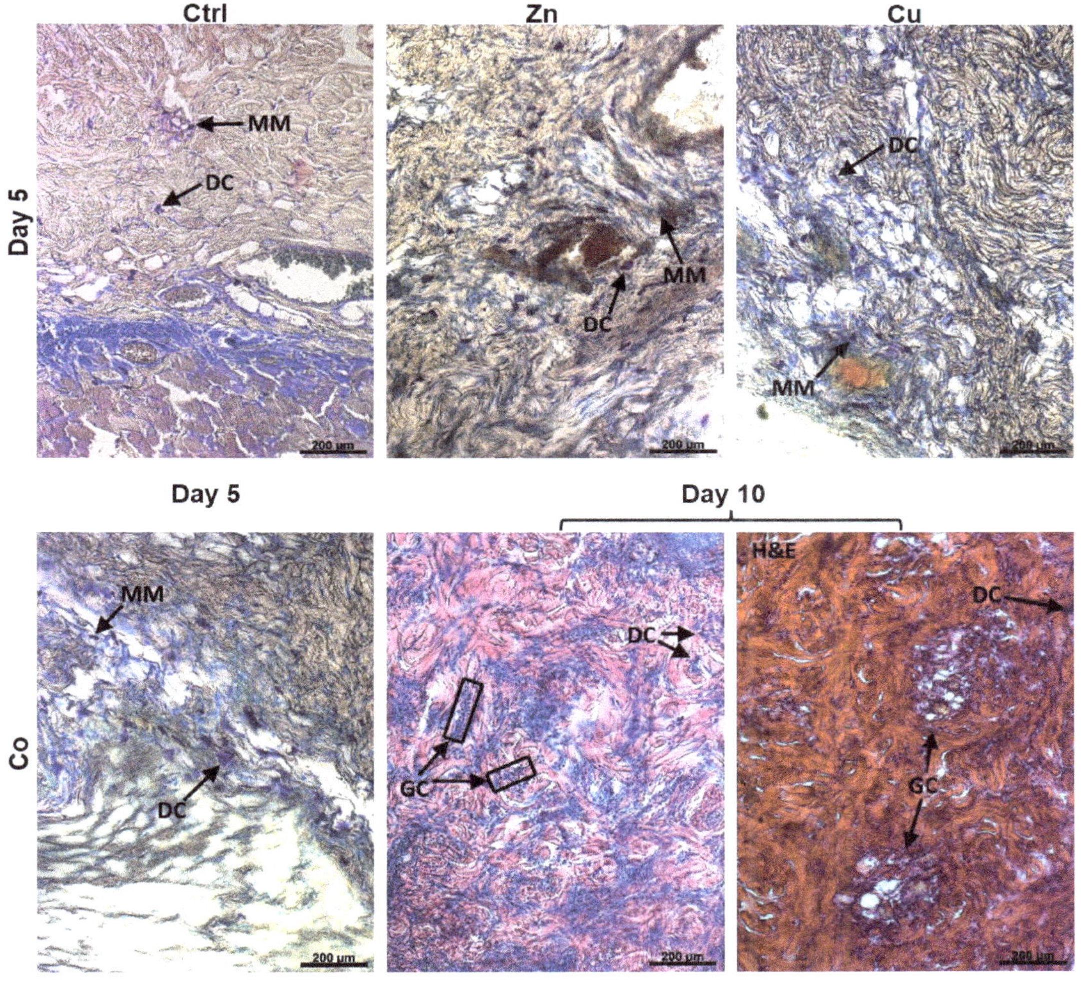

Figure 9. Representative bright-field microscopy images of Giemsa-stained cross-sections of skin explants contacted with subcutaneously implanted TM-doped cryogels. Mononuclear macrophages (**MM**), dendritic cells (**DC**), and multinucleated giant cells (**GC**) were identified. H&E-stained section was additionally shown in right image of lower panel.

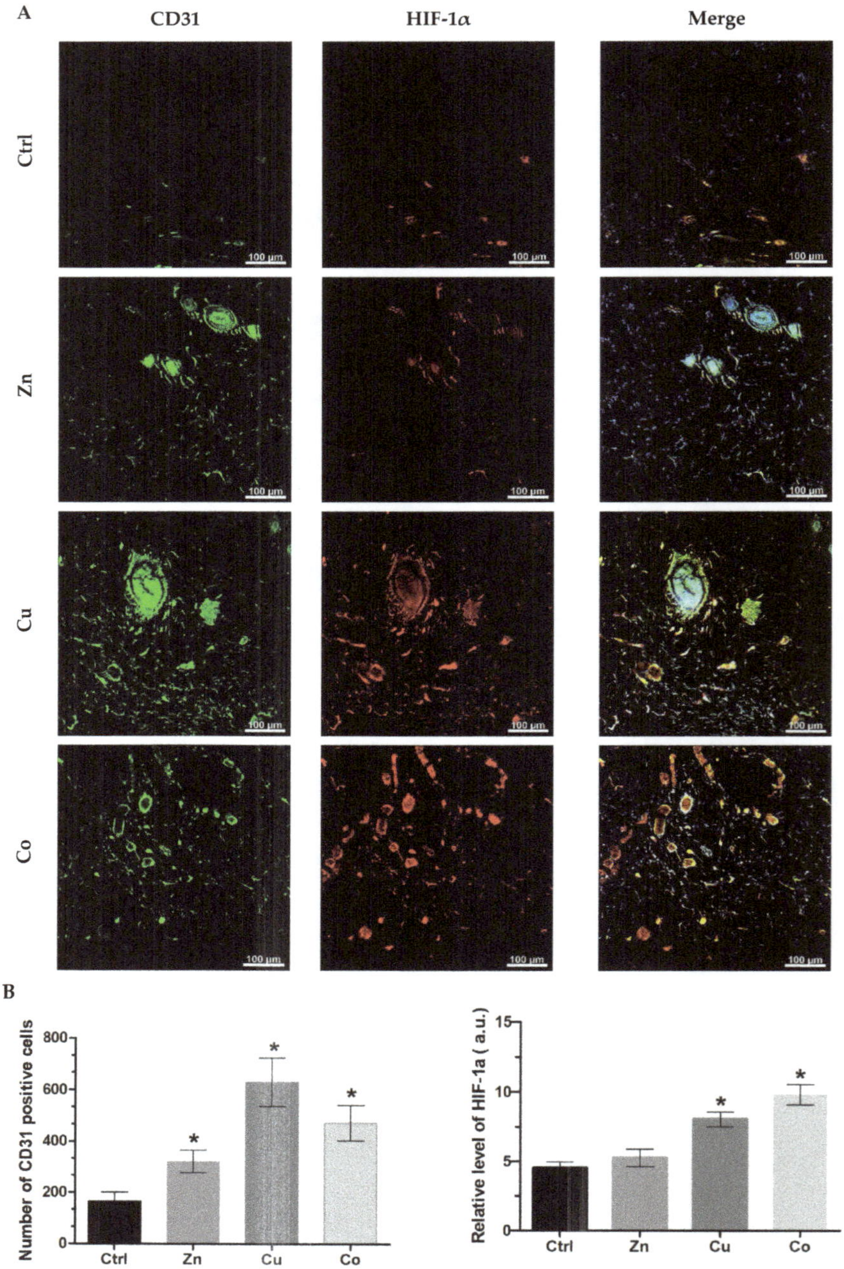

Figure 10. (**A**) Immunofluorescent analysis of cross-sections of skin explants contacted with subcutaneously implanted TM-doped cryogels. Simultaneous CD31 Alexa Fluor 488 (green), HIF-1α Alexa Fluor 647 (red) and DAPI staining was performed. (**B**) Mean number of CD31-positive cells (**left**) and relative density of HIF-1α expression (**right**) per field of view (mean ± SD, * $p < 0.05$).

Furthermore, the Zn- and Cu-doped cryogels were found to significantly increase the dermal level of MMP-2 and MPP-3 involved in the degradation of ECM proteins, the release of ECM-bound growth factors, angiogenesis and tissue remodeling. Similarly, MMP-9 level was elevated in the presence of these materials (data not shown). The effect of metal dopants decreased at day 10 over day 5, as more clearly observed for the rapidly degrading Zn-doped material (Figure S6). The Co-doped cryogel generally did not increase the content of MMPs compared to the control material (Figure S6A,B); however, at day 10, there were distinct structures with highly expressed MMP-2 in the Co group (Figure S6C) attributed to the areas containing giant cells and newly formed capillaries (Figure 9).

4. Discussion

The cross-linked gelatin cryogel was used as a biodegradable hydrogel scaffold both to examine regenerative activities of incorporated TMs and to develop improved bioinductive materials. As previously shown, different TMs can be stably incorporated into the cryogel [19], presumably involving complex formation with collagen polypeptide groups in accordance with the earlier observation [48]. The advanced macroporous structure of the cryogels favors their bulk interactions with gases, nutrients and living cells, allowing one to assess regenerative factors in three-dimensional tissue-mimicking conditions [2,34].

Cu, Co, and Zn were studied here as recognized components of solid osteogenic materials [11–13,49]; however, therapeutic effects of these TMs in cryogel scaffolds toward soft tissues have not been compared to date. The composition of TM-containing cryogels was optimized so that the metal dopants did not show a cytotoxic effect, while they were able to stimulate the proliferative and functional activity of mammalian cells (Figure 2). The corresponding effective concentrations of the incorporated TMs lay within the range of up to 1 mM, which moderately affected the structure of cryogels (Figure 1). In particular, the metal dopants comparably increased the elastic over viscous behavior of the cryogels up to ca. 2 times (Figure 1B), attributed to additional TM-mediated cross-linking. These data suggest that the incorporated TMs participate in forming a polymer network of the materials and that the stiffness of the TM-doped cryogels should fit with different soft tissues and support cell–matrix interactions [50].

Earlier, the macroporous structure of cryogels allowed us to monitor bulk affinity interactions of the materials with fluorescently labeled peptide ligands [34]. Similarly, the interaction of H_2O_2 with TM-doped cryogels was assessed using an ROS-sensitive DCFDA probe, considering H_2O_2 as both an extracellular and intracellular precursor of ROS generated in Fenton-like reactions [29,51]. The hydroxyl radical is particularly recognized as a secondary messenger involved in cell redox regulation via oxidizing thiols and activating transcriptional factors such as HIF-1α [52]. The ROS-generating ability of the Co and Cu dopants revealed in the presence of H_2O_2 (Figure 3) should reflect hypoxia-mimicking and concomitant angiogenic properties of these TMs [22,24,29]. H_2O_2 is known to be released to the site of tissue injury (for example, in association with NADPH oxidase or dual oxidase activities), playing an important regulatory role in the healing process as well as a potential pathological role [53,54]. Therefore, the Co- and Cu-doped cryogels upon tissue implantation are expected to promote the extracellular activity of H_2O_2 in contrast to the Zn-doped cryogel.

In comparison with the above reactions (Figure 3), the soluble TMs showed a different ROS-generating profile in 3T3 fibroblasts exposed to 1 or 10 μM compounds in antioxidant-free HBSS (Figure S7). Soluble Cu and, to a lesser extent, Zn, in contrast to Co, were found to induce some ROS overproduction accompanied by a weak decrease in glutathione level in cells. Together, these data support variable condition-dependent ROS-modulating activities of the TMs. The results, in particular, suggest that soluble Zn is also able to increase ROS formation in the fibroblasts, though weaker than soluble Cu (Figure S7), regardless of the ability of Zn co-dopant to inhibit the Fenton-like activity of the Co- and Cu-doped cryogels (Figure 3). The intracellular redox effect of Zn ions can be associated with their interaction with multiple thiol groups of Zn-binding cysteine-rich proteins, particularly metallothionein [55], activation of ROS-producing mitochondrial lipoamide dehydrogenase (LADH) as well as the inhibition of LADH thiol oxidoreductase [56]. Such prooxidant activity is controlled by MTF-1 transcriptional factor activated by increased Zn concentrations, leading to the overexpression of metallothionein and Zn efflux transporters [55,57,58].

The lack of the ROS-generating effect of soluble Co on 3T3 cells (Figure S7) could be explained by the decreased physiological requirement and cellular uptake of this TM in fibroblasts, noting that in cancerous cells the TM readily caused oxidative stress under similar conditions [59]. Unlike soluble Co, nano- and microparticles of this TM were shown to induce oxidative stress in 3T3 cells probably due to the internalization of the particulate formulations by endocytosis and phagocytosis [60]. Interestingly, the proliferation-stimulating effect of the TM-doped cryogels on the fibroblast cells (Figure 2) is in accordance with prooxidant cellular effects of the soluble TMs (Figure S7), suggesting that the former effect involves intracellular uptake of the metal ions released from the materials. Together, our results demonstrate that the studied TMs possess multifaceted ROS-modulating activities depending on the conditions. Variable prooxidant effects of Cu and Co, both extracellular and intracellular, could be particularly expected, whereas Zn is expected to have dual prooxidant/antioxidant effects.

Among the metal dopants, Cu exhibited the highest stimulation of proliferation of all studied cells (3T3 fibroblasts, HSFs, HUVECs) with as high as 2-fold stimulation for HUVECs (Figures 2 and 5). The Co dopant was almost inactive toward 3T3 cells and HSFs, whereas it significantly increased the proliferation of HUVECs, suggesting increased sensitivity of these ECs to the TMs compared to other types of cells in accordance with reported effects of the GHK-Cu complex [20] and metal chelators [61].

Based on the above results, the mitogenic activity of the TM-doped cryogels could be explained by the activation of ROS-mediated signaling pathways in the cells [52]. Such an effect seems to require an appropriate cell-specific level of overproduced ROS, presumably achieved in both fibroblasts (Figure 2) and HUVECs (Figure 5) grown in the Cu- and Zn-doped cryogels. This level should be essential for the stimulation of regenerative processes, particularly angiogenesis [62]. Excessive ROS levels can decrease cell viability and lead to sustained secretion of pro-inflammatory cytokines and endothelial permeability [63]. Therefore, double composition of prooxidant TMs, namely, Cu and Co, demonstrated lower or lack of stimulation toward HUVECs/HSFs compared to Cu alone (Figures 2B and 5A) attributed to excessive generation of cellular ROS. Moreover, the Zn co-dopant abolished stimulation of HUVECs in the Cu- and Co-doped cryogels (Figure 5A) and decreased it for HSFs in the Cu-doped cryogel (Figure 2B). This is attributed to the prevention of ROS production by the Zn co-dopant presumably via the inhibition of the Fenton-like reaction catalyzed by Cu and Co [52] and/or the possible reversal of HIF1-α upregulation induced by the latter TMs [64].

The mitogenic activity of the metal dopants was accompanied by the angiogenic differentiation of HUVECs in the presence of Cu and Co viewed through the morphological rearrangement and overexpression of VEGF and ICAM-2 (Figures 5B and 6). Furthermore, all the TM dopants induced the overexpression of a series of cytokines and growth factors by HSFs (Figure 4), including VEGF, FGF-2 and PDGF, the potent regulators of functional activity of ECs and blood vessels [65]. Moreover, the enhanced release of MCP-1 by the Zn and Cu dopant (Figure 4) may be essential for the recruitment of monocytes/macrophages involved in tissue repair and angiogenesis [66,67]. Of note, the aforementioned cytokines and growth factors can be produced and secreted by ECs, playing important roles in the autocrine regulation and recruitment of other cells to support neovascularization [20]. Thus, our results show that the incorporated TMs allow for the effective induction of pleiotropic growth factors and cytokines in the cryogel-grown cells. Such an activation of cryogels with TMs provides a promising alternative to the immobilization of low-stable recombinant polypeptides such as VEGF [68] and FGF-2 [69], solely or in combination [70], in order to improve regenerative and angiogenic properties of polymeric scaffolds.

The subcutaneous evaluation allows one to understand regenerative or toxicological mechanisms of (bio)materials at molecular, cellular and tissue levels [30–32,71,72], and these are relevant not only for skin but for other organs and tissues. The potential of subcutaneous implantation for the analysis of TM-containing bioactive hydrogels is poorly investigated, apart from a few reports [73,74] not dealing with TM comparison. Such a model was optimized here to establish specific localized effects of macroporous cryogel-formulated TMs on various intact skin tissues and appendages (Figure 7).

The degradation rate of implanted TM-doped cryogels was in the order: Zn > Cu > Co $\geq$ Ctrl (Figure 8, **1**). The profound promoting effect of the Zn dopant is apparently associated with this TM activity as a key cofactor of different MMPs, including MMP-1, MMP-3, MMP-8, MMP-13, MMP-2, and MMP-9 [75]. The peptidase activity of MMP-1 and MMP-9 can be also promoted by Cu [38] in accordance with some lesser effect of the latter dopant. Some delayed effect of the Co dopant at day 10 (Figure 8, **1**) may be attributed to the induction of MMP-1 and MMP-2 expression under oxidative stress conditions [76,77]. Immunofluorescence analysis confirmed a significant increase in the expression of MMP-2 and MMP-3 (as well as MMP-9) in the dermis in the presence of Zn- and Cu-doped cryogels, which, however, had a comparable effect with each other (Figure S6). Therefore, the increased resorption of the former material should be due to the overexpression of MMPs and/or a higher contribution to specific peptidase activity by Zn ions as compared to Cu ions.

The Cu dopant induced a significant thickening of the dermal layer along with an increase in the number of hair follicles recognized as an important source of stem and progenitor cells for skin repair (Figure 8). These data confirm the positive role of this metal in tissue renewal and remodeling in concordance with the reported ability of peptide-complexed Cu (AHK-Cu) [78] and Cu-doped bioactive glass [79] to promote the formation and growth of hair follicles. Such an effect could be mediated by the stimulation of angiogenesis facilitating the migration of follicle progenitor cells to the dermis [79]. A folliculogenic activity of the Zn-doped cryogel detected at day 10 is in agreement with the recently reported effect of a Zn-containing chitosan-PEG hydrogel in a full-thickness skin defect model in mice [80].

Stimulation of collagen deposition by the Cu and Zn dopants may be associated with effective vascularization or increased collagen stabilization by lysyl oxidase in the presence of Cu [81] as well as with early Zn-mediated attraction of fibroblasts that produce collagen, among other ECM components [80]. Considering some disorganization of skin layers by the Co dopant (Figure 8), its effect on skin functioning is assumed to be negative, presumably due to excessive oxidative stress. It has been previously shown that the use of the Co component at an increased concentration in hydrogels reduces cytocompatibility, collagen deposition and slows down wound closure and re-epithelization [18].

The established ability of metal dopants to increase tissue vascularization (Figures 8 and 10B) is well explained by the generation of ROS (Figure 3 and Figure S7) in combination with the increased tissue level of ROS-sensitive HIF-1α (Figure 10B). It was reported that the modification of mesoporous bioactive glasses by Co [22] and Cu [24] promoted neovascularization by inducing hypoxic cascade and the expression of HIF-1α in human bone marrow stromal cells, yet the induction of HIF-1α was not investigated in vivo. Moreover, the incorporation of Co in a gauze calcium alginate hydrogel was shown to stabilize HIF-1α expression, stimulate angiogenesis and accelerate full-thickness skin wound healing in mice [18]. HIF-1α is a major transcription regulator of VEGF [18,82], which is consistent with a considerable boost in VEGF secretion by fibroblasts (Figure 4) and expression by HUVECs (Figure 6) cultured in Cu- and Co-doped cryogels. In addition, metal-induced VEGF and FGF-2 (Figure 4) are known to stimulate the production of MMPs that degrade basement membrane and ECM, allowing ECs to migrate and form sprouts [70].

The results suggest that the prooxidant activity of the TMs may occur via both extra-cellular and intracellular reactions. The antioxidant effect of the Zn co-dopant upon the induced ROS production (Figure 3) presumably was not manifested under experimental conditions in vivo, since the TM was found to cause a weak increase in HIF-1α level and some angiogenic response, which were inferior to the effect of Cu and Co. Other Zn formulations, namely, zinc oxide (ZnO) and zinc peroxide (ZnO_2) nanoparticles, embedded into a cross-linked polymeric hydrogel promoted angiogenesis via the generation of ROS, in particular, the onsite production of H_2O_2 [83]. The Cu-doped cryogel provided a more physiological vascularization with a predominant hypodermic localization, whereas the Co-doped material, disturbing hypodermic tissues, caused the redistribution of newly formed vessel structures into the dermis. Such a distinct vascularization pattern, however, may result from the degree of prooxidant activity of the TMs (Co > Cu).

Consistent with these results, microvessel growth in the skin was earlier stimulated by Cu-containing hyaluronic acid hydrogel [73] and borate glass microfibers [84]. However, no mechanisms of Cu-mediated angiogenic activity and its comparison with other TMs were provided in these studies. According to the density of CD31-positive cells (Figure 10B), the Cu-doped cryogel should have better angiogenic potential than other proposed Cu-containing materials [73,79,84]. Furthermore, Co-containing collagen/alginate-based hydrogel was earlier demonstrated to exhibit an angiogenic effect on rat mesenchymal stem cells accompanied by the expression of CD31 and VEGF. However, this activity was not significantly affected by the Co component in vivo [85]. Likewise, implanted Co-doped sol-gel bioactive glasses promoted the appearance of blood vessel structures; however, angiogenic effects were not assessed quantitatively [86].

The angiogenic effect of the Zn-doped cryogel without a significant induction of HIF-1α (Figures 8 and 10B) could result from the activation of other pathway(s) implicating increased FGF-2 production (Figure 4), faster material degradation by MMPs promoting cell infiltration, and from the attraction of mononuclear cells (Figure 9). It is noteworthy that the in vivo effects of the Zn-doped cryogel generally decreased on day 10 compared to day 5. This suggests a transient activity of TM-doped cryogels presumably associated with the release of metal ions during degradation so that it disappears after resorption of the material. This also confirms the importance of assessing the effects of TM-doped cryogels in the early stages post-implantation.

In the presence of implanted Zn-doped cryogel, recruitment of mononuclear cells was observed (Figure 9). This process is known to be mediated by MCP-1 chemokine [87], which was stimulated by the material in vitro (Figure 4). It was reported the role of Zn in stimulating the infiltration of monocytes into the damaged tissues, where they can differentiate into macrophages [87], and regulating the transition of macrophages from pro-inflammatory to immune-regulatory tissue repair phenotypes [88]. Recruited macrophages can contribute to the angiogenic process via the degradation of ECM leading to EC migration, the release of angiogenic cytokines, and vessel wall formation by differentiating into ECs [66,67]. The incorporation of Zn into the hydroxyapatite-collagen scaffold promoted osteogenesis and angiogenesis by activating the p38 MAPK signaling pathway in the monocytes, further contributing to the release of TGF-β, VEGF, and PDGF, which stimulate the recruitment of BMSCs and ECs to the injury site [89]. The regulatory roles of Zn on immune system homeostasis have been reviewed [90]. These particularly involve a balance between normal immune response and potential tissue damage in relation with Zn distribution in the extracellular and intracellular compartments.

The detected overall number of immune cells in the case of Cu-doped cryogels was lower compared to the other implanted gels (Figure 9), which indicates that the host immune response to this material occurred smoothly. On day 10 post-implantation, the Co dopant caused the formation of multinucleated giant cells surrounding vascularization zones in dermis (Figure 9), in concordance with the reported enhanced angiogenesis by these VEGF-expressing cells within the implantation bed [47]. Subcutaneously implanted Co-containing bioactive glasses presented a significantly increased number of cell nuclei, morphologically resembling the detected giant cells [86]. This was attributed to the HIF-1α-mediated recruitment of proinflammatory cells and coordination of regeneration processes by the material [86]. Given that lymphocytes are essential for giant cell formation [91], it could be assumed that Co may trigger an adaptive immune response, the mechanism of which should be studied elsewhere.

5. Conclusions

This study proves biodegradable cryogels as advanced hydrogel scaffolds both to clarify the regeneration-related effects of incorporated TMs in vitro and in vivo and to develop improved biomaterials activated with Cu, Co, Zn or their compositions. Since these effects are assumed to be associated with ROS generation, the ROS-modulating activity of the TMs was characterized. The results suggest Cu and Co as well as their binary compositions as probable extracellular Fenton-like prooxidants, which are antagonized by Zn. All the TMs are capable of intracellular ROS generation to different extents and depending on cellular uptake. The ROS-modulating activities of the TMs and compositions were consistent with the proliferation rate of mammalian cells, the production of regeneration-related cytokines and growth factors by HSFs and the angiogenic differentiation of HUVECs in the cryogels in vitro. The subcutaneous implantation model was optimized to elucidate and compare host tissue effects of TMs containing cryogels. Biodegradation of the materials was promoted mainly by the Zn and Cu dopants, which also induced the overexpression of matrix metalloproteinases, the stimulation of collagen deposition and hair follicle growth. All the TMs, especially Cu and Co, increased the level of ROS-sensitive markers and vascularization, where more physiological vasculature pattern was observed in the case of Cu compared to Co. The latter dopant, unlike Cu and Zn, caused an obvious disturbance in the organization of skin layers along with the appearance of multinucleated giant cells attributed to the excessive prooxidant effect. The results contribute to the understanding of regenerative and adverse activities of hydrogel-formulated TMs and provide the basis for tissue engineering and regeneration applications of TM-containing cryogels.

Supplementary Materials: The following supporting information can be downloaded at: https://www.mdpi.com/article/10.3390/gels8020118/s1, Figure S1: (**A**) LSCM images of dual TM-doped cryogel sheets (top surface) visualized by autofluorescence upon argon laser excitation (488 nm). (**B**) Frequency sweep test (strain deformation $\delta = 1\%$) data for the cryogels. Figure S2: Representative bright-field microscopy images of H&E-stained cross-sections of skin explants contacted with subcutaneously implanted TM-doped cryogels. Figure S3: Visualization of epidermis of treated skin according to H&E staining. (**A**) Representative bright-field microscopy images for non-doped and TM-doped cryogels. (**B**) Localized area with dermatitis manifestations caused by Co-doped cryogel. Figure S4: (**A**) Representative polarization microscopy images of Picrosirius red-stained cross-sections of skin explants (dermal area) contacted with subcutaneously implanted TM-doped cryogels. (**B**) Relative area of mature collagen per field of view. Figure S5: Immunofluorescent analysis of cross-sections of skin explants contacted with subcutaneously implanted TM-doped cryogels at day 5 (Zn and Cu) and day 10 (Ctrl and Co). Simultaneous CD31 CruzFluor™ 488 (green), HIF-1α Alexa Fluor 647 (red) and DAPI staining was performed. Figure S6: (**A**) Immunofluorescent analysis of cross-sections of skin explants contacted with subcutaneously implanted TM-doped cryogels (MMP-2 Alexa Fluor 488 (green) and MMP-3 Alexa Fluor 647 (red)). (**B**,**C**) Relative MMP levels in the dermis per field of view. (**D**) The area with localized giant cells (MMP-2, Co-doped cryogel). Figure S7: Effect of dissolved metals on relative levels of (**A**) ROS and (**B**) glutathione in 3T3 fibroblasts according to DCFDA ($\lambda_{ex}/\lambda_{em}$ = 490/526) and monochlorobimane ($\lambda_{ex}/\lambda_{em}$ = 380/480) fluorescence.

Author Contributions: Conceptualization, T.I.A.; methodology, I.N.S. and T.I.A.; investigation, A.A.Y. and M.Z.; formal analysis: R.A.I.; writing—original draft preparation, A.A.Y., M.Z. and T.I.A.; writing—review and editing, M.Z., I.N.S. and T.I.A. All authors have read and agreed to the published version of the manuscript.

Funding: The reported study was funded by RFBR and VAST according to the research project 21-515-54003.

Institutional Review Board Statement: The animal study protocol was approved by the Ethics Committee of Kazan Federal University (protocol no. 435B from 28.02.2020).

Data Availability Statement: The data presented in this study are contained within the article.

Acknowledgments: This work was co-funded by the Russian Science Foundation (research project 20-73-10105 in studying ROS-generating activity of TM formulations) and is part of Kazan Federal University (KFU) Strategic Academic Leadership Program. Ilnur Salafutdinov (Institute of Fundamental Medicine and Biology, KFU) is greatly acknowledged for invaluable assistance. The authors thank S.V. Fedosimova and A.G. Daminova (Interdisciplinary Center for Analytical Microscopy, KFU) for LSCM analysis and I.T. Rakipov (Alexander Butlerov Institute of Chemistry, KFU) for rheological analysis.

Conflicts of Interest: The authors declare no conflict of interest.

References

1. Abbasian, M.; Massoumi, B.; Mohammad-Rezaei, R.; Samadian, H.; Jaymand, M. Scaffolding polymeric biomaterials: Are naturally occurring biological macromolecules more appropriate for tissue engineering? *Int. J. Biol. Macromol.* **2019**, *134*, 673–694. [CrossRef] [PubMed]
2. Savina, I.N.; Zoughaib, M.; Yergeshov, A.A. Design and Assessment of Biodegradable Macroporous Cryogels as Advanced Tissue Engineering and Drug Carrying Materials. *Gels* **2021**, *7*, 79. [CrossRef] [PubMed]
3. Rustad, K.C.; Wong, V.W.; Sorkin, M.; Glotzbach, J.P.; Major, M.R.; Rajadas, J.; Longaker, M.T.; Gurtner, G.C. Enhancement of mesenchymal stem cell angiogenic capacity and stemness by a biomimetic hydrogel scaffold. *Biomaterials* **2012**, *33*, 80–90. [CrossRef]
4. Yergeshov, A.A.; Siraeva, Z.Y.; Kazakova, R.R.; Mullin, R.I.; Davliev, D.M.; Zakirova, A.A.; Salikhova, T.I.; Kuznetsova, E.V.; Luong, D.T.; Savina, I.N.; et al. Effect of gelatin cryogel on proliferation and synthetic activity of fibroblasts in excision wound model. *Genes Cells* **2015**, *10*, 29–33.
5. Kamalov, M.I.; Lavrov, I.A.; Yergeshov, A.A.; Siraeva, Z.Y.; Baltin, M.E.; Rizvanov, A.A.; Kuznetcova, S.V.; Petrova, N.V.; Savina, I.N.; Abdullin, T.I. Non-invasive topical drug delivery to spinal cord with carboxyl-modified trifunctional copolymer of ethylene oxide and propylene oxide. *Colloids Surf. B Biointerfaces* **2016**, *140*, 196–203. [CrossRef]
6. Etulain, J. Platelets in wound healing and regenerative medicine. *Platelets* **2018**, *29*, 556–568. [CrossRef]
7. Zoughaib, M.; Pavlov, R.V.; Gaynanova, G.A.; Garifullin, R.; Evtugyn, V.G.; Abdullin, T.I. Amphiphilic RGD and GHK peptides synergistically enhance liposomal delivery into cancer and endothelial cells. *Mater. Adv.* **2021**, *2*, 7715–7730. [CrossRef]
8. Brochhausen, C.; Lehmann, M.; Halstenberg, S.; Meurer, A.; Klaus, G.; Kirkpatrick, C.J. Signalling molecules and growth factors for tissue engineering of cartilage—what can we learn from the growth plate? *J. Tissue Eng. Regen. Med.* **2009**, *3*, 416–429. [CrossRef]
9. Su, Y.; Cockerill, I.; Wang, Y.; Qin, Y.-X.; Chang, L.; Zheng, Y.; Zhu, D. Zinc-Based Biomaterials for Regeneration and Therapy. *Trends Biotechnol.* **2019**, *37*, 428–441. [CrossRef]
10. Mourino, V.; Cattalini, J.P.; Boccaccini, A.R. Metallic ions as therapeutic agents in tissue engineering scaffolds: An overview of their biological applications and strategies for new developments. *J. R. Soc. Interface* **2012**, *9*, 401–419. [CrossRef]
11. Ryan, E.J.; Ryan, A.J.; Gonzalez-Vazquez, A.; Philippart, A.; Ciraldo, F.E.; Hobbs, C.; Nicolosi, V.; Boccaccini, A.R.; Kearney, C.J.; O'Brien, F.J. Collagen scaffolds functionalised with copper-eluting bioactive glass reduce infection and enhance osteogenesis and angiogenesis both in vitro and in vivo. *Biomaterials* **2019**, *197*, 405–416. [CrossRef] [PubMed]
12. Deng, Z.; Lin, B.; Jiang, Z.; Huang, W.; Li, J.; Zeng, X.; Wang, H.; Wang, D.; Zhang, Y. Hypoxia-Mimicking Cobalt-Doped Borosilicate Bioactive Glass Scaffolds with Enhanced Angiogenic and Osteogenic Capacity for Bone Regeneration. *Int. J. Biol. Sci.* **2019**, *15*, 1113–1124. [CrossRef] [PubMed]
13. Zhang, F.; Zhou, M.; Gu, W.; Shen, Z.; Ma, X.; Lu, F.; Yang, X.; Zheng, Y.; Gou, Z. Zinc-/copper-substituted dicalcium silicate cement: Advanced biomaterials with enhanced osteogenesis and long-term antibacterial properties. *J. Mater. Chem. B* **2020**, *8*, 1060–1070. [CrossRef] [PubMed]
14. Flegeau, K.; Pace, R.; Gautier, H.; Rethore, G.; Guicheux, J.; Le Visage, C.; Weiss, P. Toward the development of biomimetic injectable and macroporous biohydrogels for regenerative medicine. *Adv. Colloid Interface Sci.* **2017**, *247*, 589–609. [CrossRef] [PubMed]
15. Lee, J.H.; Kim, H.W. Emerging properties of hydrogels in tissue engineering. *J. Tissue Eng.* **2018**, *9*, 2041731418768285. [CrossRef]
16. Tao, B.; Lin, C.; Deng, Y.; Yuan, Z.; Shen, X.; Chen, M.; He, Y.; Peng, Z.; Hu, Y.; Cai, K. Copper-nanoparticle-embedded hydrogel for killing bacteria and promoting wound healing with photothermal therapy. *J. Mater. Chem. B* **2019**, *7*, 2534–2548. [CrossRef]
17. Zhou, Q.; Kang, H.; Bielec, M.; Wu, X.; Cheng, Q.; Wei, W.; Dai, H. Influence of different divalent ions cross-linking sodium alginate-polyacrylamide hydrogels on antibacterial properties and wound healing. *Carbohydr. Polym.* **2018**, *197*, 292–304. [CrossRef]
18. Shi, Q.; Luo, X.; Huang, Z.; Midgley, A.C.; Wang, B.; Liu, R.; Zhi, D.; Wei, T.; Zhou, X.; Qiao, M.; et al. Cobalt-mediated multi-functional dressings promote bacteria-infected wound healing. *Acta Biomater.* **2019**, *86*, 465–479. [CrossRef]
19. Luong, D.; Yergeshov, A.A.; Zoughaib, M.; Sadykova, F.R.; Gareev, B.I.; Savina, I.N.; Abdullin, T.I. Transition metal-doped cryogels as bioactive materials for wound healing applications. *Mater. Sci. Eng. C Mater. Biol. Appl.* **2019**, *103*, 109759. [CrossRef]

20. Zoughaib, M.; Luong, D.; Garifullin, R.; Gatina, D.Z.; Fedosimova, S.V.; Abdullin, T.I. Enhanced angiogenic effects of RGD, GHK peptides and copper (II) compositions in synthetic cryogel ECM model. *Mater. Sci. Eng. C Mater. Biol. Appl.* **2021**, *120*, 111660. [CrossRef]
21. Luong, T.D.; Zoughaib, M.; Garifullin, R.; Kuznetsova, S.; Guler, M.O.; Abdullin, T.I. In Situ functionalization of Poly(hydroxyethyl methacrylate) Cryogels with Oligopeptides via β-Cyclodextrin–Adamantane Complexation for Studying Cell-Instructive Peptide Environment. *ACS Appl. Bio Mater.* **2020**, *3*, 1116–1128. [CrossRef] [PubMed]
22. Wu, C.; Zhou, Y.; Fan, W.; Han, P.; Chang, J.; Yuen, J.; Zhang, M.; Xiao, Y. Hypoxia-mimicking mesoporous bioactive glass scaffolds with controllable cobalt ion release for bone tissue engineering. *Biomaterials* **2012**, *33*, 2076–2085. [CrossRef] [PubMed]
23. Glushchenko, N.N.; Bogoslovskaya, O.A.; Shagdarova, B.T.; Il'ina, A.V.; Olkhovskaya, I.P.; Varlamov, V.P. Searching for synergistic effects of low-molecular weight chitosan derivatives, chitosan and copper nanoparticles for wound healing ointment. *Adv. Nat. Sci. Nanosci. Nanotechnol.* **2021**, *12*, 035016. [CrossRef]
24. Wu, C.; Zhou, Y.; Xu, M.; Han, P.; Chen, L.; Chang, J.; Xiao, Y. Copper-containing mesoporous bioactive glass scaffolds with multifunctional properties of angiogenesis capacity, osteostimulation and antibacterial activity. *Biomaterials* **2013**, *34*, 422–433. [CrossRef]
25. Khan, S.A.; Kanwal, S.; Rizwan, K.; Shahid, S. Enhanced antimicrobial, antioxidant, in vivo antitumor and in vitro anticancer effects against breast cancer cell line by green synthesized un-doped SnO_2 and Co-doped SnO_2 nanoparticles from Clerodendrum inerme. *Microb. Pathog.* **2018**, *125*, 366–384. [CrossRef]
26. Abdollahi, Z.; Zare, E.N.; Salimi, F.; Goudarzi, I.; Tay, F.R.; Makvandi, P. Bioactive Carboxymethyl Starch-Based Hydrogels Decorated with CuO Nanoparticles: Antioxidant and Antimicrobial Properties and Accelerated Wound Healing In Vivo. *Int. J. Mol. Sci.* **2021**, *22*, 2531. [CrossRef]
27. Szabo, R.; Bodolea, C.; Mocan, T. Iron, Copper, and Zinc Homeostasis: Physiology, Physiopathology, and Nanomediated Applications. *Nanomaterials* **2021**, *11*, 2958. [CrossRef]
28. Powell, S.R. The antioxidant properties of zinc. *J. Nutr.* **2000**, *130*, 1447S–1454S. [CrossRef]
29. Valko, M.; Jomova, K.; Rhodes, C.J.; Kuca, K.; Musilek, K. Redox- and non-redox-metal-induced formation of free radicals and their role in human disease. *Arch Toxicol.* **2016**, *90*, 1–37. [CrossRef]
30. Modulevsky, D.J.; Cuerrier, C.M.; Pelling, A.E. Biocompatibility of Subcutaneously Implanted Plant-Derived Cellulose Biomaterials. *PLoS ONE* **2016**, *11*, e0157894. [CrossRef]
31. Khorramirouz, R.; Go, J.L.; Noble, C.; Jana, S.; Maxson, E.; Lerman, A.; Young, M.D. A novel surgical technique for a rat subcutaneous implantation of a tissue engineered scaffold. *Acta Histochem.* **2018**, *120*, 282–291. [CrossRef] [PubMed]
32. Kim, M.S.; Ahn, H.H.; Shin, Y.N.; Cho, M.H.; Khang, G.; Lee, H.B. An in vivo study of the host tissue response to subcutaneous implantation of PLGA- and/or porcine small intestinal submucosa-based scaffolds. *Biomaterials* **2007**, *28*, 5137–5143. [CrossRef] [PubMed]
33. Tsepaeva, O.V.; Nemtarev, A.V.; Abdullin, T.I.; Grigor'Eva, L.R.; Kuznetsova, E.V.; Akhmadishina, R.A.; Ziganshina, L.E.; Cong, H.H.; Mironov, V.F. Design, Synthesis, and Cancer Cell Growth Inhibitory Activity of Triphenylphosphonium Derivatives of the Triterpenoid Betulin. *J. Nat. Prod.* **2017**, *80*, 2232–2239. [CrossRef] [PubMed]
34. Zoughaib, M.H.; Luong, D.T.; Siraeva, Z.Y.; Yergeshov, A.A.; Salikhova, T.I.; Kuznetsova, S.V.; Kiyamova, R.G.; Abdullin, T.I. Tumor Cell Behavior in Porous Hydrogels: Effect of Application Technique and Doxorubicin Treatment. *Bull. Exp. Biol. Med.* **2019**, *167*, 590–598. [CrossRef]
35. Akhmadishina, R.A.; Kuznetsova, E.V.; Sadrieva, G.R.; Sabirzyanova, L.R.; Nizamov, I.S.; Akhmedova, G.R.; Nizamov, I.D.; Abdullin, T.I. Glutathione salts of O, O-diorganyl dithiophosphoric acids: Synthesis and study as redox modulating and antiproliferative compounds. *Peptides* **2018**, *99*, 179–188. [CrossRef]
36. Huang, M.T.; Mason, J.C.; Birdsey, G.M.; Amsellem, V.; Gerwin, N.; Haskard, D.O.; Ridley, A.J.; Randi, A.M. Endothelial intercellular adhesion molecule (ICAM)-2 regulates angiogenesis. *Blood* **2005**, *106*, 1636–1643. [CrossRef]
37. Gustafsson, A.; Jonasson, S.; Sandstrom, T.; Lorentzen, J.C.; Bucht, A. Genetic variation influences immune responses in sensitive rats following exposure to TiO_2 nanoparticles. *Toxicology* **2014**, *326*, 74–85. [CrossRef]
38. Philips, N.; Hwang, H.; Chauhan, S.; Leonardi, D.; Gonzalez, S. Stimulation of cell proliferation and expression of matrixmetalloproteinase-1 and interluekin-8 genes in dermal fibroblasts by copper. *Connect. Tissue Res.* **2010**, *51*, 224–229. [CrossRef]
39. Hyldig, K.; Riis, S.; Pennisi, C.P.; Zachar, V.; Fink, T. Implications of Extracellular Matrix Production by Adipose Tissue-Derived Stem Cells for Development of Wound Healing Therapies. *Int. J. Mol. Sci.* **2017**, *18*, 1167. [CrossRef]
40. Naldaiz-Gastesi, N.; Goicoechea, M.; Alonso-Martin, S.; Aiastui, A.; Lopez-Mayorga, M.; Garcia-Belda, P.; Lacalle, J.; San Jose, C.; Arauzo-Bravo, M.J.; Trouilh, L.; et al. Identification and Characterization of the Dermal Panniculus Carnosus Muscle Stem Cells. *Stem Cell Rep.* **2016**, *7*, 411–424. [CrossRef]
41. Perni, S.; Alotaibi, H.F.; Yergeshov, A.A.; Dang, T.; Abdullin, T.I.; Prokopovich, P. Long acting anti-infection constructs on titanium. *J. Control. Release* **2020**, *326*, 91–105. [CrossRef] [PubMed]
42. Meruane, M.A.; Rojas, M.; Marcelain, K. The use of adipose tissue-derived stem cells within a dermal substitute improves skin regeneration by increasing neoangiogenesis and collagen synthesis. *Plast. Reconstr. Surg.* **2012**, *130*, 53–63. [CrossRef] [PubMed]

43. Muller-Rover, S.; Handjiski, B.; van der Veen, C.; Eichmuller, S.; Foitzik, K.; McKay, I.A.; Stenn, K.S.; Paus, R. A comprehensive guide for the accurate classification of murine hair follicles in distinct hair cycle stages. *J. Investig. Dermatol.* **2001**, *117*, 3–15. [CrossRef] [PubMed]
44. Oh, J.W.; Kloepper, J.; Langan, E.A.; Kim, Y.; Yeo, J.; Kim, M.J.; Hsi, T.C.; Rose, C.; Yoon, G.S.; Lee, S.J.; et al. A Guide to Studying Human Hair Follicle Cycling In Vivo. *J. Investig. Dermatol.* **2016**, *136*, 34–44. [CrossRef]
45. Jidigam, V.K.; Srinivasan, R.C.; Patthey, C.; Gunhaga, L. Apical constriction and epithelial invagination are regulated by BMP activity. *Biol. Open* **2015**, *4*, 1782–1791. [CrossRef]
46. Lugo-Villarino, G.; Balla, K.M.; Stachura, D.L.; Banuelos, K.; Werneck, M.B.; Traver, D. Identification of dendritic antigen-presenting cells in the zebrafish. *Proc. Natl. Acad. Sci. USA* **2010**, *107*, 15850–15855. [CrossRef]
47. Al-Maawi, S.; Orlowska, A.; Sader, R.; James Kirkpatrick, C.; Ghanaati, S. In vivo cellular reactions to different biomaterials-Physiological and pathological aspects and their consequences. *Semin. Immunol.* **2017**, *29*, 49–61. [CrossRef]
48. Spadaro, J.A.; Becker, R.O.; Bachman, C.H. Size-specific metal complexing sites in native collagen. *Nature* **1970**, *225*, 1134–1136. [CrossRef]
49. Hoppe, A.; Mourino, V.; Boccaccini, A.R. Therapeutic inorganic ions in bioactive glasses to enhance bone formation and beyond. *Biomater. Sci.* **2013**, *1*, 254–256. [CrossRef]
50. Zhong, J.; Yang, Y.; Liao, L.; Zhang, C. Matrix stiffness-regulated cellular functions under different dimensionalities. *Biomater. Sci.* **2020**, *8*, 2734–2755. [CrossRef]
51. Galaup, C.; Picard, C.; Couderc, F.; Gilard, V.; Collin, F. Luminescent lanthanide complexes for reactive oxygen species biosensing and possible application in Alzheimer's diseases. *FEBS J.* **2021**, 15859. [CrossRef] [PubMed]
52. Alhayaza, R.; Haque, E.; Karbasiafshar, C.; Sellke, F.W.; Abid, M.R. The Relationship Between Reactive Oxygen Species and Endothelial Cell Metabolism. *Front Chem.* **2020**, *8*, 592688. [CrossRef] [PubMed]
53. van der Vliet, A.; Janssen-Heininger, Y.M.W. Hydrogen Peroxide as a Damage Signal in Tissue Injury and Inflammation: Murderer, Mediator, or Messenger? *J. Cell. Biochem.* **2014**, *115*, 427–435. [CrossRef] [PubMed]
54. Niethammer, P.; Grabher, C.; Look, A.T.; Mitchison, T.J. A tissue-scale gradient of hydrogen peroxide mediates rapid wound detection in zebrafish. *Nature* **2009**, *459*, 996–999. [CrossRef] [PubMed]
55. Lee, S.R. Critical Role of Zinc as Either an Antioxidant or a Prooxidant in Cellular Systems. *Oxidative Med. Cell. Longev.* **2018**, *2018*, 9156285. [CrossRef]
56. Maret, W. Metallothionein redox biology in the cytoprotective and cytotoxic functions of zinc. *Exp. Gerontol.* **2008**, *43*, 363–369. [CrossRef]
57. Rodriguez-Menendez, S.; Garcia, M.; Fernandez, B.; Alvarez, L.; Fernandez-Vega-Cueto, A.; Coca-Prados, M.; Pereiro, R.; Gonzalez-Iglesias, H. The Zinc-Metallothionein Redox System Reduces Oxidative Stress in Retinal Pigment Epithelial Cells. *Nutrients* **2018**, *10*, 1874. [CrossRef]
58. Andrews, G.K. Cellular zinc sensors: MTF-1 regulation of gene expression. *Biometals* **2001**, *14*, 223–237. [CrossRef]
59. Tsepaeva, O.V.; Salikhova, T.I.; Grigor'Eva, L.R.; Ponomaryov, D.V.; Dang, T.; Ishkaeva, R.A.; Abdullin, T.I.; Nemtarev, A.V.; Mironov, V.F. Synthesis and in vitro evaluation of triphenylphosphonium derivatives of acetylsalicylic and salicylic acids: Structure-dependent interactions with cancer cells, bacteria, and mitochondria. *Med. Chem. Res.* **2021**, *30*, 925–939. [CrossRef]
60. Sabbioni, E.; Fortaner, S.; Farina, M.; Del Torchio, R.; Olivato, I.; Petrarca, C.; Bernardini, G.; Mariani-Costantini, R.; Perconti, S.; Di Giampaolo, L.; et al. Cytotoxicity and morphological transforming potential of cobalt nanoparticles, microparticles and ions in Balb/3T3 mouse fibroblasts: An in vitro model. *Nanotoxicology* **2014**, *8*, 455–464. [CrossRef]
61. Lovejoy, D.B.; Richardson, D.R. Iron chelators as anti-neoplastic agents: Current developments and promise of the PIH class of chelators. *Curr. Med. Chem.* **2003**, *10*, 1035–1049. [CrossRef] [PubMed]
62. Dunnill, C.; Patton, T.; Brennan, J.; Barrett, J.; Dryden, M.; Cooke, J.; Leaper, D.; Georgopoulos, N.T. Reactive oxygen species (ROS) and wound healing: The functional role of ROS and emerging ROS-modulating technologies for augmentation of the healing process. *Int. Wound J.* **2017**, *14*, 89–96. [CrossRef] [PubMed]
63. Cao, H.; Yu, D.; Yan, X.; Wang, B.; Yu, Z.; Song, Y.; Sheng, L. Hypoxia destroys the microstructure of microtubules and causes dysfunction of endothelial cells via the PI3K/Stathmin1 pathway. *Cell Biosci.* **2019**, *9*, 20. [CrossRef] [PubMed]
64. Nardinocchi, L.; Pantisano, V.; Puca, R.; Porru, M.; Aiello, A.; Grasselli, A.; Leonetti, C.; Safran, M.; Rechavi, G.; Givol, D.; et al. Zinc downregulates HIF-1alpha and inhibits its activity in tumor cells in vitro and in vivo. *PLoS ONE* **2010**, *5*, e15048. [CrossRef] [PubMed]
65. Bai, Y.; Bai, L.; Zhou, J.; Chen, H.; Zhang, L. Sequential delivery of VEGF, FGF-2 and PDGF from the polymeric system enhance HUVECs angiogenesis in vitro and CAM angiogenesis. *Cell. Immunol.* **2018**, *323*, 19–32. [CrossRef] [PubMed]
66. Niu, J.; Wang, K.; Zhelyabovska, O.; Saad, Y.; Kolattukudy, P.E. MCP-1-induced protein promotes endothelial-like and angiogenic properties in human bone marrow monocytic cells. *J. Pharmacol. Exp. Ther.* **2013**, *347*, 288–297. [CrossRef] [PubMed]
67. Sanberg, P.R.; Park, D.H.; Kuzmin-Nichols, N.; Cruz, E.; Hossne, N.A., Jr.; Buffolo, E.; Willing, A.E. Monocyte transplantation for neural and cardiovascular ischemia repair. *J. Cell. Mol. Med.* **2010**, *14*, 553–563. [CrossRef]
68. Ozturk, B.Y.; Inci, I.; Egri, S.; Ozturk, A.M.; Yetkin, H.; Goktas, G.; Elmas, C.; Piskin, E.; Erdogan, D. The treatment of segmental bone defects in rabbit tibiae with vascular endothelial growth factor (VEGF)-loaded gelatin/hydroxyapatite "cryogel" scaffold. *Eur. J. Orthop. Surg. Traumatol. Orthop. Traumatol.* **2013**, *23*, 767–774. [CrossRef]

69. Matsui, M.; Tabata, Y. Enhanced angiogenesis by multiple release of platelet-rich plasma contents and basic fibroblast growth factor from gelatin hydrogels. *Acta Biomater.* **2012**, *8*, 1792–1801. [CrossRef]
70. Nillesen, S.T.; Geutjes, P.J.; Wismans, R.; Schalkwijk, J.; Daamen, W.F.; van Kuppevelt, T.H. Increased angiogenesis and blood vessel maturation in acellular collagen-heparin scaffolds containing both FGF2 and VEGF. *Biomaterials* **2007**, *28*, 1123–1131. [CrossRef]
71. Joseph, B.; Augustine, R.; Kalarikkal, N.; Thomas, S.; Seantier, B.; Grohens, Y. Recent advances in electrospun polycaprolactone based scaffolds for wound healing and skin bioengineering applications. *Mater. Today Commun.* **2019**, *19*, 319–335. [CrossRef]
72. Rennert, R.C.; Sorkin, M.; Garg, R.K.; Januszyk, M.; Gurtner, G.C. Cellular response to a novel fetal acellular collagen matrix: Implications for tissue regeneration. *Int. J. Biomater.* **2013**, *2013*, 527957. [CrossRef]
73. Giavaresi, G.; Torricelli, P.; Fornasari, P.M.; Giardino, R.; Barbucci, R.; Leone, G. Blood vessel formation after soft-tissue implantation of hyaluronan-based hydrogel supplemented with copper ions. *Biomaterials* **2005**, *26*, 3001–3008. [CrossRef] [PubMed]
74. Siebert, L.; Luna-Cerón, E.; García-Rivera, L.E.; Oh, J.; Jang, J.; Rosas-Gómez, D.A.; Pérez-Gómez, M.D.; Maschkowitz, G.; Fickenscher, H.; Oceguera-Cuevas, D.; et al. Light-Controlled Growth Factors Release on Tetrapodal ZnO-Incorporated 3D-Printed Hydrogels for Developing Smart Wound Scaffold. *Adv. Funct. Mater.* **2021**, *31*, 2007555. [CrossRef]
75. Nosrati, R.; Kheirouri, S.; Ghodsi, R.; Ojaghi, H. The effects of zinc treatment on matrix metalloproteinases: A systematic review. *J. Trace Elem. Med. Biol.* **2019**, *56*, 107–115. [CrossRef]
76. Zhang, X.; Chen, L. Effects of $CoCl_2$-simulated hypoxia on the expression levels of matrix metalloproteinases in renal adenocarcinoma cells and renal tubular epithelial cells. *Exp. Ther. Med.* **2018**, *16*, 1454–1460. [CrossRef]
77. Luo, L.; Petit, A.; Antoniou, J.; Zukor, D.J.; Huk, O.L.; Liu, R.C.; Winnik, F.M.; Mwale, F. Effect of cobalt and chromium ions on MMP-1, TIMP-1, and TNF-alpha gene expression in human U937 macrophages: A role for tyrosine kinases. *Biomaterials* **2005**, *26*, 5587–5593. [CrossRef]
78. Pyo, H.K.; Yoo, H.G.; Won, C.H.; Lee, S.H.; Kang, Y.J.; Eun, H.C.; Cho, K.H.; Kim, K.H. The effect of tripeptide-copper complex on human hair growth in vitro. *Arch. Pharmacal Res.* **2007**, *30*, 834–839. [CrossRef]
79. Zhao, S.; Li, L.; Wang, H.; Zhang, Y.; Cheng, X.; Zhou, N.; Rahaman, M.N.; Liu, Z.; Huang, W.; Zhang, C. Wound dressings composed of copper-doped borate bioactive glass microfibers stimulate angiogenesis and heal full-thickness skin defects in a rodent model. *Biomaterials* **2015**, *53*, 379–391. [CrossRef]
80. Yang, Y.; Liang, Y.; Chen, J.; Duan, X.; Guo, B. Mussel-inspired adhesive antioxidant antibacterial hemostatic composite hydrogel wound dressing via photo-polymerization for infected skin wound healing. *Bioact. Mater.* **2022**, *8*, 341–354. [CrossRef]
81. Gerard, C.; Bordeleau, L.-J.; Barralet, J.; Doillon, C.J. The stimulation of angiogenesis and collagen deposition by copper. *Biomaterials* **2010**, *31*, 824–831. [CrossRef] [PubMed]
82. Park, J.H.; Yoon, J.; Park, B. Pomolic acid suppresses HIF1 α/VEGF-mediated angiogenesis by targeting p38-MAPK and mTOR signaling cascades. *Phytomedicine* **2016**, *23*, 1716–1726. [CrossRef] [PubMed]
83. Ahtzaz, S.; Nasir, M.; Shahzadi, L.; Amir, W.; Anjum, A.; Arshad, R.; Iqbal, F.; Chaudhry, A.A.; Yar, M.; ur Rehman, I. A study on the effect of zinc oxide and zinc peroxide nanoparticles to enhance angiogenesis-pro-angiogenic grafts for tissue regeneration applications. *Mater. Des.* **2017**, *132*, 409–418. [CrossRef]
84. Lin, Y.; Brown, R.F.; Jung, S.B.; Day, D.E. Angiogenic effects of borate glass microfibers in a rodent model. *J. Biomed. Mater. Res. Part A* **2014**, *102*, 4491–4499. [CrossRef]
85. Perez, R.A.; Kim, J.H.; Buitrago, J.O.; Wall, I.B.; Kim, H.W. Novel therapeutic core-shell hydrogel scaffolds with sequential delivery of cobalt and bone morphogenetic protein-2 for synergistic bone regeneration. *Acta Biomater.* **2015**, *23*, 295–308. [CrossRef]
86. Barrioni, B.R.; de Laia, A.G.S.; Valverde, T.M.; Martins, T.M.D.; Caliari, M.V.; de Sa, M.A.; de Goes, A.M.; Pereira, M.D. Evaluation of in vitro and in vivo biocompatibility and structure of cobalt releasing sol-gel bioactive glass. *Ceram. Int.* **2018**, *44*, 20337–20347. [CrossRef]
87. Lin, P.-H.; Sermersheim, M.; Li, H.; Lee, P.H.U.; Steinberg, S.M.; Ma, J. Zinc in Wound Healing Modulation. *Nutrients* **2017**, *10*, 16. [CrossRef]
88. Dierichs, L.; Kloubert, V.; Rink, L. Cellular zinc homeostasis modulates polarization of THP-1-derived macrophages. *Eur. J. Nutr.* **2018**, *57*, 2161–2169. [CrossRef]
89. Song, Y.; Wu, H.; Gao, Y.; Li, J.; Lin, K.; Liu, B.; Lei, X.; Cheng, P.; Zhang, S.; Wang, Y.; et al. Zinc Silicate/Nano-Hydroxyapatite/Collagen Scaffolds Promote Angiogenesis and Bone Regeneration via the p38 MAPK Pathway in Activated Monocytes. *ACS Appl. Mater. Interfaces* **2020**, *12*, 16058–16075. [CrossRef]
90. Gammoh, N.Z.; Rink, L. Zinc in Infection and Inflammation. *Nutrients* **2017**, *9*, 624. [CrossRef]
91. Landgraeber, S.; Jäger, M.; Jacobs, J.J.; Hallab, N.J. The Pathology of Orthopedic Implant Failure Is Mediated by Innate Immune System Cytokines. *Mediat. Inflamm.* **2014**, *2014*, 185150. [CrossRef] [PubMed]

gels 7, 180

Article

Preparation and Biocompatibility of Poly Methyl Methacrylate (PMMA)-Mesoporous Bioactive Glass (MBG) Composite Scaffolds

Irina Atkinson [1,*], Ana Maria Seciu-Grama [2,*], Oana Catalina Mocioiu [1], Ana Maria Mocioiu [3], Luminita Predoana [1], Mariana Voicescu [1], Jeanina Pandele Cusu [1], Ramona Marina Grigorescu [4], Rodica Mariana Ion [4] and Oana Craciunescu [2]

1 Romanian Academy, "Ilie Murgulescu" Institute of Physical Chemistry, 202, Spl. Independentei, 060021 Bucharest, Romania; omocioiu@icf.ro (O.C.M.); lpredoana@icf.ro (L.P.); voicescu@icf.ro (M.V.); jeaninamirea@yahoo.com (J.P.C.)
2 National Institute of Research and Development for Biological Sciences, 296, Spl. Independentei, 060031 Bucharest, Romania; oana_craciunescu2009@yahoo.com
3 National R&D Institute for Non-ferrous and Rare Metals, 102, Biruintei Blvd, 077145 Pantelimon, Ilfov, Romania; ammocioiu@imnr.ro
4 National Institute for Research & Development in Chemistry and Petrochemistry—ICECHIM Bucharest, 202, Spl. Independentei, 060021 Bucharest, Romania; ramona.grigorescu@icechim.ro (R.M.G.); rodica_ion2000@yahoo.co.uk (R.M.I.)
* Correspondence: irinaatkinson@yahoo.com (I.A.); ana.seciu@yahoo.com (A.M.S.-G.)

Citation: Atkinson, I.; Seciu-Grama, A.M.; Mocioiu, O.C.; Mocioiu, A.M.; Predoana, L.; Voicescu, M.; Cusu, J.P.; Grigorescu, R.M.; Ion, R.M.; Craciunescu, O. Preparation and Biocompatibility of Poly Methyl Methacrylate (PMMA)-Mesoporous Bioactive Glass (MBG) Composite Scaffolds. *Gels* **2021**, *7*, 180. https://doi.org/10.3390/gels7040180

Academic Editors: Arish Dasan, Filippo Rossi, Ashokraja Chandrasekar and Nupur Kohli

Received: 7 September 2021
Accepted: 19 October 2021
Published: 23 October 2021

Abstract: In recent years, the rising number of bone diseases which affect millions of people worldwide has led to an increased demand for materials with restoring and augmentation properties that can be used in therapies for bone pathologies. In this work, PMMA- MBG composite scaffolds containing ceria (0, 1, 3 mol%) were obtained by the phase separation method. The obtained composite scaffolds were characterized by X-ray diffraction, infrared spectroscopy, and scanning electron microscopy. UV–Vis measurement and EDX analysis confirmed the presence of cerium ions in the composite scaffolds. Evaluation of the in-vitro biocompatibility using MTT assay showed that composite scaffold containing 1 mol% of ceria presented higher viability than control cells (100%) for concentrations ranging between 5 and 50% after 96 h of incubation.

Keywords: biocompatibility; ceria; polymer-bioglass scaffolds

1. Introduction

Developments in tissue engineering have raised significantly the potential for treating bone defects caused by trauma, tissue resection, congenital anomalies, cancer, and osteoporosis [1]. Recently, scaffolds manufactured from natural or synthetic materials that provide structural support allowing cell proliferation upon transplantation, have become one of the important elements for regenerative medicine [2]. Moreover, the synthetic scaffolds can overcome limitations in current treatments associated with autologous bone grafting such as immunological rejection, the possibility of transmitting infectious diseases and low tissue availability [3,4]. Various types of scaffolds have been produced, but the main challenge facing us today is the selection of appropriate materials for scaffold manufacturing. To obtain scaffolds with the best properties, different types of materials have been used, such as natural or synthetic polymers, bioglasses, ceramics, metals, composites, and hydrogels. In addition, mechanical properties, biocompatibility, bioactivity, surface properties and biodegradability are essential in regenerative medicine applications and need to be considered when designing a scaffold [5].

Polymer/bioglass composite synthesis and development have played a key role in the advancement of biomedical technologies, including tissue engineering. The polymer/bioglass composite scaffolds combine two types of materials e.g., polymers and bio-glass to overcome their disadvantages and eventually produce a scaffold with superior properties [6].

Bioglasses used in synthetic scaffolds preparation for bone regeneration are attractive materials due to their ability to induce in-vitro hydroxyapatite mineralization and have excellent cytocompatibility [7]. Furthermore, bioglasses can be doped with different functional elements to enhance their biological properties [7,8].

To date, polymers have been the material of choice in the field of tissue engineering. PMMA acrylic bone cement has been widely used to repair or replace joints [9] and is used in a variety of medicinal and dentistry applications [10–13]. Polymers are extensively used in in-vivo and in-vitro biomedical applications due to their aesthetic, injecting molding ability [14]. Furthermore, PMMA is non-toxic, offers good compressive resistance and shows good versatile processing capabilities.

Several preparation techniques such as freeze-drying [15], phase separation [5], solvent casting [16], and matrix-assisted pulsed laser [17] methods have been used to produce scaffold-based polymer matrixes with adequate properties for bone tissue engineering. Among them, the phase separation method is an easy and simple way to obtain scaffolds that mimic bone morphology. Dhinasekaran D. et al. [5] obtained scaffolds by a phase separation method, with adequate properties of a bone grafting material starting from Bioglass 45S5 and PMMA using different solvents (ethanol, acetone, and chloroform). Studies regarding preparation PMMA-MBG scaffolds by phase separation method with the aid of nonionic surfactant Pluronic P123 are less frequently reported in the literature. Han X. et al. [18] obtained 3D Ti-doped meso-macroporous bioglass/PMMA scaffolds using Pluronic P123 and PMMA colloidal crystals by steam acid techniques. These scaffolds exhibit good antimicrobial properties and biocompatibility.

This study researches PMMA- Ce doped MBG composite scaffolds using the phase separation method. The influence of cerium addition on the biocompatibility of the obtained scaffolds using mouse fibroblast cells (NCTC clone L929) was investigated. In our previous study [8] good biocompatibility was obtained for Ce doped mesoporous bioglasses based on $70SiO_2$–$26CaO$–$4P_2O_5$ system prepared by sol-gel method in the presence of surfactant Pluronic P123.

Among the therapeutic elements being more recently included in research studies is cerium. It has received particular interest due to its antibacterial, anti-inflammatory, pro-osteogenesis, and pro-angiogenesis properties due to the oxidation state transition Ce^{4+} and Ce^{3+} during redox reactions in physiological fluids with the formation of free radicals [19]. In addition, compressive strength and bioactivity of the obtained composite scaffolds were also studied.

2. Results and Discussion

2.1. FTIR

FTIR analysis was used to obtain information regarding chemical bonds. All of the prepared scaffolds (Figure 1) show the absorption bands at 2992, 2952 cm^{-1} assigned to C–H stretching vibrations related to the polymer matrix. The presence of hydrocarbon is indicated by the band at 1434 cm^{-1} due to the CH asymmetric bending vibration of CH_2 [20]. Furthermore, the band at 1734 cm^{-1} correspond to the stretching vibration of the carbonyl (C=O) group while the band at 1638 cm^{-1} is due to C=C stretching vibration. The narrow and sharp band at 1384 cm^{-1} is attributed to the presence of NO^{3-} group. The absorption band at 3430 cm^{-1} corresponds to –OH stretching vibration (water or ethanol), while the band at 1638 cm^{-1} can be attributed to the interlayer stretching and bending vibration of molecular water [21].

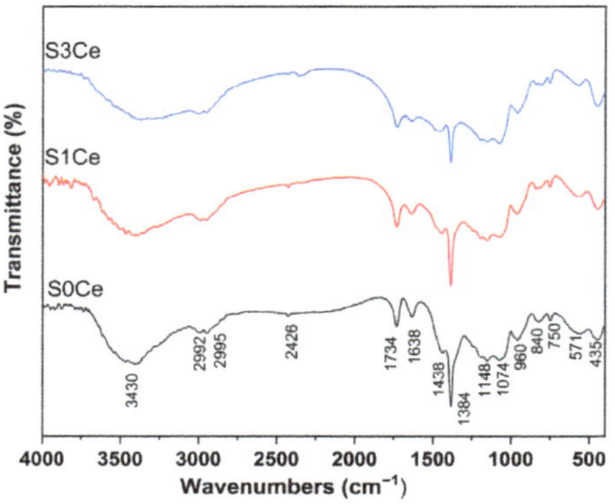

Figure 1. FTIR spectra of S0Ce-S3Ce composite scaffolds.

The absorption bands located at 1074 cm^{-1} can be ascribed to Si–O–Si asymmetric stretching vibration and are due to the vibrations of the non-bridging Si–O bonds in the structural units Q3 which are tetrahedral [SiO4], with one non-bridging oxygen atom and three bridging oxygen atoms. It can be noticed that the band located at 1074 cm^{-1} slightly intensifies with the increase of cerium concentration, suggesting the depolymerization of Si–O network to units with less bridging oxygen and cerium acting as a network modifier in glass [22,23]. The bands situated at 750 and 435 cm^{-1} are due to Si–O–Si symmetric stretching of bridging oxygen and Si–O bending vibration, respectively. The presence of silanol is indicated with the band at 960 cm^{-1} due to Si-O stretching vibration on the silanol group [24]. The absorption bands at 850 and 570 cm^{-1} are related to the P–O vibrations.

The results indicate that in the bioglass solution under investigation, almost all of the alkoxy groups are hydrolyzed into silanol groups. According to [5], the addition of a hydrolyzed silica to the polymerized PMMA solution using ethanol and water as solvents can induce the phase separation which may be considered as (i) formation of glass network in the solution containing organic polymers; (ii) parallel growth of the bioglass network and the PMMA polymer; (iii) simultaneous growth of a bioglass–PMMA interconnected polymer network; (iv) and development of a bioglass–PMMA network connected by covalent bonds.

2.2. Thermal Analysis

In order to examine the thermal stability, thermal analyses were carried out on PMMA-MBGs composite scaffolds as well as on pristine PMMA for comparison. The thermal gravimetric analysis (TG) and differential thermal analysis (DTA) data obtained from pristine PMMA and dried composite scaffolds are shown in Figure 2a,b. Both the pristine PMMA and the composite scaffolds underwent only single step degradation. The thermal decomposition for pure PMMA was completed around 400 °C. The onset of decomposition temperature decreased in the composite scaffolds, indicating the effect of cerium addition on the course of PMMA thermal degradation. No other effects were observed on the DTA curves.

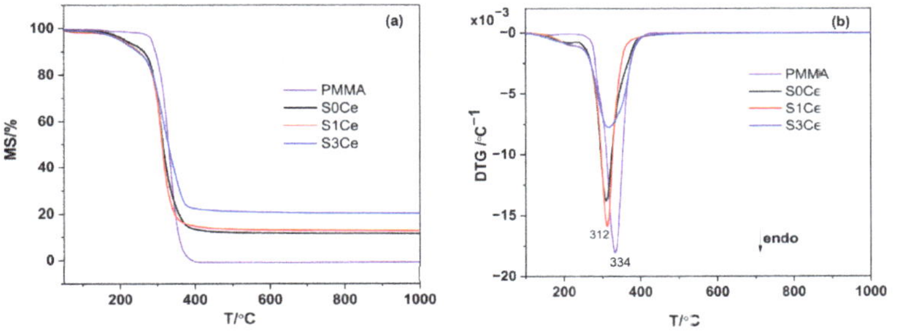

Figure 2. TG/DTA analyses of the pristine PMMA and PMMA-MBGs composite scaffolds: (**a**) TG analysis; (**b**) DTA analysis.

2.3. *UV-Vis*

UV-Vis analysis (Figure 3) was performed to obtain information regarding the oxidation state of cerium in the PMMA-MBGs composite scaffolds.

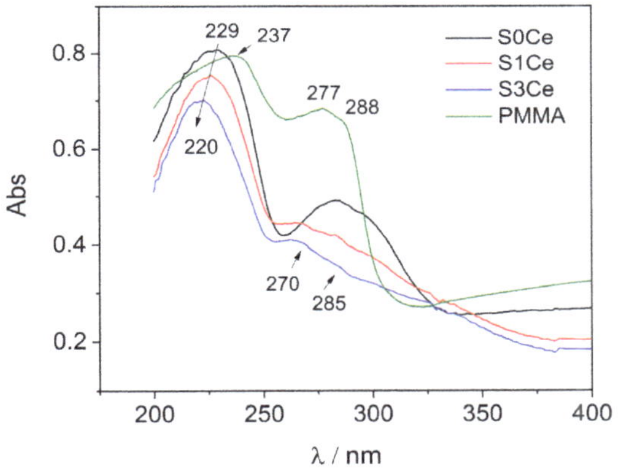

Figure 3. UV-Vis spectra of PMMA-MBGs composite scaffolds in direct comparison with pristine PMMA absorption spectrum.

Likewise, spectra of the S0Ce and cerium doped composites scaffolds show an absorption band at about 229 nm attributed to PMMA. According to Aziz et al. [25] in the UV region, a sharp absorption edge of about 270 nm due to electronic transitions n ® σ * occurs. Recently, Matamoros-Ambrocio et al. [26], observed that for PMMA microspheres in powdered form, synthesized under different conditions, all spectra in the UV region (200–400 nm) show a pronounced absorbance edge. On the other hand, in the Vis region (400–800 nm), the absorption is small and almost constant. This band is attenuated with the incorporation of cerium which can be associated with the band-gap absorption of CeO_2 and hypochromic shifted with the absorption intensity linearly decreasing from S0Ce to S3Ce. It is known that significant changes in the absorption spectra of Ce^{4+} due to the composition of glass occur, the molar extinction coefficient of Ce^{4+} being 5 to 10 times stronger than that of Ce^{3+} [27]. Moreover, Ce^{4+} produces a very strong and broad charge transfer band around 250 nm, with the intensity, half-width, and position of the absorption wavelength changing significantly with glass composition [27]. Therefore, the hypochromic shift observed (S0Ce ® S1Ce ® S3Ce) which linearly decreases with the absorption intensity may be due to the incomplete inner electronic shell of Ce^{4+}, as the scaffolds composite changed. Also, the charge transfer from O^{2-} to Ce^{4+} is taken into consideration. In the specific absorption spectra of Ce^{3+}, within the 250–350 nm range, two well-structured absorption bands appear at about 270/285 nm [28]. According with Paul et al., Ce^{3+} in glass produces a number of absorption bands in the UV region corresponding to the f ® d transitions [27]. Generally, Ce^{4+} is the stable valence in the cerium oxide due to its electronic configuration with the loss of a single electron in the 4f orbital of the Ce^{3+} ions [28]. The data from the literature reports the preparation of ceria-based oxides based on evaporation induces the self-assembly method using Pluronic P123 and ethanol as template removal, instead of the more common removal method (e.g., calcinations) [29]. Thus, the different type of spectrum pattern with the strong bleaching of the absorption band in the 270–280 nm region of S1Ce and S3Ce compared with S0Ce absorption band in this region, may be due to the fact that, by cerium doping of the composites' scaffolds, a large accumulation of Ce^{3+} defects occurs. In this context, and based on the fact that cerium absorption overlaps with PMMA-MBG's contribution, decomposition of the cerium (Ce^{4+}/Ce^{3+}) absorption band into the sample with Gaussian bands was performed (Figure 4). We conclude that, as composite scaffolds are doped with cerium, the concentration of Ce^{4+} increases while the concentration of Ce^{3+} decreases (Table 1). Moreover, the difference observed compared to that of pristine PMMA absorption may be due to porosity, defects, and oxygen vacancies in the PMMA structure of the scaffolds.

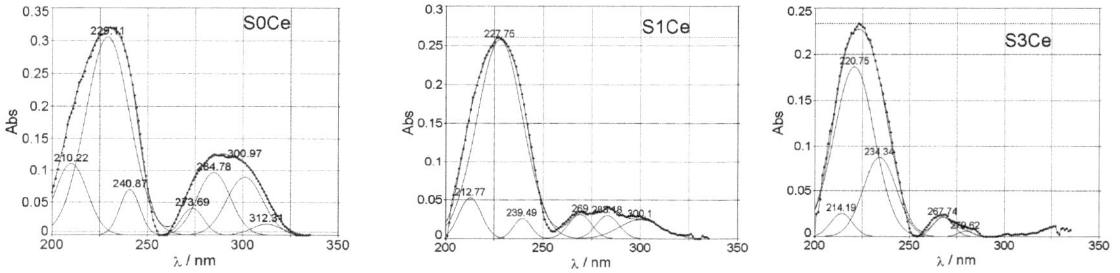

(a) (b) (c)

Figure 4. The curve fitting of the cerium (Ce^{4+}/Ce^{3+}) absorption band in PMMA-MBGs composite scaffolds: (**a**) S0Ce composite scaffold; (**b**) S1Ce composite scaffold; (**c**) S3Ce composite scaffold.

Table 1. Percentage of Ce^{4+}/Ce^{3+} in PMMA-MBGs composite scaffolds calculated from the fitting of Gaussians in the UV region, with the corresponding positions indicated in parentheses.

Sample	PMMA-MBGs	Ce^{4+}	Ce^{3+}
S0Ce	13.40% (210.22 nm)	-	-
	51.70% (229.11 nm)	-	-
	5.29% (240.87 nm)	-	-
	3.42% (273.69 nm)	-	-
	12.10% (284.78 nm)	-	-
	12.18% (300.97 nm)	-	-
	1.91% (312.31 nm)	-	-
S1Ce	-	7.72% (212.77 nm)	4.59% (269.00 nm)
	72.26% (227.75 nm)	-	4.34% (283.48 nm)
	7.43% (300.10 nm)	3.66% (239.49 nm)	-
S3Ce	-	5.26% (214.19 nm)	3.80% (267.74 nm)
	65.32% (220.75 nm)	-	1.00% (279.62 nm)
	-	24.62% (234.34 nm)	-

2.4. Morphology and Mechanical Properties Evaluation

The morphology of the composite scaffolds surface and qualitative compositional analyses were evaluated by SEM. The SEM micrographs taken from the surface area of S0Ce-S3Ce samples are presented in Figure 5. Irregular micro-pores (less than 50 µm) were observed for S0Ce composite scaffold. It can be observed that the size and distribution of pores varies within different surface regions examined and increases in the cerium containing composite scaffolds. The size of the irregular pores was estimated to be in the range of 100 µm and 300 µm for S1Ce and S3 Ce composite scaffolds, respectively (Figure 5c,e). The formed pores can lead to some porosity which is one of the most important characteristics of the materials for tissue engineering applications. They also have an important role in bone regeneration and cell migration. The obtained composite scaffolds have a porosity between 41% and 47% (Figure 6). Depending on porosity, bone tissue can be classified in two categories: cortical bone (also known as dense bone) with a porosity of 5% to 15% and trabecular (cancellouse) bone with a porosity from 40% to 95% [30].

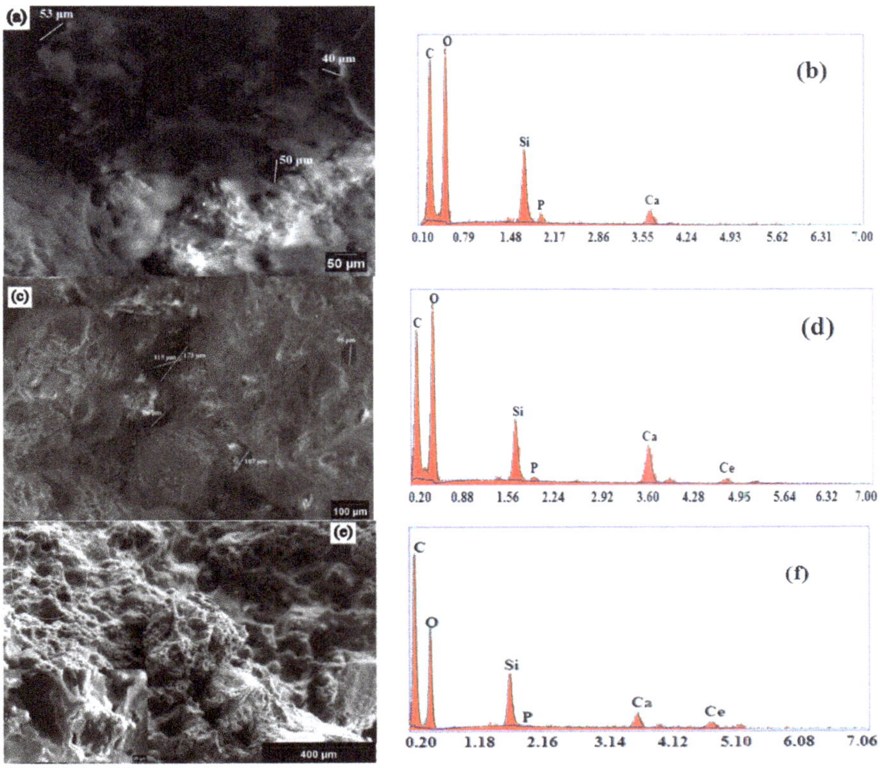

Figure 5. SEM micrographs (different magnifications) and EDX analysis of S0Ce (**a**,**b**), S1Ce(**c**,**d**), and S3Ce (**e**,**f**).

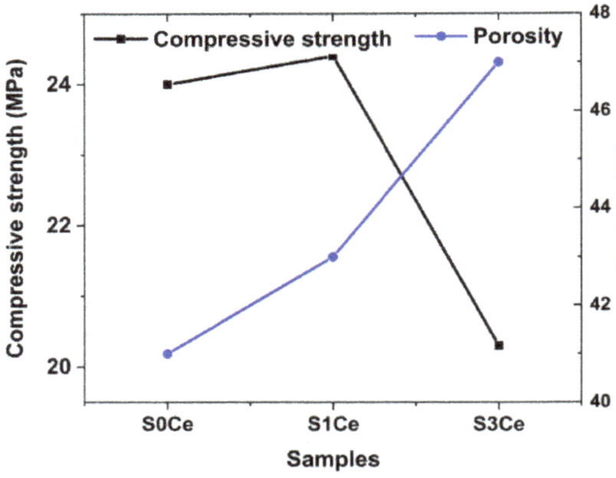

Figure 6. Porosity (blue line) and compressive strength (black line) of S0Ce-S1Ce composite scaffolds.

The results are in accordance with the data from the literature for materials with potential applications in tissue engineering (e.g., porosity in the range of 40–90%) [31]. In general, a pore size above 100 μm is required to promote the osteogenesis and angiogenesis [32]. Studies also report that a pore size under 100 μm plays a significant role in inducing the osteoinduction [33].

As can be observed in Figure 6, porosity and compressive strength are strongly correlated. The lowest compressive strength coefficient of 20 MPa is obtained for the S3Ce sample while the highest compressive strength value is found at 24.5 MPa for the S1Ce sample. The decrease of compressive strength coefficient with the increase of ceria content was reported by Zhou et al. [34]. An increase of the compressive strength of foamed glass-ceramics for a 2.5 wt% CeO_2 addition was observed. Further increases of CeO_2 content determine the decrease of the compressive strength. The higher content of Ce^{4+} as was revealed by UV-Vis analysis can lead to aggregation effect and consequently to distortion of the glass matrix. The data from the literature reported that the compressive strength of bioactive glass scaffolds ranges from 0.2 to 150 MPa depending on the scaffold's composition, microstructure, and preparation method [35,36]. It is known that while the compressive strength of human cortical bone ranges between 90 and 209 MPa, the strength of cancellous bone is between 1.5 and 45 MPa [37]. Hence, the obtained values of compressive strength in the present study classify the scaffold as being a material that has promising mechanical behavior for application as a substitute of cancellous bone.

2.5. *In Vitro Bioactivity Assessment*

The XRD patterns of PMMA-MBGs composite scaffold before and after immersion in SBF are depicted in Figure 7. The XRD patterns of the sample before immersion in SBF show broad diffraction lines of PMMA at 2θ 13.21° along with low intensity diffraction lines at 23.54° and 41.47° [38]. No diffraction lines corresponding to SiO_2 or cerium oxides were observed.

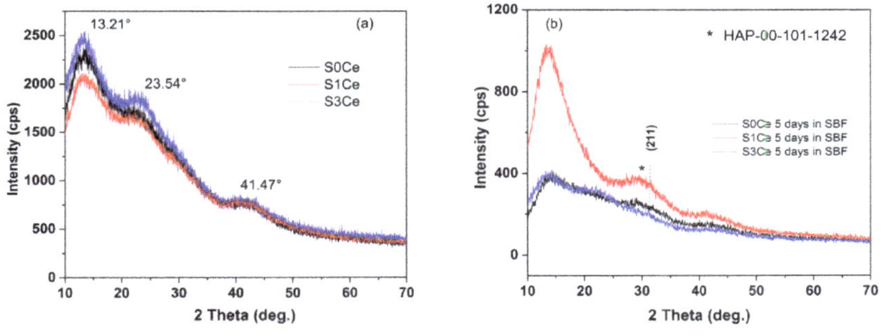

Figure 7. XRD patterns of S0Ce-S3Ce composite scaffolds (**a**) before immersion in SBF and (**b**) after immersion in SBF for five days.

After five days of immersion, XRD patterns of the S0Ce and the S1Ce samples reveal the beginning of hydroxyapatite (HAP) crystallization at 2θ around 31° (overlapping of (211), (112) and (300) reflection planes) in agreement with data from JCPDS card no. 101-1242. The overlapping of (211), (112) and (300) reflection planes is a characteristic of low crystalline HAP. Similar XRD patterns for PMMA-45S5 bioglass composites were obtained by [5] after 28 days of immersion in SBF.

Additionally, the formation of HAP layer on the surface of the composite scaffolds was also monitored by Fourier transforms infrared spectroscopic technique. Figure 8 shows the FTIR spectra of the obtained scaffolds immersed for five days in SBF solution. The characteristic absorption bands for phosphate group are observed at 552 and 602 cm^{-1} which are assigned to P–O bending vibration (ν4). The vibration band at 475 cm^{-1} corresponds to the PO_4^{-3} bending vibration (ν2) [39]. The bands, assigned to P–O bending vibration modes of the orthophosphate PO_4^{-3} group, are generally used to monitor the bioactivity [40].

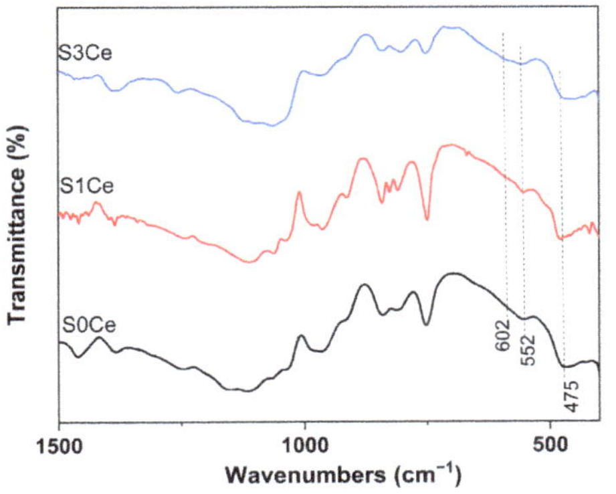

Figure 8. FTIR spectra of the PMMA-MBGs composite scaffolds immersed for five days in SBF solution.

The surface of the S1Ce and S3Ce composite scaffolds after immersion in SBF for 5 five was also examined by SEM. As can be observed in Figure 9, the surface of samples is covered by fine particles confirming the beginning of HAP crystallization (as was revealed by XRD measurements).

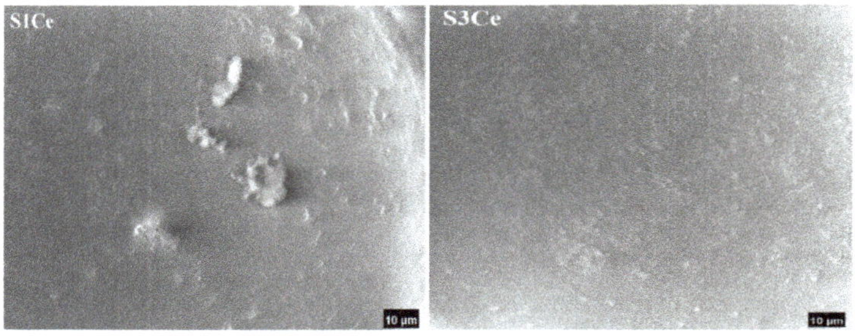

Figure 9. SEM micrographs of the S1Ce and S3Ce composite scaffolds after immersion for five days in SBF solution.

2.6. Biocompatibility Evaluation

MTT assay (Figure 10) was used to evaluate indirect toxicity and the number of metabolic-active cells. Viability of L929 cells exposed to different concentrations of PMMA MBGs composite scaffolds was evaluated after 48 h (a) and 96 h (b). Data are presented as mean ± SD (n = 3). * $p < 0.05$ compared to control (untreated cells); # $p < 0.05$ compared to scaffolds-treated cells.

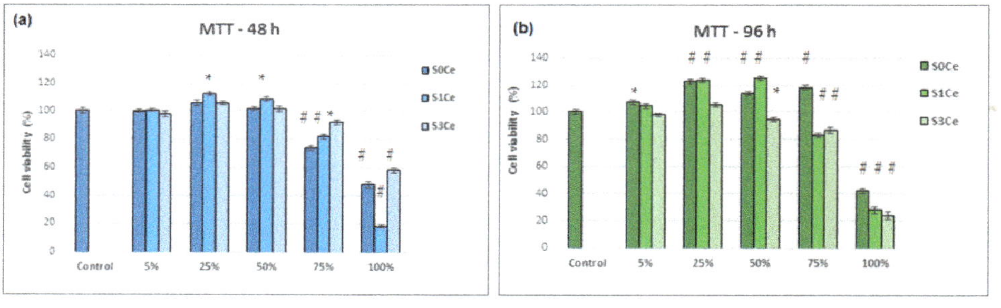

Figure 10. Viability of L929 cells exposed to different concentrations of PMMA-MBGs composite scaffolds evaluated by MTT assay after 48 h (**a**) and 96 h (**b**). Data are presented as mean ± SD (n = 3). * $p < 0.05$ compared to control (untreated cells); # $p < 0.01$ compared to control (untreated cells).

All tested samples show no cell cytotoxic activity in the concentration range between 5% and 75%, as seen in Figure 9. For all of the composite scaffolds produced during the investigation, the cell viability was above 80% (non-cytotoxic) for the aforementioned concentration range with exposure times of 96 h. At concentrations ranging between 5 and 50%, the S1Ce composite scaffold presented higher cell viability than control cells (100%) after 96 h of incubation. Good cell viability (84.73%) at a concentration of 100% was obtained for the MBGs containing 1% mole ceria in our previous study [8]. For the S0Ce composite scaffold, the cell viability was higher than control cells within concentrations ranging from 5 to 75% (Figure 10b). At 100% concentration, cell viability decreased drastically by up to 40% for the S0Ce after 96 h of incubation. The lowest cell viability after 96 h of incubation was observed for the S3Ce composite scaffold. This result can be explained based on the Ce^{4+}/Ce^{3+} ratio. Naganuma et al. [41] reported that cell proliferation and adhesion in cerium-doped materials are influenced by the oxidation state of cerium (Ce^{3+} vs. Ce^{4+}): Ce^{3+} ions inhibit cell proliferation and Ce^{4+} ions promote cell proliferation. In addition, the size and shape of CeO_2 can influence its cytotoxicity with smaller sized CeO_2 exhibiting higher toxicity [42].

3. Conclusions

PMMA-Ce doped MBG composite scaffolds with promising potential for application in tissue engineering were prepared by phase separation method by combining MBGs with addition of 0, 1, and 3 mol% ceria and PMMA.

UV-Vis measurements confirm both Ce^{3+} and Ce^{4+} oxidation states.

The compressive strength of the obtained composite scaffolds varies between 20–24.5 MPa that classify them as promising materials for application as a substitute of cancellous bone.

An in vitro biocompatibility evaluation determined using MTT assay indicated that all tested samples showed no cell cytotoxic activity on L929 cells in the concentration range of 5–75% after 96 h of incubation. Between concentration ranges of 5% and 50%, the S0Ce and S1Ce samples exhibited higher cell viability than control cells (100%).

XRD, FTIR, and SEM analyses confirmed the beginning of the hydroxyapatite layer crystallization over the sample surfaces after incubation in SBF for 5 days.

Based on the promising results, the PMMA-MBGs composite scaffolds investigated in the present study show potential for bone regeneration applications.

4. Materials and Methods

4.1. Reagents

This study used the following reagents: tetraethylorthosilicate (TEOS) (98%, Sigma-Aldrich, Darmstadt Germany), triethylphosphate (TEP) (99% Sigma-Aldrich, Darmstadt, Germany), calcium nitrate tetrahydrated ($Ca(NO_3)_2{\cdot}4H_2O$) (99% Sigma-Aldrich, Darmstadt, Germany) and cerium(III) nitrate hexahydrate (99% Sigma-Aldrich, Darmstadt, Germany) as silica, phosphate-, calcium- and cerium-oxide precursors, respectively, hydrochloric acid (HCl) (Sigma-Aldrich, Darmstadt, Germany) as a catalyst, PEG-PPG-PEG, called Pluronic® P123 (Sigma-Aldrich, Darmstadt, Germany) as structure directing agent and poly methyl methacrylate (Alfa Aesar, Ward Hill, MA, USA).

4.2. Preparation of MBG Solution

The bio-glass precursor sol was directly used to obtain the scaffolds. In brief, Ce-doped mesoporous bioglasses in the $70SiO_2$-(26-x) CaO-$4P_2O_5$-xCeO_2 system (where x stands for 0, 1, 3 mol%) were synthesized using the procedure described in paper [8]. Pluronic P123 was used as a structure directing agent.

4.3. Preparation of the Polymer-MBG Scaffolds

PMMA-MBG scaffolds were prepared by the phase separation method following the procedure described in [5]. PMMA (15%) with a molecular weight of 550,000 and a density of 1.18 g cm^3 was dissolved in an ethanol and water mix. Equal volumes of the MBG solution and the polymer/water/ethanol mixture were mixed to obtain the scaffold materials. Ethanol and water were mixed in the ratio 4:1 and preheated to 60 °C before adding PMMA. Subsequently, the obtained scaffolds were washed with ethanol to remove the Pluronic P123 structure directing agent and dried in the oven at 60 °C. The obtained scaffolds were labeled as follows: S0Ce, S1Ce, and S3Ce. The preparation chart is presented in Figure 11.

Figure 11. Composite scaffolds preparation chart.

4.4. Composite Scaffolds Characterization

Powdered X-ray patterns of the scaffolds were recorded using a RigakuUltima IV diffractometer in parallel beam geometry equipped with CuKα radiation (wavelength 1.5406 Å) in 2θ range between 10 to 70 with a speed of 2°/min and a step size of 0.02°. PDXL software (Version 1.8) from Rigaku, connected to ICDD database was used for phase identification.

Fourier transform infrared (FTIR) spectroscopy was performed with a Nicolet Spectrometer 6700 FTIR, within 400–4000 cm^{-1} range, in transmittance mode.

The thermal behavior of the obtained MBG-composite scffolds was determined by thermal gravimetric analysis and differential thermal analysis (TG/DTA) using a Mettler TOLEDO TGA/SDTA 851e equipment in flowing air atmosphere using alumina crucible. The maximum temperature was set at 1000 °C and the heating rate was 10 °C/min

The absorption measurements were recorded with a Perkin Elmer Lambda 35 Spectrometer with integrating sphere in 900–200 nm range using: data interval, 1 nm; scan speed, 60 nm/min; slit, 4 nm; sample holder at 8° wedge and a certified reflectance standard. To estimate the cerium concentration into scaffolds, a least-squares iterative curve fitting was performed with Gaussian bands using the peak fit analysis program (Sea-Solve, Framingham, MA, USA). The areas of all bands assigned to a given concentration were summed up and divided by the total area in order to obtain the contribution of cerium (Ce^{4+}/Ce^{3+}).

The morphology of the composite scaffolds was investigated by scanning electron microscopy using a microscope; Quanta FEI 200 model coupled with energy dispersive X-ray (EDX) analysis.

The compressive strength of the composite scaffolds was evaluated using a portable Schmidt hammer. The rebound number can be converted to uniaxial compressive strength (UCS-MPa) that, according to Wang R. and Yuan Z. [43] can be correlated to a qualitative evaluation of sample hardness.

The porosity of the samples was determined using Archimedes method. The porosity was calculated using Equation (1).

$$\text{Porosity, \%} = \frac{(m_1 - m_0)}{\delta V_0} \times 100 \quad (1)$$

where: m_0 and m_1 indicate the weight of the scaffold before and after immersing, respectively, V_0 is the scaffold volume before immersing, and δ is the liquid density. The experiment was performed in triplicate.

In vitro bioactivity of the composite scaffolds was assessed by immersing the samples in the simulated body fluid (SBF) as proposed by Kokubo et al. [44] at 37 °C for five days Once removed from the incubation solution, the samples were washed with deionized water and dried at 70 °C for 24 h. The presence of newly formed hydroxyapatite on the surface of composites was further examined by XRD analysis.

In vitro evaluation of cell viability was performed using the MTT assay protocol, as described in [45]. To summarize, after both 48 h and 96 h of cultivation in the presence of the scaffolds, the cells were rinsed with phosphate buffered saline solution (PBS), pH 7.4, and incubated with an MTT working solution (0.25 mg/mL) for 3 h at 37 °C to obtain crystallized formazan. Afterwards, the medium was removed and isopropyl alcohol was added to each well. After incubation at room temperature for 15 min under gently stirring, the optical density of the solution was determined at 570 nm using a Tecan Sunrise microplate reader (Tecan, Austria). The amount of converted dye directly correlates to the number of metabolically active cells. Cell viability was expressed as a percentage of the control cells (cells incubated without sample) considered being as 100% viable. The tests were performed in triplicate.

Author Contributions: Conceptualization, I.A.; investigation, O.C.M., L.P., M.V., R.M.I., R.M.G., J.P.C. and A.M.M.; writing—original draft preparation, I.A., A.M.S.-G. and O.C.; writing—review and editing, I.A., A.M.S.-G. and O.C.; supervision, I.A., A.M.S.-G.; project administration, I.A., A.M.S.-G.; funding acquisition I.A. All authors have read and agreed to the published version of the manuscript.

Funding: The research and APC were funded by Romanian Executive Agency for Higher Education Research Development and Innovation Funding (UEFISCDI), grant number PN-III-P2-2.1-PED-2019-0598, no. 258 PED/2020.

Institutional Review Board Statement: Not applicable.

Informed Consent Statement: Not applicable.

Data Availability Statement: The data presented in this study are contained within the article.

Conflicts of Interest: The authors declare no conflict of interest. The funders had no role in the design of the study; in the collection, analyses, or interpretation of data; in the writing of the manuscript; or in the decision to publish the results.

References

1. Xia, Y.; Mei, F.; Duan, Y.L.; Gao, Y.; Xiong, Z.; Zhang, T.; Zhang, H. Bone tissue engineering using bone marrow stromal cells and an injectable sodium alginate/gelatin scaffold. *J. Biomed. Mater. Res. A* **2012**, *100A*, 1044–1050. [CrossRef]
2. Dolcimascolo, A.; Calabrese, G.; Conoci, S.; Parenti, R. Innovative Biomaterials for Tissue Engineering In *Book Biomaterial-Supported Tissue Reconstruction or Regeneration*; Barbeck, M., Jung, O., Smeets, R., Eds.; IntechOpen: London, UK, 2019.
3. Funda, G.; Taschieri, S.; Bruno, G.A.; Grecchi, E.; Paolo, S.; Girolamo, D.; Del Fabbro, M. Nanotechnology Scaffolds for Alveolar Bone Regeneration. *Materials* **2020**, *13*, 201. [CrossRef]
4. Fiume, E.; Serino, G.; Bignardi, C.; Verné, E.; Baino, F. Bread-Derived Bioactive Porous Scaffolds: An Innovative and Sustainable Approach to Bone Tissue Engineering. *Molecules* **2019**, *24*, 2954. [CrossRef] [PubMed]
5. Durgalakshmi, D.; Balakumar, S. Analysis of solvent induced porous PMMA–Bioglass monoliths by the phase separation method-mechanical and in vitro biocompatible studies. *Phys. Chem. Chem. Phys.* **2015**, *17*, 1247–1256. [CrossRef] [PubMed]
6. Zhu, L.; Luo, D.; Liu, Y. Effect of the nano/microscale structure of biomaterial scaffolds on bone regeneration. *Int. J. Appl. Dent. Sci.* **2020**, *12*, 1–15. [CrossRef]
7. Atkinson, I.; Anghel, E.M.; Predoana, L.; Mocioiu, O.C.; Jecu, L.; Raut, I.; Munteanu, C.; Culita, D.; Zaharescu, M. Influence of ZnO addition on the structural, in vitro behavior and antimicrobial activity of sol–gel derived $CaO–P_2O_5–SiO_2$ bioactive glasses. *Ceram. Int.* **2016**, *42*, 3033–3045. [CrossRef]
8. Atkinson, I.; Anghel, E.M.; Petrescu, S.; Seciu, A.M.; Stefan, L.M.; Mocioiu, O.C.; Predoana, L.; Voicescu, M.; Somacescu, S.; Culita, D.; et al. Cerium-containing mesoporous bioactive glasses: Material characterization, in vitro bioactivity, biocompatibility and cytotoxicity evaluation. *Micro. Mesop. Mater.* **2019**, *276*, 76–88. [CrossRef]
9. Charnley, J. Anchorage of the femoral head prosthesis to the shaft of the femur. *Br. Vol.* **1960**, *42*, 28–30. [CrossRef]
10. Shundo, A.; Hori, K.; Penaloza, D.P.; Yoshihiro, K.; Annaka, M.; Tanaka, K. Nonsolvents-induced swelling of poly (methyl methacrylate) nanoparticles. *Phys. Chem. Chem. Phys.* **2013**, *15*, 16574–16578. [CrossRef] [PubMed]
11. Zafar, M.S. Prosthodontic Applications of Polymethyl Methacrylate (PMMA): An Update. *Polymers* **2020**, *12*, 2299. [CrossRef]
12. Manoukian, O.S.; Sardashti, N.; Stedman, T.; Gailiunas, K.; Ojha, A.; Penalosa, A.; Mancuso, C.; Hobert, M.; Kumbar, S.G. Biomaterials for Tissue Engineering and Regenerative Medicine. In *Book Encyclopedia of Biomedical Engineering*; Narayan, R., Ed.; Elsevier: Amsterdam, The Netherlands, 2019; pp. 462–482. [CrossRef]

13. Radha, G.; Balakumar, S.; Venkatesan, B.; Vellaichamy, E. A novel nano-hydroxyapatite—PMMA hybrid scaffolds adopted by conjugated thermal induced phase separation (TIPS) and wet-chemical approach: Analysis of its mechanical and biological properties. *Mater. Sci. Eng. C* **2017**, *75*, 221–228. [CrossRef]
14. Jagdale, P.; Serino, G.; Oza, G.; Audenino, A.L.; Bignardi, C.; Tagliaferro, A.; Alvarez-Gayosso, C. Physical Characterization of Bismuth Oxide Nanoparticle Based Ceramic Composite for Future Biomedical Application. *Materials* **2021**, *14*, 1626. [CrossRef]
15. Murphy, W.L.; Dennis, R.G.; Kileney, J.L.; Mooney, D.J. Salt fusion: An approach to improve pore interconnectivity within tissue engineering scaffolds. *Tissue Eng.* **2002**, *8*, 43–52. [CrossRef] [PubMed]
16. Matbouei, A.; Fathi, A.; Mahmood Rabiee, S.; Shirzad, M. Layered manufacturing of a three-dimensional polymethyl methacrylate (PMMA) scaffold used for bone regeneration. *Mater. Tech.* **2019**, *34*, 167–177. [CrossRef]
17. Han, X.; Lin, H.; Chen, X.; Li, X.; Guo, G.; Qu, F. One-step method for the preparation of poly(methyl methacrylate) modified titanium bioactive glass three-dimensional scaffolds for bone tissue engineering. *IET Nanobiotech.* **2016**, *10*, 45–53. [CrossRef] [PubMed]
18. Westhauser, F.; Rehder, F.; Decker, S.; Kunisch, E.; Moghaddam, A.; Zheng, K.; Boccaccini, A.R. Ionic dissolution products of Cerium-doped bioactive glass nanoparticles promote cellular osteogenic differentiation and extracellular matrix formation of human bone marrow derived mesenchymal stromal cells. *Biomed. Mater.* **2020**, *16*, 035028. [CrossRef] [PubMed]
19. Jabbour, J.; Calas, S.; Gatti, S.; Kribich, R.K.; Myara, M.; Pille, G.; Ettienne, P.; Moreau, Y. Characterization by IR spectroscopy of an hybrid sol–gel material used for photonic devices fabrication. *J. Non-Cryst. Solids* **2008**, *354*, 651–658. [CrossRef]
20. Chibac-Scutaru, A.L.; Melinte, V.; Coseri, S. Cellulose Acetate Incorporating Organically Functionalized CeO_2 NPs: Efficient Materials for UV Filtering Applications. *Materials* **2020**, *13*, 2955. [CrossRef]
21. Kaur, P.; Singh, G.P.; Kaur, S.; Singh, D.P. Modifier role of cerium in lithium aluminium borate glasses. *J. Molec. Struct.* **2012**, *1020*, 83–87. [CrossRef]
22. Wang, Z.; Cheng, L. Structural Features and Synthesis of CeO_2-Doped Boroaluminosilicate Oxyfluoride Transparent Glass Ceramics. *J. Chem.* **2015**, *2015*, 1–10. [CrossRef]
23. Mori, Y.; Pinnavaia, T.J. Optimizing organic functionality in mesostructured silica: Direct assembly of mercaptopropyl groups in wormhole framework structures. *Chem. Mater.* **2001**, *13*, 2173–2178. [CrossRef]
24. Nicolini, V.; Gambuzzi, E.; Malavasi, G.; Menabue, L.; Menziani, M.C.; Lusvardi, G.; Pedone, A.; Benedetti, F.; Luches, P.; D'Addato, S.; et al. Evidence of catalase mimetic activity in Ce3+/Ce4+ doped bioactive glasses. *J. Phys. Chem.* **2015**, *B119*, 4009–4019. [CrossRef]
25. Aziz, S.B.; Abdullah, O.G.; Hussein, A.M.; Ahmed, H.M. From insulating PMMA polymer to conjugated double bond behavior: Green chemistry as a novel approach to fabricate small band gap polymers. *Polymers* **2017**, *9*, 626. [CrossRef]
26. Matamoros-Ambrocio, M.; Sánchez-Mora, E.; Gómez-Barojas, E.; Luna-López, J.A. Synthesis and Study of the Optical Properties of PMMA Microspheres and Opals. *Polymers* **2021**, *13*, 2171. [CrossRef]
27. Paul, A.; Mulholland, M.; Zaman, M.S. Ultraviolet absorption of cerium (IV) in some simple glasses. *J. Mater. Sci.* **1976**, *11*, 2082–2086. [CrossRef]
28. Cicconi, M.R.; Veber, A.; Neuville, D.R.; Baudelet, F.; De Ligny, D. Cerium speciation in silicate glasses: Structure-property relationships. *J. Non-Cryst. Solids* **2021**, *563*, 120785. [CrossRef]
29. Wang, P.; Chen, S.; Gao, S.; Zhang, J.; Wang, H.; Wu, Z. Niobium oxide confined by ceria nanotubes as a novel SCR catalyst with excellent resistance to potassium, phosphorus, and lead. *Appl. Catal. B Environ.* **2018**, *231*, 299–309. [CrossRef]
30. Morgan, E.F.; Unnikrisnan, G.U.; Hussein, A.I. Bone Mechanical Properties in Healthy and Diseased States. *Annu. Rev. Biomed. Eng.* **2018**, *20*, 119–143. [CrossRef] [PubMed]
31. Loh, Q.L.; Choong, C. Three-Dimensional Scaffolds for Tissue Engineering Applications: Role of Porosity and Pore Size. *Tissue Eng. Part B Rev.* **2013**, *19*, 485–502. [CrossRef] [PubMed]
32. Zaborowska, M.; Bodin, A.; Bäckdahl, H.; Popp, J.; Goldstein, A.; Gatenholm, P. Microporous bacterial cellulose as a potential scaffoldfor bone regeneration. *Acta Biomater.* **2010**, *6*, 2540–2547. [CrossRef] [PubMed]
33. Bohner, M.; Baroud, G.; Bernstein, A.; Döbelin, N.; Galea, L.; Hesse, B.; Heuberger, R.; Meille, S.; Michel, P.; von Rechenberg, B.; et al. Characterization and distribution of mechanically competent mineralized tissue in micropores of beta-tricalcium phosphate bone substitutes. *Mater. Today* **2017**, *20*, 106–115. [CrossRef]
34. Zhou, H.-L.; Feng, K.-Q.; Chen, C.-H.; Yan, D.-L. Influence of CeO_2 addition on the preparation of foamed glass-ceramics from high-titanium blast furnace slag. *Int. J. Miner. Metall.* **2018**, *25*, 689–695. [CrossRef]
35. Hutmacher, D.W. Scaffolds in tissue engineering boneand cartilage. *Biomaterials* **2000**, *21*, 2529–2543. [CrossRef]
36. Fiume, E.; Serino, G.; Bignardi, C.; Verné, E.; Baino, F. Sintering Behavior of a Six-Oxide Silicate Bioactive Glass for Scaffold Manufacturing. *Appl. Sci.* **2020**, *10*, 8279. [CrossRef]
37. Chowdhury, S.K.R.; Mishra, A.; Pradhan, B.; Saha, D. Wear characteristic and biocompatibility of some polymer composite acetabular cups. *Wear* **2004**, *256*, 1026–1036. [CrossRef]
38. Hashem, M.; Fayez Al Rez, M.; Fouad, H.; Elsarnagawy, T.; Elsharawy, M.A.; Umar, A.; Assery, M.; Ansar, S.G. Influence of Titanium Oxide Nanoparticles on the Physical and Thermomechanical Behavior of Poly Methyl Methacrylate (PMMA): A Denture Base Resin. *Sci. Adv. Mater.* **2017**, *9*, 938–944. [CrossRef]

39. Gopi, D.; Shinyjoy, E.; Karthika, A.; Nithiya, S.; Kavitha, L.; Rajeswari, D.; Tang, T. Single walled carbon nanotubes reinforced mineralized hydroxyapatite composite coatings on titanium for improved biocompatible implant applications. *RSC Adv.* **2015**, *5*, 36766. [CrossRef]
40. Notingher, I.; Jones, J.R.; Verrier, S.; Bisson, I.; Embanga, P.; Edwards, P.; Polak, J.M.; Hench, L.L. Application of FTIR and Raman spectroscopy to characterization obioactive materials and living cells. *Spectroscopy* **2003**, *17*, 275–288. [CrossRef]
41. Naganuma, T.; Traversa, E. The effect of cerium valence states at cerium oxide nanoparticle surfaces on cell proliferation. *Biomaterials* **2014**, *35*, 4441–4453. [CrossRef] [PubMed]
42. Qi, M.; Li, W.; Zheng, X.; Li, X.; Sun, Y.; Wang, Y.; Li, C.; Wang, L. Cerium and Its Oxidant-Based Nanomaterials for Antibacterial Applications: A State-of-the-Art Review. *Front. Mater.* **2020**, *7*, 213. [CrossRef]
43. Zhang, P.; Wang, J.; Jiang, L.; Zhou, T.; Yan, X.; Yuan, L.; Chen, W. Influence Analysis and Stepwise Regression of Coal Mechanical Parameters on Uniaxial Compressive Strength Based on Orthogonal Testing Method. *Energies* **2020**, *13*, 3640. [CrossRef]
44. Kokubo, T.; Takadama, H. How useful is SBF in predicting in vivo bone bioactivity? *Biomaterials* **2006**, *277*, 2907–2915. [CrossRef] [PubMed]
45. Craciunescu, O.; Gaspar, A.; Trif, M.; Moisei, M.; Oancea, A.; Moldovan, L.; Zarnescu, O. Preparation and characterization of a collagen-liposome-chondroitin sulfate matrix with potential application for inflammatory disorders treatment. *J. Nanomater.* **2014**, *903691*, 1–9. [CrossRef]

MDPI
St. Alban-Anlage 66
4052 Basel
Switzerland
www.mdpi.com

Gels Editorial Office
E-mail: gels@mdpi.com
www.mdpi.com/journal/gels

www.ingramcontent.com/pod-product-compliance
Lightning Source LLC
Chambersburg PA
CBHW061833210926
51850CB00006B/875
* 9 7 8 3 0 3 6 5 9 4 9 5 8 *